Cognitive Science and Technology

Series editor

David M.W. Powers, Adelaide, Australia

More information about this series at http://www.springer.com/series/11554

Cowperthraite knew she had to prepare more for her class, so she settled down to read at her favorite spot on the beach

James K. Peterson

BioInformation Processing

A Primer on Computational
Cognitive Science

 Springer

James K. Peterson
Department of Mathematical Sciences
Clemson University
Clemson, SC
USA

ISSN 2195-3988   ISSN 2195-3996   (electronic)
Cognitive Science and Technology
ISBN 978-981-13-5718-3   ISBN 978-981-287-871-7   (eBook)
DOI 10.1007/978-981-287-871-7

This Springer imprint is published by SpringerNature
The registered company is Springer Science+Business Media Singapore Pte Ltd.

I dedicate this work to the many people, practicing scientists, mathematicians and software developers and computer scientists, who have who have helped me think more carefully about the whole picture. The larger problems are what interests me as my family can attest as they have listened to my ideas in the living room and over dinner for many years. I hope these notes help inspire my readers to consider the intersection of biology, mathematics, and computer science as a fertile research area.

Contents

List of Figures

List of Tables

List of Code Examples

Abstract

This book tries to show how mathematics, computer science, and science can be usefully and pleasurably intertwined. Here we begin to build a general model of cognitive processes in a network of computational nodes such as neurons using a variety of tools from mathematics, computational science, and neurobiology. We begin with a derivation of the general solution of a diffusion model from a low-level random walk point of view. We then show how we can use this idea in solving the cable equation in a different way. This will enable us to better understand neural computation approximations. Then we discuss neural systems in general and introduce a fair amount of neuroscience. We can then develop many approximations to the first and second messenger systems that occur in excitable neuron modeling. Next, we introduce specialized data for emotional content which will enable us to build a first attempt at a normal brain model. Then, we introduce tools that enable us to build graphical models of neural systems in MATLAB. We finish with a simple model of cognitive dysfunction. We also stress our underlying motto: always take the modeling results and go back to the scientists to make sure they retain relevance. We agree with the original financial modeler's manifesto due to Emanuel Derman and Paul Wilmott from January 7, 2009 available on the Social Science Research Network which takes the shape of the following Hippocratic oath. We have changed one small thing in the list of oaths. In the second item, we have replaced the original word *value* by the more generic term *variables of interest* which is a better fit for our modeling interests.

- I will remember that I did not make the world, and it does not satisfy my equations.
- Though I will use models boldly to estimate *variables of interest*, I will not be overly impressed by mathematics.
- I will never sacrifice reality for elegance without explaining why I have done so.
- Nor will I give the people who use my model false comfort about its accuracy. Instead, I will make explicit its assumptions and oversights.
- I understand that my work may have enormous effects on society and the economy, many of them beyond my comprehension.

There is much food for thought in the above lines and all of us who strive to develop models should remember them. In my own work, many times managers and others who oversee my work have wanted the false comfort the oaths warn against. We should always be mindful not to give in to these demands.

History

Based On:
Research Notes: 1992–1998
Class Notes: MTHSC 982 Spring 1997
Research Notes: 1998–2000
Class Notes: MTHSC 860 Summer Session I 2000
Class Notes: MTHSC 450, MTHSC 827 Fall 2001
Research Notes: Fall 2007 and Spring 2008
Class Notes: MTHSC 450 Fall 2008
Class Notes: MTHSC 450 Fall 2009
Class Notes: MTHSC 827 Summer Session I 2009
Class Notes: MTHSC 399 Creative Inquiry Spring 2010 and Fall 2010
Research Notes: Summer 2010 and Fall 2010
Class Notes: MTHSC 450 Spring 2010 and Fall 2010
Class Notes: MTHSC 827 Summer Session I 2010
Class Notes: MTHSC 434 Summer Session II 2010
Class Notes: MTHSC 399 Creative Inquiry Spring 2011 and Fall 2011
Class Notes: MTHSC 450 Spring 2011 and Fall 2011
Research Notes: Spring 2011 and Fall 2011
Class Notes: MTHSC 827 Summer Session I 2011
Class Notes: MTHSC 434 Summer Session II 2011
Class Notes: MTHSC 450 Spring 2012
Class Notes: BIOSC 491 Fall 2011 and Spring 2012 Research Notes: Spring 2012
Class Notes: MTHSC 827 Summer Session I 2013
Research Notes: Spring and Fall 2013
Research Notes: Spring and Fall 2014
Research Notes: Spring 2015
Class Notes: MTHSC 827 Summer Session I 2015

Part I
Introductory Matter

Chapter 1
BioInformation Processing

In this book, we will first discuss some of the principles behind models of biological information processing and then begin the process of using those ideas to build models of cognition and cognitive dysfunction. Most of the mathematical tools needed for this journey are covered in the first, second book and third book on calculus tools for cognitive scientists, Peterson (2015a, b, c, d) and so will be either assumed or covered very lightly here. Other mathematical ideas will be new and their coverage will be more complete as we try to bring everyone up a notch in mathematical sophistication.

We will discuss the principles of first and second messenger systems in terms of abstract triggering events and develop models of networks of computational nodes such as neurons that use various approximations to the action potential generation process. As usual, we therefore use tools at the interface between science, mathematics and computer science.

Our ultimate goal is to develop infrastructure components allowing us to build models of cognitive function that will be useful in several distinct arenas such as the modeling of depression and other forms of mental dysfunction that require a useful model of overall brain function. From such a model, which at the minimum involves modules for cortex, midbrain and thalamus, motor responses and sensory subsystems, we could obtain insight into how dopamine, serotonin and norepinephrine neurotransmitters interact. It is sobering to realize that as the time of this writing, June 2014, many drugs used for the treatment of cognitive dysfunction are prescribed without understanding of how they *work*. Hence, we want to develop models of neurotransmitter interaction that have predictive capability for mental disease. We are also unapologetically interested in small brain models. How much function can we obtain in a small package of neuronal elements? We want our models to be amenable to lesion analysis and so we want easily accessible scripting that will let us build and then deconstruct models as we see fit. These ideas are also needed in the development of autonomous robotics with minimal hardware.

© Springer Science+Business Media Singapore 2016
J.K. Peterson, *BioInformation Processing*, Cognitive Science and Technology,
DOI 10.1007/978-981-287-871-7_1

It is, of course, very difficult to achieve these goals and we are indeed far from making it happen. Further, to make progress, our investigations must use ideas from many fields. Above all, we must find the proper level of abstraction from the messiness of biological detail that will give us insight. Hence, in this first volume on cognitive modeling, we focus on the challenging task of how to go about creating a model of information processing in the brain sufficiently simple it can be implemented and yet sufficiently complicated to help us investigate cognition in general. We will even build a simple model of cognitive dysfunction using multiple neurotransmitters. In other volumes, we will use these abstractions of biological reality to build more interesting cognitive models.

1.1   The Proper Level of Abstraction

In the field of Biological Information Processing, the usual tools used to build models of follow from techniques that are loosely based on rather simplistic models from animal neurophysiology called artificial neural architectures. We believe we can do much better than this if we can find the *right* abstraction of the wealth of biological detail that is available; the *right* design to guide us in the development of our modeling environment. Indeed, there are three distinct and equally important areas that must be jointly investigated, understood and *synthesized* for us to make major progress. We refer to this as the Software, Wetware and Hardware **SWH** triangle, Fig. 1.1. We use double-edged arrows to indicate that ideas from these disciplines can both enhance and modify ideas from the others. The labels on the edges indicate possible intellectual pathways we can travel upon in our quest for unification and synthesis. Almost twenty years ago, an example of the **HW** leg of the triangle were the algorithms mapped into Verilog Hardware Description Language (VHDL) used to build hardware versions of a variety of low-level biological computational units. In addition, there were the new concepts of Evolvable Hardware where new hardware primitives referred to as **F**ield **P**rogrammable **G**ate **A**rrays offered the ability to program the devices Input/Output response via a bit string which could be chosen as a consequence to environmental input (see Higuchi et al. 1997; Sanchez 1996 and Sipper 1997). There were several ways to do this: *online* and *offline*. In *online*

Fig. 1.1 Software–
Hardware–Wetware
Triangle

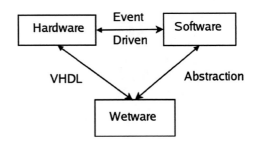

strategies, groups of FPGAs are allowed to interact and "evolve" toward an appropriate bit string input to solve a given problem and in *offline* strategies, the evolution is handled via software techniques similar to those used in genetic programming and the solution is then implemented in a chosen FPGA. In a related approach, it was even possible to perform software evolution within a pool of carefully chosen hardware primitives and generate output directly to a standard hardware language such as VHDL so that the evolved hardware can then be fabricated once an appropriate fitness level is reached. Thus, even 20 years ago, various areas of research used new relatively plastic hardware elements (their I/O capabilities are determined at *run-time*) or carefully reverse-engineered analog VLSI chipsets to provide us with a means to take abstractions of neurobiological information and implement them on silicon substrates.

Much more is possible now with the advent of synthetic biology and the construction of almost entirely artificial cells. in effect, there was and still is, a blurring between the traditional responsibilities of hardware and software for the kinds of typically event driven modeling tasks we see here. The term **Wetware** in the figure is then used to denote things that are of biological and/or neurobiological scope. And, of course, we attempt to build software to tie these threads together.

These issues from the hardware side are highly relevant to any search for useful ideas and tools for building biological models. Algorithms that can run in hardware need to be very efficient and to really build useful models of complicated biology will require large amounts of memory and computational resources unless we are *very* careful. So there are lessons we can learn by looking at how these sorts of problems are being solved in these *satellite* disciplines. Even though our example discussion here is from the arena of information processing in neurobiology, our point is that **Abstraction** is a principle tool that we use to move back and forth in the fertile grounds of software, biology and hardware.

In general, we will try try hard to *illuminate* your journey as a reader through this material. Let's begin with a little philosophy; there are some general principles to pay attention to.

1.2 The Threads of Our Tapestry

As we have mentioned, there are many threads to pull together in our research. We should not be dismayed by this as interesting things are complex enough to warrant it. You can take heart from what the historian Barbara Tuchman said in speaking about her learning the art of being a historian. She was working one of her first jobs and her editor was unhappy with her: as she says (Tuchman 1982, P. 17) in her essay *In Search of History*

> The desk editor, a newspaperman by training, grew very impatient with my work. 'Don't look up so much material,' he said. 'You can turn out the job much faster if you don't know too much.' While this was no doubt true for a journalist working against a deadline, it was not advice that suited my temperament.

There is a specific lesson for us in this anecdote. We also are in the almost unenviable position of realizing that we can never *know too much*. The problem is that in addition to mixing disciplines, languages and points of view, we must also find the *right level of abstraction* for our blended point of view to be useful for our synthesis. However, this process of finding the right level of abstraction requires reflection and much reading and thinking.

In fact, in any endeavor in which we are trying to create a road map through a body of material, to create a general theory that explains what we have found, we, in this process of creating the right level of abstraction, like the historian Tuchman, have a number of duties (Tuchman 1982, p. 17):

> The first is to distill. [We] must do the preliminary work for the reader, assemble the information, make sense of it, **select the essential**, discard the irrelevant—above all discard the irrelevant—and put the rest together so that it forms a developing ... narrative. ...To discard the unnecessary requires courage and also extra work. ...[We] are constantly being beguiled down fascinating byways and sidetracks. But the art of writing—the test of [it]—is to resist the beguilement and cleave to the subject.

Although our intellectual journey will be through the dense technical material of computer science, mathematics, biology and even neuroscience, Tuchman's advice is very relevant. It is *our* job to find this proper level of abstraction and this interesting road map that will illuminate this incredible tangle of facts that are at our disposal. We have found that this need to explain is common in most human endeavors. In music, textbooks on the art of composing teach the beginner to think of writing music in terms of musical primitives like the nouns, verbs, sentences, phrases, paragraphs of a given language where in the musical context each phoneme is a short collection of five to seven notes. Another example from history is the very abstract way that Toynbee attempted to explain the rise and fall of civilizations by defining the individual entities *city*, *state* and *civilization* (Toynbee and Caplan 1995).

From these general ideas about appropriate abstraction, we can develop an understanding of which abstractions might be of good use in an attempt to develop a working model of autonomous or cognitive behavior. We will need sophisticated models of connections between computational elements (synaptic links between classical lumped sum neuronal models are one such type of connection) and new ideas for the asynchronous learning between objects. Our challenges are great as biological systems are much more complicated than those we see in engineering or mathematics.

These are sobering thoughts aren't they? Still, we are confident we can shed some insight on how to handle these modeling problems. In this volume, we will discuss relevant abstractions that can open the doorway for you to get involved in useful and interesting biological modeling.

Unfortunately, as we have tried to show, all of the aforementioned items are very nicely (or very horribly!) intertwined (it depends on your point of view!). We hope that you the reader will persevere. As you have noticed, our style in this report is to use the "royal" *we* rather than any sort of third person narrative. We have always felt that this should be interpreted as the author asking the reader to explore the material with the author. To this end, the text is liberally sprinkled with *you* to encourage this point of view. Please take an active role!

1.3 Chapter Guide

This text covers the material in bioinformation processing in the following way.

Part I: Introductory Matter   This is the material you are reading now. Here we take the time to talk about modeling in general and use examples taken from the first three courses on calculus for Cognitive Scientists (Peterson 2015a, b, c, d).

Part II: Diffusion Models   In this portion of the text, we show you how to solve the time dependent cable equation in a different way. This gives us additional insight into what these solutions look like which helps us when we are trying to find effective ways to approximate these computations.

- In Chap. 2, we discuss the solutions to diffusion equations with a model of a particle moving randomly with right and left steps called a random walk model. We use the solution of this model as as way to motivate a solution to a general diffusion equation.
- In Chap. 3, we introduce Laplace and Fourier Transform methods so that we can solve the cable equation in a different way.
- In Chap. 4, we show how we use integral transform techniques to find a new way of expressing the solution to the time dependent cable equation under certain conditions. At this point, we understand how to approximate solutions to voltage impulses input into a dendritic model in a variety of ways.

Part III: Neural Systems   In this part, we begin the real process of abstracting biological information processing.

- In Chap. 5, we provide a top level overview of the neural structure of a typical mammalian brain.
- In Chap. 6, we work out the details of an abstract second messenger trigger system in terms of how the trigger initiates a sudden departure from dynamic equilibrium. We then use this to develop simple estimates of the size to the response to a triggering event. We also discuss a few specific examples of second messenger events so that you can see how some of the general biological information processing pathways have evolved to solve problems. We also discuss calcium ion current triggers and general ligand receptor response strategies to help us understand how action potentials are used to build responses in a network of interacting neurons.
- In Chap. 7, we discuss second messenger triggers that are diffusion based in general. Our discussion of a generic model of Calcium ion triggers uses a combined diffusion model with a finite number of Calcium binding molecules and

Calcium storage in the endoplasmic reticulum. Then, we make two simplifying assumptions and finally arrive at a model with which we can model calcium ion trigger events to first order.

- In Chap. 8, we discuss an relatively abstract way of modeling how the output characteristics of a first or second messenger trigger are computed.
- In Chap. 9, we look at the inputs and outputs to a neuron much more abstractly. We use these discussions to motivate the creation of a biological feature vector and to motivate why we think it is useful construct, we use it to map changes in the gate parameters of a typical Hodgkin–Huxley model such as we see when a toxin is introduced into a neural system to particular toxins. This leads to our discussion of how the approximation of the action potential into a biological feature vector can have useful computational advantages. We begin to see how we could use these ideas to approximate the response of a neuron without the large amounts of computation the previous approaches built on solving systems of differential equations and partial differential equations require.

Part IV: Models of Emotion and Cognition We then introduce models that are built of collections of interacting computational nodes. These network models are actually more general than ones built from excitable neurons. We discuss some basic ideas involving neurotransmitters and the salient issues, in our mind, that must be addressed to build useful software models. We finish these chapters with an introduction to how we might make a network model adapt to environment input and given target information. To do this, we introduce a variant of the cable equation for graphs using its Laplacian. In detail

- In Chap. 10, we introduce non biologically based emotional models so that we can learn how to use ideas from psycho-physics.
- In Chap. 11, we show how we can build emotionally labeled models of music which we can use to train the auditory cortex.
- In Chap. 12, we build emotionally labeled models of painting which we can use to train the visual cortex.
- In Chap. 13, we discuss generic connectionist strategies to use the emotionally labeled data to build a model of emotion and cognition although our true interest is in biologically plausible models.
- In Chap. 14, we go over some basic ideas of neurotransmitters.
- In Chap. 15, we illustrate what we could do with to simulate a simple network of neurons organized in a graph using MatLab and introduce many of the basic ideas we have to figure out how to implement.

Part V: Simple Abstract Neurons In this part, we go over the standard model of a matrix based feed forward network as a simplistic model of bioinformation processing in a neural system and then segue into the more flexible paradigm that uses a graph structure.

- In Chap. 16, we introduce the matrix feed forward network as a way to abstract neural computation. We also show how to code the training of this model in MatLab.

- In Chap. 17, we recast the matrix based networks of simple neuron processing functions as a chain, or graph, of computational nodes. At this point, it is clear the the neural processing can be much more interesting. We also introduce MatLab coding of these graphs.

Part VI: Graph Based Modeling In Matlab   We now show you how to build reasonable graph models of neural computation using MatLab. This is not a great choice of programming language but it is readily available and easy to learn compared to **C, C++** and **Python**.

- In Chap. 18, we build object oriented class models in MatLab (be warned, this is clumsy and it makes us hunger for a better computing language—but we digress....).
- In Chap. 19, we move to graph based models that are based on vector addressing: each neuron has a global address and a vector address which we can use to find it in a model.
- In Chap. 20, we actually build some simple brain models.

Part VII: Models of Cognitive Dysfunction   In Chap. 21, we finish this text with some preliminary conversations on how to build a model of normal cognitive function so that we can properly discuss cognitive dysfunction. We introduce the details of subgraph training based on the Laplacian based training briefly introduced in Chap. 15 and show how we could use those ideas and the music and painting data from Chaps. 11 and 12 to first build a model of a normal brain (yes, we know that is hard to define!) and from that a model of dysfunction such as depression. We only sketch the ideas as the implementation will take quite a long time and deserves a long treatment. This will be done in the next volume.

Part VIII: Conclusions   In Chap. 22 we talk a bit about where this volume has led us and what we can do with the material we have learned.

Part IX: Further Reading   In Chap. 23 we mention other papers and books and software tools we, and you, might find useful.

Finally, this course is just a first step in the development of the tools needed to build the graphs of computational nodes which subserve information processing in the brain. The goal is to build models of cognitive dysfunction using reasonable models of neurotransmitter action and we pursue this further in the next volume. Our style in these lectures is to assume you the reader are willing to go on this journey with us. So we have worked hard at explaining in detail all of our steps. We don't assume you have background in biophysics and the biology of neurons and so we include chapters on this material. We develop the mathematical tools you need to a large extent in house, so to speak. We also try to present the algorithm design and the code behind our software experiments in a lot of detail. The interactive nature of the MatLab/Octave integrated development environment makes it ideal for doing the exploratory exercises we have given you in these pages. The basic background we assume is contained in our companion books (Peterson 2015a, b, c, d).

Finally, as we mentioned, we are interested in developing models of cognitive dysfunction which provides insight into how cognitive function arises and how cognitive dysfunction might be ameliorated. The address based graph models (implemented here in MatLab but it is simply an exercise to reimplement in Fortran, C, C++, Python or a functional based language such as Erlang, Clojure or Haskell) provide a useful theoretical framework for modeling the interactions of neurons at varying levels of realism. We think a general graph based model, $\mathcal{G}(N, E)$, with nodes N providing computation at the neuron level and edges $E_{i \to j}$ between nodes N_i and node N_j giving neuronal connection computation provides a useful framework for exploring cognitive models theoretically. In our development of a graph model of a neural system, the neural circuitry architecture we describe is fixed, but it is clear dynamic architectures can be handled as sequence of these directed graphs. We see in Chap. 19 on the address based graphs and Chap. 20 on brain modeling how to organize the directed graph using interactions between neural modules (visual cortex, thalamus etc.) which are themselves subgraphs of the entire circuit. Once we have chosen a direct graph to represent our neural circuitry, note the addition of a new neural module is easily handled by adding it and its connections to other modules as a subgraph addition. Hence, at a given level of complexity, if we have the graph $\mathcal{G}(N, E)$ that encodes the connectivity we wish to model, then the addition of a new module or modules simply generates a new graph $\mathcal{G}'(N', E')$ for which there are straightforward equations for explaining how G' relates to G which are easy to implement. Thus, we show in Chap. 21, we can learn how to do subgraph level training which will allow us to potentially build very useful models of cognition

We find this approach intellectually liberating as we are completely agnostic as to the particular details of the nodal and edge processing functions. Indeed, once a particular level of module complexity has been chosen, it is a straightforward task to use that model as a base point for a different model. Also, it is possible to outline **longitudinal** studies which allow us to model brain function from low resolution to high keeping the same intellectual framework at all resolutions. We also are explicitly asynchronous in our modeling approach and we discuss how this impacts our computational strategies. Many of these questions will be addressed in the next volume.

We will begin our discussions with a few models from our own teaching and research to illustrate some general points about the type of modeling we are interested in. Hence, we include high level modeling challenges such as understanding painting and music, the spread of altruistic genes in a population and modeling depression as grist for our mill. We are always striving for ways to obtain insight about complicated questions. Hence, the quest here is to develop insight into how cognition might arise and also how cognitive dysfunction might occur and how such dysfunction might be ameliorated by external agents.

1.4   Theoretical Modeling Issues

The details of how proteins interact, how genes interact and how neural modules interact to generate the high level outputs we find both interesting and useful are known to some degree but it is quite difficult to use this detailed information to build models of high level function. Let's explore several models before we look at a cognitive model. In all of these models, we are asking high level questions and wondering how we might create a model that gives us some insight. We all build models of the world around us whether we use mathematical, psychological, political, biological and so forth tools to do this. All such models have built in assumptions and we must train ourselves to think abut these carefully. We must question the abstractions of the messiness of reality that led to the model and be prepared to adjust the modeling process if the world as experienced is different from what the model leads them to expect. There are three primary sources of error when we build models:

- The error we make when we abstract from reality; we make choices about which things we are measuring are important. We make further choices about how these things relate to one another. Perhaps we model this with mathematics, diagrams, words etc.; whatever we choose to do, there is error we make. This is called *Model* error.
- The error we make when we use computational tools to solve our abstract models. This error arises because we typically must replace the model we came up with in the first step with an approximate model that can be implemented on a computer. This is called *Truncation* error.
- The last error is the one we make because we can not store numbers exactly in any computer system. Hence, there is always a loss of accuracy because of this. This is called *Round Off* error.

All three of these errors are always present and so the question is how do we know the solutions our models suggest relate to the real world? We must take the modeling results and go back to original data to make sure the model has relevance. These models are not too hard but capture the tension between the need to understand a very high level question using what we know at the first principle level.

Transcription Models: In a model of model of protein transcription, we start with a simple model that lumps all kinds of things together and we end with a much more detailed one that is based on an equilibrium analysis point of view. We make many assumptions in order to build these models and by the end, we can see how to introduce positive and negative feedback and delays into the process. We end there with the statement that all of this complexity would then have to be folded into models of gene interactions called regulatory gene networks. Our high level questions such as *"What gene clusters control a phenotype?"* are difficult to answer using this level of detail and hence, useful abstractions are needed. Note the *model error* that inevitably is introduced.

The spread of a gene throughout a population: We can ask the question of how does a given phenotype become dominant in a population? To do this, we try to formulate the problem in a very abstract way for a very simple population of two phenotypes with simple assumptions about fertility etc. This simple model, nevertheless, has good explanatory power and we can use it to understand better how domesticated wheat became dominant and wild wheat became rare. Answering that high level question is difficult without some sort of abstract model to develop an equation based model we can use to see how the split between two phenotypes alters over time. The really interesting thing about this model is that all of it is developed using only algebra.

The Spread of Altruism in a Population: To develop a model for the spread of altruistic behavior in a population, we assume this behavior can be explained as a phenotype due to some combination of genes. We can build on the Viability model to some extent, but to do this properly, we have to find a way to model actions made by individuals when there is a cost to them even though there is a benefit to the good of the population. This is a very high level idea and all the detail about protein transcription, regulatory gene networks and so forth is really irrelevant. We need a better way to find insight. Here, we carefully go over the insight into the spread of altruism called *Hamilton's Rule* which was formulated by Hamilton in the early 1960's. Hamilton's Rule was based on a parameter called the *kinship* coefficient and it is quite difficult to understand what is might mean. We give several different ways of looking at it all of which shed insight in their own way.

A **Simple Insulin Model**: In diabetes there is too much sugar in the blood and the urine. This is a metabolic disease and if a person has it, they are not able to use up all the sugars, starches and various carbohydrates because they don't have enough **insulin**. Diabetes can be diagnosed by a **glucose tolerance test** (GTT). If you are given this test, you do an overnight fast and then you are given a large dose of sugar in a form that appears in the bloodstream. This sugar is called **glucose**. Measurements are made over about five hours or so of the concentration of glucose in the blood. These measurements are then used in the diagnosis of diabetes. It has always been difficult to interpret these results as a means of diagnosing whether a person has diabetes or not. Hence, different physicians interpreting the same data can come up with a different diagnosis, which is a pretty unacceptable state of affairs! The idea of this model, which we introduce in Peterson (2015b), for diagnosing diabetes from the GTT is to find a simple dynamical model of the complicated blood glucose regulatory system in which the values of two parameters would give a nice criterion for distinguishing normal individuals from those with mild diabetes or those who are pre diabetic. Of course, we have to choose what these parameters are and that is the art of the modeling process!

A **Cancer Model**: A simple model of cancer is based on a tumor suppressor gene (TSG) that occurs in two alleles. There are two pathways to cancer: one is due to point mutations that knock out first one allele and then another of the TSG. The other pathway starts with cells having both alleles of the TSG but a chromosome that is damaged in some way (copy errors in cell division etc.). The question is which pathway to cancer is dominant? This is hard to answer, of course. With our model, we can generate a prediction that there is an algebraic relationship between the chromosomal damage rate and the population of cells in the cancer model. It is quite messy to do this, but imagine how hard it would be to find this relationship using computer simulations. Knowing the relationship equation is very helpful and gives us a lot of insight.

The Classical Predator Prey Model: In the classical Predator—Prey model, we use abstraction to develop the nonlinear system of differential equations that represents the interactions between food fish and predator fish. All the food fish are lumped together and all the different types of predators are also thrown into one bin. Nevertheless, this model predicts periodic solutions whose average food fish and predator values explain very nicely puzzling data from World War I. However, the model does not take into account self interaction between food fish and predator fish. When that is added, we can still explain the data, but the trajectories for the food and predator fish converge over time to a fixed value. This hardly represents the behavior we see in the sea, so we have lost some of our insight and the model is much less explanatory even though it is more accurate because self interaction has been added. A general principle seems to be peeking out here: adding self interaction which is essentially adding damping rules out periodicity. So in a biological model which has periodicity in it, there should be a *push–pull* mechanism where the damping can become excitation. Another way to look at it is that we need positive and negative feedback in the model. This is not a proof, of course, just a reasonable hypothesis.

A West Nile Virus Infection Model: A West Nile Virus infection does something very puzzling. Normally, as the amount of virus in an animal host increases, the animal's probability of survival drops. So if you took 10 mice and exposed them to a level of virus, you could measure how many mice survived easily. You just look at the mice to see if they are still alive. If you plot host survival versus virus infection level, for most viral infections, you see a nice reversed S-curve. At low levels, you have 10 surviving and as the viral load increases, that number smoothly decays to 0 with no upticks in between. West Nile Virus infection is quite different: there is some sort of immunopathology going on that allows the number of surviving mice to oscillate up and down. Hence, even at high viral load, you can have more mice surviving than you expect. The question here is simple: *explain the survival curve data*. This is very hard. We build a model but to do it, we have to come up with abstract approximations of T-Cell interactions, the spread of various chemical messengers throughout the system and so forth. But most importantly, what constitutes host death in our simulation? Host death is a very high level idea and to compare our simulation results to the real survival data requires that our measure of host death is reasonable. So this model is quite messy and filled with abstractions and at the end of the day, it is still hard to decide what a simulated host death should be. This model is a lot like a brain model where our question is *"What is depression?"*

1.5   Code

All of the code we use in this book is available for download from the site **Biological Information Processing** (http://www.ces.clemson.edu/~petersj/CognitiveModels. html). The code samples can then be downloaded as the zipped tar ball **CognitiveCode.tar.gz** and unpacked where you wish. If you have access to MatLab, just add this folder with its sub folders to your MatLab path. If you don't have such access, download and install **Octave** on your laptop. Now Octave is more of a command line tool, so the process of adding paths is a bit more tedious. When we start up an Octave session, we use the following trick. We write up our paths in a file we call **MyPath.m**. For us, this code looks like this

Listing 1.1: How to add paths to Octave

```
function MyPath()
%
s1 = '/home/petersj/MatLabFiles/BioInfo/:';
s2 = '/home/petersj/MatLabFiles/BioInfo/GSO:';
s3 = '/home/petersj/MatLabFiles/BioInfo/HH:';
s4 = '/home/petersj/MatLabFiles/BioInfo/Integration:';
s5 = '/home/petersj/MatLabFiles/BioInfo/Interpolation:';
s6 = '/home/petersj/MatLabFiles/BioInfo/LinearAlgebra:';
s7 = '/h ome/petersj/MatLabFiles/BioInfo/Nernst:';
s8 = '/home/petersj/MatLabFiles/BioInfo/ODE:';
s9 = '/home/petersj/MatLabFiles/BioInfo/RootsOpt:';
s10 = '/home/petersj/MatLabFiles/BioInfo/Letters:';
s11 = '/home/petersj/MatLabFiles/BioInfo/Graphs:';
s12 = '/home/petersj/MatLabFiles/BioInfo/PDE:';
s13 = '/home/petersj/MatLabFiles/BioInfo/FDPDE:';
s14 = '/home/petersj/MatLabFiles/BioInfo/3DCode';
s = [s1,s2,s3,s4,s5,s6,s7,s8,s9,s12];
addpath(s);
end
```

The paths we want to add are setup as strings, here called **s1** etc., and to use this, we start up Octave like so. We copy **MyPath.m** into our working directory and then do this.

Listing 1.2: Set paths in octave

```
octave≫ MyPath();
```

We agree it is not as nice as working in MatLab, but it is free! You still have to think a bit about how to do the paths. For example, in this text, we develop two different ways to handle graphs in MatLab. The first is in the directory **GraphsGlobal** and the second is in the directory **Graphs**. They are not to be used together. So if we wanted to use the setup of **Graphs** and nothing else, we would edit the **MyPath.m** file to set **s = [s11];** only. If we wanted to use the **GraphsGlobal** code, we would edit **MyPath.m** so that **s11 ='/home/petersj/MatLabFiles/BioInfo/GraphsGlobal:';** and then set **s = [s11];**. Note the directories in the **MyPath.m** are ours: the main directory is   **'/home/petersj/MatLabFiles/BioInfo/** and of course, you will have to edit this file to put your directory information in there instead of ours.

All the code will work fine with **Octave**. So pull up a chair, grab a cup of coffee or tea and let's get started.

References

T. Higuchi, M. Iwata, W. Liu, (eds.), in *Evolvable Systems: From Biology to Hardware*, Proceedings of the First International Conference, October 1996. Lecture Notes in Computer Science, Vol. 1259 (Tsukuba, Japan, 1997)

J. Peterson, *Calculus for Cognitive Scientists: Derivatives, Integration and Modeling*, Springer Series on Cognitive Science and Technology (Springer Science+Business Media Singapore Pte Ltd., Singapore, 2015a in press)

J. Peterson, *Calculus for Cognitive Scientists: Higher Order Models and Their Analysis*, Springer Series on Cognitive Science and Technology (Springer Science+Business Media Singapore Pte Ltd., Singapore, 2015b in press)

J. Peterson, *Calculus for Cognitive Scientists: Partial Differential Equation Models*, Springer Series on Cognitive Science and Technology (Springer Science+Business Media Singapore Pte Ltd., Singapore, 2015c in press)

J. Peterson, *BioInformation Processing: A Primer On Computational Cognitive Science*, Springer Series on Cognitive Science and Technology (Springer Science+Business Media Singapore Pte Ltd., Singapore, 2015d in press)

M. Sipper, *Evolution of Parallel Cellular Machines: The Cellular Programming Approach*, Number 1194 in Lecture Notes in Computer Science (Springer, Berlin, 1997)

E. Sanchez, *Towards Evolvable Hardware: The Evolutionary Engineering Approach*. in E. Sanchez, M. Tomassini, (eds.), Proceedings of the First International Workshop, Lausanne, Switzerland, October 1995, Vol. 1259 (Springer, Berlin, 1996)

A. Toynbee, J. Caplan, *A Study of History: A New Edition* (Barnes and Noble Books, New York, 1995). Revised and abridged, Based on the Oxford University Press Edition 1972

B. Tuchman, *Practicing History: Selected Essays* (Ballantine Books, New York, 1982)

Part II
Diffusion Models

Chapter 2
The Diffusion Equation

We are now ready to study the time dependent solutions of the cable equation which are not separable. We assume you have read the derivation of the cable equation presented in Peterson (2015) so that you are comfortable with that material. Also, we have already studied how to solve the cable equation using the separation of variables technique. This method solved the cable equation with boundary conditions as a product $\hat{v}_m(\lambda, \tau) = u(\lambda)\, w(\tau)$. Then, in order to handle applied voltage pulses or current pulses, we had to look at an infinite series solution of the form $\sum_n A_n u_n(\lambda)\, w_n(\tau)$. Now, we will see how we can find a solution which is not time and space separable. This requires a fair bit of work: first, a careful discussion of how the random walk model in a limiting sense gives a probability distribution which serves as a solution to the classical diffusion model. This is done in this chapter. We follow the standard presentation of this material in for example Johnston and Wu (1995) and others.

The solution of the diffusion model is very important because we can use a clever *integral transform* technique to turn solve a general diffusion model. So in Chap. 3 we discuss the basics of the Laplace and Fourier Transform. Finally, in Chap. 4, we use a change of variables to transform a cable model into a diffusion equation. Then, we use integral transforms to solve the diffusion model. We will get the same solution we obtained using the random walk model which gives us a lot of insight into the structure of the solution.

Now these ideas are important because many second messenger systems are based on the diffusion of a signal across the cytoplasm of the excitable cell. So knowing how to work with diffusion based triggers is necessary.

2.1 The Microscopic Space–Time Evolution of a Particle

The simplest probabilistic model that links the Brownian motion of particles to the macroscopic laws of diffusion is the 1D random walk model. In this model, we assume a particle moves every τ_m seconds along the y axis a distance of either λ_c

© Springer Science+Business Media Singapore 2016

J.K. Peterson, *BioInformation Processing*, Cognitive Science and Technology,
DOI 10.1007/978-981-287-871-7_2

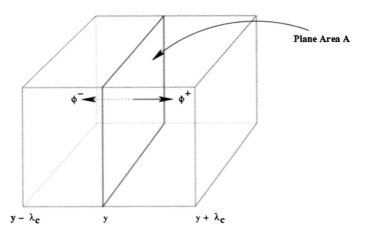

Fig. 2.1 Walking in a volume box

or $-\lambda_c$ with probability $\frac{1}{2}$. Consider the thought experiment shown in Fig. 2.1 where we see a volume element which has length $2\lambda_c$ and cross sectional area A. Since we want to do microscopic analysis of the space time evolution of the particles, we assume that $\lambda_c < y$.

Let $\phi^+(s, y)$ denote the flux density of particles crossing from left to right across the plane located at position y at time s; similarly, let $\phi^-(s, y)$ be the flux density of particles crossing from right to left. Further, $c(s, y)$ denote the concentration of particles at coordinates (s, y). What is the net number of particles that cross the plane of area A?

Since the particles randomly change their position every τ seconds by $\pm\lambda_c$, we can calculate flux as follows: first, recall that flux here is **the number of particles** per unit area and time; i.e., the units are $\frac{particles}{sec-cm^2}$. Since the walk is random, half of the particles will move to the right and half to the left. Since the distance moved is λ_c, half the concentration at $c(y - \frac{\lambda_c}{2}, s)$ will move to the right and half the concentration at $c(y + \frac{\lambda_c}{2}, s)$ will move to the left. Now the number of particles crossing the plane is concentration times the volume. Hence, the flux terms are

$$\phi^+(s, y) = \frac{1}{2} \frac{A\lambda_c \, c\left(y - \frac{\lambda_c}{2}, s\right)}{A\tau_m}$$

$$= \frac{\lambda_c}{2\tau_m} c\left(y - \frac{\lambda_c}{2}, s\right)$$

$$\phi^-(s, y) = \frac{1}{2} \frac{A\lambda_c \, c\left(y + \frac{\lambda_c}{2}, s\right)}{A\tau_m}$$

$$= \frac{\lambda_c}{2\tau_m} c\left(y + \frac{\lambda_c}{2}, s\right)$$

The net flux, $\phi(s, y)$ is thus

$$\phi(s, y) = \phi^+(s, y) - \phi^-(s, y)$$
$$= \frac{\lambda_c}{2\tau_m}\left(c\left(s, y - \frac{\lambda_c}{2}\right) - c\left(s, y + \frac{\lambda_c}{2}\right)\right)$$

Since, $\frac{\lambda_c}{y}$ is very small in microscopic analysis, we can approximate the concentration c using a first order Taylor series expansion in two variables if we assume that the concentration is sufficiently smooth. Our knowledge of the concentration functions we see in the laboratory and other physical situations implies that it is very reasonable to make such a smoothness assumption. Hence, for small perturbations $s + a$ and $y + b$ from the base point (s, y), we find

$$c(s + a, y + b) = c(s, y) + \frac{\partial c}{\partial s}(s, y) a + \frac{\partial c}{\partial y}(s, y) b + e(s, y, a, b)$$

where $e(s, y, a, b)$ is an error term which is proportional to the size of the largest of $|a|$ and $|b|$. Thus, e goes to zero as (a, b) goes to zero in a certain way. In particular, for $a = 0$ and $b = \pm\frac{\lambda_c}{2}$, we obtain

$$c\left(s, y - \frac{\lambda_c}{2}\right) = c(s, y) - \frac{\partial c}{\partial y}(s, y)\frac{\lambda_c}{2} + e\left(s, y, 0, -\frac{\lambda_c}{2}\right)$$
$$c\left(s, y + \frac{\lambda_c}{2}\right) = c(s, y) + \frac{\partial c}{\partial y}(s, y)\frac{\lambda_c}{2} + e\left(s, y, 0, \frac{\lambda_c}{2}\right)$$

and we note that the error terms are proportional to λ_c^2. Thus,

$$\phi(s, y) = -\frac{\lambda_c}{2\tau_m}\frac{\partial c}{\partial y}(s, y)\,\lambda_c + \left(e\left(s, y, 0, -\frac{\lambda_c}{2}\right) - e\left(s, y, 0, \frac{\lambda_c}{2}\right)\right)$$
$$\approx -\frac{\lambda_c^2}{2\tau_m}\frac{\partial c}{\partial y}(s, y)$$

as the difference in error terms is still proportional to λ_c^2 at worst and since λ_c is very small compared to y, the error term is negligible.

Recall from Peterson (2015) that Ficke's law of diffusion for particles across a membrane (think of our plane at y as the membrane) can be written as

$$J_{diff} = -D\frac{\partial [b]}{\partial x}$$

Equating c with $[b]$ and J_{diff} with ϕ, we see that the diffusion constant D in Ficke's Law of diffusion can be interpreted in this context as

$$D = \frac{\lambda_c^2}{2\tau_m}$$

This will give us a powerful connection between the macroscopic diffusion coefficient of Ficke's Law with the microscopic quantities that define a random walk as we will see in the next section.

2.2 The Random Walk and the Binomial Distribution

In this section, we will be following the discussion presented in Weiss (1996), but paraphrasing and simplifying it for our purposes. More details can be gleaned from a careful study of that volume. Let's assume that a particle is executing a random walk starting at position $x = 0$ and time $t = 0$. This means that from that starting point, the particle can move either $+\lambda_c$ or $-\lambda_c$ at each tick of a clock with time measured in time constant units τ_m. We can draw this as a tree as shown in Fig. 2.2.

In this figure, the labels shown in each node refer to three things: the time, in units of the time constant τ_m (hence, $t = 3$ denotes a time of $3\tau_m$); spatial position in units of the space constant λ_c (thus, $x = -2$ means a position of $-2\lambda_c$ units); and the number of possible paths that read that node (therefore, Paths equal to 6 means there are six possible paths that can be taken to arrive at that terminal node.

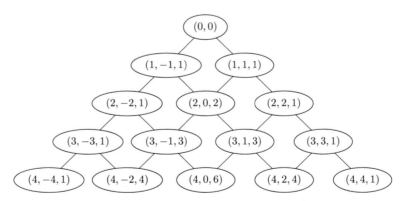

A particle starts at $t = 0, x = 0$ and can move with a given probability left or right a distance λ_c in time increments of τ_m. The triple (a, b, c) indicates the time is $t = a\tau_m$, the x location is $x = b\lambda_c$ and the number of paths from the root of the tree to the node is $P = c$.

Fig. 2.2 The random walk of a particle

Since time and space are discretized into units of time and space constants, we have a physical system where time and space are measured as integers. Thus, we can ask what is the probability, $W(m, n)$, that the particle will be at position m at time n? In the time interval of n units, let's define some auxiliary variables: n^+ is the number of time steps where the particle moves to the right—i.e., the movement is $+1$; and n^- is the number of time steps the particle moves to the left, a movement of -1. We clearly see that M is really the net displacement and

$$n = n^+ + n^-, \quad m = n^+ - n^-$$

Solving, we see

$$n^+ = \frac{n + m}{2}, \quad n^- = \frac{n - m}{2}$$

Let the usual binomial coefficients be denoted by $B_{n,j}$ where

$$B_{n,j} = \binom{n}{j} = \frac{n!}{j!\,(n - j)!}$$

Now look closely at Fig. 2.2. If you look at a node, you can count how many right hand moves are made in any given path to that node. For example, the node for time 4 and space position 2 can be reached by 4 different paths and each of them contains $3 + \lambda_c$ moves. Note three $+\lambda_c$ moves corresponds to $n^+ = 3$ and the number 4 is the same as the binomial coefficient $B_{n=4,n^+=3} = 4$. Hence, at a given node, all of the paths that can be taken to that node have the same n^+ value as shown in Table 2.1. The triangle in Fig. 2.2 has the characteristic form of Pascal's triangle: the root node is followed by two nodes which split into three nodes and so on. The pattern is typically written in terms of levels. At level zero, there is just the root node. This is time 0 for us. At level one or time 1, there are two nodes written as $1 - 1$. At level two or time 2, there are three nodes written as $1 - 2 - 1$ to succinctly capture the branching we are seeing. Continuing, we see the node pattern can be written

Level or Time	Paths
0	1
1	1 - 1
2	1 - 2 - 1
3	1 - 3 - 3 - 1
4	1 - 6 - 4 - 6 - 1

Each of the nodes at a given time then will have some paths leading to it and on each of these paths, there will be the same number of right hand moves. So counting right hand moves, the paths denoted by $1 - 1$ correspond to $n^+ = 0$ for the left node and $n^+ = 1$ for the right node. The number of paths for these are the same as $B_{n=1,n^+=0} = 1$ and

Table 2.1 Comparing paths and rightward movements

Time	Paths	n^+	Binomial coefficient B_{n,n^+}
1	1 - 1	0 - 1	$B_{1,0}$ - $B_{1,1}$
2	1 - 2 - 1	0 - 1 - 2	$B_{2,0}$ - $B_{2,1}$ - $B_{2,2}$
3	1 - 3 - 3 - 1	0 - 1 - 2 - 3	$B_{3,0}$ - $B_{3,1}$ - $B_{3,2}$ - $B_{3,3}$
4	1 - 6 - 4 - 6 - 1	0 - 1 - 2 - 3 - 4	$B_{4,0}$ - $B_{4,1}$ - $B_{4,2}$ - $B_{4,3}$ - $B_{4,4}$

$B_{n=1,n^+=1} = 1$. The next level, $1 - 2 - 1$ corresponds to $n^+ = 0$ for the left node, $n^+ = 1$ for the middle one and $n^+ = 2$ for the right one. The number of paths for these nodes are then $B_{n=2,n^+=0} = 1$ and $B_{n=2,n^+=1} = 2$ and $B_{n=2,n^+=2} = 1$. We can do this for each level giving us the information shown in Table 2.1.

2.3   Rightward Movement Has Probability 0.5 or Less

Let the probability that the particle moves to the right be p and the probability it moves to the left be q. Then we have $p + q = 1$. Let's assume that $p \leq 0.5$. This forces $q \geq 0.5$. Then, in n time units, the probability a given path of length n is taken is $\rho(n)$ where

$$\rho(n) = p^{n^+} q^{n^-} = p^{n^+} q^{n-n^+}$$

and the probability that a path will terminate at position m, $W(m, n)$, is just the number of paths that reach that position in n time units multiplied by $\rho(n)$. We know that for a given number of time steps, certain position will never be reached. Note that if the fraction $0.5(n + m)$ is not an integer, then there will be no paths that reach that position m. Let n_f be the results of the computation $n_f = 0.5(n + m)$. We know n_f need not be an integer. Define the extended binomial coefficients C_{n,n_f} by

$$C_{n,n_f} = \begin{cases} B_{n,n_f} & \text{if } n_f \text{ is an integer} \\ 0 & \text{else} \end{cases}$$

Then, we see

$$W(m, n) = C_{n,n_f} p^{n_f} q^{n-n_f}$$

From our discussions above, it is clear for any position m that is reached in n time units, that this can be rewritten in terms of n^+ as

$$W(m, n) = B_{n,n^+(m)} p^{n^+(m)} q^{n-n^+(m)}$$

where $n^+(m)$ is the value of n^+ associated with paths terminating on position m. If you think about this a bit, you'll see that for even times n, only even positions m are reached; similarly, for odd times, only odd positions are reached.

2.3.1 Finding the Average of the Particles Distribution in Space and Time

The expectation $E(m)$ is defined by

$$E(m) = \sum_{m=-n}^{n} m\, W(m, n)$$

where of course, many of the individual terms $W(m, n)$ are actually zero for a given time n because the positions are never reached. From this, we can infer that

$$E(m) = \sum_{n^+=0}^{n} m\, B_{n,n^+}\, p^{n^+} q^{n-n^+}$$

To compute this, first, switch to a simple notation. Let $n^+ = j$. Then, since $n^+ = \frac{m+n}{2}$, $m = 2j - n$ and so

$$E(m) = \sum_{j=0}^{n} m\, B_{n,j}\, p^j q^{n-j}$$

$$= \sum_{j=0}^{n} (2j - n)\, B_{n,j}\, p^j q^{n-j}$$

$$= 2\, E(j)\, n\, S$$

where

$$E(j) = \sum_{j=0}^{n} j\, B_{n,j}\, p^j q^{n-j}$$

$$S = \sum_{j=0}^{n} B_{n,j}\, p^j q^{n-j}$$

Since $p + q = 1$, we know that

$$(p + q)^n = S = \sum_{j=0}^{n} B_{n,j}\, p^j q^{n-j} = 1$$

Further, by taking derivatives with respect to p, we see that

$$p \frac{d}{dp} \left((p+q)^n \right) = p \, n \, (p+q)^{n-1} = p \, n$$

Thus,

$$E(j) = \sum_{j=0}^{n} j \, B_{n,j} \, p^j \, q^{n-j}$$

$$= \sum_{j=0}^{n} j \, p \, B_{n,j} \, p^{j-1} \, q^{n-j}$$

$$= \sum_{j=0}^{n} p \, \frac{d}{dp} \left(B_{n,j} \, p^j \, q^{n-j} \right)$$

$$= p \, \frac{d}{dp} \left(\sum_{j=0}^{n} B_{n,j} \, p^j \, q^{n-j} \right)$$

$$= p \, \frac{d}{dp} \left((p+q)^n \right)$$

$$= p \, n$$

by our calculations above. We conclude that

$$E(m) = 2 \, E(j) \, nS$$
$$= 2 \, pn - n$$

2.3.2   Finding the Standard Deviation of the Particles Distribution in Space and Time

We compute the standard deviation of our particle's movement through space and time in a similar way. First, we find the second moment of m,

$$E(m^2) = \sum_{m=-n}^{n} m^2 W(m, n)$$

Our earlier discussions still apply and we find we can rewrite this as

$$E(m^2) = \sum_{j=0}^{n} (2j - n)^2 \, B_{n,j} \, p^j \, q^{n-j}$$

$$= \sum_{j=0}^{n} (4j^2 - 4jn + n^2) \, B_{n,j} \, p^j \, q^{n-j}$$

$$= 4E(j^2) - 4n \, E(j) + n^2 \, S$$

where $E(j^2)$ is the second moment of the binomial distribution. We know S is 1 and $E(j)$ is pn. So we only have to compute $E(j^2)$. Note that

$$p^2 \, \frac{d^2}{d^2 p} \left((p + q)^n \right) = p^2 \, n \, (n - 1) \left((p + q)^{n-2} \right) = p^2 \, n \, (n - 1)$$

Also

$$p^2 \, n \, (n - 1) = p^2 \, \frac{d^2}{d^2 p} \left((p + q)^n \right)$$

$$= p^2 \, \frac{d^2}{d^2 p} \left(\sum_{j=0}^{n} B_{n,j} \, p^j \, q^{n-j} \right)$$

$$= \sum_{j=0}^{n} p^2 \, \frac{d^2}{d^2 p} \left(B_{n,j} \, p^j \, q^{n-j} \right)$$

$$= \sum_{j=0}^{n} p^2 \, j \, (j - 1) \, B_{n,j} \, p^{j-2} \, q^{n-j}$$

$$= \sum_{j=0}^{n} (j^2 - j) \, B_{n,j} \, p^j \, q^{n-j}$$

$$= E(j^2) - E(j)$$

We conclude that

$$E(j^2) = p^2 \, n \, (n - 1) + p \, n$$

Since, we also have a formula for $E(m^2)$, we see

$$E(m^2) = 4E(j^2) - 4n \, E(j) + n^2$$

Now recall the standard formula from Statistics: the square of the standard deviation of our distribution is $\sigma^2 = E(m^2) - (E(m))^2$. Hence,

$$
\begin{aligned}
\sigma^2 &= E(m^2) - (E(m))^2 \\
&= 4E(j^2) - 4\,n\,E(j) + n^2 - (2E(j) - n)^2 \\
&= 4(E(j^2) - (E(j))^2) \\
&= 4\left(p^2\,n\,(n-1) + pn - (pn)^2\right) \\
&= 4\,np(1-p) = 4\,npq.
\end{aligned}
$$

Hence, the standard deviation is

$$
\sigma = \sqrt{4pqn}.
$$

2.3.3　Specializing to an Equal Probability Left and Right Random Walk

Here, p and q are both 0.5. We see that $4pq = 1$ and

$$
E(m) = 2\,p\,n - n = 0
$$
$$
\sigma = \sqrt{n}
$$

Note, that if the random walk is skewed, with say $p = 0.1$, then we would obtain

$$
E(m) = 2\,p\,n - n = -0.8n
$$
$$
\sigma = .6\sqrt{n}
$$

so that for large n, the standard deviation of our particle's movement would be approximately $0.6n$ rather than n.

2.4　Macroscopic Scale

For a very large number of steps, the probability distribution, $W(m, n)$, will approach a limiting form. This is done by using an approximation to $k!$ that is know as the Stirling Approximation. It is known that for very large k,

$$
k! \approx \sqrt{2\pi k}\,\left(\frac{k}{e}\right)^k.
$$

The distribution of our particle's position throughout space and time can be written as

$$W(m, n) = B_{n, \frac{n+m}{2}} \, p^{\frac{n+m}{2}} \, q^{\frac{n-m}{2}}$$

using our definitions of n^+ and n^- (it is understood that $W(m, n)$ is zero for non integer values of these fractions). We can apply Stirling's approximation to $B_{n, \frac{n+m}{2}}$:

$$n! \approx \sqrt{2\pi n} \left(\frac{n}{e}\right)^n$$

$$\left(\frac{n+m}{2}\right)! \approx \sqrt{2\pi \frac{n+m}{2}} \left(\frac{n+m}{2e}\right)^{\frac{n+m}{2}}$$

$$\approx \sqrt{\pi(n+m)} \left(\frac{n}{2e}\right)^{\frac{n+m}{2}} \left(1 + \frac{m}{n}\right)^{\frac{n+m}{2}}$$

$$\left(\frac{n-m}{2}\right)! \approx \sqrt{2\pi \frac{n-m}{2}} \left(\frac{n-m}{2e}\right)^{\frac{n-m}{2}}$$

$$\approx \sqrt{\pi(n-m)} \left(\frac{n}{2e}\right)^{\frac{n-m}{2}} \left(1 - \frac{m}{n}\right)^{\frac{n-m}{2}}$$

From this, we find

$$\left(\frac{n+m}{2}\right)! \left(\frac{n-m}{2}\right)! \approx \pi \, n \sqrt{1 - \frac{m^2}{n^2}} \left(\frac{n}{2e}\right)^n \left(1 - \frac{m^2}{n^2}\right)^{\frac{n}{2}} \left(1 - \frac{m}{n}\right)^{\frac{-m}{2}} \left(1 + \frac{m}{n}\right)^{\frac{m}{2}}$$

Hence, we see

$$B_{n, \frac{n+m}{2}} \approx \frac{\sqrt{2\pi n}}{\pi n} \left(\frac{n}{e}\right)^n \left(\frac{n}{2e}\right)^{-n} \left(1 - \frac{m^2}{n^2}\right)^{-\frac{1}{2}} \left(1 - \frac{m^2}{n^2}\right)^{-\frac{n}{2}} \left(1 - \frac{m}{n}\right)^{\frac{m}{2}} \left(1 + \frac{m}{n}\right)^{-\frac{m}{2}}$$

$$\approx \sqrt{\frac{2}{\pi n}} \, 2^n \left(1 - \frac{m^2}{n^2}\right)^{-\frac{1}{2}} \left(1 - \frac{m^2}{n^2}\right)^{-\frac{n}{2}} \left(1 - \frac{m}{n}\right)^{\frac{m}{2}} \left(1 + \frac{m}{n}\right)^{-\frac{m}{2}}$$

Thus,

$$\ln\left(B_{n, \frac{n+m}{2}}\right) \approx \frac{1}{2} \ln\left(\frac{2}{\pi n}\right) + n \ln(2) + -\frac{1}{2} \ln\left(1 - \frac{m^2}{n^2}\right) + \frac{-n}{2} \ln\left(1 - \frac{m^2}{n^2}\right)$$

$$+ \frac{m}{2} \ln\left(1 - \frac{m}{n}\right) + \frac{-m}{2} \ln\left(1 + \frac{m}{n}\right)$$

Now, for small x, the standard Taylor's series approximation gives $\ln(1 + x) \approx x$; hence, for $\frac{m}{n}$ sufficiently small, we can say

$$
\ln\left(B_{n, \frac{n+m}{2}}\right) \approx \frac{1}{2} \ln\left(\frac{2}{\pi n}\right) + n \ln(2) + \frac{1}{2} \frac{m^2}{n^2} + \frac{n}{2} \frac{m^2}{n^2} - \frac{m}{2} \frac{m}{n} - \frac{m}{2} \frac{m}{n}
$$

$$
\approx \frac{1}{2} \ln\left(\frac{2}{\pi n}\right) + n \ln(2) + \frac{1}{2} \frac{m^2}{n^2} - \frac{m^2}{2n}
$$

For very large n (i.e. after a very large number of time steps $n\tau_m$), since we assume $\frac{m}{m}$ is very small, the term $\frac{m^2}{n^2}$ is negligible. Hence dropping that term and exponentiating, we find

$$
B_{n, \frac{n+m}{2}} \approx \sqrt{\frac{2}{n\pi}} \exp\left(\frac{-m^2}{2n}\right) 2^n
$$

This implies that

$$
W(m, n) \approx \sqrt{\frac{2}{n\pi}} \exp\left(\frac{-m^2}{2n}\right) 2^n \, p^{\frac{n+m}{2}} \, q^{\frac{n-m}{2}}
$$

$$
\approx \sqrt{\frac{2}{n\pi}} \exp\left(\frac{-m^2}{2n}\right) (4pq)^{\frac{n}{2}} \left(\frac{p}{q}\right)^{\frac{m}{2}}
$$

Note that if the particle moves with equal probability 0.5 to the right or the left at any time tick, this reduces to

$$
W(m, n) \approx \sqrt{\frac{2}{n\pi}} \, e^{\frac{-m^2}{2n}}
$$

and for $p = \frac{1}{3}$ and $q = \frac{2}{3}$, this becomes

$$
W(m, n) \approx \sqrt{\frac{2}{n\pi}} \exp\left(\frac{-m^2}{2n}\right) \left(\frac{8}{9}\right)^{\frac{n}{2}} \left(\frac{1}{2}\right)^{\frac{m}{2}}
$$

2.5   Obtaining the Probability Density Function

From our discrete approximations in previous sections, we can now derive the probability density function, $P(x, t)$ at position x at time t. In what follows, we will assume that $p \leq 0.5$. Let Δx be a small number which is approximately $m\lambda_c$ for some m. The probability that the particle is in an interval $[x - \frac{\Delta x}{2}, x + \frac{\Delta x}{2}]$ can then be approximated by

$$
P(x, t) \, \Delta x \approx \sum_k W(k, n)
$$

where the sum is over all indices k such that the position $k\lambda_c$ lies in the interval $[x - \frac{\Delta x}{2}, x + \frac{\Delta x}{2}]$. Hence,

$$\left[x - \frac{\Delta x}{2}, x + \frac{\Delta x}{2}\right] \equiv \{(m - j)\lambda_c, ..., m\lambda_c, ..., (m + j)\lambda_c\}$$

for some integer j. Now from the way the particle moves in a random walk, only half of these tick marks will actually be positions the particle can occupy. Hence, half of the probabilities $W(m - i, n)$ for i from $-j$ to j are zero. The number of nonzero probabilities is thus $\approx \frac{\Delta x}{2\lambda_c}$. We can therefore approximate the sum by taking the middle term $W(m, n)$ and multiplying by the number of nonzero probabilities.

$$P(x, t) \, \Delta x \approx W(m, n) \frac{\Delta x}{2\lambda_c}$$

which implies, since $x = m\lambda_c$ and $t = n\tau_m$, that for very large n,

$$P(x, t) = \frac{1}{2\lambda_c} W(m, n)$$

$$= \frac{1}{\sqrt{4\pi \frac{\lambda_c^2}{2\tau_m} t}} \exp\left(\frac{-x^2}{4\frac{\lambda_c^2}{2\tau_m} t}\right) (4pq)^{\frac{t}{2\tau_m}} \left(\frac{p}{q}\right)^{\frac{x}{2\lambda_c}}$$

Note that the term $\frac{\lambda_c^2}{2\tau_m}$ is the diffusion constant D. Thus,

$$P(x, t) = \frac{1}{\sqrt{4\pi D t}} \exp\left(\frac{-x^2}{4D t}\right) (4pq)^{\frac{t}{2\tau_m}} \left(\frac{p}{q}\right)^{\frac{x}{2\lambda_c}}$$

Next, rewrite all the power terms as exponentials:

$$P(x, t) = \frac{1}{\sqrt{4\pi D t}} \exp\left(\frac{-x^2}{4D t}\right) \exp\left(\ln(4pq) \frac{t}{2\tau_m}\right) \exp\left(\ln\left(\frac{p}{q}\right) \frac{x}{2\lambda_c}\right)$$

2.5.1 p Less Than 0.5

Note since $p + q = 1$, $4pq$ is between bigger than zero and strictly less than 1 for all nonzero p and q with p not 0.5. So for us, if we let ξ be the value $\frac{1}{4pq}$, we know that $\ln(\xi) > 0$. Let $A = \ln(\xi)$. Next, let ζ be the value $\frac{q}{p}$. Since p is less than $1/2$ here, this means $\zeta > 1$. Let $B = \ln(\zeta)$. Then, we know A and B are both positive in this case unless the probability of movement left and right is equal. In the case of equal probability, $\xi = 1$, $\zeta = 1$ and $A = B = 0$. However, with unequal probability, we have

$$(4pq)^{\frac{t}{2\tau_m}} = (\xi)^{\frac{-t}{2\tau_m}} = \exp\left(\frac{-t \ln(\xi)}{2\tau_m}\right).$$

Also,

$$\left(\frac{p}{q}\right)^{\frac{x}{2\lambda_c}} = (\zeta)^{\frac{-x}{2\lambda_c}} = \exp\left(\frac{-x \ln(\zeta)}{2\lambda_c}\right)$$

Rewriting the density function, we obtain

$$P(x, t) = \frac{1}{\sqrt{4\pi D t}} \exp\left(\frac{-x^2}{4D t} - \ln(\xi)\frac{t}{2\tau_m} - \ln(\zeta)\frac{x}{2\lambda_c}\right)$$

We thus have

$$P(x, t) = \frac{1}{\sqrt{4\pi D t}} \exp\left(\frac{-x^2}{4D t} - A\frac{t}{2\tau_m} - B\frac{x}{2\lambda_c}\right)$$

After manipulation, we can complete the square on the quadratic term and rewrite it as

$$\frac{-x^2}{4D t} - A\frac{t}{2\tau_m} - B\frac{x}{2\lambda_c} = -\frac{1}{4Dt}\left(x + \frac{BD}{\lambda_c}t\right)^2 + \frac{B^2 - 4A}{8\tau_m}t$$

and thus

$$P(x, t) = \frac{1}{\sqrt{4\pi D t}} \exp\left(-\frac{1}{4Dt}\left(x + \frac{BD}{\lambda_c}t\right)^2\right)\exp\left(\frac{B^2 - 2A}{4\tau_m}t\right)$$

The case where p is larger than $1/2$ is handled in a symmetric manner. We simply reverse the role of p and q in the argument above.

2.5.2 p and q Are Equal

The equal probability random walk has $A = B = 0$ and so the probability density function reduces to

$$P(x, t) = \frac{1}{\sqrt{4\pi D t}} \exp\left(-\frac{x^2}{4Dt}\right)$$

2.6 Understanding the Probability Distribution of the Particle

It is important to get a strong intuitive feel for the probability distribution of the particle under the random walk and skewed random walk protocols. A normal distribution has the form

$$P(x) = \frac{1}{\sqrt{2\pi}\sigma} e^{-\frac{x^2}{2E(m^2)}}$$

Hence, comparing, we see the standard deviation of our particle can be interpreted as

$$\sigma = \lambda_C^2 \frac{t}{\tau_m}.$$

Note, in general, for fixed time, we can plot the particle's position. We show this in Fig. 2.3 for three different standard deviations, σ. In the plot, the standard deviation is labeled as D. Now if we skew the distribution so that the probability of moving to the right is now $\frac{1}{6}$, we find

$$P(x, t) = \frac{1}{\sqrt{4\pi D t}} \exp\left(-\frac{1}{4Dt}\left(x - \frac{1.61Dt}{\lambda_c}\right)^2\right) \exp\left(\frac{0.35t}{\tau_m}\right)$$

which generates the plot shown in Fig. 2.4.

Fig. 2.3 Normal distribution: spread depends on standard deviation

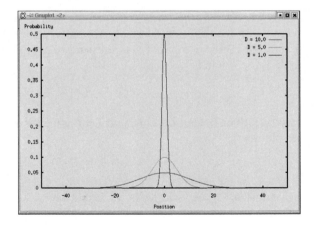

Fig. 2.4 Skewed random
walk probability distribution:
p is 0.1666

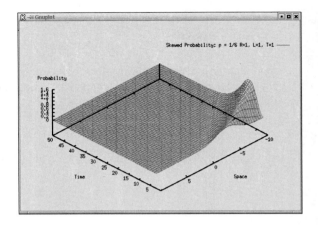

2.7   The General Diffusion Equation

We will now show that the probability density function $P(x, t)$ given by Eq. 2.1

$$P(x, t) = \frac{1}{\sqrt{4\pi D\, t}}\, e^{-\frac{x^2}{4Dt}} \tag{2.1}$$

solves the diffusion equation

$$\frac{\partial \Phi}{\partial t} = D\frac{\partial^2 \Phi}{\partial x^2}.$$

where D is the diffusion constant. A more general diffusion equation would not
assume the diffusion constant D is independent of position x. The diffusion equation
in that case would be

$$\frac{\partial u}{\partial t} = \frac{\partial}{\partial x}\left[D\frac{\partial u}{\partial x}\right]$$

However, often the term D is independent of the variable x allowing us to write the
simpler form

$$\frac{\partial u}{\partial t} = D\frac{\partial^2 u}{\partial x^2}$$

in the time and space variables (t, x). In this model, we assume x is one dimensional,
although it is easy enough to extend to higher dimensions. In 3 dimensions, u has
units of mM per liter or cm^3, but in a one dimensional setting u has units of mM per
cm. However, the diffusion coefficient D will always have units of length squared
per time unit; i.e. $\frac{cm^2}{s}$. Note in our earlier discussions, we found $P(x, t)$ had to have
units of particles per cm or simple cm^{-1}. We also assume the diffusion coefficient

D is positive. We can now show a typical solution for positive time is given by our $P(x, t)$ and let's do it by direct calculation. With a bit of manipulation, we find

$$\frac{\partial P}{\partial t} = \frac{1}{\sqrt{4\pi D}} t^{-3/2} e^{-x^2/(4Dt)} \left(-\frac{1}{2} + \frac{x^2}{4Dt} \right).$$

Next,

$$\frac{\partial P}{\partial x} = \frac{1}{\sqrt{4\pi Dt}} e^{-x^2/(4Dt)} \left(-\frac{2x}{4Dt} \right)$$

$$\frac{\partial^2 P}{\partial x^2} = \frac{1}{\sqrt{4\pi Dt}} \left(e^{-x^2/(4Dt)} \left(-\frac{2}{4Dt} \right) \right.$$

$$\left. + \left(-\frac{2x}{4Dt} \right) \left(-\frac{2x}{4Dt} \right) e^{-x^2/(4Dt)} \right)$$

$$= \frac{1}{\sqrt{4\pi Dt}} e^{-x^2/(4Dt)} \frac{2}{4Dt} \left(-1 + \frac{2x^2}{4Dt} \right)$$

$$= \frac{1}{D} \frac{1}{\sqrt{4\pi D}} t^{-3/2} e^{-x^2/(4Dt)} \left(-\frac{1}{2} + \frac{x^2}{4Dt} \right)$$

$$= \frac{1}{D} \frac{\partial P}{\partial t}.$$

Hence, we see P solves $\frac{\partial P}{\partial t} = D \frac{\partial^2 P}{\partial x^2}$. From our previous discussions, we see we can interpret the motion of our particle as that of a random walk for space constant λ_c and time constant τ_m as the number of time and position steps gets very large.

Note the behavior at $t = 0$ seems undefined. However, we can motivate the interpretation of the limiting behavior as $t \to 0$ as an impulse injection of current as follows. Note using the substitution, $y = \frac{1}{t}$, for x not zero,

$$\lim_{t \to 0} u(t, x) = \lim_{y \to \infty} \sqrt{y} \exp\left(\frac{-x^2 y}{4D} \right)$$

and so at $x = 0$, we find $u(t, 0) = \frac{1}{\sqrt{t}}$. This behaves qualitatively like an impulse. Thus, we interpret this solution as an impulse injection of essentially unbounded magnitude at $t = 0$. This corresponds to the usual delta function input $\delta(0)$. If the amount of current injected is $I\delta(0)$, the solution is

$$u(t, x) = \frac{I}{\sqrt{4\pi Dt}} \exp\left(\frac{-x^2}{4Dt} \right).$$

Since for $x \neq 0$, $\lim_{x \to \infty} u(t, x) = 0$ and $\lim_{t \to 0} u(t, x) = 0$, we know $u(t, x)$ has a maximum value at some value of t. The generic shape of the solution can be seen in Fig. 2.5. The solution at $x = 0$ behaves like the curve $\frac{1}{\sqrt{t}}$ as is shown. The solutions for non zero values of x behave like spatially spread and damped pulses. In fact, the

Fig. 2.5 The generic
diffusion solution behavior

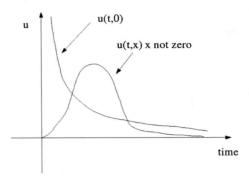

solution to the diffusion equation for nonzero x provides a nice model of the typical
excitatory pulse we see in a dendritic tree. Also, if $I < 0$, this is a good model of an
inhibitory pulse as well. Letting

$$I(t) = \frac{1}{\sqrt{4\pi Dt}} \exp\left(-\frac{x^2}{4Dt}\right)$$

we find for nonzero t, that

$$I'(t) = \frac{1}{\sqrt{4\pi D}} t^{\frac{-3}{2}} \exp\left(-\frac{x^2}{4Dt}\right)\left\{-\frac{1}{2} + \frac{x^2}{4Dt}\right\}$$

For $t \neq 0$, $f'(t) = 0$ implies $-\frac{1}{2} + \frac{x^2}{4Dt} = 0$. Thus, the maximum of the injection
pulse occurs at $t = \frac{x^2}{2D}$. The maximum value of the current is then

$$I\left(\frac{x^2}{2D}\right) = \frac{1}{\sqrt{4\pi D\,(x^2/(2D))}} e^{\left(-\frac{x^2}{4D}\frac{2D}{x^2}\right)}$$

$$= \frac{1}{\sqrt{2\pi}\,|x|}\, e^{-\frac{1}{2}}.$$

it follows that the current pulses become very sharply defined with large heights as x
approaches 0 and for large x, the pulse has a low height and is spread out significantly.
We show the qualitative nature of these pulses in Fig. 2.6. In both curves shown, the
maximum is achieved at the time point $\frac{x^2}{2D}$. These general solutions can also be
centered at the point (t_0, x_0) giving the solution

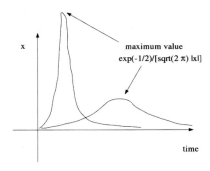

Fig. 2.6 The generic diffusion solution maximum

$$u(t, x) = \frac{I}{\sqrt{4\pi D(t - t_0)}} \exp\left(\frac{-(x - x_0)^2}{4D(t - t_0)}\right).$$

which is not defined at t_0 itself.

References

D. Johnston, S. Miao-Sin Wu, *Foundations of Cellular Neurophysiology* (MIT Press, Cambridge, 1995)

J. Peterson, *Calculus for Cognitive Scientists: Partial Differential Equation Models*, Springer Series on Cognitive Science and Technology (Springer Science+Business Media Singapore Pte Ltd., Singapore, 2015 in press)

T. Weiss, Transport, *Cellular Biophysics*, vol. 1 (MIT Press, Cambridge, 1996)

Chapter 3
Integral Transforms

We now discuss the basics of two important tools in the analysis of models: the Laplace Transform and the Fourier Transform. We will use these tools to solve the diffusion model which we obtain from the cable model by using a change of variable technique.

3.1 The Laplace Transform

The Laplace Transform acts on the time domain of a problem as follows: given the function x defined on $s \geq 0$, we define the Laplace Transform of x to be

$$\mathscr{L}(x) = \int_0^\infty x(s)\, e^{-\beta s}\, ds$$

The new function $\mathscr{L}(x)$ is defined for some domain of the new variable β. The variable β's domain is called the *frequency* domain and in general, the values of β where the transform is defined depend on the function x we are transforming. Also, in order for the Laplace transform of the function x to work, x must not grow to fast—roughly, x must decay like an exponential function with a negative coefficient. The solutions we seek to our cable equation are expected on physical grounds to decay to zero exponentially as we let t (and therefore, also s!) go to infinity and as we let the space variable z (and hence w) go to $\pm\infty$. Hence, the function Φ we seek will have a well-defined Laplace transform with respect to the s variable.

Now, what about the Laplace transform of a derivative? Consider

$$\mathscr{L}\left(\frac{dx}{ds}\right) = \int_0^\infty \frac{dx}{ds}\, e^{-\beta s}\, ds$$

© Springer Science+Business Media Singapore 2016
J.K. Peterson, *BioInformation Processing*, Cognitive Science and Technology,
DOI 10.1007/978-981-287-871-7_3

Integrating by parts, we find

$$\int_0^\infty \frac{dx}{ds} e^{-\beta s}\, ds = x(s)\, e^{-\beta s}\Big|_0^\infty + \beta \int_0^\infty x(s)\, e^{-\beta s} ds$$

$$= \lim_{R \to \infty} (x(R)\, e^{-\beta R}) - x(0) + \beta \mathcal{L}(x)$$

Now if the function x grows slower than some e^{-cR} for some constant c, the limit will be zero and we obtain

$$\mathcal{L}\left(\frac{dx}{ds}\right) = \beta \mathcal{L}(x) - x(0)$$

We often use the symbol $\hat{f}$ to denote the Laplace Transform of f.

For a simple example, consider the Laplace Transform of the function

$$f(s) = \begin{cases} e^{-as} & s \geq 0 \\ 0 & s < 0 \end{cases}$$

where a is positive. It is easy to show that for $\beta > a$,

$$\hat{f} = \mathcal{L}(f)$$

$$= \int_0^\infty e^{-as} e^{\beta s}\, ds$$

$$= \frac{1}{\beta + a}.$$

Hence, the inverse Laplace Transform, denoted by $\mathcal{L}^{-1}(\hat{f})$, here is

$$\mathcal{L}^{-1}\left(\frac{1}{\beta + a}\right) = e^{-as}.$$

3.1.1   Homework

Exercise 3.1.1 *Find the Laplace transform of $f(t) = e^t$.*

Exercise 3.1.2 *Find the Laplace transform of $f(t) = e^{-2t}$.*

Exercise 3.1.3 *Find the Laplace transform of $f(t) = e^{-6t}$.*

Exercise 3.1.4 *Find the inverse Laplace transform of $\hat{f}(s) = \frac{2}{s+4}$.*

Exercise 3.1.5 *Find the inverse Laplace transform of $\hat{f}(s) = \frac{3}{s-4}$.*

Exercise 3.1.6 *Find the inverse Laplace transform of $\hat{f}(s) = \frac{6}{2s+5}$.*

3.2 The Fourier Transform

Given a function g defined on the y axis, we define the Fourier Transform of g to be

$$\mathcal{F}(g) = \frac{1}{\sqrt{2\pi}} \int_{-\infty}^{\infty} g(y)\, e^{-j\xi y}\, dy$$

where j denotes the square root of minus 1 and the exponential term is defined to mean

$$e^{-j\xi y} = \cos(\xi y) - j\sin(\xi y)$$

This integral is well defined if g is what is called square integrable which roughly means we can get a finite value for the integral of g^2 over the y axis. Note that we can compute the Fourier Transform of the derivative of g as follows:

$$\mathcal{F}\left(\frac{dg}{dy}\right) = \frac{1}{\sqrt{2\pi}} \int_{-\infty}^{\infty} \frac{dg}{dy}\, e^{-j\xi y}\, dy$$

Integrating by parts, we find

$$\frac{1}{\sqrt{2\pi}} \int_{-\infty}^{\infty} \frac{dg}{dy}\, e^{-j\xi y}\, dy = g(y)\, e^{-j\xi y}\Big|_{-\infty}^{\infty} + j\xi\, \frac{1}{\sqrt{2\pi}} \int_{-\infty}^{\infty} g(y)\, e^{-j\xi y} dy$$

$$= \lim_{R\to\infty}\left(g(R)\, e^{-j\xi R}\right) - \lim_{R\to-\infty}\left(g(R)\, e^{-j\xi R}\right) + j\xi\, \mathcal{F}(g)$$

Since we assume that the function g decays sufficiently quickly as $y \to \pm\infty$, the first term vanishes and we have

$$\frac{1}{\sqrt{2\pi}} \int_{-\infty}^{\infty} \frac{dg}{dy}\, e^{-j\xi y}\, dy = +j\xi\, \mathcal{F}(g)$$

Applying the same type of reasoning, we can see that

$$\frac{1}{\sqrt{2\pi}} \int_{-\infty}^{\infty} \frac{d^2g}{dy^2}\, e^{-j\xi y}\, dy = \frac{dg}{dy}\, e^{-j\xi y}\Big|_{-\infty}^{\infty} + j\xi\, \frac{1}{\sqrt{2\pi}} \int_{-\infty}^{\infty} \frac{dg}{dy}\, e^{-j\xi y} dy$$

$$= \lim_{R\to\infty}\left(\frac{dg}{dy}(R)\, e^{-j\xi R}\right) - \lim_{R\to-\infty}\left(\frac{dg}{dy}(R)\, e^{-j\xi R}\right) + j\xi\, \mathcal{F}\left(\frac{dg}{dy}\right)$$

We also assume that the function g's derivative decays sufficiently quickly as $y \to \pm\infty$. Thus the the first term vanishes and we have

$$\frac{1}{\sqrt{2\pi}} \int_{-\infty}^{\infty} \frac{d^2g}{dy^2}\, e^{-j\xi y}\, dy = +j\xi\, \mathcal{F}\left(\frac{dg}{dy}\right) = j^2\xi^2\, \mathcal{F}(g) = -\xi^2\, \mathcal{F}(g)$$

because $j^2 = -1$. Hence,

$$\mathscr{F}\left(\frac{d^2g}{dy^2}\right) = -\xi^2\,\mathscr{F}(g)$$

We can then define the inverse Fourier Transform as follows. If the Fourier Transform transform of the function $g(y)$ is denoted by $\hat{g}(\xi)$, recall the inverse Fourier Transform of $\hat{g}(\xi)$ is given by

$$\mathscr{F}^{-1}(\hat{g}) = \frac{1}{\sqrt{2\pi}}\int_{-\infty}^{\infty}\hat{g}(\xi)\,e^{j\xi y}\,d\xi.$$

Now, if the Fourier Transform transform of the function $g(y)$ is denoted by $\hat{g}(\xi)$, recall the inverse Fourier Transform of $\hat{g}(\xi)$ is given by

$$\mathscr{F}^{-1}(\hat{g}) = \frac{1}{\sqrt{2\pi}}\int_{-\infty}^{\infty}\hat{g}(\xi)\,e^{j\xi y}\,d\xi.$$

Here is a standard example of such an inversion. Consider the inverse Fourier transform of

$$\frac{r_0\lambda_c I_0}{\sqrt{2\pi}}\frac{1}{\sqrt{2\pi}}\int_{-\infty}^{\infty}e^{-\xi^2 s}\,e^{j\xi y}\,d\xi = \frac{r_0\lambda_c I_0}{2\pi}\int_{-\infty}^{\infty}e^{-\xi^2 s + j\xi y}\,d\xi.$$

Now to invert the remaining part, we rewrite the exponent by completing the square:

$$-\xi^2 s + j\xi y = -s\left(\xi^2 - j\frac{\xi}{s}\xi + \left(\frac{jy}{2s}\right)^2 - \left(\frac{jy}{2s}\right)^2\right)$$

$$= -s\left(\left(\xi - \frac{jy}{2s}\right)^2 - \left(\frac{jy}{2s}\right)^2\right)$$

Hence,

$$e^{-\xi^2 s + j\xi y} = e^{-s(\xi - \frac{jy}{2s})^2}\,e^{-\frac{y^2}{4s}}$$

because $j^2 = -1$. To handle the inversion here, we note that for any positive a and positive base points x_0 and y_0:

$$\int_{-\infty}^{\infty}\int_{-\infty}^{\infty}e^{-a(x-x_0)^2}\,e^{-a(y-y_0)^2}\,dxdy = \int_{-\infty}^{\infty}\int_{-\infty}^{\infty}e^{-u^2}\,e^{-v^2}\frac{1}{a}\,dudv$$

using the change of variables $u = \sqrt{a}(x - x_0)$ and $v = \sqrt{a}(y - y_0)$. Now we convert to polar coordinates and it can be shown the resulting calculation gives

$$\int_{-\infty}^{\infty}\int_{-\infty}^{\infty} e^{-u^2} e^{-v^2} \frac{1}{a} du dv = \frac{1}{a}\int_{0}^{\infty}\int_{0}^{2\pi} e^{-r^2} r dr d\theta = \frac{\pi}{a}$$

We are doing what is called a double integration here which we have never discussed. However, our need for it never arose until this moment. For our purposes, think of it as applying our standard Riemann integral two times in succession. For example, if we had the double integral

$$\int_{1}^{2}\int_{2}^{3} (x^2 + y^2)\, dx\, dy$$

we would interpret this as two Riemann integrals working from the *inside* out. We would first find $\int_{2}^{3} (x^2 + y^2)\, dx$ by thinking of the integrand as a function of x only where the y variable is treated as a constant. This is essentially the inverse of partial differentiation! We would find

$$\int_{2}^{3} (x^2 + y^2)\, dx = \left(\frac{x^3}{3} + y^2 x\right)\Big|_{2}^{3}$$

where we use our usual one variable Cauchy Fundamental Theorem of Calculus. This gives $\frac{19}{3} + y^2$ as the answer. Then you apply the outer Riemann integral $\int_{1}^{2}$ to this result to get

$$\int_{1}^{2} \left(\frac{19}{3} + y^2\right) dy = \left(\frac{19}{3}y + \frac{y^3}{3}\right)\Big|_{1}^{2} = \frac{19}{3} + \frac{7}{3}.$$

Our situation above is more interesting than our example, of course. We have infinite limits, $\int_{-\infty}^{\infty}$ which we interpret as as usual improper integral. Naturally, since Riemann integrals can be interpreted as areas under curves, many time an integration like this over an unbounded domain ends up with an infinite answer. But the functions we are integrating in our example strongly decay as they are e^{-u^2} and e^{-v^2}. So there is hope that even in this infinite case, we will get a finite result. You should take more mathematics classes! Our volumes do not discuss two and three dimensional Riemann integration although we have discussed differentiation in the setting of more than one variable. There is always more to learn, but for our purposes, we just need this one calculation so we didn't want to belabor this point with several extra chapters. Now getting back to our problem, we see after converting to polar coordinates, we converted to convert the integral $\int_{-\infty}^{\infty}\int_{-\infty}^{\infty} du dv$ into the integral $\int_{0}^{\infty}\int_{0}^{2\pi} r dr d\theta$. This used some additional multiple integration tools we are skipping: essentially, to add up the area of the uv plane using rectangular blocks of area $du\, dv$ we need an infinite number of them stretching from $-\infty$ to ∞ in both the u and v directions. However, to add up the area of the u v plane using area chunks based on the polar coordinate variables r and θ, we divide up the two dimensional space using rays from the origin every $d\theta$ radians. Then the area pieces formed from slice the rays between radius r and $r + dr$. The resulting area is like a chopped off piece of pie and has area $r\, dr\, d\theta$. Then to cover the whole plane we need an infinite number of dr's and $d\theta$'s from 0 to 2π. Yeah, that's all there is too it. Fortunately, like we said, this background is not so important here. Our move to polar coordinates thus gives us

$$\left(\int_{-\infty}^{\infty} e^{-a(x-x_0)^2}\, dx \right)^2 = \frac{\pi}{a} \implies \int_{-\infty}^{\infty} e^{-a(x-x_0)^2}\, dx = \sqrt{\frac{\pi}{a}}$$

We can apply this result to our problem to see

$$\frac{r_0 \lambda_c I_0}{2\pi} \int_{-\infty}^{\infty} e^{-\xi^2 s + j\xi y}\, d\xi = \frac{r_0 \lambda_c I_0}{2\pi} \int_{-\infty}^{\infty} e^{-s(\xi - \frac{jy}{2s})^2}\, e^{-\frac{y^2}{4s}}\, d\xi$$

$$= \frac{r_0 \lambda_c I_0}{2\pi} \sqrt{\frac{\pi}{s}}\, e^{-\frac{y^2}{4s}} = r_0 \lambda_c I_0\, \frac{1}{\sqrt{4\pi s}}\, e^{-\frac{y^2}{4s}}$$

3.2.1   Homework

Exercise 3.2.1 *Find the inverse Fourier Transform of $e^{-2\xi^2}$ showing all of the messy steps in the discussion above.*

Exercise 3.2.2 *Find the inverse Fourier Transform of $e^{-1.5\xi^2}$ showing all of the messy steps in the discussion above.*

Exercise 3.2.3 *Find the inverse Fourier Transform of $e^{-4\xi^2}$ showing all of the messy steps in the discussion above.*

Chapter 4
The Time Dependent Cable Solution

We are now ready to solve the cable equation directly. There are lots of details here to go through, so it will take awhile. However, as we said earlier, the solutions we obtain here are essentially the same as the ones we found using random walk ideas. So we now have a good understanding of what the diffusion constant means. Note this is very important to understanding diffusion based triggers. Recall the full cable equation

$$\lambda_c^2 \frac{\partial^2 v_m}{\partial z^2} = v_m + \tau_m \frac{\partial v_m}{\partial t} - r_o \lambda_c^2 k_e$$

Recall that k_e is current per unit length. We are going to show you a mathematical way to solve the above cable equation when there is an instantaneous current impulse applied at some nonnegative time t_0 and nonnegative spatial location z_0. Essentially, we will think of this instantaneous impulse as a Dirac delta function input as we have discussed before: i.e. we will need to solve

$$\lambda_c^2 \frac{\partial^2 v_m}{\partial z^2} = v_m + \tau_m \frac{\partial v_m}{\partial t} - r_o \lambda_c^2 I_e \delta(t - t_0, z - z_0)$$

where $\delta(t - t_0, z - z_0)$ is a Dirac impulse applied at the ordered pair (t_0, z_0). We will simplify our reasoning by thinking of the impulse applied at $(0, 0)$ as we can simply translate our solution later as we did before for the idealized impulse solution to the time independent cable equation.

© Springer Science+Business Media Singapore 2016
J.K. Peterson, *BioInformation Processing*, Cognitive Science and Technology,
DOI 10.1007/978-981-287-871-7_4

4.1   The Solution for a Current Impulse

We assume that the cable is infinitely long and there is a current injection at some point z_0 on the cable which instantaneously delivers I_0 amps of current. As usual, we will model this instantaneous delivery of current using a family of pulses. For convenience of exposition, we will consider the point of application of the pulses to be $z_0 = 0$.

4.1.1   Modeling the Current Pulses

Consider the two parameter family of pulses defined on the rectangle $[-\frac{\tau_m}{n}, \frac{\tau_m}{n}] \times [-\frac{\lambda_c}{m}, \frac{\lambda_c}{m}]$ by

$$P_{nm}(t, z) = \frac{I_0 nm}{\tau_m \lambda_c \gamma^2} \, e^{-\frac{2\tau_m^2}{n^2 t^2 - \tau_m^2}} \, e^{-\frac{2\lambda_c^2}{m^2 z^2 - \lambda_c^2}}$$

and defined to be 0 off the rectangle. This family is well-known to be a C^∞ family of functions with compact support (the support of each function is the rectangle). At this point, γ is a constant to be determined. Note that P_{nm} is measured in $\frac{\text{amp}}{\text{cm}-\text{s}}$. We know the currents k_e have units of $\frac{\text{amp}}{\text{cm}}$; hence, we model the family of current impulses k_e^{nm} by

$$k_e^{nm}(t, z) = \tau_m \, P_{nm}(t, z)$$

This gives us the proper units for the current impulses. This is then a specific example of the type of pulses we have used in the past. There are several differences:

- Here we give a specific functional form for our pulses which we did not do before. It is straightforward to show that these pulses are zero off the (t, z) rectangle and are infinitely differentiable for all time and space points. Most importantly, this means the pulses are very smooth at the boundary points $t = \pm \frac{\tau_m}{n}$ and $z = \pm \frac{\lambda_c}{m}$. Note we are indeed allowing these pulses to be active for a small interval of time before zero. When we solve the actual cable problem, we will only be using the positive time portion.
- The current delivered by this pulse is obtained by the following integration:

$$J = \int_{-\infty}^{\infty} \int_{-\infty}^{\infty} P_{nm}(t, z) dt dz$$

$$= \int_{-\frac{\lambda_c}{m}}^{\frac{\lambda_c}{m}} \int_{-\frac{\tau_m}{n}}^{\frac{\tau_m}{n}} P_{nm}(t, z) dt dz$$

Since we will be interested only in positive time, we will want to evaluate

$$\frac{J}{2} = \int_{-\frac{\lambda_c}{m}}^{\frac{\lambda_c}{m}} \int_{0}^{\frac{\tau_m}{n}} P_{nm}(t, z)dt dz$$

The constant γ will be chosen so that these integrals give the constant value of $J = 2I_0$ for all n and m. If we integrate over the positive time half of the pulse, we will then get the constant I_0 instead.

- The units of our pulse are $\frac{amps}{cm-s}$. The time integration then gives us $\frac{amp}{cm}$ and the following spatial integration gives us amperes.

Let's see how we should choose γ: consider

$$J = \int_{-\frac{\lambda_c}{m}}^{\frac{\lambda_c}{m}} \int_{-\frac{\tau_m}{n}}^{\frac{\tau_m}{n}} P_{nm}(t, z)dt dz$$

Make the substitutions $\beta_1 = \frac{nt}{\tau_m}$ and $\beta_2 = \frac{mz}{\lambda_c}$. We then obtain with a bit of algebra

$$
\begin{aligned}
J &= \frac{\tau_m \lambda_c}{nm} \int_{-1}^{1} \int_{-1}^{1} \frac{I_0 nm}{\tau_m \lambda_c \gamma^2} e^{-\frac{2}{\beta_1^2 - 1}} e^{-\frac{2}{\beta_2^2 - 1}} d\beta_1 d\beta_2 \\
&= \frac{I_0}{\gamma^2} \left(\int_{-1}^{1} e^{-\frac{2}{\beta_1^2 - 1}} d\beta_1 \int_{-1}^{1} e^{-\frac{2}{\beta_1^2 - 1}} d\beta_2 \right) \\
&= \frac{I_0}{\gamma^2} \left(\int_{-1}^{1} e^{-\frac{2}{x^2 - 1}} dx \right)^2
\end{aligned}
$$

Clearly, if we choose the constant γ to be

$$\gamma = \frac{1}{\sqrt{2}} \int_{-1}^{1} e^{-\frac{2}{x^2 - 1}} dx$$

then all full pulses P_{nm} deliver $2I_0$ amperes of current when integrated over space and time and all pulses with only nonnegative time deliver I_0 amperes.

4.1.2   Scaling the Cable Equation

We can convert the full cable equation

$$\lambda_c^2 \frac{\partial^2 v_m}{\partial z^2} = v_m + \tau_m \frac{\partial v_m}{\partial t} - r_o \lambda_c^2 k_e,$$

where k_e is current per unit length into a diffusion equation with a change of variables. First, we introduce a dimensionless scaling to make it easier via the change of variables: $y = \frac{z}{\lambda_c}$ and $s = \frac{t}{\tau_m}$. With these changes, space will be measured in units of space constants and time in units of time constants. We then define the a new *voltage* variable w by

$$w(s, y) = v_m(\tau_m t, \lambda_c z)$$

It is then easy to show using the chain rule that

$$\frac{\partial^2 w}{\partial y^2} = \lambda_c^2 \frac{\partial^2 v_m}{\partial z^2}$$

$$\frac{\partial w}{\partial s} = \tau_m \frac{\partial v_m}{\partial t}$$

giving us the scaled cable equation

$$\frac{\partial^2 w}{\partial y^2} = w + \frac{\partial w}{\partial s} - r_o \lambda_c^2 k_e(\tau_m s, \lambda_c y)$$

Now to further simplify our work, let's make the additional change of variables

$$\Phi(s, y) = w(s, y) e^s$$

Then

$$\frac{\partial \Phi}{\partial s} = \left(\frac{\partial w}{\partial s} + w \right) e^s$$

$$\frac{\partial^2 \Phi}{\partial y^2} = \frac{\partial^2 w}{\partial y^2} e^s$$

leading to

$$\frac{\partial^2 \Phi}{\partial y^2} e^{-s} = \frac{\partial \Phi}{\partial s} e^{-s} - r_0 \lambda_c^2 k_e(\tau_m s, \lambda_c y)$$

After rearranging, we have the version of the transformed cable equation we need to solve:

$$\frac{\partial^2 \Phi}{\partial y^2} = \frac{\partial \Phi}{\partial s} - r_0 \lambda_c^2 \tau_m k_e(\tau_m s, \lambda_c y) e^s \tag{4.1}$$

Recall that k_e is current per unit length. We are going to show you a physical way to find a solution to the above diffusion equation when there is an instantaneous current impulse of strength I_0 applied at some nonnegative time t_0 and nonnegative spatial location z_0. Essentially, we will think of this instantaneous impulse as a Dirac delta function input as we have discussed before: i.e., we will need to solve

$$\frac{\partial^2 \Phi}{\partial y^2} = \frac{\partial \Phi}{\partial s} - r_o \, \lambda_c^2 \, I_0 \, \delta(s - s_0, y - y_0) \, e^s$$

In particular, for our family of pulses, P_{nm}, we have

$$P_{nm}(\tau_m s, \lambda_c y) = \frac{I_0 nm}{\tau_m \lambda_c \gamma^2} \, e^{\frac{2\tau_m^2}{n^2 \tau_m^2 s^2 - \tau_m^2}} \, e^{\frac{2\lambda_c^2}{m^2 \lambda_c^2 y^2 - \lambda_c^2}}$$

$$= \frac{I_0 nm}{\tau_m \lambda_c \gamma^2} \, e^{\frac{2}{n^2 s^2 - 1}} \, e^{\frac{2}{m^2 y^2 - 1}}$$

Then, we find

$$\frac{\partial^2 \Phi}{\partial y^2} = \frac{\partial \Phi}{\partial s} - r_0 \lambda_c^2 \, \tau_m \, P_{nm}(\tau_m s, \lambda_c y) \, e^s. \tag{4.2}$$

4.1.3 Applying the Laplace Transform in Time

We will apply some rather sophisticated mathematical tools to solve the Eq. (4.2). These include the Laplace and Fourier Transforms which were introduced in Chap. 3. Hence, applying the Laplace transform to both sides of Eq. 4.2, we have

$$\mathscr{L}\left(\frac{\partial^2 \Phi}{\partial y^2}\right) = \mathscr{L}\left(\frac{\partial \Phi}{\partial s}\right) - r_0 \lambda_c^2 \, \mathscr{L}(P_{nm}(\tau_m s, \lambda_c y) \, e^s)$$

$$\frac{\partial^2 \mathscr{L}(\Phi)}{\partial y^2} = \beta \, \mathscr{L}(\Phi) - \Phi(0, y) - r_0 \lambda_c^2 \, \tau_m \, \mathscr{L}(P_{nm}(\tau_m s, \lambda_c y) \, e^s)$$

This is just the transform of the time portion of the equation. The space portion has been left alone. We will now further assume that

$$\Phi(0, y) = 0, \quad y \neq 0$$

This is the same as assuming

$$v_m(0, z) = 0, \quad z \neq 0$$

which is a reasonable physical initial condition. This gives us

$$\frac{\partial^2 \mathscr{L}(\Phi)}{\partial y^2} = \beta \, \mathscr{L}(\Phi) - r_0 \lambda_c^2 \, \tau_m \, \mathscr{L}(P_{nm}(\tau_m s, \lambda_c y) \, e^s)$$

4.1.4 Applying the Fourier Transform in Space

We now apply the Fourier Transform in space to the equation we obtained after applying the Laplace Transform in time. We have

$$\mathscr{F}\left(\frac{\partial^2 \mathscr{L}(\Phi)}{\partial y^2}\right) = \beta\,\mathscr{F}(\mathscr{L}(\Phi)) - r_0\lambda_c^2\,\tau_m\,\mathscr{F}(\mathscr{L}(P_{nm}(\tau_m s, \lambda_c y)\,e^s))$$

or

$$-\xi^2\mathscr{F}(\mathscr{L}(\Phi)) = \beta\,\mathscr{F}(\mathscr{L}(\Phi)) - r_0\lambda_c^2\,\tau_m\,\mathscr{F}(\mathscr{L}(P_{nm}(\tau_m s, \lambda_c y)\,e^s))$$

For convenience, let

$$\mathscr{T}(\Phi) = \mathscr{F}(\mathscr{L}(\Phi))$$

Then, we see we have

$$-\xi^2\,\mathscr{T}(\Phi) = \beta\,\mathscr{T}(\Phi) - r_0\lambda_c^2\,\tau_m\,\mathscr{T}(P_{nm}(\tau_m s, \lambda_c y)\,e^s)$$

4.1.5 The $\mathscr{T}$ Transform of the Pulse

We must now compute the $\mathscr{T}$ transform of the pulse term $P_{nm}(\tau_m s, \lambda_c y)\,e^s$. Since the pulse is zero off the (t, z) rectangle $[-\frac{\tau_m}{n}, \frac{\tau_m}{n}] \times [-\frac{\lambda_c}{m}, \frac{\lambda_c}{m}]$, we see the (s, y) integration rectangle reduces to $[-\frac{1}{n}, \frac{1}{n}] \times [-\frac{1}{m}, \frac{1}{m}]$. Hence,

$$\mathscr{T}(P_{nm}(\tau_m s, \lambda_c y)\,e^s) = \frac{1}{\sqrt{2\pi}} \int_{-\frac{1}{m}}^{\frac{1}{m}} \int_0^{\frac{1}{n}} P_{nm}(\tau_m s, \lambda_c y)\,e^s\,e^{-\beta s}\,e^{-j\xi y}\,ds dy$$

$$= \frac{1}{\sqrt{2\pi}} \int_{-\frac{1}{m}}^{\frac{1}{m}} \int_0^{\frac{1}{n}} \frac{I_0 nm}{\tau_m \lambda_c \gamma^2}\,e^{-\frac{2}{n^2 s^2 - 1}}\,e^{-\frac{2}{m^2 y^2 - 1}}\,e^s\,e^{-\beta s}\,e^{-j\xi y}\,ds dy$$

Now use the change of variables $\zeta = ns$ and $u = my$ to obtain

$$\mathscr{T}(P_{nm}(\tau_m s, \lambda_c y)\,e^s) = \frac{1}{\sqrt{2\pi}} \int_{-1}^{1} \int_0^1 \frac{I_0 nm}{\tau_m \lambda_c \gamma^2}\,e^{-\frac{2}{\zeta^2 - 1}}\,e^{-\frac{2}{u^2 - 1}}\,e^{\frac{\zeta}{n}}\,e^{-\frac{\beta\zeta}{n}}\,e^{-\frac{j\xi u}{m}}\,\frac{d\zeta}{n}\,\frac{du}{m}$$

$$= \frac{1}{\sqrt{2\pi}} \frac{1}{2} \int_{-1}^{1} \int_{-1}^{1} \frac{I_0}{\tau_m \lambda_c \gamma^2}\,e^{-\frac{2}{\zeta^2 - 1}}\,e^{-\frac{2}{u^2 - 1}}\,e^{\frac{\zeta}{n}}\,e^{-\frac{\beta\zeta}{n}}\,e^{-\frac{j\xi u}{m}}\,d\zeta\,du$$

Note that this implies that

$$\lim_{n,m\to\infty} \mathscr{T}(P_{nm}(\tau_m s, \lambda_c y) e^s) = \frac{1}{\sqrt{8\pi}} \int_{-1}^{1} \int_{-1}^{1} \frac{I_0}{\tau_m \lambda_c \gamma^2} e^{\frac{2}{\zeta^2-1}} e^{\frac{2}{u^2-1}} d\zeta \, du \lim_{n,m\to\infty} \left(e^{\frac{\zeta}{n}} e^{-\frac{\beta\zeta}{n}} e^{-\frac{j\xi u}{m}}\right)$$

$$= \frac{1}{\sqrt{8\pi}} \int_{-1}^{1} \int_{-1}^{1} \frac{I_0}{\tau_m \lambda_c \gamma^2} e^{\frac{2}{\zeta^2-1}} e^{\frac{2}{u^2-1}} d\zeta \, du$$

$$= \frac{I_0}{\sqrt{8\pi}\tau_m \lambda_c \gamma^2} 2\gamma^2$$

$$= \frac{I_0}{\sqrt{2\pi}\tau_m \lambda_c}$$

4.1.6 The Idealized Impulse $\mathscr{T}$ Transform Solution

For a given impulse P_{nm}, we have the $\mathscr{T}$ transform solution satisfies

$$-\xi^2 \mathscr{T}(\Phi) = \beta \, \mathscr{T}(\Phi) - r_0 \lambda_c^2 \tau_m \, \mathscr{T}(P_{nm}(\tau_m s, \lambda_c y) e^s)$$

and so the idealized solution we seek is obtained by letting n and m go to infinity to obtain

$$-\xi^2 \mathscr{T}(\Phi) = \beta \, \mathscr{T}(\Phi) - r_0 \lambda_c^2 \tau_m \frac{I_0}{\sqrt{2\pi}\tau_m \lambda_c}$$

$$= \beta \, \mathscr{T}(\Phi) - r_0 \lambda_c I_0 \implies (\xi^2 + \beta)\mathscr{T}(\Phi) = \frac{r_0 \lambda_c I_0}{\sqrt{2\pi}}$$

Thus, we find

$$\Phi^* = \mathscr{T}(\Phi) = \frac{r_0 \lambda_c I_0}{\sqrt{2\pi}(\beta + \xi^2)}$$

where for convenience, we denote the $\mathscr{T}$ transform of Φ by Φ^*.

4.1.7 Inverting the $\mathscr{T}$ Transform Solution

To move back from the transform (β, ξ) space to our original (s, y) space, we apply the inverse of our $\mathscr{T}$ transform. For a function $h(\beta, \xi)$, this is defined by

$$\mathscr{T}^{-1}(h) = \frac{1}{\sqrt{2\pi}} \int_{-\infty}^{\infty} \int_{0}^{\infty} h(\beta, \xi) \, e^{\beta s} \, e^{j\xi y} \, d\beta d\xi = \mathscr{F}^{-1}\left(\mathscr{L}^{-1}(h)\right)$$

To find the solution to our cable equation, we will apply this inverse transform to Φ. First, we can compute the inner inverse Laplace transform to obtain

$$\mathscr{L}^{-1}(\Phi^*) = \frac{r_0 \lambda_c I_0}{\sqrt{2\pi}} \, \mathscr{L}^{-1}\left(\frac{1}{\beta + \xi^2}\right) = r_0 \lambda_c I_0 \, e^{-\xi^2 s}.$$

Recall the inverse Fourier Transform of $\hat{g}(\xi)$ is given by

$$\mathscr{F}^{-1}(\hat{g}) = \frac{1}{\sqrt{2\pi}} \int_{-\infty}^{\infty} \hat{g}(\xi) \, e^{j\xi y} \, d\xi.$$

Hence, we need to calculate

$$\frac{r_0 \lambda_c I_0}{2\pi} \int_{-\infty}^{\infty} e^{-\xi^2 s} \, e^{j\xi y} \, d\xi = \frac{r_0 \lambda_c I_0}{2\pi} \int_{-\infty}^{\infty} e^{-\xi^2 s + j\xi y} \, d\xi.$$

This inversion has been done in Chap. 3. Hence, we find

$$\frac{r_0 \lambda_c I_0}{2\pi} \int_{-\infty}^{\infty} e^{-\xi^2 s + j\xi y} \, d\xi = \frac{r_0 \lambda_c I_0}{2\pi} \int_{-\infty}^{\infty} e^{-s(\xi - \frac{jy}{2s})^2} \, e^{-\frac{y^2}{4s}} \, d\xi$$

$$= \frac{r_0 \lambda_c I_0}{2\pi} \sqrt{\frac{\pi}{s}} \, e^{-\frac{y^2}{4s}} = r_0 \lambda_c I_0 \frac{1}{\sqrt{4\pi s}} \, e^{-\frac{y^2}{4s}}$$

Hence, our idealized solution is

$$\Phi(s, y) = r_0 \lambda_c I_0 \frac{1}{\sqrt{4\pi s}} \, e^{-\frac{y^2}{4s}}$$

Note, if we think of the diffusion constant as $D_0 = \frac{\lambda_c^2}{\tau_m}$, we can rewrite $\Phi(s, y)$ as follows:

$$\Phi(s, y) = r_0 \lambda_c I_0 \frac{1}{\sqrt{4\pi (\lambda_c^2/\tau_m)(t/\lambda_c^2)}} \, e^{-\frac{x^2}{4(\lambda_c^2/\tau_m)t}}$$

$$= r_0 \lambda_c I_0 \frac{1}{\sqrt{4\pi D_0(t/\lambda_c^2)}} \, e^{-\frac{x^2}{4D_0 t}} = r_0 \lambda_c^2 I_0 \frac{1}{\sqrt{4\pi D_0 t}} \, e^{-\frac{x^2}{4D_0 t}}$$

It is exciting to see the term

$$P_0(x, t) = \frac{1}{\sqrt{4\pi D_0 t}} \, e^{-\frac{x^2}{4D_0 t}} \tag{4.3}$$

which is the usual probability density function for a random walk with space constant $\lambda_c/\sqrt{2}$ and time constant τ_m! Thus,

$$\Phi(s, y) = (r_0 \lambda_c) I_0 \left(\lambda_c P_0(x, t) \right)$$

The term $\lambda_c P_0(x, t)$ is the probability we are in an interval of width $\lambda_c/2$ around x and so is a scalar without units. The term $r_0 \lambda_c I_0$ has units *ohms amps* or *volts*. We can then find the full solution w since

$$w(s, y) = \Phi(s, y) e^{-s}$$
$$= r_0 \lambda_c I_0 \frac{1}{\sqrt{4\pi s}} e^{-\frac{y^2}{4s}} e^{-s}$$

We can write this in the unscaled form at pulse center (t_0, z_0) as

$$v_m(t, z) = r_0 \lambda_c I_0 \frac{1}{\sqrt{4\pi ((t - t_0)/\tau_m)}} e^{-\frac{((z-z_0)/\lambda_c)^2}{4((t-t_0)/\tau_m)}} e^{-(t-t_0)/\tau_m} \tag{4.4}$$

4.1.8 A Few Computed Results

We can use the scaled solutions to generate a few surface plots. In Fig. 4.1 we see a pulse applied at time zero and spatial location 1.0 of magnitude 4. We can use the linear superposition principle to sum two applied pulses: in Fig. 4.2, we see the

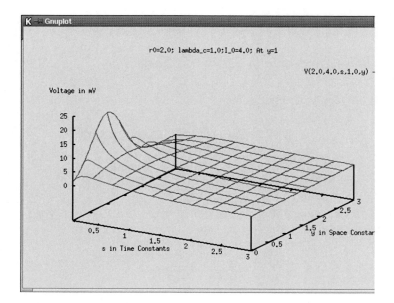

Fig. 4.1 One time dependent pulse

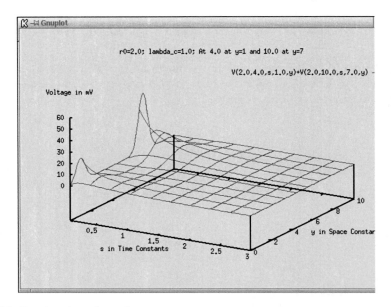

Fig. 4.2 Two time dependent pulses

effects of a pulse applied at space position 1.0 of magnitude 4.0 add to a pulse of strength 10.0 applied at position 7.0. Finally, we can take the results shown in Fig. 4.2 and simply plot the voltage at position 10.0 for the first three seconds. This is shown in Fig. 4.3.

4.1.9   Reinterpretation in Terms of Charge

Note that our family of impulses are

$$
k_e^{nm}(t, z) = \tau_m P_{nm}(t, z) = \tau_m \left(\frac{I_0 nm}{\tau_m \lambda_c \gamma^2} \, e^{\frac{2\tau_m^2}{n^2 t^2 - \tau_m^2}} \, e^{\frac{2\lambda_c^2}{m^2 z^2 - \lambda_c^2}} \right)
$$
$$
= \frac{(\tau_m I_0) nm}{\tau_m \lambda_c \gamma^2} \, e^{\frac{2\tau_m^2}{n^2 t^2 - \tau_m^2}} \, e^{\frac{2\lambda_c^2}{m^2 z^2 - \lambda_c^2}} = \frac{(Q_0) nm}{\tau_m \lambda_c \gamma^2} \, e^{\frac{2\tau_m^2}{n^2 t^2 - \tau_m^2}} \, e^{\frac{2\lambda_c^2}{m^2 z^2 - \lambda_c^2}}
$$

where Q_0 is the amount of charge deposited in one time constant. The rest of the analysis is quite similar. Thus, we can also write our solutions as

$$
v_m(t, z) = \frac{r_0 \lambda_c Q_0}{\tau_m} \frac{1}{\sqrt{4\pi \left((t - t_0)/\tau_m \right)}} \, e^{-\frac{((z - z_0)/\lambda_c)^2}{4((t - t_0)/\tau_m)}} \, e^{-(t - t_0)/\tau_m} \tag{4.5}
$$

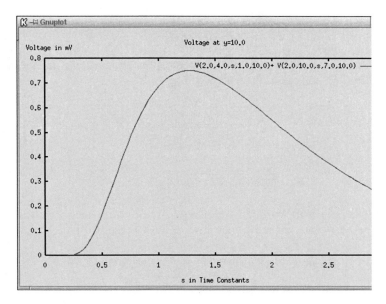

Fig. 4.3 Summed voltage at 3 time and 10 space constants

4.2 The Solution to a Constant Current

Now we need to attack a harder problem: we will apply a constant external current i_e which is defined by

$$i_e(t) = \begin{cases} I_e, & t > 0 \\ 0, & t \le 0 \end{cases}$$

Recall that we could rewrite this as $i_e = I_e \, u(t)$ for the standard unit impulse function u. Now in a time interval h, charge $Q_e = I_e \, h$ is delivered to the cable through the external membrane. Fix the positive time t. Now divide the time interval $[0, t]$ into K equal parts using h equals $\frac{t}{K}$. This gives us a set of $K + 1$ points $\{t_i\}$

$$t_i = i \, \frac{t}{K}, \quad 0 \le i \le K$$

where we note that t_0 is 0 and t_K is t. This is called a partition of the interval $[0, t]$ which we denote by the symbol $\mathscr{P}_K$. Here h is the fraction $\frac{t}{K}$. Now let's think of the charge deposited into the outer membrane at t_i as being the full amount $I_e \, h$ deposited between t_i and t_{i+1}. Then the time dependent solution due to the injection of this charge at t_i is given by

$$v_m^i(t,z) = \frac{r_0 \lambda_c I_e\, h}{\tau_m} \frac{1}{\sqrt{4\pi\,((t-t_i)/\tau_m)}}\, e^{-\frac{(z/\lambda_c)^2}{4((t-t_i)/\tau_m)}}\, e^{-((t-t_i)/\tau_m)}$$

and since our problem is linear, by the superposition principle, we find the solution due to charge $I_e h$ injected at each point t_i is given by

$$v_m(t,z,\mathscr{P}_K) = \frac{r_0 \lambda_c I_e\, h}{\tau_m} \sum_{j=0}^{K} \frac{1}{\sqrt{4\pi\,((t-t_i)/\tau_m)}}\, e^{-\frac{(z/\lambda_c)^2}{4((t-t_i)/\tau_m)}}\, e^{-((t-t_i)/\tau_m)}$$

$$= \frac{r_0 \lambda_c I_e}{\tau_m} \sum_{j=0}^{K} \frac{1}{\sqrt{4\pi\,((t-t_i)/\tau_m)}}\, e^{-\frac{(z/\lambda_c)^2}{4((t-t_i)/\tau_m)}}\, e^{-((t-t_i)/\tau_m)}\, (t_{i+1} - t_i)$$

Now we can reorder the partition $\mathscr{P}_K$ using $u_{K-i} = t - t_i$ which gives $u_K = t$ and $u_0 = 0$; hence, we are just moving backwards through the partition. Note that

$$t_{i+1} - t_i = u_{K-i} - u_{K-i-1} = h$$

This relabeling allows us to rewrite the solution as

$$v_m(t,z,\mathscr{P}_K) = \frac{r_0 \lambda_c I_e}{\tau_m} \sum_{j=0}^{K} \frac{1}{\sqrt{4\pi\,(u_i/\tau_m)}}\, e^{-\frac{(z/\lambda_c)^2}{4(u_i/\tau_m)}}\, e^{-(u_i/\tau_m)}\, (u_{i+1} - u_i)$$

We can do this for any choice of partition $\mathscr{P}_K$. Since all of the functions involved here are continuous, we see that as $K \to \infty$, we obtain the Riemann Integral of the idealized solution for an impulse of size I_e applied at the point u

$$v_m^I(u,z) = \frac{r_0 \lambda_c I_e}{\tau_m} \frac{1}{\sqrt{4\pi\,(u/\tau_m)}}\, e^{-\frac{(z/\lambda_c)^2}{4(u/\tau_m)}}\, e^{-(u/\tau_m)}$$

leading to

$$v_m(t,z) = \int_0^t v_m^I(u,z)\, du = \int_0^t \frac{r_0 \lambda_c I_e}{\tau_m} \frac{1}{\sqrt{4\pi\,(u/\tau_m)}}\, e^{-\frac{(z/\lambda_c)^2}{4(u/\tau_m)}}\, e^{-(u/\tau_m)}\, du$$

We can rewrite this in terms of the probability density function associated with this model. Using Eq. 4.3, we have

$$v_m(t,z) = \int_0^t v_m^I(u,z)\, du = \int_0^t \frac{r_0 \lambda_c^2 I_e}{\tau_m}\, P_0(z,u)\, e^{-(u/\tau_m)}\, du$$

which is a much more convenient form. Much more can be said about this way of looking at the solutions, but that leads us into much more advanced mathematics!

4.3 Time Dependent Solutions

We now know the solution to the time dependent infinite cable model for an idealized applied charge pulse has the form

$$v_m(z, t) = \frac{r_o \lambda_C Q_e}{\tau_M} \frac{1}{\sqrt{4\pi(\frac{t}{\tau_M})}} \exp\left(-\frac{(\frac{z}{\lambda_C})^2}{4(\frac{t}{\tau_M})}\right) e^{-\frac{t}{\tau_M}}. \tag{4.6}$$

If we apply a current I_e for all t greater than zero, it is also possible to apply the superposition principle for linear partial differential equations and write the solution in terms of a standard mathematical function, the error function, $erf(x) = \frac{2}{\sqrt{\pi}} \int_0^x e^{-y^2} dy$. We will not go through these details as we feel you have been abused enough at this point. However, the work to verify these comments is not that different than what we have already done, so if you have followed us so far, you can go read up on this and be confident you can follow the arguments. Note that this situation is the equivalent of the steady state solution for the infinite cable with a current impulse applied at $z = 0$ which can be shown to be

$$v_m(z, t) = \frac{r_o \lambda_C I_e}{4} \left[erf\left(\sqrt{\frac{t}{\tau_M}} + \frac{|\frac{z}{\lambda_C}|}{2\sqrt{\frac{t}{\tau_M}}}\right) - 1 \right] e^{|\frac{z}{\lambda_C}|}$$

$$+ \frac{r_o \lambda_C I_e}{4} \left[erf\left(\sqrt{\frac{t}{\tau_M}} - \frac{|\frac{z}{\lambda_C}|}{2\sqrt{\frac{t}{\tau_M}}}\right) + 1 \right] e^{-|\frac{z}{\lambda_C}|} \tag{4.7}$$

Although this solution is much more complicated in appearance, note that as $t \to \infty$, we obtain the usual steady state solution given by equation given in Peterson (2015). We can use this solution to see how quickly the voltage due to current applied to the fiber decays. We want to know when the voltage v_m decays to one half of its starting value. This is difficult to do analytically, but if you plot the voltage v_m vs. the t for various cable lengths ℓ, you can read off the time at which the voltage crosses the one half value. This leads to the important empirical result that gives a relationship that tells us the position on the fiber z at which the voltage v_m has dropped to one half of its starting value. This relationship is linear and satisfies

$$z = 2\left(\frac{\lambda_C}{\tau_M}\right) t_{\frac{1}{2}} - \frac{\lambda_C}{2}. \tag{4.8}$$

The slope of this line can then be interpreted as a *velocity*, the rate at which the fiber position for half-life changes; this is a *conduction velocity* and is given by

$$v_{\frac{1}{2}} = 2\left(\frac{\lambda_C}{\tau_M}\right), \tag{4.9}$$

having units of (cm/s). Using our standard assumption that $r_i \gg r_o$ and our equations for λ_C and g_m in terms of membrane parameters, we find

$$v_{\frac{1}{2}} = \sqrt{\frac{2G_M}{\rho_i C_M^2}}\, a^{\frac{1}{2}}, \tag{4.10}$$

indicating that the induced voltage attenuates proportional to the square root of the fiber radius a. Hence, the double the ratio of the *fundamental space to time* constant is an important heuristic measure of the propagation speed of the voltage pulse. Good to know.

Reference

J. Peterson, *Calculus for Cognitive Scientists: Partial Differential Equation Models*, Springer Series on Cognitive Science and Technology (Springer Science+Business Media Singapore Pte Ltd, Singapore, 2015 in press)

Part III
Neural Systems

Chapter 5
Mammalian Neural Structure

In this chapter, we will discuss the basic principles of the neural organization that subserves cognitive function. A great way to get a handle on this material is in the neuroanatomy books (Pinel and Edwards 2008) and (Diamond et al. 1985) which we have spent many hours working with using colored pencils and many sheets of paper. We encourage you to do this as our simple introduction below is just to get you started.

5.1 The Basic Model

Our basic neural model is based on abstractions from neurobiology. The two halves or hemispheres of the brain are connected by the corpus callosum which is like a cap of tissue that sits on top of the brain stem. The structures in this area are very old in an evolutionary sense. The outer surface of the brain is the cortex which is a thin layer organized into columns. There is too much cortex to fit comfortably inside the human skull, so as the human species evolved and the amount of cortical tissue expanded, the cortex began to develop folds. Imagine a deep canyon in the earth's surface. The walls of the canyon are called a *gyrus* and are the cortical tissue. The canyon itself called a *sulcus* or *fissure*. There are many such *sulci* with corresponding *gyri*. Some of the gyri are deep enough to touch the corpus callosum. One such gyrus is the longitudinal cerebral gyrus fissure which contains the cingulate gyri at the very bottom of the fissure touching the corpus callosum.

Consider the simplified model of information processing in the brain that is presented in Fig. 5.3. This has been abstracted out of much more detail (Nolte 2002). In Brodal (1992) and Diamond et al. (1985), we can trace out the details of the connections between cortical areas and deeper brain structures near the brain stem and construct a very simplified version which will be suitable for our modeling purposes. Raw visual input is sent to area 17 of the occipital cortex where is is further processed by the occipital association areas 18 and 19. Raw auditory input is sent to area 41 of the Parietal cortex. Processing continues in areas 5, 7 (the parietal association

© Springer Science+Business Media Singapore 2016
J.K. Peterson, *BioInformation Processing*, Cognitive Science and Technology,
DOI 10.1007/978-981-287-871-7_5

areas) and 42. There are also long association nerve fiber bundles starting in the cingulate gyrus which connect the temporal, parietal and occipital cortex together. The Temporal—Occipital connections are labeled C2, the superior longitudinal fasciculus; and B2, the inferior occipitofrontal nerve bundles, respectively. The C2 pathway also connects to the cerebellum for motor output. The Frontal—Temporal connections labeled A, B and C are the cingular, uncinate fasciculus and arcuate fasciculus bundles and connect the Frontal association areas 20, 21, 22 and 37.

Area 37 is an area where inputs from multiple sensor modalities are fused into higher level constructs. The top boundary of area 17 in the occipital cortex is marked by a fold in the surface of the brain called the lunate sulcus. This sulcus occurs much higher in a primate such as a chimpanzee. Effectively, human like brains have been reorganized so that the percentage of cortex allotted to vision has been reduced. Comparative studies show that the human area 17 is 121 % smaller than it should be if its size was proportionate to other primates. The lost portion of area 17 has been reallocated to area 7 of the parietal cortex. There are special areas in each cortex that are devoted to secondary processing of primary sensory information and which are not connected directly to output pathways. These areas are called associative cortex and there are primarily defined by function, not a special cell structure. In the parietal cortex, the association areas are 5 and 7; in the temporal cortex, areas 20, 21, 22 and 37; and in the frontal, areas 6 and 8. Hence, human brains have evolved to increase the amount of associative cortex available for what can be considered symbolic processing needs. Finally, the same nerve bundle A connects the Parietal and Temporal cortex. In addition to these general pathways, specific connections between the cortex association areas are shown as bidirectional arrows. The box labeled *cingular gyrus* is essentially a simplified model of the processing that is done in the limbic system. Note the bidirectional arrows connecting the cingulate gyrus to the septal nuclei inside the subcallosal gyrus, the anterior nuclei inside the thalamus and the amygdala. There is also a two way connection between the anterior nuclei and the mamillary body inside the hypothalamus. Finally, the cingulate gyrus connects to the cerebellum for motor output. We show the main cortical areas of the brain in Fig. 5.1. Our ability to process symbolic information is probably due to changes in the human brain that have occurred over evolutionary time. A plausible outline of the reorganizations in the brain that have occurred in human evolution has been presented in Holloway (1999, p. 96), which we have paraphrased somewhat in Table 5.1.

In the table presented, we use some possible unfamiliar terms which are the abbreviation *mya*, this refers to *millions of years ago*; the term *petalias*, which are asymmetrical projections of the occipital and frontal cortex; and *endocast*, which is a cast of the inside of a hominid fossil's skull. An endocast must be carefully planned and examined, of course, as there are many opportunities for wrong interpretations. Holloway (1999, p. 77), believes that there are no new evolutionarily derived structures in the human brain as compared to other animals—nuclear masses and the fiber systems interconnecting them are the same. The differences are in the quantitative relationships between and among these nuclei and fiber tracts, and the organization of cerebral cortex structurally, functionally and in integration.

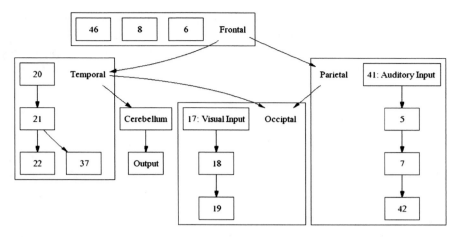

Fig. 5.1 The main cortical subdivisions of the brain

Table 5.1 Brain reorganizations

Brain changes (Reorganization)	Hominid fossil	Time (mya)	Evidence
Reduction of Area 17	A. afarensis	3.5–3.0	Brain endocast
Increase in posterior parietal cortex reorganization of frontal cortex third inferior frontal convolution Broca's area	Homo habilis	2.0–1.8	Brain endocast
Cerebral asymmetries left occipital, right frontal petalias	Homo habilis	2.0–1.8	Brain endocast
Refinements in cortical organization to a modern human pattern	Homo erectus	1.5–0.0	Brain endocast

We make these prefatory remarks to motivate why we think a reasonable model of cognition (and hence, cognitive dysfunction) needs a working model of cortical processing. Our special symbolic processing abilities appear to be closely linked to the reorganizations of our cortex that expanded our use of associative cortex. Indeed, in Holloway (1999, p. 97), we find a statement of one of the guiding principles in our model building efforts.

> our humanness resides mostly in our brain, endowed with symbolic abilities, which have permitted the human animal extraordinary degrees of control over its adaptive powers in both the social and material realms.

As might be expected, clear evidence of our use of symbolic processing does not occur until humans began to leave archaeological artifacts which were well enough preserved. Currently, such evidence first appears in painted shells used for adornment found in African sites from approximately seventy thousand years ago. Previous to those discoveries, the strongest evidence for abstract and symbolic abilities came in the numerical examples of Paleolithic art from the Europe and Russia in the form of cave paintings and carved bone about twenty-five thousand years ago (25 kya). Paleolithic art comprises thousands of images made on bone, antler, ivory and limestone cave walls using a variety of techniques, styles and artistic conventions. In Conkey (1999, p. 291), it is asserted that

> If,...,we are to understand the 'manifestations and evolution of complex human behavior',
> there is no doubt that the study of paleolithic 'art' has much to offer.

In the past, researchers assumed that the Paleolithic art samples that appeared so suddenly 25 kya were evidence of the emergence of a newly reorganized human brain that was now capable of symbolic processing of the kind needed to create *art*. Conkey argues persuasively that this is not likely. Indeed, the table of brain reorganization presented earlier, we noted that the increase in associative parietal cortex in area 7 occurred approximately 3 mya. Hence, the capability of symbolic reasoning probably steadily evolved even though the concrete evidence of cave art and so forth does not occur until really quite recently. However, our point is that the creation of 'art' is intimately tied up with the symbolic processing capabilities that must underlies any model of cognition. Further, we assert that the abstractions inherent in mathematics and optimization are additional examples of symbolic processing.

In addition, we indicate in simplified form, three major neurotransmitter pathways. In the brain stem, we focus on the groups of neurons called the raphe and the locus coeruleus as well as dopamine producing cells which are labeled by their location, the substantia nigra in the brain stem. These cell groups produce neurotransmitters of specific types and send their outputs to a collection of neural tissues that surround the thalamus called the basal ganglia. The basal ganglia are not shown in Fig. 5.3 but you should imagine them as another box surrounding the thalamus. The basal ganglia then sends outputs to portions of the cerebral cortex; the cerebral cortex in turn sends connections back to the basal ganglia. These connections are not shown explicitly; instead, for simplicity of presentation, we use the thalamus to cingulate gyrus to associative connections that are shown. The raphe nuclei produce serotonin, the locus coeruleus produce norepinephrine and the substantia nigra (and other cells in the brain stem) produce dopamine. There are many other neurotransmitters, but the model we are presenting here is a deliberate choice to focus on a few basic neurotransmitter pathways.

The limbic processing presented in Fig. 5.3 is shown in more detail in Fig. 5.2. Cross sections of the brain in a plane perpendicular to a line through the eyes and back of head shown that the cingulate gyrus sits on top of the corpus callosum. Underneath the corpus callosum is a sheath of connecting nerve tissue known as the fornix which is instrumental in communication between these layers and the structures that lie below. The arrows in Fig. 5.2 indicate structural connections only and should not be used to infer information transfer. A typical model of biological information processing that can be abstracted from what we know about brain function is thus shown in the simplified brain model of Fig. 5.3. This shows a chain of neural modules which subserve cognition. It is primarily meant to illustrate the hierarchical complexity of

Fig. 5.2 The major limbic
system structures with
arrows indicating structural
connections

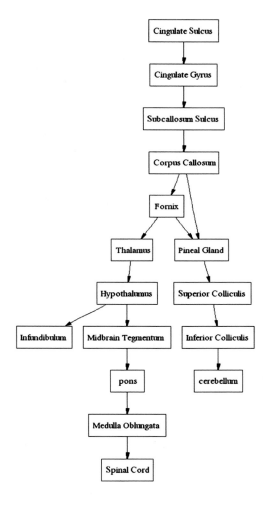

the brain structures we need to become familiar with. There is always cross-talk and feedback between modules which is not shown.

Some of the principle components of the information processing system of the brain are given in Table 5.2. The central nervous system (CNS) is divided roughly into two parts; the brain and the spinal cord. The brain can be subdivided into a number of discernible modules. For our purposes, we will consider the brain model to consist of the cerebrum, the cerebellum and the brain stem. Finer subdivisions are then shown in Table 5.2 where some structures are labeled with a corresponding number for later reference in other figures. The numbers we use here bear no relation to the numbering scheme that is used for the cortical subdivisions shown in Figs. 5.1 and 5.3. The numbering scheme there has been set historically. The numbering scheme shown in Table 5.2 will help us locate brain structures deep in the brain that can

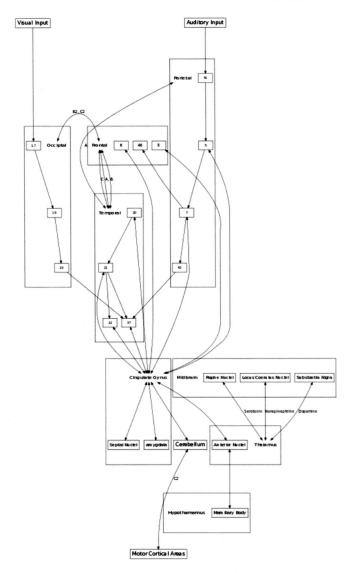

Fig. 5.3 A simplified path of information processing in the brain. *Arrows* indicate information processing pathways

Table 5.2 Information processing components

Brain	→	Cerebrum	Cerebellum **1**	Brain Stem
Cerebrum	→	Cerebral Hemisphere	Diencephalon	
Brain Stem	→	Medulla **4**	Pons **3**	Midbrain **2**
Cerebral Hemisphere	→	Amygdala **6**	Hippocampus **5**	
		Cerebral Cortex	Basal Ganglia	
Diencephalon	→	Hypothalamus **8**	Thalamus **7**	
Cerebral Cortex	→	Limbic **13**	Temporal **12**	Occipital **11**
		Parietal **10**	Frontal **9**	
Basal Ganglia	→	Lenticular Nucleus **15**	Caudate Nucleus **14**	
Lenticular Nucleus	→	Global Pallidus **16**	Putamen **15**	

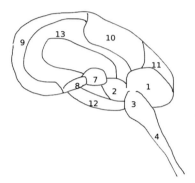

Fig. 5.4 Brain structures: the numeric labels correspond to structures listed in Table 5.2

only be seen by taking slices. These numbers thus correspond to the brain structures shown in Fig. 5.4 (modules that can be seen on the surface) and the brain slices of Figs. 5.5a, b and 5.6a, b. A useful model of the processing necessary to combine disparate sensory information into higher level concepts is clearly built on models of cortical processing.

There is much evidence that cortex is initially uniform prior to exposure to environmental signal and hence a good model of such generic cortical tissue, isocortex, is needed. A model of isocortex is motivated by recent models of cortical processing outlined in Raizada and Grossberg (2003). This article uses clues from visual processing to gain insight into how virgin cortical tissue (isocortex) is wired to allow for its shaping via environmental input. Clues and theoretical models for auditory cortex can then be found in the survey paper of Merzenich (2001). For our purposes, we will use the terms auditory cortex for area 41 of the Parietal cortex and visual cortex for area 17 of the occipital cortex. These are the areas which receive primary sensory input with further processing occurring first in cortex where the input is received and then in the temporal cortex in areas 20, 21, 22 and 37 as shown in Fig. 5.3. The first layer of auditory cortex is bathed in an environment where sound

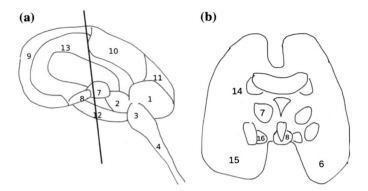

Fig. 5.5 Brain slice 1 details. **a** Slice 1 orientation. **b** Neural slice 1 cartoon

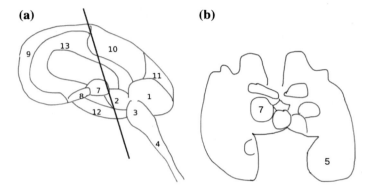

Fig. 5.6 Brain slice 2 details. **a** Slice 2 orientation. **b** Neural slice 2 cartoon

is chunked or batched into pieces of 200 ms length which is the approximate size of the phonemes of a person's native language. Hence, the first layer of cortex develops circuitry specialized to this time constant.

The second layer of cortex then naturally develops a chunk size focus that is substantially larger, perhaps on the order of 1000–10,000 ms. Merzenich details how errors in the imprinting of these cortical layers can lead to cognitive impairments such as dyslexia. As processing is further removed from the auditory cortex via mylenated pathways, additional meta level concepts (tied to even longer time constants) are developed. One way to do this is through what is called *Hebbian* learning. This training method strengthens the strength of a connection between a pre and post neuron on the basis of coincident activity. It has many variations, of course, but that is the gist of it and it is based on the neurobiology of long term learning.

While it is clear that a form of such Hebbian learning is used to set up these circuits, the pitfalls of such learning are discussed clearly in the literature (McClelland 2001). For example, the inability of adult speakers of Japanese to distinguish the sound of an *ell* and an *r* is indicative of a bias of Hebbian learning that makes it difficult to learn new strategies by unlearning previous paradigms. For this reason, we will not use strict Hebbian learning protocols; instead, we will model auditory and visual cortex with three layers using techniques based on information flow through graphs in conjunction with versions of Hebbian learning. Our third layer of cortex is then an abstraction of the additional anatomical layers of cortex as well as appropriate myelinated pathways which conduct upper layer processing results to other cognitive modules.

5.2 Brain Structure

The connections from the limbic core to the cortex are not visible from the outside. The outer layer of the cortex is deeply folded and wraps around an inner core that consists of the limbic lobe and corpus callosum. Neural pathways always connect these structures. In Fig. 5.5a, we can see the inner structures known as the amygdala and putamen in a brain cross-section. Figure 5.5a show the orientation of the brain cross-section while Fig. 5.5b displays a cartoon of the slice itself indicating various structures.

The thalamus is located in a portion of the brain which can be seen using the cross-section indicated by Fig. 5.6a. The structures the slice contains are shown in Fig. 5.6b.

5.3 The Brain Stem

Cortical columns interact with other parts of the brain in sophisticated ways. The patterns of cortical activity (modeled by the Folded Feedback Pathways (FFP) and On-Center Off-surround (OCOS) circuits as discussed in Sect. 5.4.2) are modulated by neurotransmitter inputs that originate in the *reticular formation* or *RF* of the brain. The outer cortex is wrapped around an inner core which contains, among other structures, the midbrain. The midbrain is one of the most important information processing modules. Consider Fig. 5.8 which shows the midbrain in cross section. A number of integrative functions are organized at the brainstem level. These include complex motor patterns, respiratory and cardiovascular activity and some aspects of consciousness. The brain stem has historically been subdivided into functional groups starting at the spinal cord and moving upward toward the middle of the brain. The groups are, in order, the **caudal and rostral medulla**, the **caudal, mid and rostral pons** and the **caudal and rostral midbrain**. We will refer to these as the brainstem layers **1–7**. In this context, the terms *caudal* and *rostral* refer to whether a slice is closest to the spinal cord or not. Hence, the pons is divided into the *rostral pons* (farthest from the spinal cord) and the *caudal pons* (closest to the spinal cord).

Fig. 5.7 The brainstem
layers

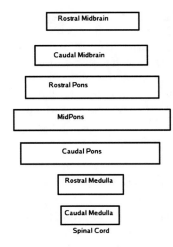

Fig. 5.8 The brainstem
structure

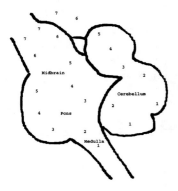

The seven brain stem layers can be seen by taking cross-sections through the brain as indicated by repeated numbers 1–7 shown in Fig. 5.8. Each shows interesting structure which is not visible other than in cross section. Each slice of the midbrain also has specific functionality and shows interesting structure which is not visible other than in cross section. The layers are shown in Fig. 5.7. Slice 1 is the *caudal medulla* and is shown in Fig. 5.9a. The Medial Longitudinal Fasciculus (MLF) controls head and eye movement. The *Medial Lemniscus* is not shown but is right below the MLF. Slice 2 is the *rostral medulla*. The Fourth Ventricle is not shown in Fig. 5.9b but would be right at the top of the figure. The *Medial Lemniscus* is still not shown but is located below the MLF like before. In slice 2, a lot of the *inferior olivary nucleus* can be seen. The caudal pons is shown in Slice 3, Fig. 5.10a, and the mid pons in Slice 4, Fig. 5.10b.

The rostral pons and the caudal and rostral midbrain are shown in Figs. 5.11a, b and 5.12, respectively. Slice 4 and 5 contain pontine nuclei and Slice 7 contains the

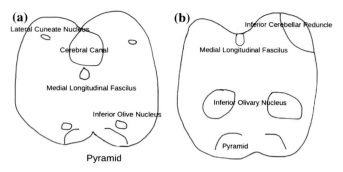

Fig. 5.9 The medulla cross-sections. **a** The caudal medulla: slice 1. **b** The rostral medulla: slice 2

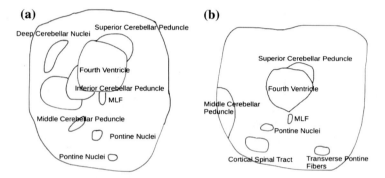

Fig. 5.10 The caudal and mid pons slices. **a** The caudal pons: slice 3. **b** The mid pons: slice 4

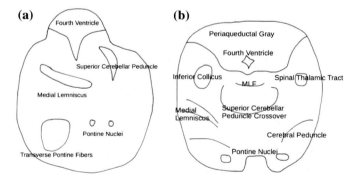

Fig. 5.11 The rostral pons and caudal midbrain cross-sections. **a** The rostral pons: slice 5. **b** The caudal midbrain: slice 6

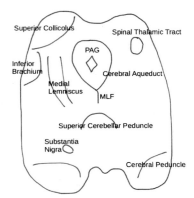

Fig. 5.12 The rostral midbrain: slice 7

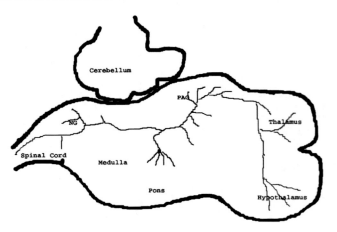

Fig. 5.13 Typical neuron in the reticular formation

cerebral peduncle, a massive nerve bundle containing cortico-pontine and cortico-spinal fibers. Fibers originating in the frontal, parietal, and temporal cortex descend to pontine nuclei. The Pontine also connects to cerebral peduncles.

The *Reticular Formation* or *RF* is this central core of the brain stem which contains neurons whose connectivity is characterized by huge fan-in and fan-out. Reticular neurons therefore have extensive and complex axonal projections. Note we use the abbreviation *PAG* to denote the cells known as periaqueductal gray and *NG* for the Nucleus Gracilis neurons. The extensive nature of the afferent connections for a RF neuron are shown in Fig. 5.13. The indicated neuron in Fig. 5.13 sends its axon to many CNS areas. If one cell has projections this extensive, we can imagine the complexity of reticular formation as a whole. Its neurons have ascending projections that terminate in the thalamus, subthalamus, hypothalamus, cerebral cortex and the basal ganglia (caudate nucleus, putamen, globus pallidus, substantia nigra). The midbrain and rostral pons *RF* neurons thus collect sensory modalities and project this information to intralaminar nuclei of thalamus. The intralaminar nuclei project to widespread areas of cortex causes heightened arousal in response to sensory stimuli; e.g. attention. Our cortical column models must therefore be able to accept modulatory inputs from the *RF* formation. It is also know that some RF neurons release monoamines which are essential for maintenance of consciousness. For example, bilateral damage to the midbrain RF and the fibers through it causes prolonged coma. Hence, even a normal intact cerebrum can not maintain consciousness. It is clear that input from brainstem RF is needed. The monoamine releasing RF neurons release norepinephrine, dopamine and serotonin. The noradrenergic neurons are located in the pons and medulla (locus coeruleus). This is Slice 5 of midbrain and it connects to cerebral cortex. This system is inactive in sleep and most active in startling situations, or those calling for watchful situations. Dopaminergic neurons in slice 7 of midbrain are are located in the Substantia Nigra (SG). Figure 5.14 shows the brain areas that are influenced by dopamine. Figure 5.15a also shows the SC (superior collicus), PAG (periaqueductal gray) and RN (red nucleus) structures.

These neurons project in overlapping fiber tracts to other parts of the brain. The nigrostriatal sends information to the substantia nigra and then to the caudate, putamen and midbrain. The medial forebrain bundle projects from the substantia nigra to the frontal and limbic lobes. The indicated projections to the motor cortex are consistent with initiation of movement. We know that there is disruption to cortex function due to dopamine neurons ablation in Parkinson's disease. The projections to other frontal cortical areas and limbic structures imply there is a motivation and cognition role. Hence, imbalances in these pathways will play a role in mental dysfunction. Furthermore, certain drugs cause dopamine release in limbic structures which implies a pleasure connection. Serotonergic neurons occur in in most levels of brain stem, but concentrate in raphe nuclei in slice 5 of midbrain—also near the locus coeruleus (Fig. 5.15b). Their axons innervate many areas of the brain as shown

Fig. 5.14 The dopamine
innervation pathway

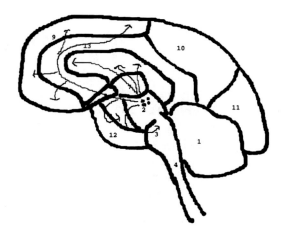

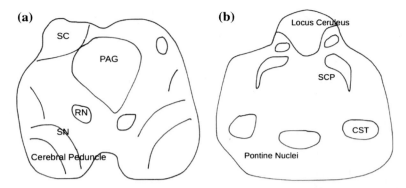

(a) **(b)**

Fig. 5.15 The dopamine and serotonin releasing neurons sites. **a** Location of dopamine sites. **b** Location of serotonin sites

Fig. 5.16 The serotonin innervation pathway

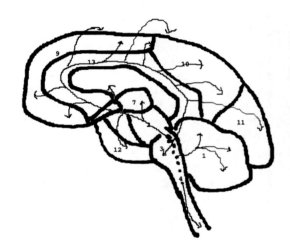

in Fig. 5.16. It is known that serotonin levels determine the set point of arousal and influence the pain control system.

5.4   Cortical Structure

The cortex consists of the frontal, occipital, parietal, temporal and limbic lobes and it is folded as is shown in Fig. 5.17. It consists of a number of regions which have historically been classified as illustrated in Fig. 5.18a. The cortical layer is thus subdivided into several functional areas known as the *Frontal*, *Occipital*, *Parietal*, *Temporal* and *Limbic* lobes. Most of what we know about the function of these lobes comes from the analysis of the diminished abilities of people who have unfortunately

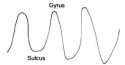

Fig. 5.17 Cortical folding

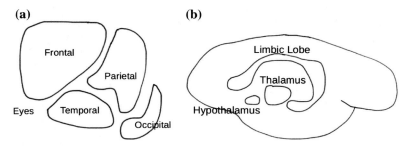

Fig. 5.18 Cortical lobes. **a** Cortical lobes. **b** The limbic lobe inside the cortex

had strokes or injuries. By studying these brave people very carefully we can discern what functions they have lost and correlate these losses with the areas of their brain that has been damaged. This correlation is generally obtained using a variety of brain imaging techniques such as Computer Assisted Tomography (CAT) and Functional Magnetic Resonance Imagery (fMRI) scans.

The **Frontal** lobe has a number of functionally distinct areas. It contains the primary motor cortex which is involved in initiation of voluntary movement. Also, it has an specialized area known as Broca's area which is important in both written and spoken language ability. Finally, it has the prefrontal cortex which is instrumental in the maintenance of our personality and is involved in the critical abilities of insight and foresight. The **Occipital** lobe is concerned with visual processing and visual association. In the **Parietal** lobe, primary somato-sensory information is processed in the area known as the primary somato-sensory cortex. There is also initial cortical processing of tactile and proprioceptive input. In addition, there are areas devoted to language comprehension and complex spatial orientation and perception. The **Temporal** lobe contains auditory processing circuitry and develops higher order auditory associations as well. For example, the temporal lobe contains Wernicke's area which is involved with language comprehension. It also handles higher order visual processing and learning and memory. Finally, the **Limbic** system lies beneath the cortex as in shown in Fig. 5.18b and is involved in emotional modulation of cortical processing.

Fig. 5.19 Generic overview

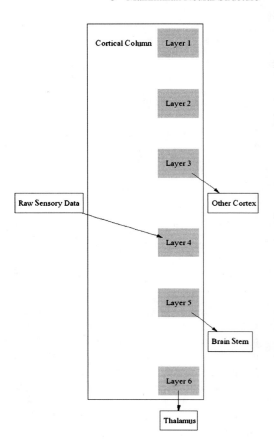

5.4.1   Cortical Processing

In order to build a useful model of cognition, we must be able to model interactions of cortical modules with the limbic system. Further, these interactions must be modulated by a number of monoamine neurotransmitters. It also follows that we need a reasonable model of cortical processing. There is much evidence that cortex is initially uniform prior to exposure to environmental signal. Hence, a good model of generic cortical tissue, called isocortex, is needed. A model of isocortex is motivated by recent models of cortical processing outlined in Raizada and Grossberg (2003). There are other approaches, of course, but we will focus on this model here. A good discussion of auditory processing is given in Nelken (2004) and how information from multiple cortical areas can be combined into a useful signal is discussed in Beauchamp (2005). But for the moment, let's stick to one cortex at a time. The Raizada article uses clues from visual processing to gain insight into how virgin cortical tissue (isocortex) is wired to allow for its shaping via environmental input.

Clues and theoretical models for auditory cortex can then be found in the survey paper of Merzenich (2001). We begin with a general view of a typical cortical column taken from the standard references of Brodal (1992), Diamond et al. (1985) and Nolte (2002) as shown in Fig. 5.19. This column is oriented vertically with layer one closest to the skull and layer six furthest in. We show layer four having a connection to primary sensory data. The details of some of the connections between the layers are shown in Fig. 5.20. The six layers of the cortical column consist of specific cell types and mixtures described in Table 5.3.

We can make some general observations about the cortical architecture. First, layers three and five contain pyramidal cells which collect information from layers above themselves and send their processed output for higher level processing. Layer three outputs to motor areas and layer five, to other parts of the cerebral cortex. Layer six contains cells whose output is sent to the thalamus or other brain areas. Layer four is a collection layer which collates input from primary sensory modalities or from other cortical and brain areas. We see illustrations of the general cortical column structure in Figs. 5.19 and 5.20. The cortical columns are organized into larger vertical structures following a simple stacked protocol: sensory data → cortical column 1 → cortical column 2 → cortical column 3 and so forth. For convenience, our models will be shown with three stacked columns. The output from the last column is then sent to other cortex, thalamus and the brain stem.

5.4.2 Isocortex Modeling

A useful model of generic cortex, isocortex, is that given in Grossberg (2003), Grossberg and Seitz (2003) and Raizada and Grossberg (2003). Two fundamental cortical circuits are introduced in these works: the *on-center, off-surround* (OCOS) and the *folded feedback pathway* (FFP) seen in Fig. 5.21a, b. In Fig. 5.21a, we see the *On-Center, Off-Surround* control structure that is part of the cortical column control circuitry. Outputs from the thalamus (perhaps from the nuclei of the Lateral Geniculate Body) filter upward into the column at the bottom of the picture. At the top of the figure, the three circles that are not filled in represent neurons in layer four whose outputs will be sent to other parts of the column.

There are two thalamic output lines: the first is a direct connection to the input layer four, while the second is an indirect connection to layer six itself. This connection then connects to a layer of inhibitory neurons which are shown as circles filled in with black. The middle layer four output neuron is thus innervated by both inhibitory and excitatory inputs while the left and right layer four output neurons only receive inhibitory impulses. Hence, the *center* is excited and the part of the circuit that is off the center, is inhibited. We could say the *surround is off*. It is common to call this type of activation the *off surround*. Next consider a stacked cortical column consisting of two columns, column one and column two. There are cortico-cortical feedback axons originating in layer six of column two which input into layer one of column one. From layer one, the input connects to the dendrites of layer five pyramidal neurons which

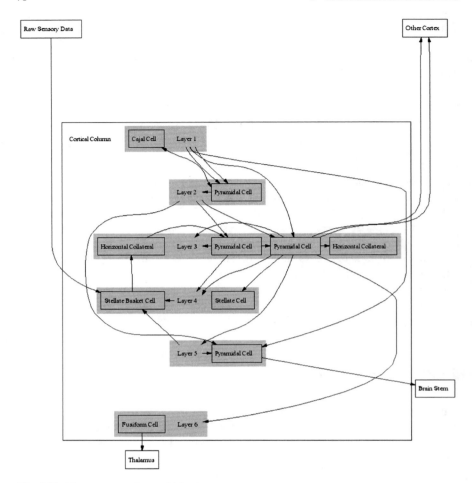

Fig. 5.20 The structure of a cortical column

Table 5.3 Cortical column cell types

Layer	Description	Use
One	Molecular	
Two	External granule layer	
Three	External pyramidal layer	Output to other cortex areas
Four	Internal granule layer	Collect primary sensory input or input from other brain areas
Five	Internal pyramidal layer	Output to motor cortex
Six	Multiform layer	Output to thalamus brain areas

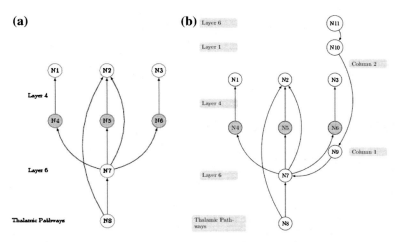

Fig. 5.21 OCOS and FFP circuits. **a** The on-center, off-surround control structure. **b** The folded feedback pathway control structure

connects to the thalamic neuron in layer six. Hence, the *higher level cortical input* is fed back into the previous column layer six and then can excite column one's fourth layer via the on-center, off-surround circuit discussed previously. This description is summarized in Fig. 5.21b. We call this type of feedback a *folded feedback pathway*. For convenience, we use the abbreviations OCOS and FFP to indicate the on-center, off-surround and folded feedback pathway, respectively. The layer six–four OCOS is connected to the layer two–three circuit as shown in Fig. 5.22a. Note that the layer four output is forwarded to layer two–three and then sent back to layer six so as to contribute to the standard OCOS layer six–layer four circuit. Hence, we can describe this as another FFP circuit. Finally, the output from layer six is forwarded into the thalamic pathways using a standard OCOS circuit. This provides a way for layer six neurons to modulate the thalamic outputs which influence the cortex. This is shown in Fig. 5.22b.

The FFP and OCOS cortical circuits can also be combined into a multi-column model (Raizada and Grossberg 2003) as seen in Fig. 5.23. It is known that cortical outputs dynamically assemble into spatially and temporally localized phase locked structures. A review of such functional connectivity appears in Fingelkurts et al. (2005). We have abstracted a typical snapshot of the synchronous activity of participating neurons from Fingelkurts' discussions which is shown in Fig. 5.24. Each burst of this type of synchronous activity in the cortex is measured via skull-cap EEG equipment and Fingelkurts presents a reasoned discussion why such coordinated ensembles are high level representations of activity. A cognitive model is thus inherently multi-scale in nature. The cortex uses clusters of synchronous activity as shown in Fig. 5.24 acting on a sub second to second time frame to successively transform raw data representations to higher level representations. Also, cortical transformations from different sensory modalities are combined to fuse representations into new

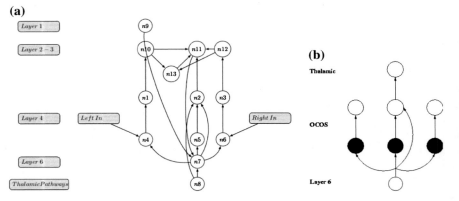

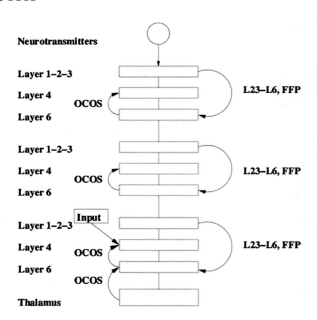

Fig. 5.22 Layer six connections. **a** The layer six–four connections to layer two–three are another FFP. **b** The layer six to thalamic OCOS

Fig. 5.23 The OCOS/FFP cortical model

Fig. 5.24 Synchronous cortical activity

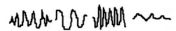

Fig. 5.25 The
norepinephrine innervation
pathway

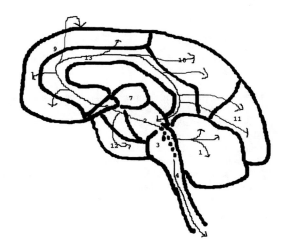

representations at the level of *cortical modules*. These higher level representations
both arise and interact on much longer time scales. Further, monoamine inputs from
the brain stem modulate the cortical modules as shown in Figs. 5.25, 5.14 and 5.16.
Additionally, reticular formation neurons modulate limbic and cortical modules. It
is therefore clear that cognitive models require abstract neurons whose output can be
shaped by many modulatory inputs. To do this, we need a theoretical model of the
input/output (I/O) process of a single excitable cell that is as simple as possible yet
still computes the salient characteristics of our proposed system.

References

M. Beauchamp, See me, hear me, touch me: multisensory integration in lateral occipital—temporal
   cortex. Curr. Opin. Neurobiol. **15**, 145–153 (2005)
P. Brodal, *The Central Nervous System: Structure and Function* (Oxford University Press, New
   York, 1992)
M. Conkey, A history of the interpretation of European 'paleolithic art': magic, mythogram, and
   metaphors for modernity, in *Handbook of Human Symbolic Evolution*, ed. by A. Lock, C. Peters
   (Blackwell Publishers, Massachusetts, 1999)
M. Diamond, A. Scheibel, L. Elson, *The Human Brain Coloring Book* (Barnes and Noble Books,
   New York, 1985)
A. Fingelkurts, A. Fingelkurts, S. Kähkönen, Functional connectivity in the brain—is it an elusive
   concept? Neurosci. Biobehav. Rev. **28**, 827–836 (2005)
S. Grossberg, How does the cerebral cortex work? development, learning, attention and 3D vision
   by laminar circuits of visual cortex, Technical Report TR-2003-005, Boston University, CAS/CS,
   2003
S. Grossberg, A. Seitz, Laminar development of receptive fields, maps, and columns in visual cortex:
   the coordinating role of the subplate, Technical Report 02-006, Boston University, CAS/CS, 2003
R. Holloway, Evolution of the human brain, in *Handbook of Human Symbolic Evolution*, ed. by
   A. Lock, C. Peters (Blackwell Publishers, Massachusetts, 1999), pp. 74–116

J. McClelland, Failures to learn and their remediation, in *Failures to learn and their remediation*, ed. by J. McClelland, R. Siegler (Lawrence Erlbaum Associates Publishers, USA, 2001), pp. 97–122

M. Merzenich, Cortical plasticity contributing to child development, in *Mechanisms of Cognitive Development: Behavioral and Neural Perspectives*, ed. by J. McClelland, R. Siegler (Lawrence Erlbaum Associates Publishers, USA, 2001), pp. 67–96

I. Nelken, Processing of complex stimuli and natural scenes in the auditory cortex. Curr. Opin. Neurobiol **14**, 474–480 (2004)

J. Nolte, *The Human Brain: An Introduction to Its Functional Anatomy* (Mosby, A Division of Elsevier Science, 2002)

J. Pinel, M. Edwards, *A Colorful Introduction to the Anatomy of the Human Brain: A Brain and Psychology Coloring Book* (Pearson, New York, 2008)

R. Raizada, S. Grossberg, Towards a theory of the laminar architecture of cerebral cortex: computational clues from the visual system. Cereb. Cortex **13**, 100–113 (2003)

Chapter 6
Abstracting Principles of Computation

Classical computing via traditional hardware starts with computing primitives such as **AND** and **OR** gates and then assembles large numbers of such gates into circuits using microfabrication techniques to implement boolean functions. More complicated functions and data other than boolean and/or integer can then be modeled using this paradigm. The crucial feature here is that simple building blocks are realized in hardware primitives which are then assembled into large structures. Quantum computing is taking a similar approach; the primitive unit of information is no longer a boolean **bit** but instead is a more complicated entity called a **qubit**. The quantum nature of the qubit then allows the primitive functional **AND/OR** gates to have unusual properties, but the intellectual pathway is still clear. Hardware primitives are used to assemble logic gates' logic gate ensembles are used to implement computable numbers via functions. Is this a viable model of biological computation?

In the thoughts that follow, we will argue that this is the incorrect approach and biological computation is instead most profitably built from a blurring of the hardware/ software distinction. Following Gerhart and Kirschner (1997), we present a generic biological primitive based on signal transduction into a second messenger pathway. We will look at specific cellular mechanisms to give the flavor of this approach—all based on the presentation in Gerhart and Kirschner (1997) but placed in very abstract terms for our purposes. We will show how they imply an abstract computational framework for basic triggering events. The trigger model is essentially a good approximation to a general second messenger input. We can use these ideas in future development of a useful approximation to a biological neuron for use in cognitive models.

6.1 Cellular Triggers

A good example of a cellular trigger is the transcriptional control of the factor $NF\kappa B$ which plays a role in immune system response. This mechanism is discussed in

© Springer Science+Business Media Singapore 2016

J.K. Peterson, *BioInformation Processing*, Cognitive Science and Technology,

DOI 10.1007/978-981-287-871-7_6

a semi-abstract way in Gerhart and Kirschner (1997); we will discuss even more abstractly.

Consider a trigger T_0 which activates a cell surface receptor. Inside the cell, there are always protein kinases that can be activated in a variety of ways. Here we denote a protein kinase by the symbol PK. A common mechanism for such an activation is to add to PK another protein subunit $\mathcal{U}$ to form the complex $PK/\mathcal{U}$. This chain of events looks like this:

$$T_0 \rightarrow \text{Cell Surface Receptor} \rightarrow PK/\mathcal{U}$$

$PK/\mathcal{U}$ then acts to *phosphorylate* another protein. The cell is filled with large amounts of a transcription factor we will denote by T_1 and an inhibitory protein for T_1 we label as $T_1^{\sim}$. This symbol, $T_1^{\sim}$, denotes the *complement* or *anti* version of T_1. In the cell, T_1 and $T_1^{\sim}$ are generally joined together in a complex denoted by $T_1/T_1^{\sim}$. The addition of $T_1^{\sim}$ to T_1 prevents T_1 from being able to access the genome in the nucleus to transcribe its target protein.

The trigger T_0 activates our protein kinase PK to $PK/\mathcal{U}$. The activated $PK/\mathcal{U}$ is used to add a phosphate to $T_1^{\sim}$. This is called phosphorylation. Hence,

$$PK/\mathcal{U} + T_1^{\sim} \rightarrow T_1^{\sim}P$$

where $T_1^{\sim}P$ denotes the phosphorylated version of $T_1^{\sim}$. Since T_1 is bound into the complex $T_1/T_1^{\sim}$, we actually have

$$PK/\mathcal{U} + T_1/T_1^{\sim} \rightarrow T_1/T_1^{\sim}P$$

In the cell, there is always present a collection of proteins which tend to bond with the phosphorylated form $T_1^{\sim}P$. Such a system is called a **tagging** system. The protein used by the tagging system is denoted by $\mathcal{V}$ and usually a chain of n such $\mathcal{V}$ proteins is glued together to form a polymer $\mathcal{V}_n$. The tagging system creates the new complex

$$T_1/T_1^{\sim}P\mathcal{V}_n$$

This gives the following event tree at this point:

$$
\begin{aligned}
T_0 & \rightarrow \text{Cell Surface Receptor} \rightarrow PK/\mathcal{U} \\
PK/\mathcal{U} + T_1/T_1^{\sim} & \rightarrow T_1/T_1^{\sim}P \\
T_1/T_1^{\sim}P + \text{tagging system} & \rightarrow T_1/T_1^{\sim}P\mathcal{V}_n
\end{aligned}
$$

Also, inside the cell, the tagging system coexists with a complimentary system whose function is to destroy or remove the tagged complexes. Hence, the combined system

$$\text{Tagging} \longleftrightarrow \text{Removal} \rightarrow T_1/T_1^{\sim}P$$

is a **regulatory** mechanism which allows the transcription factor T_1 to be freed from its bound state $T_1/T_1^{\sim}$ so that it can perform its function of protein transcription in the genome. The removal system is specific to $\mathscr{V}_n$ molecules; hence although it functions on $T_1^{\sim} P \mathscr{V}_n$, it would work just as well on $Q\mathscr{V}_n$ where Q is any other tagged protein. We will denote the removal system which destroys $\mathscr{V}_n$ tagged proteins Q from a substrate S by the symbol

$$f \, SQ\mathscr{V}_n$$

This symbol means the system **acts** on $SQ\mathscr{V}_n$ units and outputs S via mechanism f. Note the details of the mechanism f are largely irrelevant here. Thus, we have the reaction

$$T_1/T_1^{\sim} P \mathscr{V}_n + f \, SQ\mathscr{V}_n \rightarrow T_1$$

which releases T_1 into the cytoplasm. The full event chain for a cellular trigger is thus

$$
\begin{array}{lll}
T_0 & \rightarrow \text{Cell Surface Receptor} \rightarrow & PK/\mathscr{U} \\
PK/\mathscr{U} & + \; T_1/T_1^{\sim} & \rightarrow T_1/T_1^{\sim} P \\
T_1/T_1^{\sim} P & + \; \text{tagging system} & \rightarrow T_1/T_1^{\sim} P \mathscr{V}_n \\
T_1/T_1^{\sim} P \mathscr{V}_n & + \; f \, SQ\mathscr{V}_n & \rightarrow T_1 \\
T_1 & \rightarrow \text{nucleus} & \rightarrow \text{tagged protein transcription } P(T_1)
\end{array}
$$

where $P(T_1)$ indicates the protein whose construction is initiated by the trigger T_0. Without the trigger, we see there are a variety of ways transcription can be stopped:

- T_1 does not exist in a free state; instead, it is always bound into the complex $T_1/T_1^{\sim}$ and hence can't be activated until the $T_1^{\sim}$ is removed.
- Any of the steps required to remove $T_1^{\sim}$ can be blocked effectively killing transcription:

  - phosphorylation of $T_1^{\sim}$ into $T_1^{\sim} P$ is needed so that tagging can occur. So anything that blocks the phosphorylation step will also block transcription.
  - Anything that blocks the tagging of the phosphorylated $T_1^{\sim} P$ will thus block transcription.
  - Anything that stops the removal mechanism $f \, SQ\mathscr{V}_n$ will also block transcription.

The steps above can be used therefore to further regulate the transcription of T_1 into the protein $P(T_1)$. Let T_0', T_0'' and T_0''' be inhibitors of the steps above. These inhibitory proteins can themselves be regulated via triggers through mechanisms just like the ones we are discussing. In fact, $P(T_1)$ could itself serve as an inhibitory trigger—i.e. as any one of the inhibitors T_0', T_0'' and T_0'''. Our theoretical pathway is now:

$$
\begin{array}{lll}
T_0 & \rightarrow \text{Cell Surface Receptor} \rightarrow & PK/\mathscr{U} \\
PK/\mathscr{U} & + \; T_1/T_1^{\sim} & \rightarrow \text{step i } \; T_1/T_1^{\sim} P \\
T_1/T_1^{\sim} P & + \; \text{tagging system} & \rightarrow \text{step ii } \; T_1/T_1^{\sim} P \mathscr{V}_n \\
T_1/T_1^{\sim} P \mathscr{V}_n & + \; f SQ\mathscr{V}_n & \rightarrow \text{step iii } \; T_1 \\
T_1 & \rightarrow \text{nucleus} & \rightarrow \qquad \text{tagged protein transcription } P(T_1)
\end{array}
$$

where the **step i**, **step ii** and **step iii** can be inhibited as shown below:

$$
\begin{aligned}
&T_0 && \rightarrow \text{Cell Surface Receptor} \rightarrow && PK/\mathscr{U} \\
&PK/\mathscr{U} && + \; T_1/T_1^{\sim} && \rightarrow \text{step i} \;\; T_1/T_1^{\sim}P \\
& && && \uparrow T_0'\text{kill} \\
&T_1/T_1^{\sim}P && + \; \text{tagging system} && \rightarrow \text{step ii} \;\; T_1/T_1^{\sim}P\mathscr{V}_n \\
& && && \uparrow T_0''\text{kill} \\
&T_1/T_1^{\sim}P\mathscr{V}_n && + \; \textit{fSQ}\mathscr{V}_n && \rightarrow \text{step iii} \;\; T_1 \\
& && && \uparrow T_0'''\text{kill} \\
&T_1 && \rightarrow \text{nucleus} && \rightarrow \quad \text{tagged protein transcription } P(T_1)
\end{aligned}
$$

Note we have expanded to a system of four triggers which effect the outcome of $P(T_1)$. Also, note that **step i** is a phosphorylation step. Now, let's refine our analysis a bit more. Usually, reactions are paired: we typically have the competing reactions

$$
\begin{aligned}
PK/\mathscr{U} \; + \; T_1/T_1^{\sim} &\rightarrow T_1/T_1^{\sim}P \\
T_1/T_1^{\sim}P &\rightarrow T_1/T_1^{\sim} \; + \; PK/\mathscr{U}
\end{aligned}
$$

Hence, we can imagine that **step i** is a system which is in dynamic equilibrium. The amount of $T_1/T_1^{\sim}P$ formed and destroyed forms a stable loop with no net $T_1/T_1^{\sim}P$ formed. The trigger T_0 introduces additional $PK/\mathscr{U}$ into this stable loop and thereby effects the net production of $T_1/T_1^{\sim}P$. Thus, a new trigger T_0' could profoundly effect phosphorylation of $T_1^{\sim}$ and hence production of $P(T_1)$. We can see from the above comments that very fine control of $P(T_1)$ production can be achieved if we think of each step as a dynamical system in flux equilibrium.

Note our discussion above is a first step towards thinking of this mechanism in terms of interacting objects.

6.2 Dynamical Loop Details

Our dynamical loop consists of the coupled reactions

$$
\begin{aligned}
PK/\mathscr{U} \; &+ \; T_1/T_1^{\sim} \xrightarrow{\; k_1 \;} T_1/T_1^{\sim}P \\
& \;\; k_{-1} \\
T_1/T_1^{\sim}P \; &\rightarrow \; T_1/T_1^{\sim} \; + \; PK/\mathscr{U}
\end{aligned}
$$

where k_1 and k_{-1} are the forward and backward reaction rate constants and we assume the amount of $T_1/T_1^{\sim}$ inside the cell is constant and maintained at the equilibrium concentration $[T_1/T_1^{\sim}]_e$. Since one unit of $PK/\mathscr{U}$ combines with one unit of $T_1/T_1^{\sim}$ to phosphorylate $T_1/T_1^{\sim}$ to $T_1/T_1^{\sim}P$, we see

$$
[T_1/T_1^{\sim}](t) = [T_1/T_1^{\sim}]_e - [PK/\mathscr{U}](t). \tag{6.1}
$$

Hence,

$$\frac{d[PK/\mathcal{U}]}{dt} = -k_1 \, [PK/\mathcal{U}] \, [T_1/T_1^{\sim}] + k_{-1} \, [T_1/T_1^{\sim}P]$$

For this reaction, we have $[PK/\mathcal{U}] = [T_1/T_1^{\sim}]$; thus, we find

$$\frac{d[PK/\mathcal{U}]}{dt} = -k_1 \, [PK/\mathcal{U}]^2 + k_{-1} \, [T_1/T_1^{\sim}P] \tag{6.2}$$

From Eq. 6.1, we see

$$\frac{d[T_1/T_1^{\sim}]}{dt} = -\frac{d[PK/\mathcal{U}]}{dt}$$

Hence,

$$\frac{d[T_1/T_1^{\sim}]}{dt} = k_1 \, [PK/\mathcal{U}]^2 - k_{-1} \, [T_1/T_1^{\sim}P] \tag{6.3}$$

However, kinetics also tells us that

$$\frac{d[T_1/T_1^{\sim}]}{dt} = -k_1 \, [PK/\mathcal{U}]^2 + k_{-1} \, [T_1/T_1^{\sim}P] \tag{6.4}$$

and so, equating these two expressions, we find

$$2k_1 \, [PK/\mathcal{U}]^2 = 2k_{-1}[T_1/T_1^{\sim}P]$$

which implies

$$\frac{k_1}{k_{-1}} \, [PK/\mathcal{U}]^2 = [T_1/T_1^{\sim}P] \tag{6.5}$$

Using Eq. 6.5 in Eq. 6.3, we find that

$$\frac{d[T_1/T_1^{\sim}]}{dt} = k_1 \, [PK/\mathcal{U}]^2 - k_{-1} \, [T_1/T_1^{\sim}P] = 0 \tag{6.6}$$

Now let $[PK/\mathcal{U}]_e$ denote the equilibrium concentration established by Eq. 6.5. Then if the trigger T_0 increases $[PK/\mathcal{U}]$ by $\delta_{PK/\mathcal{U}}$, from Eq. 6.5, we see

$$[T_1/T_1^{\sim}P]\text{new} = \frac{k_1}{k_{-1}} \, \left([PK/\mathcal{U}]_e + \delta_{PK/\mathcal{U}}\right)^2 \tag{6.7}$$

which implies the percentage increase from the equilibrium level is

$$100\left(1 + \frac{\delta_{PK/\mathscr{U}}}{[PK/\mathscr{U}]_e}\right)^2$$

This *back of the envelope* calculation can be done not only at **step i** but at the other steps as well. Letting $\delta_{T_1/T_1^{\sim}P}$ and $\delta_{T_1/T_1^{\sim}P\mathscr{V}_n}$ denote changes in critical molecular concentrations, lets examine the **stage ii** equilibrium loop. We have

$$\begin{array}{c} k_2 \\ T_1/T_1^{\sim}P \;\rightarrow\; T_1/T_1^{\sim}P\mathscr{V}_n \end{array} \qquad (6.8)$$

$$\begin{array}{c} k_{-2} \\ T_1/T_1^{\sim}P\mathscr{V}_n \;\rightarrow\; T_1/T_1^{\sim}P \end{array} \qquad (6.9)$$

The kinetic equations are then

$$\frac{d[T_1/T_1^{\sim}P]}{dt} = -k_2\,[T_1/T_1^{\sim}P] + k_{-2}[T_1/T_1^{\sim}P\mathscr{V}_n] \qquad (6.10)$$

$$\frac{d[T_1/T_1^{\sim}P\mathscr{V}_n]}{dt} = k_2[T_1/T_1^{\sim}P] - k_{-2}[T_1/T_1^{\sim}P\mathscr{V}_n] \qquad (6.11)$$

Dynamic equilibrium then implies that

$$\frac{d[T_1/T_1^{\sim}P]}{dt} = \frac{d[T_1/T_1^{\sim}P\mathscr{V}_n]}{dt} \qquad (6.12)$$

and hence

$$2k_2[T_1/T_1^{\sim}P] = 2k_{-2}[T_1/T_1^{\sim}P\mathscr{V}_n]$$

or

$$\frac{k_2}{k_{-2}}[T_1/T_1^{\sim}P] = [T_1/T_1^{\sim}P\mathscr{V}_n] \qquad (6.13)$$

Equation 6.13 defines the equilibrium concentrations of $[T_1/T_1^{\sim}P]_e$ and $[T_1/T_1^{\sim}P\mathscr{V}_n]_e$. Now if $[T_1/T_1^{\sim}P]$ increased to $[T_1/T_1^{\sim}P] + \delta_{T_1/T_1^{\sim}P}$, the percentage increase would be

$$100\left(1 + \frac{\delta_{T_1/T_1^{\sim}P}}{[T_1/T_1^{\sim}P]_e}\right)^2$$

If the increase in $[T_1/T_1^{\sim}P]$ is due to **step i**, we know

$$\delta_{T_1/T_1^\sim P} = [T_1/T_1^\sim P]\text{new} - [T_1/T_1^\sim P]_e$$

$$= \frac{k_1}{k_{-1}} \left([PK/\mathscr{U}]_e + \delta_{PK/\mathscr{U}}\right)^2 - \frac{k_1}{k_{-1}}[PK/\mathscr{U}]_e^2$$

$$= \frac{k_1}{k_{-1}} \left(2[PK/\mathscr{U}]\delta_{PK/\mathscr{U}} + \delta_{PK/\mathscr{U}}^2\right)$$

We also know from Eq. 6.5 that

$$[T_1/T_1^\sim P]_e = \frac{k_1}{k_{-1}} [PK/\mathscr{U}]_e^2$$

and hence,

$$\frac{\delta_{T_1/T_1^\sim P}}{[T_1/T_1^\sim P]_e} = \frac{\frac{k_1}{k_{-1}}\left(2[PK/\mathscr{U}]_e\delta_{PK/\mathscr{U}} + \delta_{PK/\mathscr{U}}^2\right)}{\frac{k_1}{k_{-1}}[PK/\mathscr{U}]_e^2}$$

$$= 2\frac{\delta_{PK/\mathscr{U}}}{[PK/\mathscr{U}]_e} + \left(\frac{\delta_{PK/\mathscr{U}}}{[PK/\mathscr{U}]_e}\right)^2$$

For convenience, let's define the relative change in a variable x as $r_x = \frac{\delta_x}{x}$. Thus, we can write

$$r_{T_1/T_1^\sim P} = \frac{\delta_{T_1/T_1^\sim P}}{[T_1/T_1^\sim P]_e}$$

$$r_{PK/\mathscr{U}} = \frac{\delta_{PK/\mathscr{U}}}{[PK/\mathscr{U}]_e}$$

which allows us to recast the change in $[T_1/T_1^\sim]$ equation as

$$r_{T_1/T_1^\sim P} = 2r_{PK/\mathscr{U}} + r_{PK/\mathscr{U}}^2$$

Hence, it follows that

$$\delta_{T_1/T_1^\sim P\mathscr{V}_n} = [T_1/T_1^\sim P\mathscr{V}_n]\text{new} - [T_1/T_1^\sim P\mathscr{V}_n]_e$$

$$= \frac{k_2}{k_{-2}} \left([T_1/T_1^\sim P]_e + \delta_{T_1/T_1^\sim P}\right) - \frac{k_2}{k_{-2}} \left([T_1/T_1^\sim P]_e\right)$$

$$= \frac{k_2}{k_{-2}}\delta_{T_1/T_1^\sim P}$$

and so

$$r_{T_1/T_1^\sim P\mathscr{V}_n} = r_{T_1/T_1^\sim P}$$

$$= 2r_{PK/\mathscr{U}} + r_{PK/\mathscr{U}}^2$$

From this, we see that trigger events which cause $2r_{PK/\mathcal{U}} + r^2_{PK/\mathcal{U}}$ to exceed one $(r_{PK/\mathcal{U}} > \sqrt{2} - 1)$, create an explosive increase in $[T_1/T_1^{\sim} P\mathcal{V}_n]$. Finally, in the third step, we have

$$T_1/T_1^{\sim} P\mathcal{V}_n \xrightarrow[k_{-3}]{k_3} T_1 \tag{6.14}$$

$$T_1 \to T_1/T_1^{\sim} P\mathcal{V}_n \tag{6.15}$$

This dynamical loop can be analyzed just as we did in **step ii**. We see

$$\frac{k_3}{k_{-3}} [T_1/T_1^{\sim} P\mathcal{V}_n]_e = [T_1]_e$$

and the triggered increase in $[PK/\mathcal{U}]_e$ by $\delta_{PK/\mathcal{U}}$ induces the relative change

$$\begin{aligned}
\delta_{T_1} &= \frac{k_3}{k_{-3}} \delta_{T_1/T_1^{\sim} P\mathcal{V}_n} \\
&= \frac{k_3}{k_{-3}} \frac{k_2}{k_{-2}} \delta_{T_1/T_1^{\sim} P} \\
&= \frac{k_3}{k_{-3}} \frac{k_2}{k_{-2}} \frac{k_1}{k_{-1}} \left(2\, \delta_{PK/\mathcal{U}}\, [PK/\mathcal{U}]_e + \delta^2_{PK/\mathcal{U}} \right) \\
&= \frac{k_3 k_2 k_1}{k_{-3} k_{-2} k_{-1}} \left(2\, r_{PK/\mathcal{U}} + r^2_{PK/\mathcal{U}} \right) [PK/\mathcal{U}]^2_e
\end{aligned}$$

We can therefore clearly see the multiplier effects of trigger T_0 on protein production T_1 which, of course, also determines changes in the production of $P(T_1)$.

The mechanism by which the trigger T_0 creates activated kinase $PK/\mathcal{U}$ can be complex; in general, each unit of T_0 creates λ units of $PK/\mathcal{U}$ where λ is quite large—perhaps 10,000 or more times the base level of $[PK/\mathcal{U}]_e$. Hence, if $r_{PK/\mathcal{U}} = \beta$ and $\mathcal{K} = \frac{k_1 k_2 k_3}{k_{-1} k_{-2} k_{-3}}$, we have

$$\delta_{T_1} = (2\beta + \beta^2)\, \mathcal{K} [PK/\mathcal{U}]^2_e >> [PK/\mathcal{U}]_e$$

for $\beta >> 1$. From this quick analysis, we can clearly see the potentially explosive effect changes in T_0 can have on $PK/\mathcal{U}$!

6.3   An Implication for Biological Computation

Our first biological primitive is thus the trigger pathway we have seen before:

Fig. 6.1 Protein
transcription tree

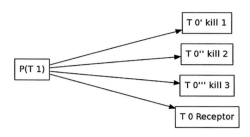

$$
\begin{array}{llll}
T_0 & \rightarrow \text{Cell Surface Receptor} \rightarrow & & PK/\mathscr{U} \\
PK/\mathscr{U} & + \ T_1/T_1^{\sim} & \rightarrow \text{step i} & T_1/T_1^{\sim}P \\
& & \uparrow T_0' \text{ kill} & \\
T_1/T_1^{\sim}P & + \ \text{tagging system} & \rightarrow \text{step ii} & T_1/T_1^{\sim}P\mathscr{V}_n \\
& & \uparrow T_0'' \text{ kill} & \\
T_1/T_1^{\sim}P\mathscr{V}_n & + \ fSQ\mathscr{V}_n & \rightarrow \text{step iii} & T_1 \\
& & \uparrow T_0''' \text{ kill} & \\
T_1 & \rightarrow \text{nucleus} & \rightarrow & \text{tagged protein transcription } P(T_1)
\end{array}
$$

We know from our discussions that δ_{T_1} is $(2\beta + \beta^2)\,\mathscr{K}\,[PK/\mathscr{U}]_e^2$ and we can interpret the output protein $P(T_1)$ as any of the molecules Q needed to **kill** a pathway. If $P(T_1)$ was a T_0 receptor, then we would have that the expression of the target protein $P(T_1)$ generates $\delta_{PK/\mathscr{U}}$!

Let us note two every important points now: There is **richness** to this pathway and the target $P(T_1)$ can **alter hardware** or **software** easily. If $P(T_1)$ was a K^+ voltage activated gate, then we see an increase of δ_{T_1} (assuming $1-1$ conversion of T_1 to $P(T_1)$) in the concentration of K^+ gates. This corresponds to a change in the characteristics of the axonal pulse. Similarly, $P(T_1)$ could create Na^+ gates thereby creating change in axonal pulse characteristics. $P(T_1)$ could also create other proteins whose impact on the axonal pulse is through indirect means such as the kill T_0' etc. pathways. There is also the positive feedback pathway via $\delta_{PK/\mathscr{U}}$ through the T_0 receptor creation. In effect, we have a forking or splitting of possibilities as shown in Fig. 6.1. Note that all of these pathways are essentially modeled by this primitive.

6.4   Transport Mechanisms and Switches

Another common method of cellular information processing is based on substances which are transported into a cell and then, depending on their concentration level, are either stored in binding sites or used to initiate the construction of a protein. The abstract model we discuss here is based on the cellular mechanism for dealing with free iron Fe in the cellular cytosol as detailed in Gerhart and Kirschner (1997). A substance M is to be carried into the cell. M can be a single atom of some type of a molecular unit. M is transported by a transport protein, tM. The transport protein binds to a specialized receptor *rM* to create *rtM*. The resulting complex initiates the transport through the membrane wall culminating in the release of *M* into the cytoplasm. Hence, letting *M/tM* denote the transported complex, we see

$$M + tM \rightarrow M \ tM \ \text{(transport)} + rM \ \text{(receptor)}$$
$$= M/rtM \ \text{(receptor complex)}$$
$$\rightarrow M \ \text{(freed M)}$$

There are many other details, of course, but the diagram above captures the basic information flow. In skeletal outline

$$M + tM \ \rightarrow \ M \ tM \ \rightarrow \ M/rtM \ \rightarrow \ M$$

Once inside the cell, *M* can be bound into a storage protein *sP* or *M* can be used in other processes. The concentration of *M*, [*M*], usually determines which path the substance will take. Let's assume there are two states, **low** and **high** for [*M*] which are determined by a threshold, *th*. A simple threshold calculation can be performed by a sigmoidal unit as shown in Fig. 6.2.

The standard mathematical form of such a sigmoid which provides a transfer between the low value of 0 and high value of 1 is given by

$$h(x) = \frac{1}{2}\left(1 + \tanh\left(\frac{x - o}{g}\right)\right) \tag{6.16}$$

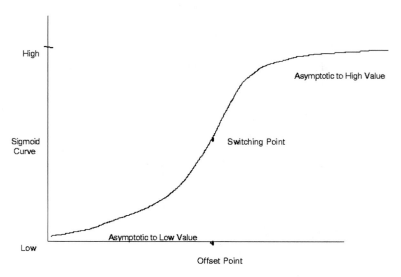

Fig. 6.2 A simple sigmoidal threshold *curve*

where x is our independent variable (here $[M]$), o is an *offset* (the center point of the sigmoid), and g is called a *gain* parameter as it controls the slope of the transfer from low signal to the high signal via the value $h'(o)$. Note,

$$h'(x) = \frac{1}{2g}\left(sech^2\left(\frac{x-o}{g}\right)\right) \qquad (6.17)$$

which gives $h'(0) = \frac{1}{2g}$. In general, the sigmoid which transfers us from a low value of L to a high value of H is given by

$$h(x) = \frac{1}{2}\left((H+L) + (H-L)\tanh\left(\frac{x-o}{g}\right)\right) \qquad (6.18)$$

This is the same sigmoidal transformation that will be used when we discuss neural computation in nervous systems in Chap. 14 and subsequent chapters. If we used a transition between low and high that was instantaneous, such an abrupt transition would not be differentiable which would cause some technical problems. Hence, we use a smooth transfer function as listed above and we interpret a low gain g as giving us a *fuzzy* or *slow* transition versus the *sharp* transition we obtain from a high gain g. Let's assume M binds to a switching complex denoted swM when its concentration is in the high state. The bound complex swM/M is then inactivated. In the low state, M does not bind to swM and swM is then able to bind to two sites on the messenger RNA, *mRNA*, for the two proteins rM and sPM.

The machinery by which *mRNA* is transformed into a protein can be thought of as a factory: hence, we let frM and $fspM$ denote the rM and sPM factories respectively. The *mRNAs* for these proteins will have the generic appearance as shown in Fig. 6.3. If you look at Fig. 6.3 closely, you'll see that there are special structures attached to the *mRNA* amino acid sequence. These look like tennis rackets and are called *stem loops*. The *mRNA* for both rM and sPM have these structures. The unbound switching protein swM can dock with any stem loop **if the conditions are right**. We know from our basic molecular biology, that binding near a 3′ end stabilizes *mRNA* against degradation and thus effectively increases protein creation. On the other hand, binding near the 5′ end inhibits translation of *mRNA* into protein and

Fig. 6.3 The mRNA object

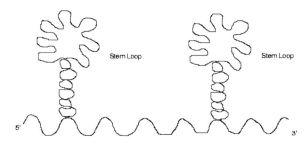

Fig. 6.4 *rM* and *sPM*
factory controls

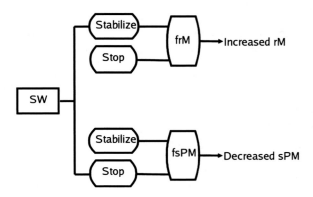

effectively serves as a stop button. Thus, for *rM* and *sPM*, we can imagine their factories possessing two control buttons—**stop** and **stabilize**. The switch binds to the 3′ end on the *mRNA* for *rM* and the 5′ end for the *sPM mRNA*. As you can see in Fig. 6.4, where *SM* denotes *swM*, Increased *rM* implies more *M* enters the cell; hence, [*M*] increases. Also, since decreased *sPM* implies there are fewer binding sites available, less *M* is bound, implying [*M*] increases also. The switch *swM* thus effects two primitive components of the cell; the number of *M* receptors in the cell wall and the number of storage sites for *M* in the cytosol. This mechanism could work for any substance *M* which enters a cell through a receptor via a transport protein whose concentration [*M*] is controlled by a storage/free cytosol interaction. The generic mechanism is illustrated in Fig. 6.5. This mechanism is based on how iron is handled in a cell. In human cells, there is no mechanism for removing excess iron; hence, the high pathway can only show inactivation of the switch and additional binding into the binding protein. From our discussion, it follows we effectively control [*M*] by a switch which can directly activate the genome!

6.5   Control of a Substance via Creation/Destruction Patterns

Another way we can control the concentration of a substance N is indirectly through a creation/destruction cycle. Assume N is bound into a complex B_1N and there are two competing reactions:

$$B_1N \rightarrow B_1 + N \quad \text{Breakdown of } B_1N$$
$$B_1 + N \rightarrow B_1N \quad \text{Creation of } B, N$$

These reactions are made possible by enzymes. Hence letting E_B be the breakdown enzyme, we have

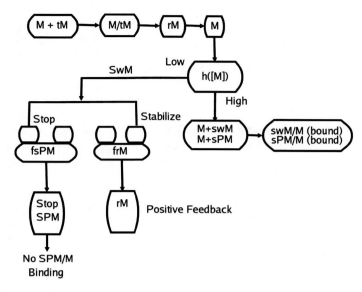

Fig. 6.5 Generic substance M control mechanism

$$B_1N + E_B \rightarrow B_1 + NR$$

where NR is the N complex that is temporarily formed. The enzyme which breaks down NR is labeled E_B^{NR} and we have the reaction

$$NR + E_B^{NR} \rightarrow N + R$$

The enzyme controlling the creation of B_1N is called E_C with reaction

$$B_1 + NR + E_C \rightarrow B_1N$$

Finally, to create NR, we need the enzyme E_C^{NR}; the reaction is

$$N + R + E_C^{NR} \rightarrow NR$$

Combining, we have four reactions:

$$\left.\begin{array}{l} B_1N + E_B \rightarrow B_1 + NR \\ NR + E_B^{NR} \rightarrow N + R \end{array}\right\} \text{ Creates } N \quad \left.\begin{array}{l} B_1 + NR + E_C \rightarrow B_1N \\ N + R + E_C^{NR} \rightarrow NR \end{array}\right\} \text{ Binds } N$$

Note that the binding of N effectively removes it from the cytosol. Now, the above four equations tell us that the creation or destroying of the complex B_1N is tantamount to the binding or creating of N itself.

We can control which of these occurs as follows. Let's assume there is a protein T_1 whose job is to phosphorylate a protein S_1. The protein S_1 is bound with another protein S_2 and the complex, denoted S_1/S_2 will function as a switch. Let the phosphorylated protein S_1 be denoted by S_{1p}; then we have the reaction

$$T_1 + S1/S2 \rightarrow S_{1p}/S_2$$

Now imagine our four enzymatically controlled reactions as being toggled on and off by an activating substance. We illustrate this situation in Fig. 6.6a where the action of S_1/S_2 and S_{1p}/S_2 are as indicated. The S_{1p}/S_2 switch increases $[N]$ and the S_1/S_2 switch decreases $[N]$. In detail, S_{1p}/S_2 increases the breakdown of NR, more effectively decreasing $[B_1N]$; the S_1/S_2 increases $[NR]$, effectively increasing $[B_1N]$. Hence, $[B_1N]$ and N work opposite one another. The control of the phosphorylation of S_1 to S_{1p} is achieved via a $T_1/T_1^\sim$ mechanism. Specifically, if rB_1N is the B_1N receptor, then we have the control architecture shown in Fig. 6.6b. In this figure, the protein $P(T_1)$ is actually T_1 itself. We see we can combine mechanisms to generate control of a substances concentration in the cytosol. Note, we are still seeing a blurring of the hardware/software distinction in this example, just as in the others. These paired reactions are often called futile cycles.

The trigger systems we are modeling must include a mechanism for deactivation. Indeed, much of the richness of biological computation comes from negative feedback. Consider the generic trigger system of Fig. 6.7. This system allows for protein creation, but it is equally important to allow for deactivation of the product if its concentration becomes too high. We can use the generic switch method discussed in Sect. 6.4 in the case where the trigger T_0 exists in both a free and bound state upon entry into the cell. The trigger T_0 is processed by the usual trigger mechanism into the protein $P(T_1)$ which is sent into a sigmoid switch h. If the concentration of $P(T_1)$ is low, the usual stop/ stabilize control is called which increases ΔT_0. If the concentration is high, we bind T_0 to a buffer complex. This mechanism is applicable to the triggers that control free Ca^{++} in the cell such as neurotransmitters.

6.6   Calcium Ion Signaling

Adding or removing a phosphate ion, PO_3^- to a protein or other molecular complex is important to the control of many reaction pathways. To add such a phosphate ion PO_3^- to a protein, however, requires energy and a helper molecule so that the reaction can proceed at a useful speed. The specialized protein which help make these reactions possible are called *protein kinases*. To understand this a little better, we need to look at the ATP–ADP reaction. Adenosine Triphosphate (ATP) is the nucleotide adenine plus a ribose sugar with three phosphates, terminated in a methyl ion, linked to it as a linear chain, $PO_3^- - PO_3^- - PO_3^- CH_2$. Adenosine Diphosphate (ADP) has a linear chain of just two phosphates $PO_3^- - PO_3^- CH_2^-$ added. When we remove another phosphate, we have only a single $PO_3^- CH_2^-$ bound to the adenosine

Fig. 6.6 Control strategies. **a** Futile control. **b** Switch control

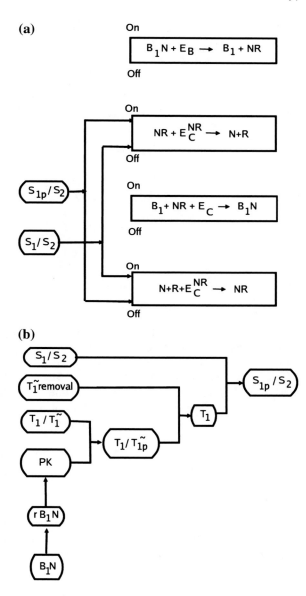

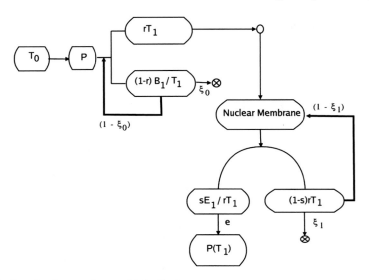

Fig. 6.7   Generic trigger pathway with feedback

core. This molecule is then called Adenosine Monophosphate or AMP. In these three versions, the PO_3^- chains are bound to the ribose sugar directly and do not have an attachment to the adenine nucleotide itself. There is a fourth version which binds a PO_3^- both to the adenine and the ribose both. This bond looks cyclic in nature and so this last version is called Cyclic Adenosine Monophosphate or CAMP. The formation and breakdown of these compounds are controlled by enzymes. In reaction form we have

$$2\ ADP \xrightarrow{\text{adenylate kinase}} ATP + AMP$$
$$AMP \quad \rightarrow \quad Adenosine$$

There are 4 phosphate ions on the 2 ADP molecules; adenylate kinase facilitates the move one of them from one ADP to the other ADP to make a molecule ATP. The remaining ADP then becomes an AMP molecule. The enzyme adenylate kinase maintains the concentration of ATP, ADP and AMP at approximately 1 : 1 : 1 ratios. Hence,

$$k_{eq} = \frac{[AMP][ATP]}{[ADP]^2} \approx 1$$

and

$$[AMP] \approx \frac{[ADP]^2}{[ATP]}$$

Now in a fully energized cell, $[ATP]$ is approximately $4\,mM$ and $[ADP]$ is on the order of $.4\,mM$. This is a $10:1$ ratio. We also know the reaction $ATP \rightarrow ADP + PO_3^-$ and its reverse maintains a $1:1:1$ ratio. Thus, the ATP to ADP reaction implies that the concentration of the two should be the same. However, the AMP reaction tells us that in a fully energized cell, ATP and ADP concentrations differ by a factor of 10 from this equilibrium ratio. From the AMP equation, we see that the equilibrium concentration of AMP must be $\frac{(0.4)^2}{4}$ or $.04\,mM$. There is also another way to break down ATP to ADP which involves muscle contraction under anaerobic conditions. This gives $[ATP] \approx [ADP] \approx 2\,mM$. In this case, when adenylate cyclase is at equilibrium, we find $[AMP]$ is approximately $\frac{2^2}{2}$ or $2\,mM$. Hence, with this second mechanism, the equilibrium concentration of ATP is half the concentration of ATP from the first reaction pathway. The equilibrium concentration of AMP however varies significantly more: it is $0.4\,mM$ in the first and $2\,mM$ in the secon—a fifty fold increase. Hence, the concentration of AMP is a very sensitive indicator of which reaction is controlling the energy status of the cell.

Now, inside the cytosol, free Ca^{++} exists in $.1\,\mu M$ to $.2\,\mu M$ concentrations. Outside the cytosol, this ion's concentration is much higher—$1.8\,mM$. Thus, the concentration gradient between inside and outside is on the order of 10^4. This very steep gradient is maintained by surface membrane proteins, a sodium/Ca^{++} exchange protein and a membrane calcium pump. In addition, the endoplasmic reticulum (ER) stores free calcium in pools within its membrane. The movement of calcium in and out of these pools is controlled by a pump in the ER which uses the reaction between adenosine triphosphate (ATP) and adenosine diphospate (ADP) to drive calcium in or out of the pools. The ER pump is therefore called the ATPase pump.

Rapid increases in cystolic Ca^{++} are used as regulatory signals. Hormones, neuro-transmitter and electrical activity control Ca^{++} movement into the cytosol. Inositol-1,4,5 triphosphate or IP_3 controls movement of the ions in and out of the storage pools in the ER. There is probably 50–$100\,\mu M$ of Ca^{++} inside the cell but most of this is bound into proteins such as calmodulin (CAM). Calmodulin *relays* calcium signals to other protein targets. The other proteins serve usually as Ca^{++} buffers which therefore limit the amount of Ca^{++} that is free in the cell. Different neurons have different levels of Ca^{++} buffering proteins. This implies that there are large differences in physiological behavior between these classes of neurons.

Ca^{++} gets into the cytosol from the extracellular fluid via Ca^{++} gates or channels. These ports can be voltage dependent like a classical sodium or potassium voltage gated channel or they can be *ligand gated*. This means that some protein must bind to the port in some fashion to initiate further mechanisms that control the amount of free Ca^{++} in the cell. These mechanisms could be as direct as the ligand opens the port and allows Ca^{++} to come in or could be indirect as we have already discussed for generic T_0 triggers and switches. The important thing to remember is that the amount of free calcium ion in the cytosol is controlled by four things:

- The ATP based pump in the ER: one calcium ion goes out for each ATP to ADP reaction. The rate of this pump is controlled by calmodulin,
- calcium ion in through a voltage gated channel,

- calcium ion in through a ligand gated channel,
- the calcium/ sodium exchange pump: one calcium ion out for each three sodium ions coming in.

Calmodulin is a Ca^{++} receptor protein. One CAM molecule has four Ca^{++} binding domains. At resting Ca^{++} concentrations, little Ca^{++} is bound to CAM. As the concentration of Ca^{++} increases, the binding sites of CAM are occupied and CAM becomes a multifunctional activator. Roughly speaking, here is what happens.

- A ligand such as a neurotransmitter or hormone binds to its receptor or gate in the cell membrane. Such a ligand is often called an *agonist*.
- In physical proximity with the receptor is a protein called the *G-protein*. It generally consists of three subunits, G_α, G_β and G_γ. The G protein is called a coupling protein as it transduces the extracellular signal into chemical pathway action inside the cytosol.
- At rest, the coupling protein is bound into a complex with *Guanine Diphosphate* or GDP. They exist diffusively within the cytosol. If the bound complex G/GDP encounters a receptor with its bound ligand or agonist, the GDP pops off the complex. Inside the cytosol, there is free GTP. Once the GDP is off, GTP from the cytosol takes its place. The coupling protein G then splits into two pieces: the complex G_α/GTP and the complex $G_\beta G_\gamma$. These two complexes are called activated forms of the coupling protein G. The bound GDP can be released and replaced by *Guanine Triphosphate* or GTP. Hence, there is a mechanism to convert the complex, G/GDP to G/GTP.
- The enzyme *adenyl cyclase* or *AC* then combines with *ATP* to create *CAMP* as we have discussed. *Phosphodiesterase* or *PDE* can then deactivate *CAMP* back to *AMP*. However, *CAMP* dependent protein kinase or *PKA* then adds a phosphate ion to a target protein. Adding a phosphate ion to a protein is called phosphorylation. Any protein that puts a phosphate ion on another protein is called a *kinase*. Kinases works by breaking a phosphate group off of *ATP* (which releases energy) and taking that phosphate and adding it to the protein. Once this is done, the protein is said to be *activated*. An activated protein can then create other molecules and complexes via reaction pathways of its own. The amount of protein that is phosphorylated at any time is determined by the balance between the rates of phosphorylation by its kinase and the rate of dephosphorylation by what is called a *phosphotase*. This is a classic example of a futile cycle we have discussed.
- In general, the activated protein from the step above will eventually activate protein transcription from the genome.

It is clear then that the coupling protein mechanism above allows an extracellular signal to increase [*CAMP*] which in turn alters the activity level of the cell via the *PKA* pathway. There are very important information processing consequences here: a small number of occupied receptors can have a major influence on the cell. First, the nature of the G protein mediated signals is that there is amplification of the message.

First, the nature of the G protein mediated signals is that there is amplification of the message. An agonist can activate many different G protein—*CAMP* pathways. Note the specificity of the extracellular signal via the agonist docking into the receiver is multiplied many fold by the fact that the G/GDP complexes exist diffusively in the cell. The extracellular signal binds to multiple receptors and each bound receptor can interact with a different G/GDP complex. Hence, out of this soup of different G/GDP complexes, multiple G/GDP complexes could interact with the same signal. Hence, the signal can be enormously amplified. Roughly speaking, this amplification is on the order of $(2\beta + \beta^2)\mathscr{K}[G]$ for suitable β and $\mathscr{K}$ as discussed in Sect. 6.2. Also, one activated G_α/GTP can interact with many $CAMP$. Each $CAMP$ can create activated proteins via PKA which could all be different due to concentration differences in the created $[CAMP]$, and the time and spatial activity envelope of the $[CAMP]$ pulse. Further, PKA can activate any protein that has a certain pattern of amino acids around serine or threonine residues. Thus, any such protein will be called by PKA to participate in the extracellular signals effects.

Second, more than one agonist can increase $[CAMP]$ concentration via these mechanisms. Thus, a general feature of this kind of agonist stimulation is that the effect is felt away from the site of the initial docking of the extracellular signal. Also, the full effect of the signal is the creation of a new protein which can completely rewrite the hardware and software of the excitable nerve cell. We conclude that there is both *amplification* and *divergence* of action. Further, an essential feature of this sort of activation is that the signaling pathway can be turned off. The G protein system is self-timing. The G_α/GTP is converted back to G/GDP automatically, so if there is not a stimulus to create new G/GTP, the system reverts to it rest state. Also, $CAMP$ is always cleaved to AMP by PDE so without the creation of new $CAMP$, the phosphorylation of the target proteins could not continue. Finally, the phosphate ions on the activated target proteins are always being removed by appropriate phosphatases. It follows that if a cell has the three mechanism mentioned above operating a a high state, then extracellular signals will not last long, approximately less than 2 s.

6.7 Modulation Pathways

The monoamine neurotransmitters norepinephrine, dopamine and serotonin modulate the use of Ca^{++} within a cell in a variety of ways. In Fig. 6.8, we have used the discussions in Hille (1992), to indicate some of the complexity of these pathways. In the figure, we use the following legends: AC, adenyl cyclase; CAMP, cyclic adenosine monophosphate; G proteins G_i, G_s, G_o and G_p; PDE, phosphodiesterase; PKA, CAMP dependent kinase; CamK, calmodulin dependent kinase; PLC, phospholipase C; PKC, C kinase; IP3, inositol-1,4,5 triphosphate; DAG, diacylglycerol; IN1, PKA inhibitor; PP, generic protein phosphotase; S_1, S_2, S_3, S_4, S_6 and S_7 are various surface proteins. Hence, creation of these proteins alters the hardware of the cell. These proteins S_i could be anything used in the functioning of the cell. We illustrate the pathways used by the three major monoamine neurotransmitters as well, DP, dopamine; S, serotonin and NE, norepinephrine. NE has two gates shown subscripted by 1 and 2 which initiate different G pathways. The examples we have discussed give us a way to unify our understanding of cell signaling. Following Bray

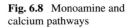 **Fig. 6.8** Monoamine and calcium pathways

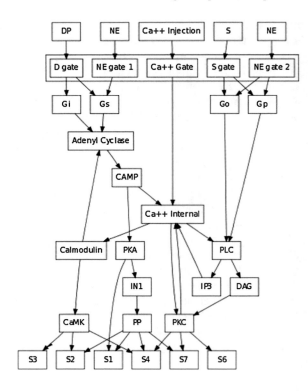

(1998), we can infer protein kinases, protein phosphatases, transmembrane receptors etc. that carry and process messages inside the cell are associated with compact clusters of molecules attached to the cell membrane which *operate as functional units* and *signaling complexes provide an intermediate level of organization like the integrated circuits used in VLSI. However, this interpretation is clearly flawed as these circuits are self-modifying.* The general organization of the action of the extracellular agonists here is thus:

$$Agonist \rightarrow Receptor \rightarrow G$$
$$\rightarrow Enzyme \rightarrow Second\ Messenger$$
$$\rightarrow Kinase \rightarrow Effector\ Protein$$

6.7.1 Ligand—Receptor Response Strategies

The *TAR* complex is a cluster of proteins associated with the chemotaxic receptor of the coliform bacteria. It is a large protein which is inserted through the bacteria's membrane. It has three portions: the extracellular part where the external signals bind, the transmembrane part which is inside the membrane and the intracellular part which is in the cellular cytosol. When the bacteria is moving in an aspartate concentration gradient, signals are generated to the motor assemblies that drive its flagella movement. This allows the bacteria to move toward the food source. The extracellular signal is the aspartate current. The way this works is as follows:

- Aspartate binds to the extracellular domain of the receptor.
- This binding causes a change in the three dimensional appearance of the intracellular portion of *TAR*. This is called a conformational change.
- The intracellular part of *TAR* is made up of 4 proteins. The two closest to the membrane are labeled *CheZ* which can either by methylated or de methylated by the enzymes *CheR* or *CheB*, respectively. Below them is the protein *CheW* and below that is the protein *CheA*. *CheA* is an enzyme which phosphorylates both *CheY* and *CheB*. *CheY* and *CheB* exist free in the cytosol.

The *TAR* control action is based on several states. First, if no ligand is bound to *TAR* and *CheZ* is highly methylated, *CheA* is active and it phosphorylates both *CheB* and *CheY*. The high level of phosphorylated *CheYp* increases the frequency of the motor's switch to clockwise flagella rotation which causes tumbling—essentially, a random search pattern for food. The phosphorylated *CheBp slowly* removes methyl from *CheZ* which resets the signaling state to that of no ligand and no *CheZ* methylation. If *TAR* does not have a ligand bound and *CheZ* is not highly methylated, *CheA* is inactive and the level of *CheYp* is reduced. This increases the rate of counterclockwise flagella rotation as there are fewer switches to clockwise. This results in the bacteria swimming in a specific direction and hence, this is not a random search pattern. Finally, if the aspartate signal binds *TAR* whether or not *CheZ* is methylated, *CheA* becomes inactive which leads to directional swimming as outlined above.

The phosphorylation of *CheY* occurs at a rate which is proportional to the rate of external signal activation of *TAR*. The *TAR* complex thus operates as a self-contained solid state processing unit which "integrates" extracellular inputs to produce a certain rate of phosphorylation of *CheA*. This influences the rate of change of *CheYp* and thus produces alterations in the way the flagella handles rotations. This switches the bacteria from random search to directed swimming. Control of the *CheYp* creation rate effectively controls the ratio of counterclockwise to clockwise rotations. This sets the direction of the swimming or if the switching is too frequent, sets the motion to be that of a random walk. The current state of the motor can thus be inferred from the methylation patterns on the cytosol side of the *TAR* complex. The general aspartate pathway thus consists of a single freely diffusing molecule *CheYp*, a protein complex which is only concerned with the external stimulus and a second protein complex which handles the behavioral response.

Next, consider how a cell uses a signal for growth called the *platelet derived growth factor* or *PDGF*. This signal must be noticed by a cell before it can be used. A single *PDGF* receptor diffuses freely in the cellular membrane. When it encounters a *PDGF* extracellularly, it combines with another receptor to form what is called a *dimer* or *double PDGF* receptor. The paired receptors have tails that extend through the membrane into the cytosol. Once paired, multiple sites on these tails phosphorylate. This phosphorylation triggers the assembly of six signaling complexes on the tails. Some fraction of the six signaling complexes are then activated. The activated complexes then initiate broadcast signals using multiple routes to many destinations inside the cell. These signals activate and coordinate the many biochemical changes needed in cell growth. As usual, the response is removed by phosphatases which take away phosphates from activated complexes. This is, therefore, a cellular strategy to make a large number of receptors catch as many of the *PDGF* molecules as possible.

Let's look at the differences in the handling of the extracellular signal in the *TAR* and *PDGF* cases. *PDGF* requires a sequence of reactions: receptor clustering into dimers, phosphorylation of critical sites and the binding of multiple signaling molecule complexes. Hence, the response here is slower than the conformational change that occurs within a preformed *TAR* complex due to the aspartate signal. Notice though that the *TAR* response needs to be quick as it alters swimming patterns for finding food. There is not a need for such speed in the *PDGF* response. Also, not every *PDGF* receptor carries a signaling complex as only those that encounter a *PDGF* molecule undergo the changes that result in the addition of signaling complexes. Thus, when the cell makes a large number of receptors and casts a wide net to catch *PGDF* signals, there is not a large cost. This is because there is no need to make expensive signaling complexes on all receptors—only the ones which need them.

TAR complex however must react quickly. In fact, there is even a real possibility that all *TAR* receptors could be occupied at the same time. Since the bacteria must respond to an aspartate concentration gradient on the order of 10^5, the response to 90 and 95 % *TAR* receptor occupancy must be different. Hence, *all* receptors must carry the signaling complexes necessary to alter the flagella motor.

The *PGDF* receptor complex produces multiple outputs which implies the initial extracellular signal is *divergent* with effects spreading throughout the cell. The *TAR* complex has only two outputs: *CheYp* and *CheBp* which both effect motor response. However, there are also other pathways that effect *CheYp* and *CheBp*. Hence, there are multiple controls here that *converge* on a single output pair.

In general, the concentration of proteins in the cytosol is high enough that there is competition among macromolecules for water. This effects the diffusion rate of the macromolecules and changes the tendency of proteins to associate with other proteins. Hence, the formation of a signaling complex requires a sequence of binding steps in which proteins come together in a diffusion limited interaction governed by mass action. This implies that association among macromolecules is strongly favored in these crowded conditions. Now, the *TAR* complex serves a focused purpose and drives changes in swimming action directly and quickly. Signaling complexes as seen in the *PDGF* receptor are favored for different reasons. Clusters of molecules in permanent association with each other are well suited to perform reactions that integrate, codify and transduce extracellular signals. These solid state computations are more rapid, efficient and noise free than a system of diffusing molecules in which encounters are subject to random fluctuations of thermal energy. Also, complexes are made only once. So even though the construction process uses diffusion, once completed, the complex can repeatedly and accurately perform the same task.

A macromolecule can switch its conformational state 10^4–10^6 times a second at the cost of 1 *ATP* per switch. So we might infer from this that biological information has a cost of 1 *ATP* per bit. However, in a real system the major expense is not in registering conformational change. The true cost is the amount of energy used to distribute, integrate and process the information of the extracellular signal within and between systems. Hence, if a switch in conformational state gives a rough indication of information storage rate, we might think the biological information storage rate is about 10^4–10^6 bits per second. However, based on the distribution, integration and processing costs, it is more like 10^3–10^5 per second at a cost of 10^6 *ATP* per bit.

Hence, a guiding principle is that the second messenger triggers will store information at a rate of about 10^3 per second at a cost of 10^6 energy units per bit.

Although signaling complexes are indeed powerful tools, there is also a strong need for diffusion based mechanisms. Most extracellular stimuli are not passed in a *linear chain* of cause and effect from one receptor in the membrane to one target molecule in the nucleus or cytosol. If such a linear chain could be used, the signal could be passed very efficiently as a chain of conformational changes along a protein filament. In fact, signals such as we see in the axon of an excitable neuron are essentially passed along in this manner. However, most extracellular stimuli are spread as a rapidly diverging influence from the receptor to multiple chemical targets in the cell at often many different spatial locations. This divergence allows the possibility of signal amplification and cross-talk between different signal processes chains. The machinery of the signal spread in this way always requires at least one molecule that diffuses as a free element. Note that always associating one receptor with one effector (as in the *CheYp* to flagella motor effector) is faster and more efficient than the general diffusion mechanism. However, it is a *much smaller and more localized* effect!

The *TAR* receptor approach does use a freely diffusing molecule, *CHeYp*, but the action of this pathway is restricted to the flagella motor. Hence there is no divergence of action. In the *PDGF* receptor approach, there is a large cost associated with the construction of the signaling complex, but once built, diffusion is used as the mechanism to send the *PDGF* signal in a divergent manner across the cell. Hence, one extracellular event triggers one change or one extracellular event triggers 10^3–10^5 changes! This is a large increase in the sensitivity of the cell to the extracellular input. We see a cell pays a price in time and consumption of *ATP* molecules for a huge reward in sensitivity.

References

D. Bray, Signalling complexes: biophysical constraints on intracellular communication. Annu. Rev. Biophys. Biomol. Struct. **27**, 59–75 (1998)

J. Gerhart, M. Kirschner, *Cells, Embryos and Evolution: Towards a Cellular and Developmental Understanding of Phenotypic Variation and Evolutionary Adaptability* (Blackwell Science, New Jersey, 1997)

B. Hille, *Ionic Channels of Excitable Membranes* (Sinauer Associates Inc, Sunderland, 1992)

Chapter 7
Second Messenger Diffusion Pathways

Now that we have discussed some of the extracellular trigger mechanisms used in bioinformation processing, it is time to look at the fundamental process of diffusion within a cell. Our second messenger systems often involve Ca^{++} ion movement in and out of the cell. The amount of free Ca^{++} ion in the cell is controlled by complicated mechanisms, but some is stored in buffer complexes. The release of calcium ion from these buffers plays a big role in cellular regulatory processes and which protein $P(T_1)$ is actually created from a trigger T_0. Following discussions in Berridge (1997), Wagner and Keizer (1994) and Höffer et al. (2001), we will now describe a general model of calcium ion movement and buffering. These ideas will give us additional insight into how to model second messenger triggering events.

7.1 Calcium Diffusion in the Cytosol

Calcium is bound to many proteins and other molecules in the cytosol. This binding is fast compared to calcium release and diffusion. From Wagner and Keizer (1994), we can develop a rapid-equilibrium approximation to calcium buffering. Let's assume there are M different calcium binding sites; in effect, there are different *calcium species*. We label these sites with the index j. Each has a binding rate constant k_j^+ and a disassociation rate constant k_j^-. We assume each of these binding sites is homogeneously distributed throughout the cytosol with concentration B_j. We let the concentration of free binding sites for species j be denoted by $b_j(t, x)$ and the concentration of occupied binding sites is $c_j(t, x)$; where t is our time variable and x is the spatial variable. The units of b_j are (mM B_j)/liter and c_j has units (mM B_j)/liter. We let $u(t, x)$ be the concentration of free calcium ion in the cytosol at (t, x). Hence, we have the reactions

© Springer Science+Business Media Singapore 2016
J.K. Peterson, *BioInformation Processing*, Cognitive Science and Technology,
DOI 10.1007/978-981-287-871-7_7

$$Ca + B_j \rightarrow_{k_j^+} Ca/B_j$$
$$Ca/B_j \rightarrow_{k_j^-} Ca + B_j$$

The corresponding dynamics are

$$\frac{d[Ca]}{dt} = -k_j^+[Ca][B_j] + k_j^-[Ca/B_j]$$
$$\frac{d[Ca/B_j]}{dt} = k_j^+[Ca][B_j] - k_j^-[Ca/B_j]$$

From this, we have the unit analysis

$$\frac{\text{mM Ca}}{\text{liter} - \text{sec}} = \text{units of } k_j^+ \times \frac{\text{mM Ca}}{\text{liter}} \frac{\text{mM } B_j}{\text{liter}} + \text{units of } k_j^- \times \frac{\text{mM Ca}/B_j}{\text{liter}}$$

This tells us the units of k_j^+ are liter/(mM B_j-sec) and k_j^- has units (mM of Ca)/
((mM Ca/B_j)-sec). Hence, k_j^- c_j has units of (mM Ca)/(liter-sec), k_j^+ b_j, 1/s and
k_j^+ $b_j u(t, x)$, (mM Ca)/(liter-sec). Let the diffusion constant for each site be D_j. We
assume D_j is independent of the bound calcium molecules; that is, it is independent
of c_j. We also assume a homogeneous distribution of the total concentration B_j.
This implies a local conservation law

$$b_j(t, x) + c_j(t, x) = B_j$$

For simplicity, we will use a $1D$ model here; thus the spatial variable is just the
coordinate x rather than the vector (x, y, z). For the one dimensional domain of
length L, we have $0 \leq x \leq L$. In a cell, some of the calcium ion is bound in storage
pools in the endoplasmic reticulum or ER. The amount of release/uptake could be a
nonlinear function of the concentration of calcium ion. Hence, we model this effect
with the nonlinear mapping $f_0(u)$. The diffusion dynamics are then

$$\frac{\partial u}{\partial t} = \text{rate of ER release and uptake}$$

$$+ \text{sum over all calcium species(amount freed from binding site } - \text{ amount bound)}$$

$$+ D_0 \frac{\partial^2 u}{\partial x^2}$$

Thus, we find

$$\frac{\partial u}{\partial t} = f_0(u) + \sum_{j=1}^{M} \left(k_j^- c_j - k_j^+(B_j - c_j)u \right) + D_0 \frac{\partial^2 u}{\partial x^2} \qquad (7.1)$$

Also, we see the dynamics for c_j are given by

$$\frac{\partial c_j}{\partial t} = -k_j^- c_j + k_j^+ (B_j - c_j) u + D_j \frac{\partial^2 c_j}{\partial x^2} \tag{7.2}$$

where D_0 is diffusion coefficient for free calcium ion. Our domain is a one dimensional line segment, so there should be diffusion across the boundaries at $x = 0$ and $x = L$ for calcium but not for binding molecules. Let J_0 and J_L denote the flux of calcium ion across the boundary at 0 and L. We will usually just say $J_{0,L}$ for short. Our boundary conditions then become

$$-D_0 \left.\frac{\partial u}{\partial x}\right|_0 = J_0 \tag{7.3}$$

$$-D_0 \left.\frac{\partial u}{\partial x}\right|_L = J_L \tag{7.4}$$

$$D_j \left.\frac{\partial c_j}{\partial x}\right|_0 = 0 \tag{7.5}$$

$$D_j \left.\frac{\partial c_j}{\partial x}\right|_L = 0 \tag{7.6}$$

The concentration of total calcium is the sum of the free and the bound. We denote this by $w(t, x)$ and note that

$$w = u + \sum_{j=1}^{M} c_j$$

$$\frac{\partial w}{\partial t} = f_0\left(w - \sum_{j=1}^{M} c_j\right) + D_0 \frac{\partial^2 w}{\partial x^2} + \sum_{j=1}^{M} (D_j - D_0) \frac{\partial^2 c_j}{\partial x^2}$$

Note that

$$\frac{\partial w}{\partial t} = \frac{\partial u}{\partial t} + \sum_{j=1}^{M} \frac{\partial c_j}{\partial t}$$

$$= f_0(u) + \sum_{j=1}^{M} \left[k_j^- c_j - k_j^+ (B_j - c_j) u \right] + D_0 \frac{\partial^2 u}{\partial x^2}$$

$$+ \sum_{j=1}^{M} \left[-k_j^- c_j + k_j^+ (B_j - c_j) u \right] + \sum_{j=1}^{M} D_j \frac{\partial^2 c_j}{\partial x^2}$$

$$= f_0(u) + D_0 \frac{\partial^2 u}{\partial x^2} + \sum_{j=1}^{M} D_j \frac{\partial^2 c_j}{\partial x^2}$$

$$= f_0\left(w - \sum_{j=1}^{M} c_j\right) + D_0\frac{\partial^2 u}{\partial x^2} + D_0\sum_{j=1}^{M}\frac{\partial^2 c_j}{\partial x^2} - D_0\sum_{j=1}^{M}\frac{\partial^2 c_j}{\partial x^2} + \sum_{j=1}^{M} D_j\frac{\partial^2 c_j}{\partial x^2}$$

$$= f_0\left(w - \sum_{j=1}^{M} c_j\right) + D_0\left(\frac{\partial^2 u}{\partial x^2} + \sum_{j=1}^{M}\frac{\partial^2 c_j}{\partial x^2}\right) - \sum_{j=1}^{M}(D_0 - D_j)\frac{\partial^2 c_j}{\partial x^2}$$

$$= f_0\left(w - \sum_{j=1}^{M} c_j\right) + D_0\frac{\partial^2 w}{\partial x^2} + \sum_{j=1}^{M}(D_j - D_0)\frac{\partial^2 c_j}{\partial x^2}$$

The boundary conditions are 0 and L can then be computed using Eqs. 7.3–7.6.

$$-D_0\frac{\partial w}{\partial x}\bigg|_{0,L} = -D_0\frac{\partial u}{\partial x}\bigg|_{0,L} - \sum_{j=1}^{M} D_j\frac{\partial c_j}{\partial x}\bigg|_{0,L} - (D_0 - D_j)\sum_{j=1}^{M}\frac{\partial c_j}{\partial x}\bigg|_{0,L} = J_{0,L}$$

Thus, the total calcium equation is

$$\frac{\partial w}{\partial t} = f_0\left(w - \sum_{j=1}^{M} c_j\right) + D_0\frac{\partial^2 w}{\partial x^2} + \sum_{j=1}^{M}(D_j - D_0)\frac{\partial^2 c_j}{\partial x^2} \tag{7.7}$$

$$J_{0,L} = -D_0\frac{\partial w}{\partial x}\bigg|_{0,L} - (D_0 - D_j)\sum_{j=1}^{M}\frac{\partial c_j}{\partial x}\bigg|_{0,L} \tag{7.8}$$

7.1.1   Assumption One: Calcium Binding Is Fast

It seems reasonable to assume that calcium binding, determined by k_j^- and $k_j^+ B_j$ is *fast* compared to the release and uptake rate function f_0. This is a key assumption of Wagner and Keizer (1994). Hence, we will assume that Eq. 7.2 is at equilibrium (that is all c_j terms are no longer changing and hence all partials are zero) giving

$$k_j^- c_j - k_j^+(B_j - c_j)u = 0$$

Solving, we find

$$c_j = \frac{B_j u}{\frac{k_j^-}{k_j^+} + u} = \frac{B_j u}{K_j + u} \tag{7.9}$$

where K_j is the ratio $\frac{k_j^-}{k_j^+}$. Now, we also know $u = w - \sum_{j=1}^{M} c_j$. Substituting this in for u, we have

$$c_j \left[K_j + \left(w - \sum_{j=1}^{M} c_j \right) \right] = B_j \left[w - \sum_{j=1}^{M} c_j \right] \tag{7.10}$$

This simplifies to

$$B_j w = -c_j^2 + (K_j + w + B_j)c_j + (B_j - 1) \sum_{k \neq j} c_k$$

This shows that the concentration of occupied binding sites for species j is a function of the total calcium concentration w. Let $c_j(w)$ denote this functional dependence. From the chain rule, we have $\frac{\partial c_j}{\partial x} = \frac{\partial c_j}{\partial w} \frac{\partial w}{\partial x}$. Then, the calcium dynamics become

$$\frac{\partial w}{\partial t} = f_0 \left(w - \sum_{j=1}^{M} c_j(w) \right) + D_0 \frac{\partial^2 w}{\partial x^2} + \frac{\partial}{\partial x} \left[\sum_{j=1}^{M} (D_j - D_0) \frac{\partial c_j}{\partial w} \frac{\partial w}{\partial x} \right]$$

$$= f_0 \left(w - \sum_{j=1}^{M} c_j(w) \right) + \frac{\partial}{\partial x} \left(\left[D_0 + \sum_{j=1}^{M} (D_j - D_0) \frac{\partial c_j}{\partial w} \right] \frac{\partial w}{\partial x} \right)$$

The boundary conditions then become

$$\left[-D_0 - \sum_{j=1}^{M} (D_j - D_0) \frac{\partial c_j}{\partial w} \bigg|_{0,L} \right] \frac{\partial w}{\partial x} \bigg|_{0,L} = J_{0,L}$$

To summarize, the dynamics for calcium, with the assumption that calcium binding into the buffers is fast compared to release and uptake from the calcium pools in the ER, are

$$\frac{\partial w}{\partial t} = f_0 \left(w - \sum_{j=1}^{M} c_j(w) \right) + \frac{\partial}{\partial x} \left(\left[D_0 + \sum_{j=1}^{M} (D_j - D_0) \frac{\partial c_j}{\partial w} \right] \frac{\partial w}{\partial x} \right) \tag{7.11}$$

$$J_{0,L} = \left[-D_0 - \sum_{j=1}^{M} (D_0 - D_j) \frac{\partial c_j}{\partial w} \bigg|_{0,L} \right] \frac{\partial w}{\partial x} \bigg|_{0,L} \tag{7.12}$$

Notice that if we define a new diffusion coefficient, $\mathscr{D}$ for the diffusion process that governs w by

$$\mathscr{D} = D_0 + \sum_{j=1}^{M} (D_j - D_0) \frac{\partial c_j}{\partial w} \tag{7.13}$$

we can rewrite Eqs. 7.11 and 7.12 as

$$\frac{\partial w}{\partial t} = f_0\left(w - \sum_{j=1}^{M} c_j(w)\right) + \frac{\partial}{\partial x}\left(\mathcal{D}\frac{\partial w}{\partial x}\right) \tag{7.14}$$

$$-\mathcal{D}\Big|_{0,L}\frac{\partial w}{\partial x}\Big|_{0,L} = J_{0,L} \tag{7.15}$$

7.1.2 Assumption Two: Binding Rate Is Much Less Than Disassociation Rate

Many calcium binding molecules have low calcium affinity in μM concentration range and above. Recall that k_j^+ has units of liter/(mM B_j-sec) and k_j^- has units of (mM of Ca)/((mM Ca/B_j)-sec). Hence, the ratio K_j has units of (mM B_j/(mM Ca/B_j)) (mM Ca)/liter. The ratio of concentration of B_j to Ca/B_j can thus also be interpreted as the (number of free binding sites/liter)/(number of bound sites/liter). Consider the equation $u < K_j$. This would imply $k_j^+ u < k_j^-$. From the units discussion above, this occurs if the number of bound sites is less than the number of free sites. The data in Wagner and Keizer (1994) suggests that typically $K_j \approx 0.2\,\mu$M per liter calcium concentration. From Chap. 6, we know that inside the cell, the free calcium concentration is $0.1 - 0.2\,\mu$ moles. Thus, it is reasonable to assume that the rate of binding is much less than the rate of disassociation. This is the second key assumption of Wagner and Keizer (1994). Hence, we further assume in our model that $u \ll K_j$. It then follows that we can simplify Eq. 7.10 to

$$c_j = \frac{B_j}{K_j}u \tag{7.16}$$

Thus, $w = \left(1 + \sum_{j=1}^{M}\frac{B_j}{K_j}\right)u$ and solving for c_j, we find

$$c_j = \frac{\frac{B_j}{K_j}w}{1 + \sum_{k=1}^{M}\frac{B_k}{K_k}} \implies \frac{\partial c_j}{\partial w} = \frac{\frac{B_j}{K_j}}{1 + \sum_{k=1}^{M}\frac{B_k}{K_k}}$$

Then, letting $\gamma_j = \frac{B_j}{K_j}$,

$$\frac{\partial w}{\partial t} = f_0\left(w - \sum_{j=1}^{M}\left(\frac{\gamma_j w}{1 + \sum_{k=1}^{M}\gamma_k}\right)\right) + \frac{\partial}{\partial x}\left[D_0 + \sum_{j=1}^{M}(D_j - D_0)\frac{\gamma_j}{1 + \sum_{k=1}^{M}\gamma_k}\frac{\partial w}{\partial x}\right]$$

$$= f_0\left(w - \frac{\gamma_j w}{1 + \sum_{k=1}^{M}\gamma_k}\right) + \left[D_0 + \sum_{j=1}^{M}(D_j - D_0)\frac{\gamma_j}{1 + \sum_{k=1}^{M}\gamma_k}\right]\frac{\partial^2 w}{\partial x^2}$$

Letting Λ denote the term $1 + \sum_{k=1}^{M} \gamma_k$, then $w = \Lambda u$ and so we have $\frac{\partial w}{\partial t} = \Lambda \frac{\partial u}{\partial t}$. Thus,

$$\Lambda \frac{\partial u}{\partial t} = f_0 \left(\Lambda u - \sum_{j=1}^{M} \frac{\gamma_j \Lambda u}{\Lambda} \right) + \left[D_0 + \sum_{j=1}^{M} (D_j - D_0) \frac{\gamma_j}{\Lambda} \right] \Lambda \frac{\partial^2 u}{\partial x^2}$$

$$= f_0 \left(\Lambda u - (\Lambda - 1)u \right) + \left[\Lambda D_0 + \sum_{j=1}^{M} (D_j - D_0) \gamma_j \right] \frac{\partial^2 u}{\partial x^2}$$

This can then be written as

$$\frac{\partial u}{\partial t} = \frac{f_0(u)}{\Lambda} + \left[\frac{\Lambda D_0 + \sum_{j=1}^{M} D_j \gamma_j}{\Lambda} \right] \frac{\partial^2 u}{\partial x^2}$$

Next, define the new release uptake function f by

$$f(u) = \frac{f_0(u)}{\Lambda}. \tag{7.17}$$

We note that, in this approximation of binding is much less than disassociation, the diffusion constant $\mathscr{D}$ has a new form

$$\mathscr{D} \rightarrow D_0 + \sum_{j=1}^{M} (D_j - D_0) \frac{\gamma_j}{\Lambda} = \frac{\Lambda D_0 + \sum_{j=1}^{M} D_j \gamma_j - (\Lambda - 1) D_0}{\Lambda} = \frac{D_0 + \sum_{j=1}^{M} D_j \gamma_j}{\Lambda}$$

This suggests we define the new diffusion constant $\hat{\mathscr{D}}$ by

$$\hat{\mathscr{D}} = \frac{D_0 + \sum_{j=1}^{M} D_j \gamma_j}{\Lambda} \tag{7.18}$$

The free calcium dynamics are thus

$$\frac{\partial u}{\partial t} = f(u) + \hat{\mathscr{D}} \frac{\partial^2 u}{\partial x^2} \tag{7.19}$$

We call the function f, the effective calcium release/uptake rate and $\hat{\mathscr{D}}$, the effective diffusion coefficient. What about the new boundary conditions? This is just another somewhat unpleasant calculation:

$$- \left(D_0 + \sum_{j=1}^{M} (D_j - D_0) \frac{\partial c_j}{\partial w} \right) \Big|_{0,L} \frac{\partial w}{\partial x} \Big|_{0,L} = J_{0,L}$$

$$-\left(D_0 + \sum_{j=1}^{M} \frac{(D_j - D_0)\gamma_j}{\Lambda}\right)\Lambda \frac{\partial u}{\partial x}\bigg|_{0,L} = J_{0,L}$$

$$-\left(\Lambda D_0 + \sum_{j=1}^{M}(D_j - D_0)\gamma_j\right)\frac{\partial u}{\partial x}\bigg|_{0,L} = J_{0,L}$$

$$-\left(D_0 + \sum_{j=1}^{M}D_j\gamma_j\right)\frac{\partial u}{\partial x}\bigg|_{0,L} = J_{0,L}$$

$$-\hat{\mathscr{D}}\Lambda \frac{\partial u}{\partial x}\bigg|_{0,L} = J_{0,L}$$

We conclude the appropriate dynamics are then

$$\frac{\partial u}{\partial t} = \frac{f_0(u)}{\Lambda} + \hat{\mathscr{D}}\frac{\partial^2 u}{\partial x^2} \tag{7.20}$$

$$-\hat{\mathscr{D}}\frac{\partial u}{\partial x}\big|_{0,L} = \frac{J_{0,L}}{\Lambda} \tag{7.21}$$

7.2   Transcriptional Control of Free Calcium

The critical review on the control of free calcium in cellular processing in Carafoli et al. (2001) notes the concentration of Ca^{++} in the cell is controlled by the reversible binding of calcium ion to the buffer complexes we have been discussing. These buffer molecules therefore act as calcium ion sensors that, in a sense, decode the information contained in the calcium ion current injection and then pass on a decision to a target. Many of these targets can be proteins transcribed by accessing the genome. Hence, the $P(T_1)$ we have discussed in Chap. 6 could be a buffer molecule B_j. The boundary condition $J_{0,L}$ plays the role of our entry calcium current. Such a calcium ion input current through the membrane could be due to membrane depolarization causing an influx of calcium ions through the port or via ligand binding to a receptor which in turn indirectly increases free calcium ion in the cytosol. Such mechanisms involve the interplay between the release/uptake ER function and the storage buffers as the previous sections have shown. This boundary current determines the $u(t, x)$ solution through the diffusion equations Eqs. 7.20 and 7.21. The exact nature of this solution is determined by the receptor types, buffers and storage sites in the ER. Differences in the period and magnitude of the calcium current $u(t, x)$ resulting from the input $J_{0,L}$ trigger different second messenger pathways. Hence, there are many possible outcomes due to a given input current $J_{0,L}$.

Let's do a back of the envelope calculation to see what might happen if a trigger event T_0 initiated an increase in a buffer B_j which then initiates a complex trigger mechanism culminating in a protein transcription. Let's assume that the buffer B_{j_0} is increased to $B_{j_0} + \epsilon$. It is reasonable to assume that both k_i^+ and k_i^- are independent of the amount of B_i that is present. Hence, we can assume all the γ_i's are unaffected by the change in B_{j_0}. If binding activity goes up, we would also expect that the release/uptake activity would also change. There are then changes in the terms that define the diffusion constant

$$\Lambda \rightarrow 1 + \sum_{k \neq j_0}^{M} \frac{B_k}{K_k} + \frac{B_{j_0} + \epsilon}{K_{j_0}}$$

$$f(u) \rightarrow f(u) + \eta$$

$$\hat{\mathscr{D}} \rightarrow \frac{D_0 + D_{j_0} \frac{B_{j_0} + \epsilon}{K_{j_0}} + \sum_{k \neq j_0}^{M} D_k \gamma_k}{1 + \sum_{k \neq j_0}^{M} \frac{B_k}{K_k} + \frac{B_{j_0} + \epsilon}{K_j}}$$

for some nonzero η and positive ϵ. Letting $\Theta = 1 + \sum_{k=1}^{M} D_k \gamma_k$, $\xi_j = \frac{1}{K_j}$, we find

$$\hat{\mathscr{D}}^{new} = \frac{D_0 + \sum_{k=1}^{M} D_k \gamma_k + \epsilon \frac{D_{j_0}}{K_{j_0}}}{1 + \sum_{k=1}^{M} \gamma_j + \epsilon \frac{1}{K_{j_0}}} = \frac{D_0 + \sum_{k=1}^{M} D_k \gamma_k + \epsilon \frac{D_{j_0}}{K_{j_0}}}{\Lambda + \epsilon \frac{1}{K_{j_0}}} = \frac{(D_0 - 1) + \Theta + \epsilon D_{j_0} \xi_{j_0}}{\Lambda + \epsilon \xi_{j_0}}$$

Thus, noting $\hat{\mathscr{D}} = \frac{D_0 + \Theta - 1}{\Lambda}$, we find

$$\Delta \hat{\mathscr{D}} = \hat{\mathscr{D}}^{new} - \hat{\mathscr{D}}$$

$$= \frac{(D_0 - 1) + \Theta + \epsilon D_{j_0} \xi_{j_0}}{\Lambda + \epsilon \xi_{j_0}} - \frac{(D_0 - 1) + \Theta}{\Lambda}$$

$$= \frac{[(D_0 - 1) + \Theta + \epsilon D_{j_0} \xi_{j_0}]\Lambda - [(D_0 - 1) + \Theta][\Lambda + \epsilon \xi_{j_0}]}{[\Lambda + \epsilon \xi_{j_0}][\Lambda]}$$

$$= \epsilon \xi_{j_0} \frac{D_{j_0} \Lambda - (D_0 - 1 + \Theta)}{\Lambda [\Lambda + \epsilon \xi_{j_0}]} = \epsilon \xi_{j_0} \frac{D_{j_0} \Lambda - \Lambda \mathscr{D}}{\Lambda [\Lambda + \epsilon \xi_{j_0}]} = \epsilon \xi_{j_0} D_{j_0} \frac{1}{\Lambda + \epsilon \xi_{j_0}} \left(1 - \frac{\mathscr{D}}{D_{j_0}}\right)$$

We see that

$$\Delta \hat{\mathscr{D}} = \frac{\epsilon}{\Lambda + \epsilon \xi_{j_0}} \xi_{j_0} D_{j_0} \left(1 - \frac{\mathscr{D}}{D_{j_0}}\right) \propto \frac{\epsilon}{\Lambda + \epsilon \xi_{j_0}}$$

Hence, to first order

$$\Delta \hat{\mathscr{D}} \propto \frac{\epsilon}{\Lambda}$$

and we see that the new diffusion dynamics are on the order of

$$\frac{\partial u}{\partial t} = \frac{f(u) + \eta}{\Lambda} + \left(\mathscr{D} + C \frac{\epsilon}{\Lambda}\right) \frac{\partial^2 u}{\partial x^2}$$

for some constant C. The alteration of the release/uptake function and the diffusion constant imply a change in the solution. This change in the solution $u(t, x)$ then can initiate further second messenger changes culminating in altered $P(T_1)$ protein production.

References

M. Berridge, Elementary and global aspects of calcium signalling. J. Physiol. **499**, 291–306 (1997)

E. Carafoli, L. Santella, D. Branca, M. Brini, Generation, control and processing of cellular calcium signals. Crit. Rev. Biochem. Mol. Biol. **36**(2), 107–260 (2001)

T. Höffer, A. Politi, R. Heinrich, Intracellular ca^{+2} wave propagation through gap - junctional ca^{+2} diffusion: A theoretical study. Biophys. J. **80**(1), 75–87 (2001)

J. Wagner, J. Keizer, Effects of rapid buffers on ca^{+2} diffusion and ca^{+2} oscillations. Biophys. J. **67**, 447–456 (1994)

Chapter 8
Second Messenger Models

In Sect. 6.7, we discussed some of the basic features of the pathways used by extracellular triggers. We now look at these again but very abstractly as we want to design principles by which we can model second messenger effects. Let T_0 denote a second messenger trigger which moves though a port P to create a new trigger T_1 some of which binds to B_1. A schematic of this is shown in Fig. 8.1. In the figure, r is a number between 0 and 1 which represents the fraction of the trigger T_1 which is free in the cytosol. Hence, $100r\%$ of T_1 is free and $100(1-r)$ is bound to B_1 creating a storage complex B_1/T_1. For our simple model, we assume rT_1 is transported to the nuclear membrane where some of it binds to the enzyme E_1. Let s in $(0, 1)$ denote the fraction of rT_1 that binds to E_1. We illustrate this in Fig. 8.2. We denote the complex formed by the binding of E_1 and T_1 by E_1/T_1. From Fig. 8.2, we see that the proportion of T_1 that binds to the genome (DNA) and initiates protein creation $P(T_1)$ is thus srT_1.

8.1 Generic Second Messenger Triggers

The protein created, $P(T_1)$, could be many things. Here, let us assume that $P(T_1)$ is a sodium, Na^+, gate. Thus, our high level model is

$$sE_1/rT_1 + DNA \rightarrow Na^+ gate$$

We therefore increase the concentration of Na^+ gates, $[Na^+]$ thereby creating an increases in the sodium conductance, g_{Na}. The standard Hodgkin–Huxley conductance model (details are in Peterson (2015)) is given by

$$g_{Na}(t, V) = g_{Na}^{max} \mathcal{M}_{Na}^p(t, v) \mathcal{H}_{Na}^q(t, V)$$

where t is time and V is membrane voltage. The variables $\mathcal{M}_{Na}$ and $\mathcal{H}_{Na}$ are the activation and inactivation functions for the sodium gate with p and q appropriate

© Springer Science+Business Media Singapore 2016
J.K. Peterson, *BioInformation Processing*, Cognitive Science and Technology,
DOI 10.1007/978-981-287-871-7_8

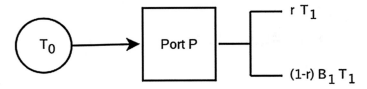

Fig. 8.1 Second messenger trigger

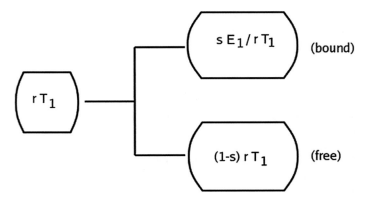

Fig. 8.2 Some T_1 binds to the genome

positive powers. Finally, G_{Na}^{max} is the maximum conductance possible. These models generate $\mathcal{M}_{Na}$ and $\mathcal{H}_{Na}$ values in the range (0, 1) and hence,

$$0 \le g_{Na}(t, V) \le g_{Na}^{max}$$

We can model increases in sodium conductances as increases in g_{Na}^{max} with efficiency e, where e is a number between 0 and 1. We will not assume all of the sE_1/rT_1 + DNA to sodium gate reaction is completed. It follows that e is similar to a Michaelson–Mentin kinetics constant. We could also alter activation, $\mathcal{M}_{Na}$, and/or inactivation, $\mathcal{H}_{Na}$, as functions of voltage, V in addition to the change in the maximum conductance. However, we are interested in a simple model at present. Our full schematic is then given in Fig. 8.3. We can model the choice process, rT_1 or $(1 - r)B_1/T_1$ via a simple sigmoid,

$$f(x) = 0.5 \left(1 + \tanh \left(\frac{x - x_0}{g} \right) \right)$$

where the transition rate at x_0 is $f'(x_0) = \frac{1}{2g}$. Hence, the "gain" of the transition can be adjusted by changing the value of g. We assume g is positive. This function can be interpreted as switching from of "low" state **0** to a high state **1** at speed $\frac{1}{2g}$. Now the function $h = rf$ provides an output in (r, ∞). If x is larger than the

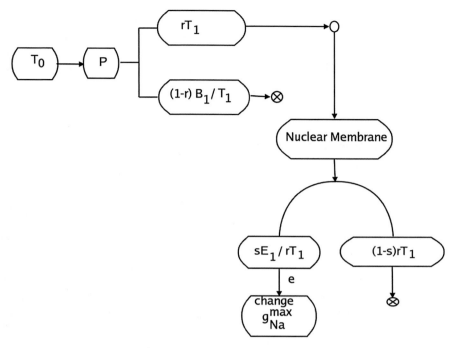

Fig. 8.3 Maximum sodium conductance control pathway

threshold x_0, h rapidly transitions to a high state r. On the other hand, if x is below threshold, the output remains near the low state 0.

We assume the trigger T_0 does not activate the port P unless its concentrations is past some threshold $[T_0]_b$ where $[T_0]_b$ denotes the *base* concentration. Hence, we can model the port activity by

$$h_p([T_0]) = \frac{r}{2}\left(1 + \tanh\left(\frac{[T_0] - [T_0]_b}{g_p}\right)\right)$$

where the two shaping parameters g_p (transition rate) and $[T_0]_b$ (threshold) must be chosen. We can thus model the schematic of Fig. 8.1 as $h_p([T_0])\,[T_1]_n$ where $[T_1]_n$ is the nominal concentration of the induced trigger T_1. In a similar way, we let

$$h_e(x) = \frac{s}{2}\left(1 + \tanh\left(\frac{x - x_0}{g_e}\right)\right)$$

Thus, for $x = h_p([T_0])\,[T_1]_n$, we have h_e is a switch from 0 to s. Note that $0 \le x \le r[T_1]_n$ and so if $h_p([T_0])\,[T_1]_n$ is close to $r[T_1]_n$, h_e is approximately s. Further, if $h_p([T_0])\,[T_1]_n$ is small, we will have h_e is close to 0. This suggests a threshold value for h_e of $\frac{r[T_1]_n}{2}$. We conclude

$$h_e\left(h_p([T_0])\ [T_1]_n\right) = \frac{s}{2}\left(1 + \tanh\left(\frac{h_p([T_0])[T_1]_n - \frac{r[T_1]_n}{2}}{g_e}\right)\right)$$

which lies in $[0, s)$. This is the amount of activated T_1 which reaches the genome to create the target protein $P(T_1)$. It follows then that

$$[P(T_1)] = h_e\left(h_p([T_0])\ [T_1]_n\right)[T_1]_n$$

The protein is created with efficiency e and so we model the conversion of $[P(T_1)]$ into a change in g_{Na}^{max} as follows. Let

$$h_{Na}(x) = \frac{e}{2}\left(1 + \tanh\left(\frac{x - x_0}{g_{Na}}\right)\right)$$

which has output in $[0, e)$. Here, we want to limit how large a change we can achieve in g_{Na}^{max}. Hence, we assume there is an upper limit which is given by $\Delta\ g_{Na}^{max} = \delta_{Na}\ g_{Na}^{max}$. Thus, we limit the change in the maximum sodium conductance to some percentage of its baseline value. It follows that $h_{Na}(x)$ is about δ_{Na} if x is sufficiently larges and small otherwise. This suggests that x should be $[P(T_1)]$ and since translation to $P(T_1)$ occurs no matter how low $[T_1]$ is, we can use a switch point value of $x_0 = 0$. We conclude

$$h_{Na}([P(T_1)]) = \frac{e}{2}\ \delta_{Na}\ g_{Na}^{max}\left(1 + \tanh\left(\frac{[P(T_1)]}{g_{Na}}\right)\right) \tag{8.1}$$

Our model of the change in maximum sodium conductance is therefore $\Delta\ g_{Na}^{max} = h_{Na}([P(T_1)])$. We can thus alter the action potential via a second messenger trigger by allowing

$$g_{Na}(t, V) = (\ g_{Na}^{max} + h_{Na}([P(T_1)]))\mathcal{M}_{Na}^p(t, V)\mathcal{H}_{Na}^q(t, V)$$

for appropriate values of p and q within a standard Hodgkin–Huxley model.

Next, if we assume a modulatory agent acts as a trigger T_0 as described above, we can generate action potential pulses using the standard Hodgkin–Huxley model for a large variety of critical sodium trigger shaping parameters. We label these with a Na to indicate their dependence on the sodium second messenger trigger.

$$\left[r^{Na}, [T_0]_b{}^{Na}, g_p^{Na}, s^{Na}, g_e^{Na}, e^{Na}, g_{Na}, \delta_{Na}\right]'$$

We can follow the procedure outlined in this section for a variety of triggers. We therefore can add a potassium gate trigger with shaping parameters

$$\left[r^K, [T_0]^K{}_b, g_p^K, s^K, g_e^K, e^K, g_K, \delta_K\right]'$$

8.1.1 Concatenated Sigmoid Transitions

In the previous section, we found how to handle alterations in g_{Na}^{max} due to a trigger T_0. We have

$$g_{Na}(t, V) = (g_{Na}^{max} + h_{Na}([P(T_1)])) \mathcal{M}_{Na}^p(t, V) \mathcal{H}_{Na}^q(t, V)$$

where

$$h_{Na}([P(T_1)]) = \frac{e}{2} \delta_{Na} \, g_{Na}^{max} \left(1 + \tanh \left(\frac{[P(T_1)]}{g_{Na}} \right) \right)$$

with e and δ_{Na} in $(0, 1)$. Using the usual transition function $\sigma(x, x_0, g_0)$, we can then write the sodium conductance modification equation more compactly as

$$h_{Na}([P(T_1)]) = e \delta_{Na} g_{Na}^{max} \sigma([P(T_1)], 0, g_{Na}).$$

Using this same notation, we see

$$h_p([T - 0]) = r\sigma([T_0], [T_0]_b, g_p)$$

$$h_e(h_p([T_0])[T_1]_n) = s\sigma \left(h_p([T_0])[T_1]_n, \frac{r[T_1]_n}{2}, g_e \right)$$

$$= s\sigma \left(r\sigma([T_0], [T_0]_b, g_p)[T_1]_n, \frac{r[T_1]_n}{2}, g_e \right)$$

Note the concatenation of the sigmoidal processing. Now $[P(T_1)] = h_e \left(h_p([T_0]) [T_1]_n \right) [T_1]_n$ Thus,

$$[P(T_1)] = s\sigma \left(r\sigma([T_0], [T_0]_b, g_p)[T_1]_n, \frac{r[T_1]_n}{2}, g_e \right) [T_1]_n.$$

Finally,

$$g_{Na}(t, V) = g_{Na}^{max}(1 + e \delta_{Na} \, \sigma([P(T_1)], 0, g_{Na})) \mathcal{M}_{Na}^p(t, V) \mathcal{H}_{Na}^q(t, V)$$

Implicit is this formula is the *cascade* "$\sigma(\sigma(\sigma$" as $\sigma([P(T_1)], 0, g_{Na})$ uses two concatenated sigmoid calculations itself. We label this as a Sigma Three Transition, σ_3, and use the notation

σ_3 ($[T_0]$, $[T_0]_b$, g_p;   inner most sigmoid r;   scale innermost calculation by r
    $[T_1]_n$;   scale again by $[T_1]_n$ this is input to next sigmoid
    $\frac{r[T_1]_n}{2}$, g_e;   offset and gain of next sigmoid s;   scale results by s
    $[T_1]_n$;   scale again by $[T_1]_n$ this is $[P(T_1)]$
                    this is input into last sigmoid
    0, g_{Na};   offset and gain of last sigmoid)

Thus, the g_{Na} computation can be written as

$$g_{Na}(t, V) = g_{Na}^{max}\left(1 + e\delta_{Na}\, h_3\left([T_0], [T_0]_b, g_p; r; [T_1]_n; \frac{r[T_1]_n}{2}, g_e; s; [T_1]_n; 0, g_{Na}\right)\right)$$
$$\mathcal{M}_{Na}^p(t, V)\mathcal{H}_{Na}^q(t, V)$$

This implies a trigger T_0 has associated with it a data vector

$$W_{T_0} = \left[[T_0], [T_0]_b, g_p, r, [T_1]_n, \frac{r[T_1]_n}{2}, g_e, s, [T_1]_n, 0, g_T\right]'$$

where g_T denotes the final gain associated with the third level sigmoidal transition to create the final gate product. We can then rewrite our modulation equation as

$$g_{Na}(t, V) = g_{Na}^{max}\, (1 + e\delta_{Na}\, h_3(W_{Na}))\; \mathcal{M}_{Na}^p(t, V)\mathcal{H}_{Na}^q(t, V)$$

8.2   A Graphic Model Computation Model

Although third order sigmoidal transformations certainly occur in our models, hiding all the details obscures what is really going on. We will now recast the model into computational graph structure. This makes it easier to see how the calculations will be performed as asynchronous agents. Consider the typical standard sigmoid transformation

$$h_p([T_0]) = \frac{r}{2}\left(1 + \tanh\left(\frac{[T_0] - [T_0]_b}{g_p}\right)\right)$$

We can draw this as a graph as is shown in Fig. 8.4 where h denotes the standard sigmoidal state transition function. We also have

$$h_e(h_p([T_0])\,[T_1]_n) = \frac{s}{2}\left(1 + \tanh\left(\frac{h_p([T_0])[T_1]_n - \frac{r[T_1]_n}{2}}{g_e}\right)\right)$$

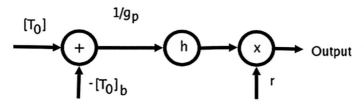

Fig. 8.4 The first level sigmoid graph computation

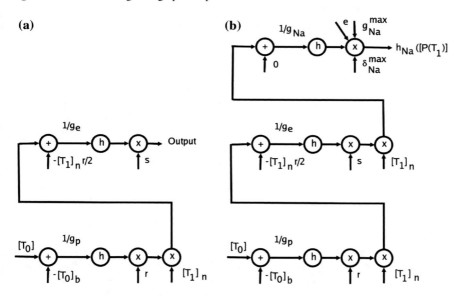

Fig. 8.5 Sigmoid graph computations. **a** Second level. **b** Third level

which for a given output Y becomes

$$h_e(Y) = \frac{s}{2}\left(1 + \tanh\left(\frac{Y - \frac{r[T_1]_n}{2}}{g_e}\right)\right)$$

and the computation can be represented graphically by Fig. 8.5a. Finally, we have used

$$[P(T_1)] = h_e(h_p([T_0])[T_1]_n)[T_1]_n$$

$$h_{Na}([P(T_1)]) = e\delta_{Na}g_{Na}^{max}h([P(T_1)], 0, g_{Na})$$

which can be shown diagrammatically as in Fig. 8.5b. These graphs are, of course, becoming increasingly complicated. However, they are quite useful in depicting how feedback pathways can be added to our computations. Let's add feedback pathways

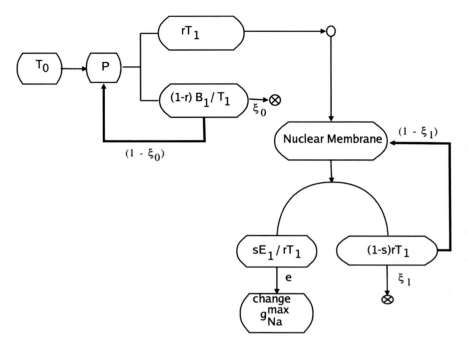

Fig. 8.6 Adding feedback to the maximum sodium conductance control pathway

to the original maximum conductance control pathway we illustrated in Fig. 8.3.
Figure 8.6 is equivalent to that of the computational graph of Fig. 8.5b but shows
more of the underlying cellular mechanisms. The feedback pathways are indicated
by the variables ξ_0 and ξ_1. The feedback ξ_0 is meant to suggest that some of the
protein kinase T_1 which is bound onto $(1 - r)B_1/T_1$ is recycled (probably due to
other governing cycles) back to *free* T_1. We show this with a line drawn back to the
port P. Similarly, some of the protein kinase not used for binding to the enzyme E_1,
to start the protein creation process culminating in $P(T_1)$, is allowed to be freed for
reuse. This is shown with the line labeled with the legend $1 - \xi_1$ leading back to
the nuclear membrane. It is clear there are many other feedback possibilities. Also
the identity of the protein $P(T_1)$ is fluid. We have already discussed the cases where
$P(T_1)$ can be voltage dependent gates for sodium and potassium and calcium second
messenger proteins. However, there are other possibilities:

- B_1: thereby increasing T_1 binding
- T_1: thereby increasing rT_1 and $P(T_1)$
- E_1: thereby increasing $P(T_1)$.

and so forth.

8.3 Ca^{++} Triggers

We now specialize to the case of the Ca^{++} triggers. Then $[T_0]$ represents $[Ca^{++}]$ and $[T_1]$ will denote some sort of Ca^{++} binding complex. We will use the generic name *calmodulin* for this complex and use the symbol C^* as its symbol. Hence, $[T_1] = [C^*]$ with $[C^*]_n$ the nominal calcineuron concentration. Thus,

$$W_{Ca^{++}} = \left[[Ca^{++}], [Ca^{++}]_b, g_{Ca^{++}}, r_{Ca^{++}}, [C^*]_n, \frac{r_{Ca^{++}}[C^*]_n}{2}, g_e^{Ca^{++}}, s^{Ca^{++}}, [C^*]_n, 0, g_{Ca^{++}} \right]'$$

Further, the sodium Ca^{++} conductance model is then

$$g_{Na}(t, V) = g_{Na}^{max}(1 + e_{Ca^{++}} \delta_{Na}^{Ca^{++}} h_3(W_{Ca^{++}})) \mathcal{M}_{Na}^p(t, V) \mathcal{H}_{Na}^q(t, V)$$

where the term $\delta_{Na}^{Ca^{++}}$ is the fraction of the sodium maximum conductance that the Ca^{++} second messenger pathway can effect. We know the situation is actually quite complicated.

- $[Ca^{++}]$ is an injected current—i.e. a current pulse—whose temporal and spatial shape are very important to the protein production $P(C^*)$.
- C^* is a Ca^{++} binding complex which serves as an intermediary; i.e. as our T_1 protein. It will eventually be translated into the protein $P(C^*)$ via mechanisms similar to what we have described. There is a lot going on here and we must remember that we are trying for a simple model which will capture some of the core ideas behind second messenger Ca^{++} signaling systems.
- We know $[Ca^{++}]$ is determined internally by a complicated diffusion model which has a damped time and spatial solution.

To enable our model to handle more of the special characteristics of the Ca^{++} triggers, we must let the time independent Ca^{++} data vector from $W_{Ca^{++}}$ become time dependent:

$$W_{Ca^{++}}(t) = \begin{bmatrix} [Ca^{++}](t) \\ [Ca^{++}]_b(t) \\ g_{Ca^{++}} \\ r_{Ca^{++}} \\ [C^*]_n \exp(\frac{-t}{\tau_c}) \\ \frac{r_{Ca^{++}}[C^*]_n}{2} \exp(\frac{-t}{\tau_c}) \\ g_e^{Ca^{++}} \\ s^{Ca^{++}} \\ [C^*]_n \\ 0 \\ g_{Ca^{++}} \end{bmatrix}$$

This gives the new model

$$g_{Na}(t, V) = g_{Na}^{max}(1 + e_{Ca^{++}} \delta_{Na}^{Ca^{++}} h_3(W_{Ca^{++}(t)})) \, \mathcal{M}_{Na}^p(t, V) \mathcal{H}_{Na}^q(t, V)$$

8.4   Spatially Dependent Calcium Triggers

Using the standard Ball and Stick model, the calcium injection current can be inserted at any electrotonic distance x_0 with $0 \le x_0 \le 4$. For convenience, we usually will only allow x_0 to be integer valued. The previous trigger equations without spatial dependence are given below:

$$h_p([Ca^{++}(t)]) = \frac{r}{2}\left(1 + \tanh\left(\frac{[Ca^{++}(t)] - [Ca^{++}]_b}{g_p}\right)\right)$$

$$h_e(h_p([Ca^{++}(t)]) [C^*]_n) = \frac{s}{2}\left(1 + \tanh\left(\frac{h_p([Ca^{++}(t)])[C^*]_n - \frac{r[C^*]_n}{2}}{g_e}\right)\right)$$

$$[P([C^*](t)] = h_e\left(h_p([Ca^{++}(t)]) [C^*]_n\right) [C^*]_n$$

$$h_{Na}([P([C^*](t))]) = \frac{e}{2}\, \delta_{Na}\, g_{Na}^{max}\left(1 + \tanh\left(\frac{[P(C^*(t))]}{g_{Na}}\right)\right)$$

where for convenience of discussion, we have refrained from labeling all of the relevant variables with Ca^{++} tags. Now given the injection current $[Ca^{++}(t)]$, only a portion of it is used to form the C^* complex. The initial fraction is $h_p([Ca^{++}(t)])[C^*]_n$ and it begins the diffusion across the cellular cytosol. Hence, following the discussions in Chap. 7, if the injection is at site x_0, the solution to the diffusion equation for this initial pulse is given by

$$I(t, x) = \frac{h_p([Ca^{++}(t)])[C^*]_n}{\sqrt{t}} \exp\left(\frac{-(x - x_0)^2}{4Dt}\right)$$

Then the amount g_{Na}^{max} changes depends on the amount of the above response that gets to the nuclear membrane. This is given by

$$h_e(I(t, x)) = \frac{s}{2}\left(1 + \tanh\left(\frac{I(t, x) - \frac{r[C^*]_n}{2}}{g_e}\right)\right)$$

Note that $I(t, x)$ explicitly contains time and spatial attenuation. Combining, we see the full spatial dependent equations are:

$$h_p([Ca^{++}(t)]) = \frac{r}{2}\left(1 + \tanh\left(\frac{[Ca^{++}(t)] - [Ca^{++}]_b}{g_p}\right)\right)$$

$$I(t, x)) = \frac{h_p([Ca^{++}(t)])[C^*]_n}{\sqrt{t}} \exp\left(\frac{-(x - x_0)^2}{4Dt}\right)$$

$$h_e(I(t, x)) = \frac{s}{2}\left(1 + \tanh\left(\frac{I(t, x) - \frac{r[C^*]_n}{2}}{g_e}\right)\right)$$

$$[P([C^*](t)) = h_e(h_p([Ca^{++}](t)) [C^*]_n)[C^*]_n$$

$$h_{Na}([P(C^*(t))]) = \frac{e}{2}\,\delta_{Na}\,g_{Na}^{max}\left(1 + \tanh\left(\frac{[P(C^*(t))]}{g_{Na}}\right)\right)$$

$$g_{Na}(t, V) = (g_{Na}^{max} + h_{Na}([P(C^*(t)0]\,)\mathcal{M}_{Na}^p(t, V)\mathcal{H}_{Na}^q(t, V)$$

Now consider the actual feedback paths shown in Fig. 8.6 for a Ca^{++} trigger. In a calcium second messenger event, the calcium current enters the cell through the port and some remains free and some is bound into the calcineuron complex. In this context, we interpret the feedback as follows: $1 - \xi_0$ of the bound $[Ca^*](t)$ is fed back to the port as free Ca^{++}. This happens because the bound complex disassociates back into Ca^{++} plus other components. This amount is fed back into the input of h_p and the feedback response $J(t)$ is given by

$$J(t) = \sum_{j=0}^{\infty} h_p((1 - \xi_0)^j\,[Ca^{++}(t)])$$

$$= \sum_{j=0}^{\infty} \frac{r}{2}\left(1 + \tanh\left(\frac{(1 - \xi_0)^j([Ca^{++}(t)] - [Ca^{++}]_b)}{g_p}\right)\right)$$

To estimate this value, note for $u = 1 - \xi_0$, u is in $(0, 1)$. It follows from the mean value theorem that for any positive integer j, we have for some point c between x and x_0

$$\frac{r}{2}\left(1 + \tanh\left(\frac{u^j(x - x_0)}{g}\right)\right) = \frac{r}{2}\frac{u^j}{g}\,sech^2\left(\frac{u^j(c_x - x_0)}{g}\right)(x - x_0)$$

Thus, since c_x is bounded by x, we have

$$\frac{r}{2}\left(1 + \tanh\left(\frac{u^j(x - x_0)}{g}\right)\right) \leq \frac{r}{2}\frac{u^j}{g}\,sech^2\left(\frac{u^j(x - x_0)}{g}\right)(x - x_0)$$

Similarly, to first order, there is a constant d_x between x_0 and x so that

$$\frac{r}{2}\left(1 + \tanh\left(\frac{x - x_0}{g}\right)\right) = \frac{r}{2}\frac{1}{g}\,sech^2\left(\frac{(d_x - x_0)}{g}\right)(x - x_0)$$

It follows then that

$$\frac{\frac{r}{2}\left(1 + \tanh\left(\frac{u^j(x-x_0)}{g}\right)\right)}{\frac{r}{2}\left(1 + \tanh\left(\frac{x-x_0}{g}\right)\right)} \approx \frac{u^j\,sech^2\left(\frac{u^j(c_x-x_0)}{g}\right)}{sech^2\left(\frac{(d_x-x_0)}{g}\right)}$$

$$\approx \frac{u^j\,sech^2\left(\frac{(c_x-x_0)}{g}\right)}{sech^2\left(\frac{(d_x-x_0)}{g}\right)}$$

Thus, to first order, letting $\gamma_x = \dfrac{u^j\,sech^2\left(\frac{(c_x-x_0)}{g}\right)}{sech^2\left(\frac{(d_x-x_0)}{g}\right)}$, we have

$$\frac{r}{2}\left(1 + \tanh\left(\frac{u^j(x-x_0)}{g}\right)\right) \approx \gamma_x u^j \frac{r}{2}\left(1 + \tanh\left(\frac{x-x_0}{g}\right)\right)$$

Applying this to our sum J we find It then follows, using $\gamma(t)$ for the term $\gamma_{Ca^{++}}(t)$,

$$J(t) \approx \gamma_x\, h_p([Ca^{++}(t)]) \sum_{j=0}^{\infty} (1 - \xi_0)^j$$

$$= \gamma(t)\, h_p([Ca^{++}(t)]) \frac{1}{\xi_0}$$

Thus, the multiplier applied to the g_{Na} computation due to feedback is of order $\frac{1}{\xi_0}$; i.e. $J \le \gamma_x \frac{1}{\xi_0}$ The constant γ_x is difficult to assess even though we have first order bounds for it. It is approximately given by

$$\gamma(t) \approx \frac{r}{2g}\, sech^2\left(\frac{(1 - \xi_0)([Ca^{++}(0)] - [Ca^{++}]_b)}{g}\right) ([Ca^{++}(t)] - [Ca^{++}]_b).$$

8.5   Calcium Second Messenger Pathways

From the discussions so far, we can infer that the calcium second messenger pathways can influence the shape of the action potential in many ways. For our purposes, we will concentrate on just a fraction of these possibilities. If a calcium current is injected into the dendrite at time and spatial position (t_0, x_0), it creates a protein response from the nucleus following the equations below for various ions. We use the label a to denote the fact that these equations are for protein $P(C^*) = a$. The calcium injection current is also labeled with the protein a as it is known that calcium currents are specifically targeted towards various protein outputs.

$$h_p^a([Ca^{a,++}(t)]) = \frac{r^a}{2}\left(1 + \tanh\left(\frac{[Ca^{a,++}(t)] - [Ca^{a,++}]_b}{g_p^a}\right)\right)$$

$$I^a(t,x)) = \frac{h_p^a([Ca^{a,++}(t)])[C^{a,*}]_n}{\sqrt{t - t_0}}\exp\left(\frac{-(x - x_0)^2}{4D^a(t - t_0)}\right)$$

$$h_e^a(I(t,x)) = \frac{s^a}{2}\left(1 + \tanh\left(\frac{I^a(t,x) - \frac{r^a[C^{a,*}]_n}{2}}{g_e^a}\right)\right)$$

$$[P^a([C^{a,*}](t))] = h_e^a(h_p^a([Ca^{++}(t)])[C^{a,*}]_n)[C^{a,*}]_n$$

$$h^a([P^a([C^{a,*}](t))]) = \frac{e^a}{2}\delta_a g_a^{max}\left(1 + \tanh\left(\frac{[P^a(C^{a,*}(t))]}{g^a}\right)\right)$$

$$g_a(t,V) = (g_a^{max} + h_a([P^a([C^{a,*}](t))))\mathcal{M}_a^p(t,V)\mathcal{H}_a^q(t,V)$$

Thus, we have

$$g_{Na}(t,V) = (g_{Na}^{max} + h_{Na}([P^{Na}([C^*](t)]))\alpha_{Na}^p(t - t_0, V)\beta_{Na}^q(t - t_0, V)$$

$$g_K(t,V) = (g_K^{max} + h_K([P^K([C^*](t)]))\mathcal{M}_a^p(t - t_0, V)\mathcal{H}_a^q(t - t_0, V)$$

implying that the calcium second messenger activity alters the maximum ion conductances as follows:

$$\delta g_{Na}(t_0, x_0) = h_{Na}\left(\left[P^{Na}([C^{Na,*}]_n \exp\left(\frac{-(t - t_0)}{\tau_c^{Na}}\right)\right]\right)$$

$$\delta g_K(t_0, x_0) = h_K\left(\left[P^K([C^{K,*}]_n \exp\left(\frac{-(t - t_0)}{\tau_c^K}\right)\right]\right)$$

We see that $[Ca^{++}]$ currents that enter the dendrite initiate cellular changes that effect potassium conductance and hence the hyperpolarization portion of the action potential curve. Also, these currents can effect the sodium conductance which modifies the depolarization portion of the action potential. Further, these changes in the maximum sodium and potassium conductances also effect the generation of the action potential itself. If feedback is introduced, we know there is a multiplier effect for each ion a:

$$J^a(t) = \gamma^a(t)\, h_p^a([Ca^{++}(t)])\,\frac{1}{\xi_0^a}$$

which alters the influence equations by adding a multiplier at the first step:

$$h_p^a([Ca^{a,++}(t)]) = \gamma^a(t)\frac{1}{\xi_0^a}\frac{r^a}{2}\left(1 + \tanh\left(\frac{[Ca^{a,++}(t)] - [Ca^{a,++}]_b}{g_p^a}\right)\right)$$

$$I^a(t,x)) = \frac{h_p^a([Ca^{a,++}(t)])[C^{a,*}]_n}{\sqrt{t - t_0}}\exp\left(\frac{-(x - x_0)^2}{4D^a(t - t_0)}\right)$$

$$h_e^a(I(t,x)) = \frac{s^a}{2}\left(1 + \tanh\left(\frac{I^a(t,x) - \frac{r^a[C^{a,*}]_n}{2}}{g_e^a}\right)\right)$$

$$[P^a([C^{a,*}](t)] = h_e^a(h_p^a([Ca^{++}(t)])\,[C^{a,*}]_n)[C^{a,*}]_n$$

$$h^a([P^a([C^{a,*}](t)]) = \frac{e^a}{2}\,\delta_a\,g_a^{max}\left(1 + \tanh\left(\frac{[P^a(C^{a,*}(t))]}{g^a}\right)\right)$$

$$g_a(t,V) = (g_a^{max} + h_a([P^a([C^{a,*}](t))]))\mathcal{M}_a^p(t,V)\mathcal{H}_a^q(t,V)$$

The full effect of the multiplier is nonlinear as it is fed forward through a series of sigmoid transformations. There are only five integer values for the electrotonic distance which can be used for the injection sites. These are $x_0 = 0$ (right at the soma) to $x_0 = 4$ (farthest from the soma). At each of these sites, calcium injection currents can be applied at various times. Hence given a specific time t_0, the total effect on the sodium and potassium conductances is given by

$$\delta g_{Na}(t_0) = \sum_{i=0}^{4} \delta g_{Na}(t_0, i), \quad \delta g_K(t_0, x_0) = \sum_{i=0}^{4} \delta g_K(t_0, i)$$

where the attenuation due to the terms $\exp\left(\frac{-(x-i)^2}{4D^a(t-t_0)}\right)$ for the ions a is built into the calculations.

8.6   General Pharmacological Inputs

When two cells interact via a synaptic interface, the electrical signal in the pre-synaptic cell in some circumstances triggers a release of a neurotransmitter (NT) from the pre-synapse which crosses the synaptic cleft and then by docking to a port on the post cell, initiates a post-synaptic cellular response. The general pre-synaptic mechanism consists of several key elements: one, NT synthesis machinery so the NT can be made locally; two, receptors for NT uptake and regulation; three, enzymes that package the NT into vesicles in the pre-synapse membrane for delivery to the cleft. There two general pre-synaptic types: *monoamine* and *peptide*. In the *monoamine* case, all three elements for the pre-cell response are first manufactured in the pre-cell using instructions contained in the pre-cell's genome and shipped to the pre-synapse. Hence, the monoamine pre-synapse does not require further instructions from the pre-cell genome and response is therefore fast. The *peptide* pre-synapse can only manufacture a *peptide* neurotransmitter in the pre-cell genome; if a peptide neurotransmitter is needed, there is a lag in response time. Also, in the peptide case, there is no re-uptake pump so peptide NT can't be reused.

On the post-synaptic side, the fast response is triggered when the bound NT/ Receptor complex initiates an immediate change in ion flux through the gate thereby altering the electrical response of the post cell membrane and hence, ultimately its action potential and spike train pattern. Examples are glutumate (excitatory) and GABA (inhibitory) neurotransmitters. The slow response occurs when the initiating NT triggers a second messenger response in the interior of the cell. There are two general families of receptors we are interested in: *family 1:7 transmembrane regions* and *family 2:4 transmembrane regions*. The responses mediated by these families are critical to the design of a proper abstract cell capable of interesting biological activity.

8.6.1   7 Transmembrane Regions

The 7 transmembrane regions are arranged in a circle in the membrane with a central core. The first messenger, FM, docks with the receptor, R_7, creating a complex, NT/R_7. The complex undergoes a conformational change allowing it to bind to a class of proteins (G-proteins) creating a new complex, $NT/R_7/G$. This causes a conformational change in the G protein allowing it to bind to an intracellular enzyme E inducing another conformational change on E. At this point, we have a final complex, $NT/R_7/G/E$, in which the E has been activated so that it can release the substance we call the second messenger, SM. The substance SM can then initiate many intracellular processes by triggering a cascade of reactions that culminate in the transcription of a protein from the genome in the post-cell nucleus. A typical pathway has SM activating an intracellular enzyme E_2 which creates a transcription factor, TF. The transcription factor TF crosses into the nucleus to transcribe a protein P. The protein P may then bind to other genes in the post genome to initiate an entire cascade of gene activity, the protein P may transport out of the post nucleus as the voltage activated gate protein which is inserted into the post cell membrane which effectively increases the maximum ion conductance for some ion or many other possibilities. These possibilities can be modeled with our abstract feature vector approach discussed in Chap. 9 where we associate a ten dimensional vector with the axonal output of the post neuron.

- The protein P transports out of the post nucleus as the voltage activated gate protein which is inserted into the post cell membrane which effectively increases the maximum ion conductance for some ion. We model this effect by alterations in the ten feature vector parameters.
- The protein P operates in the synaptic cleft increasing or decreasing neurotransmitter uptake and neurotransmitter creation. These effects are somewhat intertwined with the first possibility sketched out above, but these have an intrinsic time delay before they take effect.

The docking of the FM therefore initiates a complete repertoire of response that can completely reshape the biological structure of the post-cell and even surrounding cells. There are several time periods involved here: the protein P may be built and transported to its new site for use in minutes (6×10^4 ms), hours (3.6×10^5 ms), days (8.6×10^6 ms) or weeks (6×10^7 ms). Note the rough *times 10* magnitude increases here.

8.6.2 4 Transmembrane Regions

In this family, 4 transmembrane regions are arranged in a circle and five copies of this circle are organized to form a pore in the membrane. The central pore of these channels can be opened to increase or closed to decrease substance flux by two means: one, the gate is voltage activated (the standard Hodgkin–Huxley ion gate) and two, the gate is *ligand* activated which means substances bind to specialized regions of the gate to cause alterations in the flux of substances through the channel. The substance going through the channel could be a sodium ion or a first messenger. The ligand gated channels are very interesting as they allow very complicated control possibilities to emerge. Each 4 transmembrane region circle can have its own specific regulatory domains and hence can be denoted as a type α circle, C_4^{α}. The full 5 circle ion channel is thus a concatenation of 5 such circles and can be labeled $C_4^{\alpha(1)} \cdot C_4^{\alpha(2)} \cdot C_4^{\alpha(3)} \cdot C_4^{\alpha(4)} \cdot C_4^{\alpha(5)}$ or more simply as $C_4(\alpha_1\alpha_2\alpha_3\alpha_4\alpha_5)$. This is much more conveniently written as $C_4(\vec{\alpha})$ where the arrow over the α indicates that it has multiple independent components. There is much complexity here since a given $\alpha(i)$ module can have multiple regulatory docking sites for multiple substances *and* hence, the full R_4 type receptor can have a very sophisticated structure. The amount by which a channel is open or closed is thus capable of being controlled in an exquisitely precise fashion.

8.6.3 Family Two: The Agonist Spectrum

There is a spectrum of response which is called the *agonist spectrum* in which the channel response can be altered from full open to full closed. Call the original FM for this gate NT. Then if NT is an excitatory substance, the docking of NT to a circle opens the channel. In this case, we call NT an *agonist* and denote it by AG. On the other hand, NT might be an inhibitor and close the channel; in that case, NT is called an *antagonist* or AT. We can even have agonists and antagonists that open or close a channel partially by $r\%$, where r is something less that 100%. These work by replacing a docked agonist or antagonist and hence reducing the channel action in either case: full open goes to partially open, full closed goes to partially open. Hence, there is a spectrum of agonist activity ranging from full open to full closed. Since each circle can be of a different type, the resulting behavior can be quite complicated. Further, the C_4 gates can control first messengers that initiate second messenger cascades as well as immediate ion flux alterations. Thus, the production of the second messenger and its possible protein target can also be controlled with great precision.

8.6.4 Allosteric Modulation of Output

There is also *allosteric modification* where a first messenger FM_1 binding to receptor $C_4(\alpha)$ can have its activity modulated by another first messenger FM_2 binding to a second receptor $C_4(\beta)$ to either amplify or block the FM_1 effect. In this case, FM_2 has no action of its own; only if FM_1 is present and docked is there a modulation. This type of modification can also occur on the same receptor if $C_4(\alpha)$ has a second binding site on one of its C_4 regions for FM_2. Finally, there are cotransmitter pairings where two first messengers can both operate independently but also together to either enhance or diminish response. The ligand gated channels can often use drugs or pharmacological inputs to mimic NT/receptor binding and hence generate modulated response. To model cognition, it is thus clear that we eventually need to model pharmacological effects of the types discussed above.

8.7 Neurotransmitter Effects

We now consider how we might model the effects of neurotransmitter modulators as discussed in the previous section. Consider the general model shown in Fig. 8.7. The dendrite is modeled as a Rall cable of electrotonic length L_1 and the soma is a cylinder of length L_2. Both the Rall cable and the soma can receive excitatory and/or inhibitory current pulses generally denoted by the letters ESP and ISP. When the output of one neuron is sent into the input system of another, we typically call the neuron providing the input the *pre-neuron* and the neuron generating the output, the *post-neuron*. The axon of the pre-neuron interacts with the dendrite of the post-neuron via a structure called the *post synaptic density* or *PSD*. The pre-neuron generates an axonal pulse which is the input to the PSD structure. The PSD is really a computational object which transduces the axonal voltage signal on the pre-axon into a ESP or ISP on the post-dendrite cable. The pre-neuron's action potential influences the release of the contents of synaptic vesicles into the fluid contained in the region between the neurons. Remember, the brain is a 3D organ and all neurons are enclosed by a liquid soup of water and many other chemicals. The vesicles contain *neurotransmitters*. For convenience, we focus on one such neurotransmitter, labeled ζ. The vesicle containing ζ is inside a structure called a spine on the surface of the pre-axon. The vesicle migrates to the wall of the spine and then through the wall itself so that it is exposed to the fluid between the pre-axon and post-dendrite (the synaptic cleft). The vesicle ruptures and spreads the λ neurotransmitter into the synaptic cleft. The ζ neurotransmitter then acts like the trigger T_0 we have already

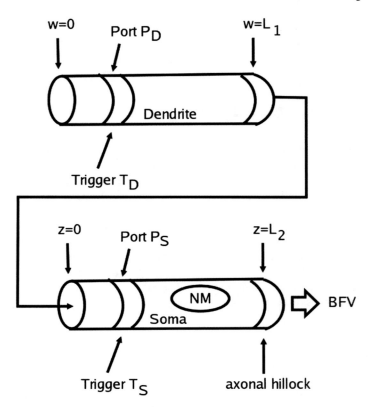

Fig. 8.7 The Dendrite–Soma BFV Model

discussed. It binds in some fashion to a port or gate specialized for the ζ neurotransmitter. The neurotransmitter ζ then initiates a cascade of reactions:

- It passes through the gate, entering the interior of the cable. It then forms a complex, $\hat{\zeta}$.
- Inside the post-dendrite, $\hat{\zeta}$ influences the passage of ions through the cable wall. For example, it may increase the passage of Na^+ through the membrane of the cable thereby initiating an ESP. It could also influence the formation of a calcium current, an increase in $K+$ and so forth.
- The influence via $\hat{\zeta}$ can be that of a second messenger trigger.

Hence, each neuron creates a brew of neurotransmitters specific to its type. A trigger of type T_0 can thus influence the production of neurotransmitters with concomitant changes in post-neuron activity. Thus, in addition to modeling Ca^{++} or other triggers as mechanisms to alter the maximum sodium and potassium conductance, we can also model triggers that provide ways to increase or decrease neurotransmitter. Since the neurotransmitter λ is a second messenger trigger, our full equations with multiplier are

$$h_p^\zeta([\lambda(t)]) = \gamma^\zeta(t) \frac{1}{\xi_0^\zeta} \frac{r^\zeta}{2} \left(1 + \tanh\left(\frac{[\zeta(t)] - [\zeta]_b}{g_p^\zeta}\right)\right)$$

$$I^\zeta(t, x)) = \frac{h_p^\zeta([\zeta(t)])[\hat\zeta]_n}{\sqrt{t - t_0}} \exp\left(\frac{-(x - x_0)^2}{4D^\zeta(t - t_0)}\right)$$

$$h_e^\zeta(I^\zeta(t, x)) = \frac{s^\zeta}{2} \left(1 + \tanh\left(\frac{I^\zeta(t, x) - \frac{r^\zeta[\hat\zeta]_n}{2}}{g_e^\zeta}\right)\right)$$

$$[P^\zeta([\hat\zeta](t)] = h_e^\zeta(h_p^\zeta([\zeta(t)]) \, [\hat\zeta]_n)[\hat\zeta]_n$$

$$h^\zeta([P^\zeta([\hat\zeta](t)]) = \frac{e^\zeta}{2} \delta_\zeta \, g_\zeta^{max} \left(1 + \tanh\left(\frac{[P^\zeta(\hat\zeta(t))]}{g^\zeta}\right)\right)$$

where x is the spatial variable for the post-dendritic cable. The final step is to interpret what function the protein P^ζ has in the cellular system. If it is a sodium or potassium gate, the modification of the conductance of those ions is as before:

$$g_a(t, V) = (g_a^{max} + h_\zeta([P^\zeta([\hat\zeta](t)]))\mathcal{M}_a^p(t, V)\mathcal{H}_a^q(t, V)$$

However, the protein could increase calcium current by adding more calcium gates. We can model the calcium current alterations as

$$Ca^{++}(t, V) = (Ca^{++})_\zeta^{max} + h_\zeta([P^\zeta([\hat\lambda](t)])$$

This gives the sodium and potassium influence equations

$$\delta g_\zeta^{max}(t, x) = \sum_{\zeta_s} \frac{h_\zeta([P^\zeta([\hat\zeta](t))])}{t - t_{\zeta_s}} \exp\left(\frac{-(x - x_{\zeta_s})^2}{4D_{\zeta_s}(t - t_{\zeta_s})}\right)$$

where the index ζ_s denotes the sites where the ζ gates are located. If the neurotransmitter modifies the calcium currents, we have a different type of influence:

$$\delta(Ca^{++})_\zeta^{max}(t, x) = \sum_{\zeta_s} h_\zeta([P^\zeta([\hat\zeta])(t)])$$

We will be focusing on only a few neurotransmitters. We will use ζ_0 as the designation for the neurotransmitter *serotonin*; ζ_1, for *dopamine* and ζ_2, for *norepinephrine*. In Fig. 8.7, we show a typical computational neuron assembly. The dendritic cable is electronic distance L_1 and we will use the variable w to denote distance along the dendrite. The soma is also modeled as a cylinder and we assume it can receive input at electronic distances up to L_2 away from the axon hillock. The inputs to the dendrite occur through ports P_D and the soma entry ports are labeled by P_S. We have a variety of inputs that enter the dendrite. For each time t_0, we sum over the dendrite distance, the effects of all the inputs. These include

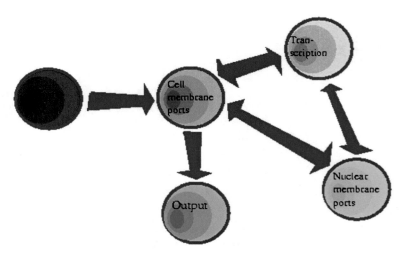

Fig. 8.8 A high level view

- a trigger T_0 which enters the dendrite port P_D and is a simple modulator of the voltage activated gates for sodium and potassium. So we directly modify the maximum ion conductance.
- a trigger T_0 which is a second messenger whose effects on sodium and ion conductance utilize a sigmoid—3 transition.
- a Ca^{++} injection current which can alter the sodium and potassium conductance functions via second messenger pathways.
- neurotransmitters λ_0 to λ_4 which can modify sodium and potassium conductance via second messenger effects or directly. In addition, they can alter calcium currents via second messenger pathways.

   In addition, there are inputs at time t_0 at various electronic distances along the soma. We sum these effects over the soma electronic distance in the same way. These values are then combined to give the voltage at the action hillock that could be used in a general Hodgkin–Huxley model to generate the resulting action potential. In summary, we can model top level computations using the cedllular graph given in Fig. 8.8, with the details of the computational agents coming from our abstractions of the relevant biology.

Reference

J. Peterson, *Calculus for Cognitive Scientists: Partial Differential Equation Models*, Springer Series on Cognitive Science and Technology (Springer Science+Business Media Singapore Pte Ltd, Singapore, 2015 in press)

Chapter 9
The Abstract Neuron Model

Let's look at neuronal inputs and outputs more abstractly. We have learned quite a bit about first and second messenger systems and how action potentials are generated for input families. Now is the time to put all the to good use!

9.1 Neuron Inputs

It is clear neuron classes can have different trigger characteristics. First, consider the case of neurons which create the monoamine neurotransmitters. Neurons of this type in the Reticular Formation of the midbrain produce a monoamine neurotransmitter packet at the synaptic junction between the axon of the pre-neuron and the dendrite of the post-neuron. The monoamine neurotransmitter is then released into the synaptic cleft and it induces a second messenger response as shown in Fig. 6.8. The strength of this response is dependent on the input from pre-neurons that form synaptic connections with the post-neuron. In addition, the size of this input determines the strength of the monoamine trigger into the post-neuron dendrite. Let the strength be given by the weighting term $c_{pre,post}^{\zeta}$ where as usual we use the values $\zeta = 0$ to denote the neurotransmitter serotonin, $\zeta = 1$, for dopamine and $\zeta = 2$, for norepinephrine. In our model, recall that time is discretized into integer values as time ticks 1, 2 and so forth via our global simulation clock. Also, spatial values are discretized into multiples of various electrotonic scaling distances. With that said, the trigger at time t and dendrite location w is therefore

$$T_0(t, w) = \frac{c_{pre,post}^{\zeta}}{\sqrt{t - t_0}} \exp\left(\frac{-(w - w_0)^2}{4D_0^{\zeta}(t - t_0)}\right).$$

where D_0^{ζ} is the diffusion constant associated with the trigger. Hence, $w = j\hat{L}_E$ for some scaling $\hat{L}_E$. The trigger T_0 has associated with it the protein T_1. We let

© Springer Science+Business Media Singapore 2016
J.K. Peterson, *BioInformation Processing*, Cognitive Science and Technology,
DOI 10.1007/978-981-287-871-7_9

$$T_1(t, w) = \frac{d^\zeta_{pre,post}}{\sqrt{t - t_0}} \exp\left(\frac{-(w - w_0)^2}{4D^\zeta_1(t - t_0)} \right).$$

where $d^\zeta_{pre,post}$ denotes the strength of the induced T_1 response and D^ζ_1 is the diffusion constant of the T_1 protein. This trigger will act through the usual pathway. Also, we let T_2 denote the protein $P(T_1)$. T_2 transcribes a protein target from the genome with efficiency e.

$$h_p([T_0(t, w)]) = \frac{r}{2} \left(1 + \tanh \left(\frac{[T_0(t, w)] - [T_0]_b}{g_p} \right) \right)$$

$$I(t, w)) = \frac{h_p([T_0(t, w)])[T_1]_n}{\sqrt{t - t_0}} \exp \left(\frac{-(w - w_0)^2}{4D^\lambda_1(t - t_0)} \right)$$

$$h_e(I(t, w)) = \frac{s}{2} \left(1 + \tanh \left(\frac{I(t, w) - \frac{r[T_1]_n}{2}}{g_e} \right) \right)$$

$$[P(T_1)](t, w) = h_e(I(t, w))$$

$$h_{T_2}(t, w) = \frac{e}{2} \left(1 + \tanh \left(\frac{[T_2](t, w)}{g_{T_2}} \right) \right)$$

$$[T_2](t, w) = h_{T_2}(t, w)[T_2]_n$$

Note $[T_2](t, w)$ gives the value of the protein T_2 concentration at some discrete time t and spatial location $j\hat{L}_E$. This response can also be modulated by feedback. In this case, let ξ denote the feedback level. Then, the final response is altered to $h^f_{T_2}$ where the superscript f denotes the feedback response and the constant ω is the strength of the feedback.

$$h^f_{T_2}(t, w) = \omega \frac{1}{\xi} h_{T_2}(t, w)$$

$$[T_2](t, w) = h^f_{T_2}(t, w)[T_2]_n$$

There are a large number of shaping parameters here. For example, for each neurotransmitter, we could alter the parameters due to calcium trigger diffusion as discussed in Sect. 7.2. These would include D^ζ_0, the diffusion constant for the trigger, and D^ζ_1, the diffusion constant for the gate induced protein T_1. In addition, transcribed proteins could alter—we know their first order quantitative effects due to our earlier analysis—$d^\zeta_{pre,post}$, the strength of the T_1 response, r, the fraction of T_1 free, g_p, the trigger gain, $[T_0]_b$, the trigger threshold concentration, s, the fraction of active T_1 reaching genome, g_e, the trigger gain for active T_1 transition, $[T_1]_n$, the threshold for T_1, $[T_2]_n$, the threshold for $P(T_1) = T_2$, g_{T_2}, the gain for T_2, ω, the feedback strength, and ξ, the feedback amount for $T_1 = 1 - \xi$. Note $d^\zeta_{pre,post}$ could be simply $c^\lambda_{pre,post}$. The neurotransmitter triggers can alter many parameters important to the creation of the action potential. The maximum sodium and potassium conductances can be altered via the equation for T_2. For sodium we have

$$T_2(t, w) = h_{T_2}(t, w)[T_2]_n$$

becomes

$$[T_2]_n = \delta_{Na} \, g_{Na}^{max}$$
$$h_{T_2}^f(t, w) = \omega \frac{1}{\xi} h_{T_2}(t, w)$$
$$[T_2](t, w) = h_{T_2}^f(t, w)[T_2]_n$$
$$g_{Na}(t, w, V) = (\, g_{Na}^{max} + [T_2](t, w) \,)\mathcal{M}_{Na}^p(t, V)\mathcal{H}_{Na}^q(t, V)$$

For potassium, the change is

$$[T_2]_n = \delta_K \, g_K^{max}$$
$$h_{T_2}^f(t, w) = \omega \frac{1}{\xi} h_{T_2}(t, w)$$
$$[T_2](t, w) = h_{T_2}^f(t, w)[T_2]_n$$
$$g_K(t, w, V) = (\, g_{Na}^{max} + [T_2](t, w) \,)\mathcal{M}_K^p(t, V)\mathcal{H}_K^q(t, V)$$

Finally, neurotransmitters and other second messenger triggers have delayed effects in general. So if the trigger T_0 binds with a port P at time t_0, the changes in protein levels $P(T_1)$ need to delayed by a factor τ^ς. The soma calculations are handled exactly like this except we are working on the soma cable and so electronic distance is measured by the variable z instead of w.

9.2 Neuron Outputs

We now understand in principle how to compute the axon hillock voltage that will be passed to the Hodgkin–Huxley engine to calculate the resulting action potential. The prime purpose of the incoming voltage is to provide the proper depolarization of the excitable cell membrane. Hence, we know that we will generate an action potential is the incoming signal exceeds the neuron's threshold. For our purposes, the shape of the action potential will be primarily determined by the alteration of the g_{Na}^{max} and g_K^{max} conductance parameters. In our model, these values are altered by the second messenger triggers which create or destroy the potassium and sodium gates in the membrane. Other triggers alter the essential hardware of the neuron and potentially the entire neuron class $\mathcal{N}$ in other ways. The protein T_2 due to a trigger u can

- directly alter the synaptic coupling weight $c^u_{pre,post}$ according to the strength of $T_2(t, w)$ by making changes in the extracellular side of the membrane in a variety of ways,
- can directly impact the maximum conductances for the potassium and sodium ions,
- can alter second messenger channels in many ways as we have discussed in the chapters on calcium and generic second messenger triggers. These alterations can affect the coupling weight or maximum ion conductances as well.

However, they can also effect more global parameters. Consider if the protein alters the ration ρ. This is an fundamental coupling parameter for the dendrite and soma system we use in our modeling. Hence, an alteration in ρ is a global change to the entire neuron class $\mathcal{N}$. This mechanism is very useful as it allows us to use monoamine modulation to alter every neuron in a particular class no matter what neural module it is a part of. This is a useful tool in implementing true RF type core modulation of cortical output. Such a global change could be as simple as an alteration to L_{DE} or L_{SE} without changing ρ. However, if rho is changed to $rho \pm \epsilon$, this changes the eigenvalue problem that the neuron class is associated with to

$$\tan(\alpha L) = -\frac{\tanh(L)}{(\rho \pm \epsilon)L}(\alpha L),$$

Rather than performing such a numerical computation every time a global monoamine modulation request is sent, it is easier to pre-compute the eigenvalue problems solution for a spectrum of ρ values. We then keep them in storage and access them as needed to update the neuron class when required. The determination of the action potential for a given axon hillock input is handled as follows. Once we know the incoming voltage is past threshold, we generate an action potential whose parameters are shaped by both the size of the pulse and the changes in ion conductances. If we want to replace the action potential generation as the solution of a complicated partial differential equation with an approximation of some kind, we need to look at how the Hodgkin–Huxley equations are solved in more detail even though we already introduced these equations in Peterson (2015). Our replacement for the action potential will be called the **Biological Feature Vector** or **BFV**.

The individual neural objects in our cognitive model (and in particular, in our cortical columns) will be abstractions of neural ensembles, their behaviors and outputs gleaned from both low level biological processing and high level psychopharmacology data. The low level information must include enough detail of how inputs are processed (spike train generation) to be useful and enough detail of second messenger pathways to see clearly that the interactions between a pre and a post neural ensemble are really communications between their respective genomes. Clearly, this implies that an appropriate abstract second messenger and genome model is needed. In this section, we modify the general *ball and stick* model to use a very simple low dimensional representation of the action potential.

9.3 Abstract Neuron Design

We can see the general structure of a typical action potential is illustrated in Fig. 9.1.

This wave form is idealized and we are interested in how much information can be transferred from one abstract neuron to another using a low dimensional biologically based feature vector Biological Feature Vector of BFV. We can achieve such an abstraction by noting that in a typical excitable neuron response, Fig. 9.1, the action potential exhibits a combination of cap-like shapes. We can use the following points on this generic action potential to construct a low dimensional feature vector of Eq. 9.1.

$$\zeta = \begin{cases} (t_0, V_0) & \text{start point} \\ (t_1, V_1) & \text{maximum point} \\ (t_2, V_2) & \text{return to reference voltage} \\ (t_3, V_3) & \text{minimum point} \\ (g, t_4, V_4) & \text{sigmoid model of tail} \\ & V_3 + (V_4 - V_3)\, \tanh(g(t - t_3)) \end{cases} \tag{9.1}$$

where the model of the tail of the action potential is of the form $V_m(t) = V_3 + (V_4 - V_3)\, \tanh(g(t - t_3))$. Note that $V'_m(t_3) = (V_4 - V_3)\, g$ and so if we were using real voltage data, we would approximate $V'_m(t_3)$ by a standard finite difference. This wave form is idealized and actual measured action potentials will be altered by noise and the extraneous transients endemic to the measurement process in the laboratory.

To study the efficacy of this BFV for capturing useful information, we performed a series of computational experiments on the recognition of toxins introduced into the input side of a cell from their effects on the cells action potential. We know

Fig. 9.1 Prototypical action potential

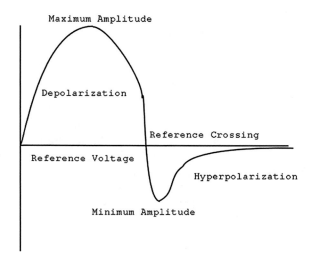

biotoxins alter the shape of this action potential in many ways. The toxin guide of Adams and Swanson (1996) shows quite clearly the variety of ways that a toxin can alter ion gate function in the cell membrane, second messenger cascades and so forth. Older material from Kaul and Daftari (1986), Wu and Narahashi (1988) and Schiavo et al. (2000) focus on the details of specific classes of toxins. Kaul and Wu focus on pharmacological active substances from the sea. Schiavo investigates the effects of toxins that interfere with the release of neurotransmitters by altering the exocytosis process. The effects of toxins on the presynaptic side are analyzed in Harvey (1990), Strichartz et al. (1987) presents a summary of how toxins act on sodium channels.

Let's focus on a single action potential and examine how its characteristics change when we introduce various alterations in the parameters that effect its shape. We know that toxins cause many such effects and we can classify toxins into families based on how they alter the parameters that can effect the action potential. A simulation is then possible that collects information from a family of generated action potentials due to the introduction of various toxin families. The BFV abstracts the characteristics into a low dimensional vector and in Peterson and Khan (2006), we showed the BFV was capable of determining which toxin family was used to alter the action potential and our ability to discern the toxin family used compared nicely to other methods of feature vector extraction such as the classic covariance technique, a total variation method and an associated spline method that uses knots determined by the total variation approach.

In order to demonstrate the efficacy of the BFV approach for capturing information, we generated families of action potentials from a classic Hodgkin–Huxley model. We also assume a toxin in a given family alters the action potential in a specific way which we will call the *toxin family signature*. We studied two type of signatures: first, families that alter the maximum sodium and potassium conductance parameters by a given percentage and second, families that perturb the standard Hodgkin–Huxley α–β values.

Now, an input into an artificial neuron generates an abstract BFV response. Larger ensembles of artificial neurons that can then be created to serve as modules or neural objects in graphs.

The particular values on the components of a single artificial neuron's or a module's feature vector, BFV, are then amenable to alteration via neurotransmitter action such as serotonin, norepinephrine and dopamine coming from the midbrain module of Fig. 5.3. The feature vector output of a neural object is thus due to the cumulative effect of second messenger signaling to the genome of this object which influences the action potential and thus feature vector of the object by altering its complicated mixture of ligand and voltage activated ion gates, enzymes and so forth. For example, the G protein-linked receptor superfamily second messenger system would consist of a receptor with 7 transmembrane regions with links to G proteins that uses a second messenger system activated by an enzyme—cAMP and PI and because a second messenger system is used, response to a signal is delayed and hence, these are *slow response* systems. Another family uses receptors with 4 transmembrane regions. In this family, the ion channel is surrounded by multiple copies of multiple different receptors and ion flow is directly controlled by a given particular mixture of neurotransmitters and receptors. Clearly, there are many control possibilities that arise here due to the combinatorial nature of this family's channels. It is difficult to find appropriate high level descriptions of such events so that the algorithmic structures are evident and not obscured by the detail. Two important resources that have been prime influences and have helped us develop generic models of neural objects which have feature vector outputs which can be shaped by such pharmacological inputs have been Stahl (2000, psychopharmacology) and Gerhart and Kirschner (1997, evolution of cellular physiology). They have been out for awhile now, but they are still packed with useful information for approximation schemes. The action potential is generated by an input on the dendritic side of an excitable nerve cell. We wish to analyze this signal and from its properties, discern whether or not a toxin/ligand has been introduced into the input stream. further, we want to be able to recognize the toxin as belonging to a certain family. Effectively, this means we can label the output as due to a certain ligand input. For our purposes, we are interested in a single action potential with general structure as illustrated in Fig. 9.1. We concentrate for the moment on single output voltage pulses produced by toxins introduced into the input side of the nerve cell. The wave form in Fig. 9.1 is, of course, idealized and the actual measured action potentials will be altered by noise and the extraneous transients endemic to the measurement process in the laboratory.

In order to show that the abstract feature vector, BFV, is capable of capturing some of the information carried by the action potential, we studied how to use this kind of feature vector to determine whether or not an action potentials has been influenced by a toxin introduced into the dendritic system of an excitable neuron. The role of the toxin is of course similar to the role of the monoamine neurotransmitters we wish to use to modulate cortical output. Recall the full information processing here is quite complex as seen in Fig. 9.2 and we are approximating portions of it. We explain the toxin studies in some detail in the sections that follow. This provides additional background on the reasons for our choice of low dimensional feature vector. The basic Hodgkin–Huxley model depends on a large number of parameters and we will be using perturbations of these as a way to model families of toxins. Of course, more sophisticated action potential models can be used, but the standard two ion gate Hodgkin–Huxley model is sufficient for our needs in this paper. In Sect. 9.3.1, we present the toxin recognition methodology for the first toxin family that modify the maximal sodium and potassium conductances. We believe that toxins of this sort include some second messenger effects. Our general classification methodology is then discussed and applied to the this collection of toxin families. We show that we can design a reasonable recognizer engine using a low dimensional biologically based feature vector. In Sect. 9.3.1.2, we introduce the second class of toxin families whose effect on the action potential is more subtle. We show that the biological feature vector performs well in this case also.

9.3.1 Toxin Recognition

Recall, when two cells interact via a synaptic interface, the electrical signal in the pre-synaptic cell in some circumstances triggers a release of a neurotransmitter (NT)

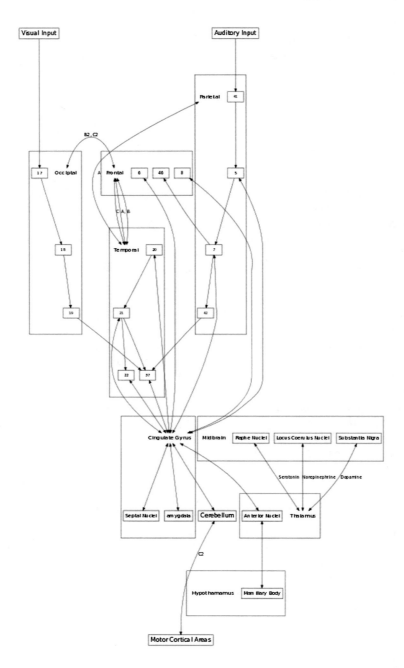

Fig. 9.2 A simplified path of information processing in the brain. *Arrows* indicate information processing pathways

from the pre-synapse which crosses the synaptic cleft and then by docking to a port on the post cell, initiates a post-synaptic cellular response. The general pre-synaptic mechanism consists of several key elements: one, NT synthesis machinery so the NT can be made locally; two, receptors for NT uptake and regulation; three, enzymes that package the NT into vesicles in the pre-synapse membrane for delivery to the cleft. On the post-synaptic side, we will focus on two general responses. The fast response is triggered when the bound NT/Receptor complex initiates an immediate change in ion flux through the gate thereby altering the electrical response of the post cell membrane and hence, ultimately its action potential and spike train pattern. The slow response occurs when the initiating NT triggers a second response in the interior of the cell. In this case, the first NT is called the first messenger and the intracellular response (quite complex in general) is the second messenger system. Further expository details of first and second messenger systems can be found in Stahl (2000). From the above discussion, we can infer that a toxin introduced to the input system of an excitable cell influences the action potential produced by such a cell in a variety of ways. For a classic Hodgkin Huxley model, there are a number of critical parameters which influence the action potential.

The nominal values of the maximum sodium and potassium conductance G_{Na}^0 and G_K^0 can be altered. We view these values as a measure of the density of a classic Hodgkin Huxley voltage activated gates per square cm of biological membrane. Hence, changes in this parameter require the production of new gates or the creation of enzymes that control the creation/destruction balance of the gates. This parameter is thus related to *second messenger* activity as access to the genome is required to implement this change. We can also perturb the parameters that shape the α and β functions which we introduced in Peterson (2015). These functions are all special cases of the of the general mapping $F(V_m, p, q)$ with $p \in \mathfrak{R}^4$ and $q \in \mathfrak{R}^2$ defined by

$$F(V_m, p, q) = \frac{p_0(V_m + q_0) + p_1}{e^{p_2(V_m + q_1)} + p_3}.$$

For ease of exposition, here we will denote $\mathcal{M}_{Na}$ by m, $\mathcal{H}_{Na}$ by h and $\mathcal{M}_K$ by n. We have used h for sigmoid type transitions elsewhere, but historically, m, h and n have been used for these Hodgkin–Huxley models. The α and β pairs are thus described using the generic F mapping by

$$\alpha_m = F(V_m, p_m^\alpha = \{-0.10, 0.0, -0.1, -1.0\}, q_m^\alpha = \{35.0, 35.0\}),$$
$$\beta_m = F(V_m, p_m^\beta = \{0.0, 4.0, 0.0556, 0.0\}, q_m^\beta = \{60.0, 60.0\}),$$
$$\alpha_h = F(V_m, p_h^\alpha = \{0.0, 0.07, 0.05, 0.0\}, q_h^\alpha = \{60.0, 60.0\}),$$
$$\beta_h = F(V_m, p_h^\beta = \{0.0, 1.0, -0.1, 1.0\}, q_h^\beta = \{30.0, 30.0\}),$$
$$\alpha_n = F(V_m, p_n^\alpha = \{-0.01, 0.0, -0.1, -1.0\}, q_n^\alpha = \{50.0, 50.0\}),$$
$$\beta_n = F(V_m, p_n^\beta = \{0.0, 0.125, 0.0125, 0.0\}, q_n^\beta = \{60.0, 60.0\}).$$

The p and q parameters control the shape of the action potential in a complex way. From our discussions about the structure of ion gates, we could think of alterations in the (p, q) pair associated with a given α and/or β as a way of modeling how passage of ions through the gate are altered by the addition of various ligands. These effects may or may not be immediate. For example, the alterations to the p and q parameters may be due to the docking of ligands which are manufactured through calls to the genome in the cell's nucleus. In that case, there is a long delay between the initiation of the second messenger signal to the genome and the migration of the ligands to the outside of the cell membrane. In addition, proteins can be made which bind to the inside of a gate and thereby alter the ion flow. We understand that this type of modeling is not attempting to explain the details of such interactions. Instead, we are exploring an approach for rapid identification and differentiation of the signals due to various toxins.

Now assume that the standard Hodgkin Huxley model for $g_{Na}^0 = 120$, $g_K^0 = 36.0$ and the classical α, β functions are labeled as the nominal values. Hence, there is a nominal vector Λ_0 given by

$$
\Lambda_0 =
\begin{bmatrix}
G_{Na}^0 = 120.0 & G_K^0 = 36.0 \\
(p_m^\alpha)^0 \ \{-0.10, 0.0, -0.1, -1.0\} & (p_m^\beta)^0 \ \{0.0, 4.0, 0.0556, 0.0\} \\
(p_h^\alpha)^0 \ \{0.0, 0.07, 0.05, 0.0\} & (p_h^\beta)^0 \ \{0.0, 1.0, -0.1, 1.0\} \\
(p_n^\alpha)^0 \ \{-0.01, 0.0, -0.1, -1.0\} & (p_n^\beta)^0 \ \{0.0, 0.125, 0.0125, 0.0\} \\
(q_m^\alpha)^0 \ \{35.0, 35.0\} & (q_m^\beta)^0 \ \{60.0, 60.0\} \\
(q_h^\alpha)^0 \ \{60.0, 60.0\} & (q_h^\beta)^0 \ \{30.0, 30.0\} \\
(q_n^\alpha)^0 \ \{50.0, 50.0\} & (q_n^\beta)^0 \ \{60.0, 60.0\}
\end{bmatrix}
$$

A toxin G thus has an associated toxin signature, $\mathcal{E}(G)$ which consists of deviations from the nominal classical Hodgkin Huxley parameter suite: $\mathcal{E}(G) = \Lambda_0 + \delta$, where δ is a vector, or percentage changes from nominal, that we assume the introduction of the toxin initiates. As you can see, if we model the toxin signature in this way, we have a rich set of possibilities we can use for parametric studies.

9.3.1.1 Simple Second Messenger Toxins

We begin with toxins whose signatures are quite simple as they only cause a change in the nominal sodium and potassium maximum conductance. These are therefore second messenger toxins. This simulation will generate five toxins whose signatures are distinct. Using C++ (not MatLab!) as our code base, we designed a *TOXIN* class whose constructor generates a family of distinct signatures using a signature as a base. Here, we will generate 20 sample toxin signatures clustered around the given signature using a neighborhood size of 0.02. First, we generate five toxins using the toxin signatures of Table 9.1.

Table 9.1 Toxin conductance signature

Toxin	Signature $[\delta g_{Na}^0, \delta g_K^0]$
A	$[0.45, -0.25]$
B	$[0.05, -0.35]$
C	$[0.10, 0.45]$
D	$[0.55, 0.70]$
E	$[0.75, -0.75]$

Fig. 9.3 Applied synaptic pulse

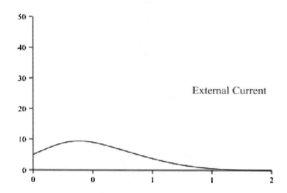

These percentage changes are applied to the base conductance values to generate the sodium, potassium and leakage conductances for each simulation. We use these toxins to generate 100 simulation runs using the 20 members of each toxin family. We believe that the data from this parametric study can be used to divide up action potentials into disjoint classes each of which is generated by a given toxin. Each of these simulation runs use the synaptic current injected as in Fig. 9.3 which injects a modest amount of current over approximately 2 s. The current is injected at four separate times to give the gradual rise seen in the picture. In Fig. 9.4, we see all 100 generated action potentials. You can clearly see that the different toxin families create distinct action potentials.

The Biological Feature Vector Based Recognizer One can easily observe that for a typical response, as in Fig. 9.1, the potential exhibits a combination of cap-like shapes. We can use the following points on this generic action potential to construct the low dimensional feature vector given earlier as Eq. 9.1 which we reproduce for convenience.

$$\xi = \begin{bmatrix} (t_0, V_0) & \text{start point} \\ (t_1, V_1) & \text{maximum point} \\ (t_2, V_2) & \text{return to reference voltage} \\ (t_3, V_3) & \text{minimum point} \\ (g, t_4, V_4) & \text{sigmoid model of tail} \end{bmatrix},$$

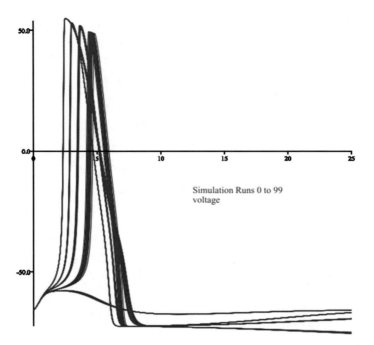

Fig. 9.4 The toxin families

with the model of the tail of the action potential is of the form

$$V_m(t) = V_3 + (V_4 - V_3) \tanh(g(t - t_3)),$$

Note that

$$V'_m(t_3) = (V_4 - V_3) g.$$

We approximate $V'_m(t_3)$ by a standard finite difference. We pick a data point (t_5, V_5) that occurs after the minimum—typically we use the voltage value at the time t_5 that is 5 time steps downstream from the minimum and approximate the derivative at t_3 by

$$V'_m(t_3) \approx \frac{V_5 - V_3}{t_5 - t_3}$$

The value of g is then determined to be

$$g = \frac{V_5 - V_3}{(V_4 - V_3)(t_5 - t_3)}$$

which reflects the asymptotic nature of the hyperpolarization phase of the potential. Note that ξ is in $\mathfrak{R}^{11}$. We see the rudimentary feature vector extraction we have called the BFV is quite capable of generating a functional recognizer; i.e. distinguishing between toxin inputs. Hence, we are confident we can use the BFV is developing approximations to nodal computations . In all of the recognizers constructed in Peterson and Khan (2006), the classification of the toxin is determined by finding the toxin class which is minimum distance from the sample.

9.3.1.2   Toxins that Reshape α–β Parameters

Our earlier discussions focused on toxins that cause a change in the nominal sodium and potassium maximum conductance, and so are effectively second messenger toxins. However, the 36 additional α–β parameters that we have listed all can be altered by toxins to profoundly affect the shape of the action potential curve. In this section, we will focus on parameter changes that effect a small subset of the full range of α–β possibilities.

First, we generate five toxins as shown below. Note that Toxin A perturbs the q parameters of the α–β for the sodium activation m only; Toxin B does the same for the q parameters of the sodium h inactivation; Toxin C alters the q parameters of the potassium activation n; Toxin D changes the $p[2]$ value of the α–β functions for the sodium inactivation h; and Toxin E, does the same for the potassium activation n. This is just a sample of what could be studied. These particular toxin signatures were chosen because the differences in the generated action potentials for various toxins from family A, B, C, D or E will be subtle. For example, in Toxin A, a perturbation of the given type generates the α–β curves for m_{NA} as shown in Fig. 9.5. Note all we are doing is changing the voltage values of 35.0 slightly. This small change introduces a significant ripple in the α curve. Note, we are only perturbing

Fig. 9.5 m perturbed
alpha–beta curves

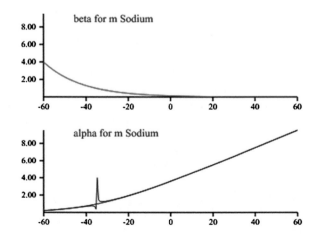

Table 9.2 Toxin α–β
signatures

Toxin A	$\delta q_m^\alpha = \{0.2, -0.1\}$
	$\delta q_m^\beta = \{-0.1, 0.1\}$
Toxin B	$\delta q_h^\alpha = \{-0.1, 0.1\}$
	$\delta q_h^\beta = \{0.1, -0.2\}$
Toxin C	$\delta q_n^\alpha = \{-0.2, 0.1\}$
	$\delta q_n^\beta = \{-0.1, 0.1\}$
Toxin D	$\delta p_h^\alpha \{0.0, 0.0, 3.0, 0.0\}$
	$\delta p_h^\beta \{0.0, 0.3, 0.0, 0.0\}$
Toxin E	$\delta p_n^\alpha \{0.0, 0.3, 0.0, 0.0\}$
	$\delta p_n^\beta \{0.0, 0.3, 0.0, 0.0\}$

two parameters at a time in a given toxin family. In all of these toxins, we will be leaving the maximum ion conductances the same. We use these toxins to generate 100 simulation runs using the 20 members of each toxin family. We believe that the data from this parametric study can be used to divide up action potentials into disjoint classes each of which is generated by a given toxin. Each of these simulation runs use the synaptic current injected as in Fig. 9.3 which injects a modest amount of current over approximately 2 s. We generated five toxins whose signatures were distinct. Again, using C++ as our code base, we designed a *TOXIN* class whose constructor generates a family of distinct signatures using a given signature as a base. Here, we generated twenty sample toxin signatures clustered around the given signature using a neighborhood size for each of five toxin families. First, we generate five toxin families. The types of perturbations each toxin family uses are listed in Table 9.2. In this table, we only list the parameters that are altered. Note that Toxin A perturbs the q parameters of the α–β for the sodium activation m only; Toxin B does the same for the q parameters of the sodium h inactivation; Toxin C alters the q parameters of the potassium activation n; Toxin D changes the $p[2]$ value of the α–β functions for the sodium inactivation h; and Toxin E, does the same for the potassium activation n. This is just a sample of what could be studied. These particular toxin signatures were chosen because the differences in the generated action potentials for various toxins from family A, B, C, D or E will be subtle. For example, in Toxin A, a perturbation of the given type generates the α–β curves for m_{NA} as shown in Fig. 9.5. Note all we are doing is changing the voltage values of 35.0 slightly. This small change introduces a significant ripple in the α curve. We use these toxins to generate 100 simulation runs using the 20 members of each toxin family and the BFV approach can easily distinguish between these families.

   Each of these simulation runs use the synaptic current injection protocol as described before. The current is injected at four separate times to give the gradual rise seen in the picture. In Fig. 9.6, we see all 100 generated action potentials. You can clearly see that the different toxin families create distinct action potentials and as shown in Peterson and Khan (2006), the BFV is very capable at as a functional recognizer which determines the toxin family that has perturbed the dendritic inputs.

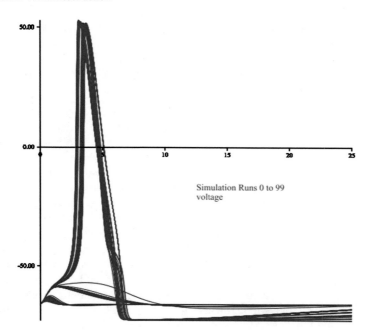

50.00

0.00

10   15   20   25

Simulation Runs 0 to 99
voltage

-50.00

Fig. 9.6 Generated voltage traces for the $\alpha-\beta$ toxin families

9.4 Feature Vector Abstraction

In the work of this chapter, we have indicated a low dimensional feature vector based on biologically relevant information extracted from the action potential of an excitable nerve cell is capable of subserving biological information processing. We can use such a BFV to extract information about how a given toxin influences the shape of the output pulse of an excitable neuron. Hence, we expect modulatory inputs into our abstract neuron can be modeled as alterations to the components of the BFV.

The biological feature vector stores many of the important features of the action potential is a low dimensional form. We note these include

- The interval $[t_0, t_1]$ is the duration of the rise phase. This interval can be altered or modulated by neurotransmitter activity on the nerve cell's membrane as well as second messenger signaling from within the cell.
- The height of the pulse, V_1, is an important indicator of excitation.
- The time interval between the highest activation level, V_1 and the lowest, V_3, is closely related to spiking interval. This time interval, $[t_1, t_3]$, is also amenable to alteration via neurotransmitter input.
- The "height" of the depolarizing pulse, V_4, helps determine how long it takes for the neuron to reestablish its reference voltage, V_0.

- The neuron voltage takes time to reach reference voltage after a spike. This is the time interval by the interval $[t_3, \infty]$.
- The exponential rate of increase in the time interval $[t_3, \infty]$ is also very important to the regaining of nominal neuron electrophysiological characteristics.

Clearly, we can model an inhibitory pulse in essentially the same way, *mutatis mutandi*. We will assume all of the data points in our feature vector are potentially mutable due to neurotransmitter activity, input pulses into the neuron's dendritic system and alteration of the neuron hardware via genome access with the second messenger system.

Although it is possible to do detailed modeling of biological systems using GENESIS and NEURON, we do not believe that they are useful tools in modeling the kind of information flow between cortical modules that is needed for a cognitive model. Progress in building large scale models that involve many cooperating neurons will certainly involve making suitable abstractions in the information processing that we see in the neuron. Neurons transduce and integrate information on the dendritic side into wave form pulses and there are many models involving filtering and transforms which attempt to "see" into the action potential and find its informational core so to speak. However, all of these methods are hugely computationally expensive and even a simple cognitive model will require ensembles of neurons acting together locally to create global effects. For reasons outlined above, we believe alterations in the parameters of the simple biological feature vector (BFV) can serve as modulatory agents in ensembles of abstract neurons. The kinds of changes one should use for a given neurotransmitters modulatory effect can be estimated from the biophysical and toxin literature. For example, an increase in sodium ion flow, Ca^{++} gated second messenger activity can be handled at a high level as a suitable change in one of the 11 parameters of the BFV.

9.4.1 The BFV Functional Form

In Fig. 9.7, we indicated the three major portions of the biological feature vector and the particular data points chosen from the action potential which are used for the model. These are the two parabolas f_1 and f_2 and the sigmoid f_3. The parabola f_1 is treated as the two distinct pieces f_{11} and f_{12} given by

$$f_{11}(t) = a^{11} + b^{11}(t - t_1)^2 \tag{9.2}$$
$$f_{12}(t) = a^{12} + b^{12}(t - t_1)^2 \tag{9.3}$$

Thus, f_1 consists of two joined parabolas which both have a vertex at t_1. The functional form for f_2 is a parabola with vertex at t_3:

$$f_2(t) = a^2 + b^2(t - t_3)^2 \tag{9.4}$$

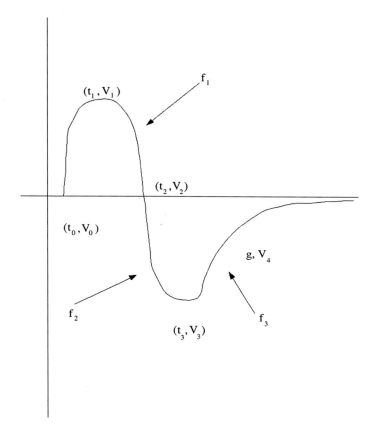

Fig. 9.7 The BFV functional form

Finally, the sigmoid portion of the model is given by

$$f_3(t) = V_3 + (V_4 - V_3)\tanh(g(t - t_3)) \tag{9.5}$$

We have also simplified the BFV even further by dropping the explicit time point t_4 and modeling the portion of the action potential after the minimum voltage by the sigmoid f_3. From the data, it follows that

$$
\begin{aligned}
f_{11}(t_0) &= V_0 = a^2 + b^{11}(t_0 - t_1)^2 \\
f_{11}(t_1) &= V_1 = a^{11} \\
f_{12}(t_1) &= V_1 = a^{12} \\
f_{11}(t_2) &= V_2 = a^{12} + b^{12}(t_2 - t_1)^2
\end{aligned}
$$

This implies

$$a^{11} = V_1$$
$$b^{11} = \frac{V_0 - V_1}{(t_0 - t_1)^2}$$
$$a^{12} = V_1$$
$$b^{12} = \frac{V_2 - V_1}{(t_2 - t_1)^2}$$

In a similar fashion, the f_2 model is constrained by

$$f_2(t_2) = V_2 = a^2 + b^2(t_2 - t_3)^2$$
$$f_2(t_3) = V_3 = a^2$$

We conclude that

$$a^2 = V_3$$
$$b^2 = \frac{V_2 - V_3}{(t_2 - t_3)^2}$$

Hence, the functional form of the BFV model can be given by the mapping f of Eq. 9.6.

$$f(t) = \begin{cases} V_1 + \frac{V_0 - V_1}{(t_0 - t_1)^2}(t - t_1)^2, & t_0 \le t \le t_1 \\ V_1 + \frac{V_2 - V_1}{(t_2 - t_1)^2}(t - t_1)^2, & t_1 \le t \le t_2 \\ V_3 + \frac{V_2 - V_3}{(t_2 - t_3)^2}(t - t_3)^2, & t_2 \le t \le t_3 \\ V_4 + (V_4 - V_3)\tanh(g(t - t - 3)), & t_3 \le t < \infty \end{cases} \qquad (9.6)$$

All of our parabolic models can also be written in the form

$$p(t) = \pm\frac{1}{4\beta}(t - \alpha)$$

where 4β is the width of the line segment through the focus of the parabola. The models f_{11} and f_{12} point down and so use the "minus" sign while f_2 uses the "plus". By comparing our model equations with this generic parabolic equation, we find the width of the parabolas of f_{11}, f_{12} and f_2 is given by

$$4\beta_{11} = \frac{(t_0 - t_1)^2}{V_1 - V_0} = \frac{-1}{b^{11}}$$
$$4\beta_{12} = \frac{(t_2 - t_1)^2}{V_1 - V_2} = \frac{-1}{b^{12}}$$
$$4\beta_2 = \frac{(t_2 - t_3)^2}{V_2 - V_3} = \frac{1}{b^2}$$

9.4.2   Modulation of the BFV Parameters

We want to modulate the output of our abstract neuron model by altering the BFV. The BFV itself consists of 10 parameters, but better insight, into how alterations of the BFV introduce changes in the "action potential" we are creating, comes from studying changes in the mapping f given in Sect. 9.4.1. In addition to changes in timing, t_0, t_1, t_2 and t_3, we can also consider the variations of Eq. 9.7.

$$
\begin{bmatrix}
\Delta a^{11} \\
\Delta b^{11} \\
\Delta a^{12} \\
\Delta b^{12} \\
\Delta a^{2} \\
\Delta b^{2}
\end{bmatrix}
=
\begin{bmatrix}
\Delta V_1 \\
\Delta \left(\frac{V_0 - V_1}{(t_0 - t_1)^2} \right) \\
\Delta V_1 \\
\Delta \left(\frac{V_2 - V_1}{(t_2 - t_1)^2} \right) \\
\Delta V_3 \\
\Delta \left(\frac{V_2 - V_3}{(t_2 - t_3)^2} \right)
\end{bmatrix}
=
\begin{bmatrix}
\Delta \text{ Maximum Voltage} \\
\Delta \left(\frac{-1}{4\beta_{11}} \right) \\
\Delta \text{ Maximum Voltage} \\
\Delta \left(\frac{-1}{4\beta_{12}} \right) \\
\Delta \text{ Minimum Voltage} \\
\Delta \left(\frac{1}{4\beta_{2}} \right)
\end{bmatrix}
\tag{9.7}
$$

It is clear that modulatory inputs that alter the cap shape and hyperpolarization curve of the BFV functional form can have a profound effect on the information contained in the "action potential". For example, a hypothetical neurotransmitter that alters V_1 will also alter the latis rectum distance across the cap f_1. Further, direct modifications to the latis rectum distance in any of the two caps f_{11} and f_{12} can induce corresponding changes in times t_0, t_1 and t_2 and voltages V_0, V_1 and V_2. A similar statement can be made for changes in the latis rectum of cap f_2. For example, if a neurotransmitter induced a change of, say 1% in $4\beta_{11}$, this would imply that $\Delta(\frac{V_1 - V_0}{(t_0 - t_1)^2}) = .04\beta_{11}^0$ where β_{11}^0 denotes the original value of β_{11}^0. Thus, to first order

$$
.04\beta_{11}^0 = \left(\frac{\partial \beta_{11}}{\partial V_0} \right)^* \Delta V_0 + \left(\frac{\partial \beta_{11}}{\partial V_1} \right)^* \Delta V_1 + \left(\frac{\partial \beta_{11}}{\partial t_0} \right)^* \Delta t_0 + \left(\frac{\partial \beta_{11}}{\partial t_1} \right)^* \Delta t_1
\tag{9.8}
$$

where the superscript $*$ on the partials indicates they are evaluated at the base point (V_0, V_1, t_0, t_1). Taking partials we find

$$
\left(\frac{\partial \beta_{11}}{\partial V_0} \right) = 2 \frac{(t_0 - t_1)^2}{(V_1 - V_0)^2} = \frac{2}{V_1 - V_0} \beta_{11}^0
$$

$$
\left(\frac{\partial \beta_{11}}{\partial V_1} \right) = -2 \frac{(t_0 - t_1)^2}{(V_1 - V_0)^2} = -\frac{2}{V_1 - V_0} \beta_{11}^0
$$

$$
\left(\frac{\partial \beta 11}{\partial t_0} \right) = 2 \frac{t_0 - t_1}{V_1 - V_0} = \frac{2}{t_0 - t_1} \beta_{11}^0
$$

$$
\left(\frac{\partial \beta 11}{\partial t_1} \right) = -2 \frac{t_0 - t_1}{V_1 - V_0} = -\frac{2}{t_0 - t_1} \beta_{11}^0
$$

Thus, Eq. 9.8 becomes

$$.04\beta_{11}^0 = \frac{2\Delta V_0}{V_1 - V_0}\beta_{11}^0 - \frac{2\Delta V_1}{V_1 - V_0}\beta_{11}^0 + 2\Delta t_0 \frac{1}{t_0 - t_1}\beta_{11}^0 - 2\Delta t_1 \frac{1}{t_0 - t_1}\beta_{11}^0$$

This simplifies to

$$.02(V_1 - V_0)(t_0 - t_1) = (\Delta V_0 - \Delta V_1)(t_0 - t_1) - (\Delta t_0 - \Delta t_1)(V_1 - V_0)$$

Since we can do this analysis for any percentage r of β_{11}^0, we can infer that a neurotransmitter that modulates the action potential by perturbing the "width" or latis rectum of the cap of f_{11} can do so satisfying the equation

$$2r(V_1 - V_0)(t_1 - t_0) = (\Delta V_0 - \Delta V_1)(t_0 - t_1) - (\Delta t_0 - \Delta t_1)(V_1 - V_0)$$

Similar equations can be derived for the other two width parameters for caps f_{12} and f_3. These sorts of equations give us design principles for complex neurotransmitter modulations of a BFV.

9.4.3   Modulation via the BFV Ball and Stick Model

The BFV model we build consists of a dendritic system and a computational core which processed BFV input sequence to generate a BFV output.

9.4.3.1   The Modulation of the Action Potential

We know that

$$C_m \frac{dV_M}{dt} = I_E - g_K^{max}(\mathscr{M}_K)^4(V_m, t)(V_m - E_K)$$
$$- g_{Na}^{max}(\mathscr{M}_{NA})^3(V_m, t)(\mathscr{H}_{NA})(V_m, t)(V_m - E_{Na}) - g_L(V_m - E_L).$$

Since the BFV is structured so that the action potential has a maximum at t_1 of value V_1 and a minimum at t_3 of value V_3, we have $V_m'(t_1) = 0$ and $V_m'(t_3) = 0$. This gives

$$I_E(t_1) = g_K^{max}(\mathscr{M}_K)^4(V_1, t_1)(V_1 - E_K)$$
$$+ g_{Na}^{max}(\mathscr{M}_{NA})^3(V_1, t_1)(\mathscr{H}_{NA})(V_1, t_1)(V_1 - E_{Na}) + g_L(V_1 - E_L)$$
$$I_E(t_3) = g_K^{max}(\mathscr{M}_K)^4(V_3, t_3)(V_3 - E_K)$$
$$+ g_{Na}^{max}(\mathscr{M}_{NA})^3(V_3, t_3)(\mathscr{H}_{NA})(V_3, t_3)(V_3 - E_{Na}) + g_L(V_3 - E_L)$$

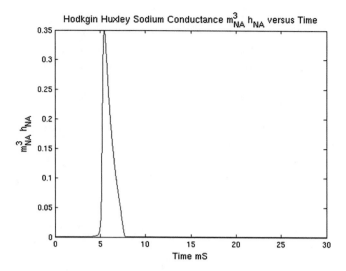

Fig. 9.8 The product $\mathscr{M}_{NA}{}^3 \, \mathscr{H}_{NA}$: sodium conductances during a pulse 1100 at 3.0

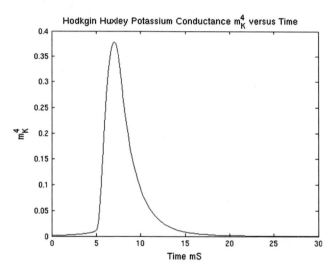

Fig. 9.9 The product $\mathscr{M}_K{}^4$: potassium conductances during a pulse 1100 at 3.0

From Figs. 9.8 and 9.9, we see that $m^3(V_1, t_1)h(V_1, t_1) \approx 0.35$ and $n^4(V_1, t_1) \approx 0.2$. Further, $m^3(V_3, t_3)h(V_3, t_3) \approx 0.01$ and $n^4(V_3, t_3) \approx 0.4$.

Thus,

$$I_E(t_1) = 0.20 g_K^{max}(V_1 - E_K) + 0.35 g_{Na}^{max}(V_1 - E_{Na}) + g_L(V_1 - E_L)$$
$$I_E(t_3) = 0.40 g_K^{max}(V_3 - E_K) + 0.01 g_{Na}^{max}(V_3 - E_{Na}) + g_L(V_3 - E_L)$$

Reorganizing,

$$I_E(t_1) = \left(0.20g_K^{max} + 0.35g_{Na}^{max} + g_L\right) V_1 - \left(0.20g_K^{max}E_K + 0.35g_{Na}^{max}E_{Na} + g_LE_L\right)$$
$$I_E(t_3) = \left(0.40g_K^{max} + 0.01g_{Na}^{max} + g_L\right) V_3 - \left(0.40g_K^{max}E_K + 0.01g_{Na}^{max}E_{Na} + g_LE_L\right)$$

Solving for the voltages, we find

$$V_1 = \frac{I_E(t_1) + 0.20g_K^{max}E_K + 0.35g_{Na}^{max}E_{Na} + g_LE_L}{0.20g_K^{max} + 0.35g_{Na}^{max} + g_L}$$

$$V_3 = \frac{I_E(t_3) + 0.40g_K^{max}E_K + 0.01g_{Na}^{max}E_{Na} + g_LE_L}{0.40g_K^{max} + 0.01g_{Na}^{max} + g_L}$$

Thus,

$$\frac{\partial V_1}{\partial g_K^{max}} = 0.20E_K \frac{1}{0.20g_K^{max} + 0.35g_{Na}^{max} + g_L}$$
$$+ \frac{I_E(t_1) + 0.20g_K^{max}E_K + 0.35g_{Na}^{max}E_{Na} + g_LE_L}{0.20g_K^{max} + 0.35g_{Na}^{max} + g_L}$$
$$\frac{-1.0}{0.20g_K^{max} + 0.35g_{Na}^{max} + g_L}.20$$

This simplifies to

$$\frac{\partial V_1}{\partial g_K^{max}} = \frac{0.20}{0.20g_K^{max} + 0.35g_{Na}^{max} + g_L}(E_K - V_1) \tag{9.9}$$

Similarly, we find

$$\frac{\partial V_1}{\partial g_{Na}^{max}} = \frac{0.35}{0.20g_K^{max} + 0.35g_{Na}^{max} + g_L}(E_{Na} - V_1) \tag{9.10}$$

$$\frac{\partial V_3}{\partial g_K^{max}} = \frac{0.40}{0.40g_K^{max} + 0.01g_{Na}^{max} + g_L}(E_K - V_3) \tag{9.11}$$

$$\frac{\partial V_3}{\partial g_{Na}^{max}} = \frac{0.40}{0.40g_K^{max} + 0.01g_{Na}^{max} + g_L}(E_{Na} - V_3) \tag{9.12}$$

We also know that as t goes to infinity, the action potential flattens and V_m' approaches 0. Also, the applied current, I_E is zero and so we must have

$$0 = -g_K^{max}(\mathcal{M}_K)^4(V_\infty, \infty)(V_\infty - E_K)$$
$$- g_{Na}^{max}(\mathcal{M}_{NA})^3(V_\infty, \infty)(\mathcal{H}_{NA})(V_\infty, \infty)(V_\infty - E_{Na}) - g_L(V_\infty - E_L)$$

Our hyperpolarization model is

$$Y(t) = V_3 + (V_4 - V_3) \tanh\left(g(t - t_3)\right)$$

We have V_∞ is V_4. Thus,

$$0 = -g_K^{max}(\mathcal{M}_K)^4(V_4, \infty)(V_4 - E_K)$$
$$- g_{Na}^{max}(\mathcal{M}_{NA})^3(V_4, \infty)(\mathcal{H}_{NA})(V_4, \infty)(V_4 - E_{Na}) - g_L(V_4 - E_L)$$

This gives, letting $(\mathcal{M}_{NA})^3(V_4, \infty)(\mathcal{H}_{NA})(V_4, \infty)$ and $(\mathcal{M}_K)^4(V_4, \infty)$ be denoted by $((\mathcal{M}_{NA})^3(\mathcal{H}_{NA}))^*$ and $((\mathcal{M}_K)^4)^*$ for simplicity of exposition,

$$\left(g_K^{max}((\mathcal{M}_K)^4)^* + g_{Na}^{max}((\mathcal{M}_{NA})^3(\mathcal{H}_{NA}))^* + g_L\right)V_4$$
$$= \left(g_K^{max}((\mathcal{M}_K)^4)^*E_K + g_{Na}^{max}((\mathcal{M}_{NA})^3(\mathcal{H}_{NA}))^*E_{Na} + g_L E_L\right)$$

Hence, letting $((\mathcal{M}_{NA})^3(\mathcal{H}_{NA}))^* \equiv (m^3 h)^*$ and $((\mathcal{M}_K)^4)^* \equiv (n^4)^*$, we have

$$V_4 = \frac{g_K^{max}n^4(V_4, \infty)E_K + g_{Na}^{max}m^3(V_4, \infty)h(V_4, \infty)E_{Na} + g_L E_L}{g_K^{max}n^4(V_4, \infty) + g_{Na}^{max}m^3(V_4, \infty)h(V_4, \infty) + g_L}$$

We see

$$\frac{\partial V_4}{\partial G_K^{max}} = \frac{n^4(V_4, \infty)}{g_K^{max}n^4(V_4, \infty) + g_{Na}^{max}m^3(V_4, \infty)h(V_4, \infty) + g_L}\left(E_K - V_4\right) \qquad (9.13)$$

$$\frac{\partial V_4}{\partial G_{Na}^{max}} = \frac{m^3(V_4, \infty)h(V_4, \infty)}{g_K^{max}n^4(V_4, \infty) + g_{Na}^{max}m^3(V_4, \infty)h(V_4, \infty) + g_L}\left(E_{Na} - V_4\right) \qquad (9.14)$$

We can also assume that the area under the action potential curve from the point (t_0, V_0) to (t_1, V_1) is proportional to the incoming current applied. If V_{In} is the axon-hillock voltage, the impulse current applied to the axon-hillock is $g_{In} V_{In}$ where g_{In} is the ball stick model conductance for the soma. Thus, the approximate area under the action potential curve must match this applied current. We have

$$\frac{1}{2}(t_1 - t_0)(V_1 - V_0) \approx g_{In} V_{In}$$

We conclude

$$(t_1 - t_0) = \frac{2g_{In}V_{In}}{V_1 - V_0}$$

Thus

$$\frac{\partial(t_1 - t_0)}{\partial g_K^{max}} = -\frac{t_1 - t_0}{V_1 - V_0} \frac{\partial V_1}{\partial g_K^{max}} \tag{9.15}$$

$$\frac{\partial(t_1 - t_0)}{\partial g_{Na}^{max}} = -\frac{t_1 - t_0}{V_1 - V_0} \frac{\partial V_1}{\partial g_{Na}^{max}} \tag{9.16}$$

Also, we know that during the hyperpolarization phase, the sodium current is off and the potassium current is slowly bringing the membrane potential back to the reference voltage. Now, our BFV model does not assume that the membrane potential returns to the reference level. Instead, by using

$$Y(t) = V_3 + (V_4 - V_3) \tanh\left(g(t - t_3)\right)$$

we assume the return is to voltage level V_4. At the midpoint, $Y = \frac{1}{2}(V_3 + V_4)$, we find

$$\frac{1}{2}(V_4 - V_3) = (V_4 - V_3) \tanh\left(g(t - t_3)\right)$$

Thus, letting $u = g(t - t_3)$,

$$\frac{1}{2} = \frac{e^{2u} - 1}{e^{2u} + 1}$$

and we find $u = \frac{\ln(3)}{2}$. Solving for t, we then have

$$t^* = t_3 + \frac{\ln(3)}{2g}$$

From t_3 on, the Hodgkin–Huxley dynamics are

$$C_m \frac{dV_M}{dt} = -g_K^{max}(\mathcal{M}_K)^4(V_m, t)(V_m - E_K) - g_L(V_m - E_L).$$

We want the values of the derivatives to match at t^*. This gives

$$g(V^* - V_3) \, \text{sech}^2\left(g(t^* - t_3)\right) = -\frac{g_K^{max}}{C_m}(\mathcal{M}_K)^4(V^*, t^*)(V^* - E_K) - g_L(V^* - E_L)$$

where $V^* = \frac{1}{2}(V_3 + V_4)$. Now $g(t^* - t_3) = \frac{\ln(3)}{2}$ and thus we find

$$\frac{g}{2}(V_4 - V_3)\,\text{sech}^2\left(g(t^* - t_3)\right) = -\frac{g_K^{max}}{C_m}(\mathscr{M}_K)^4(V^*, t)(V^* - E_K) - \frac{g_L}{C_m}(V^* - E_L)$$

$$\frac{g}{2}(V_4 - V_3)\,\frac{9}{64} = -\frac{g_K^{max}}{C_m}(\mathscr{M}_K)^4(V^*, t^*)(V^* - E_K) - g_L(V^* - E_L)$$

Next, consider the magnitude of $(\mathscr{M}_K)^4(V^*, t^*)$. We know at t^*, $(\mathscr{M}_K)^4$ is small from Fig. 9.9. Thus, we will replace it by the value 0.01. This gives

$$\frac{g}{2}(V_4 - V_3)\,\frac{9}{64} = -0.01\frac{g_K^{max}}{C_m}\left(\frac{1}{2}(V_4 + V_3) - E_K\right) - \frac{g_L}{C_m}\left(\frac{1}{2}(V_4 + V_3) - E_L\right)$$

Simplifying, we have

$$\frac{9g}{128}(V_4 - V_3) = \left(0.01\frac{g_K^{max}}{C_m}E_K + \frac{g_L}{C_m}E_L\right) - \frac{1}{2}\left(0.01\frac{g_K^{max}}{C_m} + \frac{g_L}{C_m}\right)(V_4 + V_3)$$

$$\frac{9g}{64} = \left(0.01\frac{g_K^{max}}{C_m}E_K + \frac{g_L}{C_m}E_L\right)\frac{1}{V_4 - V_3} - \left(0.01\frac{g_K^{max}}{C_m} + \frac{g_L}{C_m}\right)\frac{V_4 + V_3}{V_4 - V_3}$$

We can see clearly from the above equation, that the dependence of g on g_K^{max} and g_{Na}^{max} is quite complicated. However, we can estimate this dependence as follows. We know that $V_3 + V_4$ is about the reference voltage, $-65.9\,\text{mV}$. If we approximate V_3 by the potassium battery voltage, $E_k = -72.7\,\text{mV}$ and V_4 by the reference voltage, we find $\frac{V_3 + V_4}{V_4 - V_3} \approx \frac{-138.6}{6.8} = -20.38$ and $\frac{1}{V_4 - V_3} \approx \frac{1}{6.8} = 0.147$. Hence,

$$\frac{9C_m g}{64} = 0.147\left(0.01g_K^{max}E_K + g_L E_L\right) + 20.38\left(0.01g_K^{max} + g_L\right)$$

$$= \left(0.0147E_K + 2.038E_L\right)g_K^{max} + g_L\left(0.0147E_L + 20.38\right)$$

Thus, we find

$$\frac{\partial g}{\partial g_K^{max}} = \frac{64}{9C_m}\left(0.0147E_K + 2.038E_L\right) \tag{9.17}$$

This gives $\frac{\partial g}{\partial g_K^{max}} \approx -710.1$ Eq. 9.17 shows what our intuition tells us: *if g_K^{max} increases, the potassium current is stronger and the hyperpolarization phase is shortened. On the other hand, if g_K^{max} decreases, the potassium current is weaker and the hyperpolarization phase is lengthened.*

9.4.3.2 Multiple Inputs

Consider a typical input $V(t)$ which is determined by a BFV vector. Without loss of generality, we will focus on excitatory inputs in our discussions. The input consists

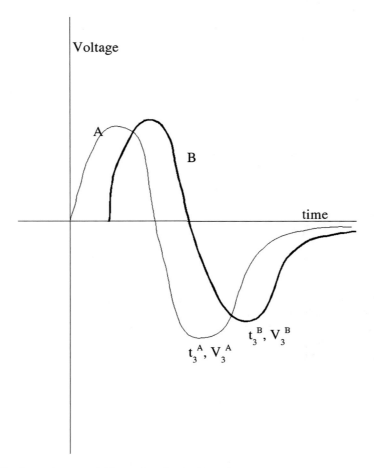

Fig. 9.10 Two action potential inputs into the dendrite subsystem

of a three distinct portions. First, a parabolic cap above the equilibrium potential determined by the values (t_0, V_0), (t_1, V_1), (t_2, V_2). Next, the input contains half of another parabolic cap dropping below the equilibrium potential determined by the values (t_2, V_2) and (t_3, V_3). Finally, there is the hyperpolarization phase having functional form $H(t) = V_3 + (V_4 - V_3)\tanh(g(t - t_3))$. Now assume two inputs arrive at the same electronic distance L. We label this inputs as A and B as is shown in Fig. 9.10. For convenience of exposition, we also assume $t_3^A < t_3^B$, as otherwise, we just reverse the roles of the variables in our arguments. In this figure, we note only the minimum points on the A and B curves. We merge these inputs into a new input V^N prior to the hyperpolarization phase as follows:

$$t_0^N = \frac{t_0^A + t_0^B}{2}$$

$$V_0^N = \frac{V_0^A + V_0^B}{2}$$

$$t_1^N = \frac{t_1^A + t_1^B}{2}$$

$$V_1^N = \frac{V_1^A + V_1^B}{2}$$

$$t_2^N = \frac{t_2^A + t_2^B}{2}$$

$$V_2^N = \frac{V_2^A + V_2^B}{2}$$

This constructs the two parabolic caps of the new resultant input by averaging the caps of V^A and V^B. The construction of the new hyperpolarization phase is more complicated. The shape of this portion of an action potential has a profound effect on neural modulation, so it is very important to merge the two inputs in a reasonable way. The hyperpolarization phases of V^A and V^B are given by

$$H^A(t) = V_3^A + (V_4^A - V_3^A) \tanh\left(g^A(t - t_3^A)\right)$$

$$H^B(t) = V_3^B + (V_4^B - V_3^B) \tanh\left(g^B(t - t_3^B)\right)$$

We will choose the 4 parameters V_3, V_4, g, t_3 so as to minimize

$$E = \int_{t_3^A}^{\infty} \left(H(t) - H^A(t)\right)^2 + \left(H(t) - H^B(t)\right)^2 dt$$

For optimality, we find the parameters where $\frac{\partial E}{\partial V_3}$, $\frac{\partial E}{\partial V_4}$, $\frac{\partial E}{\partial g}$ and $\frac{\partial E}{\partial t_3}$ are 0. Now,

$$\frac{\partial E}{\partial V_3} = \int_{t_3^A}^{\infty} 2\left\{\left(H(t) - H^A(t)\right) + \left(H(t) - H^B(t)\right)\right\} \frac{\partial H}{\partial V_3} dt$$

Further,

$$\frac{\partial H}{\partial V_3} = 1 - \tanh\left(g(t - t_3)\right)$$

so we obtain

$$0 = \int_{t_3^A}^{\infty} 2\left\{\left(H(t) - H^A(t)\right) + \left(H(t) - H^B(t)\right)\right\} \left(1 - \tanh\left(g(t - t_3)\right)\right) dt \quad (9.18)$$

We also find

$$\frac{\partial E}{\partial V_4} = \int_{t_3^A}^{\infty} 2\left\{\left(H(t) - H^A(t)\right) + \left(H(t) - H^B(t)\right)\right\}\left(\tanh\left(g(t - t_3)\right)\right) dt$$

as

$$\frac{\partial H}{\partial V_4} = \tanh\left(g(t - t_3)\right)$$

The optimality condition then gives

$$0 = \int_{t_3^A}^{\infty} 2\left\{\left(H(t) - H^A(t)\right) + \left(H(t) - H^B(t)\right)\right\}\tanh\left(g(t - t_3)\right) dt \quad (9.19)$$

Combining Eqs. 9.18 and 9.19, we find

$$0 = \int_{t_3^A}^{\infty} \left\{\left(H(t) - H^A(t)\right) + \left(H(t) - H^B(t)\right)\right\} dt$$
$$- \int_{t_3^A}^{\infty} \left\{\left(H(t) - H^A(t)\right) + \left(H(t) - H^B(t)\right)\right\}\tanh\left(g(t - t_3)\right) dt$$
$$0 = \int_{t_3^A}^{\infty} \left\{\left(H(t) - H^A(t)\right) + \left(H(t) - H^B(t)\right)\right\}\tanh\left(g(t - t_3)\right) dt.$$

It follows after simplification, that

$$0 = \int_{t_3^A}^{\infty} \left\{\left(H(t) - H^A(t)\right) + \left(H(t) - H^B(t)\right)\right\} dt \quad (9.20)$$

The remaining optimality conditions give

$$\frac{\partial E}{\partial g} = \int_{t_3^A}^{\infty} 2\left\{\left(H(t) - H^A(t)\right) + \left(H(t) - H^B(t)\right)\right\}\frac{\partial H}{\partial g} dt = 0$$
$$\frac{\partial E}{\partial t_3} = \int_{t_3^A}^{\infty} 2\left\{\left(H(t) - H^A(t)\right) + \left(H(t) - H^B(t)\right)\right\}\frac{\partial H}{\partial t_3} dt = 0$$

We calculate

$$\frac{\partial H}{\partial g} = (V_4 - V_3)(t - t_3)\, sech^2\left(g(t - t_3)\right)$$
$$\frac{\partial H}{\partial t_3} = -(V_4 - V_3)\, g\, sech^2\left(g(t - t_3)\right)$$

Thus, we find

$$0 = \int_{t_3^A}^{\infty} \left\{ \left(H(t) - H^A(t) \right) + \left(H(t) - H^B(t) \right) \right\} (V_4 - V_3)(t - t_3) \, sech^2 \left(g(t - t_3) \right) dt$$

$$0 = \int_{t_3^A}^{\infty} \left\{ \left(H(t) - H^A(t) \right) + \left(H(t) - H^B(t) \right) \right\} (V_4 - V_3) \, g \, sech^2 \left(g(t - t_3) \right) dt.$$

This implies

$$0 = \int_{t_3^A}^{\infty} \left\{ \left(H(t) - H^A(t) \right) + \left(H(t) - H^B(t) \right) \right\} t \, sech^2 \left(g(t - t_3) \right) dt$$

$$- t_3 \int_{t_3^A}^{\infty} \left\{ \left(H(t) - H^A(t) \right) + \left(H(t) - H^B(t) \right) \right\} sech^2 \left(g(t - t_3) \right) dt$$

$$0 = \int_{t_3^A}^{\infty} \left\{ \left(H(t) - H^A(t) \right) + \left(H(t) - H^B(t) \right) \right\} sech^2 \left(g(t - t_3) \right) dt.$$

This clearly can be simplified to

$$0 = \int_{t_3^A}^{\infty} \left\{ \left(H(t) - H^A(t) \right) + \left(H(t) - H^B(t) \right) \right\} t \, sech^2 \left(g(t - t_3) \right) dt \quad (9.21)$$

We can satisfy Eqs. 9.20 and 9.21 by making

$$(H(t) - H^A(t)) + (H(t) - H^B(t)) = 0. \quad (9.22)$$

Equation 9.22 can be rewritten as

$$0 = \left(V_3 - \frac{V_3^A + V_3^B}{2} \right) + (V_4 - V_3) \tanh \left(g(t - t_3) \right) \quad (9.23)$$

$$- \frac{V_4^B - V_3^B}{2} \tanh \left(g^B(t - t_3^B) \right) - \frac{V_4^A - V_3^A}{2} \tanh \left(g^A(t - t_3^A) \right) \quad (9.24)$$

This equation is true as $t \to \infty$. Thus, we obtain the identity

$$0 = \left(V_3 - \frac{V_3^A + V_3^B}{2} \right) + \left(V_4 - V_3 \right) - \frac{V_4^B - V_3^B}{2} - \frac{V_4^A - V_3^A}{2}$$

Upon simplification, we find

$$0 = V_3 - \frac{V_3^A + V_3^B}{2}$$

$$0 = V_4 - \frac{V_4^A + V_4^B}{2}$$

This leads to our choices for V_3 and V_4.

$$V_3 = \frac{V_3^A + V_3^B}{2} \tag{9.25}$$

$$V_4 = \frac{V_4^A + V_4^B}{2} \tag{9.26}$$

Equation 9.24 is also true at $t = t_3^A$ and $t = t_3^B$. This gives

$$0 = \left(V_3 - \frac{V_3^A + V_3^B}{2}\right) + (V_4 - V_3)\, \tanh\left(g(t_3^A - t_3)\right)$$
$$- \frac{V_4^B - V_3^B}{2}\, \tanh\left(g^B(t_3^A - t_3^B)\right) \tag{9.27}$$

$$0 = \left(V_3 - \frac{V_3^A + V_3^B}{2}\right) + (V_4 - V_3)\, \tanh\left(g(t_3^B - t_3)\right)$$
$$- \frac{V_4^A - V_3^A}{2}\, \tanh\left(g^A(t_3^B - t_3^A)\right) \tag{9.28}$$

For convenience, define $w_{34}^A = \frac{V_4^A - V_3^A}{2}$ and $w_{34}^B = \frac{V_4^B - V_3^B}{2}$. Then, using Eqs. 9.25, 9.28 and Eq. 9.28 become

$$0 = (V_4 - V_3)\tanh\left(g(t_3^A - t_3)\right) - w_{34}^B \tanh\left(g^B(t_3^A - t_3^B)\right)$$

$$0 = (V_4 - V_3)\tanh\left(g(t_3^B - t_3)\right) - w_{34}^A \tanh\left(g^A(t_3^B - t_3^A)\right)$$

This is then rewritten as

$$\tanh(g(t_3^A - t_3)) = \frac{w_{34}^B \, \tanh\left(g^B(t_3^A - t_3^B)\right)}{(V_4 - V_3)}$$

$$\tanh(g(t_3^B - t_3)) = \frac{w_{34}^A \, \tanh\left(g^A(t_3^B - t_3^A)\right)}{(V_4 - V_3)}$$

Defining

$$z_A = \frac{w_{34}^B \, \tanh\left(g^B(t_3^A - t_3^B)\right)}{V_4 - V_3}$$

$$z_B = \frac{w_{34}^A \, \tanh\left(g^A(t_3^B - t_3^A)\right)}{V_4 - V_3}$$

we find that the optimality conditions have led to the two nonlinear equations for g and t_3 given by

$$\tanh\left(g(t_3^A - t_3)\right) = z_A \tag{9.29}$$

$$\tanh\left(g(t_3^B - t_3)\right) = z_B \tag{9.30}$$

Note that using Eqs. 9.29 and 9.30, we have

$$z_A = \frac{w_{34}^B \tanh\left(g^B(t_3^A - t_3^B)\right)}{V_4 - V_3} = -\frac{w_{34}^B \tanh\left(g^B(t_3^B - t_3^A)\right)}{w_{34}^A + w_{34}^B}$$

$$z_B = \frac{w_{34}^A \tanh\left(g^A(t_3^B - t_3^A)\right)}{V_4 - V_3} = \frac{w_{34}^A \tanh\left(g^A(t_3^B - t_3^A)\right)}{w_{34}^A + w_{34}^B}$$

Hence,

$$z_A > -\frac{w_{34}^B}{w_{34}^A + w_{34}^B} > -1$$

$$z_B < \frac{w_{34}^A}{w_{34}^A + w_{34}^B} < 1$$

so that $z_A < 0 < z_B$. It seems reasonable that the optimal value of t_3 should lie between t_3^A and t_3^B. Note Eqs. 9.29 and 9.30 preclude the solutions $t_3 = t_3^A$ or $t_3 = t_3^B$. To solve the nonlinear system for g and t_3, we will approximate tanh by its first order Taylor Series expansion. This seems reasonable as we don't expect $g(t_3^A - t_3)$ and $g(t_3^A - t_3)$ to be far from 0. This gives the approximate system

$$g(t_3^A - t_3) \approx z_A \tag{9.31}$$
$$g(t_3^B - t_3) \approx z_B \tag{9.32}$$

From Eq. 9.32, we find

$$g = \frac{z_B}{t_3^B - t_3}$$

Substituting this into Eq. 9.32, we obtain

$$\frac{z_B}{t_3^B - t_3}(t_3^A - t_3) = z_A$$

This can be simplified as follows:

$$\frac{t_3^A - t_3}{t_3^B - t_3} = \frac{z_A}{z_B}$$

$$(t_3^A - t_3)\, z_B = (t_3^B - t_3)\, z_A$$

$$t_3^A z_B - t_3^B z_A = t_3\, (z_B - z_A)$$

Thus, we find the optimal value of t_3 is approximately

$$t_3 = \frac{t_3^A z_B - t_3^B z_A}{z_B - z_A} \tag{9.33}$$

Using the approximate value of t_3, we find the optimal value of g can be approximated as follows:

$$g = \frac{z_B}{t_3^B - \frac{t_3^A z_B - t_3^B z_A}{z_B - z_A}}$$

$$= \frac{z_B(z_B - z_A)}{t_3^B(z_B - z_A) - (t_3^A z_B - t_3^B z_A)}$$

$$= \frac{z_B(z_B - z_A)}{t_3^B z_B - t_3^A z_B}$$

$$= \frac{z_B - z_A}{t_3^B - t_3^A}$$

Hence, we find the approximate optimal value of g is

$$g = \frac{z_B - z_A}{t_3^B - t_3^A} \tag{9.34}$$

It is easy to check that this value of t_3 lies in (t_3^A, t_3^B) as we suspected it should and that g is positive. We summarize our results. Given two input BFVs, the sigmoid portions of the incoming BFVs combine into the new sigmoid given by

$$H(t) = V_3 + (V_4 - V_3)\, \tanh\!\left(g(t - t_3) \right)$$

$$H(t) = \frac{V_3^A + V_3^B}{2} + \left(\frac{V_4^A - V_3^A}{2} + \frac{V_4^B - V_3^A}{2} \right) \tanh\!\left(\frac{z_B - z_A}{t_3^B - t_3^A}\left(t - \frac{t_3^A z_B - t_3^B z_A}{z_B - z_A} \right) \right)$$

Given an input sequence of BFV's into a port on the dendrite of an accepting neuron

$$\{V_n, V_{n-1}, \ldots, V_1\}$$

the procedure discussed above computes the combined response that enters that port at a particular time. The inputs into the dendritic system are combined pairwise; V_2 and V_1 combine into a V_{new} which then combines with V_3 and so on. We can do this at each electrotonic location.

9.5 The Full Abstract Neuron Model

Pre-neurons can supply input to the dendrite cable at electronic positions $w = 0$ to $w = 4$. These inputs generate an ESP or ISP via many possible mechanisms or they alter the structure of the dendrite cable itself by the transcription of proteins. The output of a pre-neuron is a BFV which must then be associated with an abstract trigger as we have discussed in earlier chapters. The strength of a BFV output will be estimated as follows: The area under the first parabolic cap of the BFV can be approximated by the area of the triangle, A, formed by the vertices (t_0, V_0), (t_1, V_1), (t_2, V_2). This area is shown in Fig. 9.11. The area is given by

$$A = \frac{1}{2}(V_2 - V_0)(t_2 - t_0)$$

The pre-neuron signals that come into the dendritic cable of the post-neuron are the initial conditions that determined the voltage at the axon hillock of the post-neuron. We are modeling the dendritic arbor and the soma as finite length cables. At position λ on the cable, the voltage is given by

$$\hat{v}_m(\lambda, \tau) = A_0 e^{-\tau} + \sum_{n=1}^{\infty} A_n \cos(\alpha_n(4 - \lambda))e^{-(1+\alpha_n^2)\tau}$$

as discussed in Peterson (2015). This voltage arrives at the far end, $z = 7$ in Fig. 9.12, of the soma cable and the voltage we have at $z = 0$ is then the axon hillock voltage. This figure shows a general computational architecture for an artificial neuron consisting of

Fig. 9.11 The EPS triangle approximation

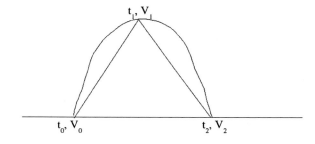

Fig. 9.12 Cellular agent computations

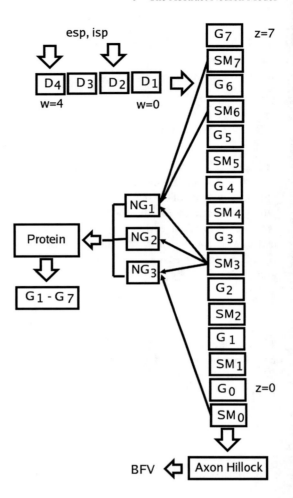

- a dendrite system of electronic length 4. Four dendrite ports are shown labeled as D_i. Each accepts ISP and ESP inputs which are generated and modulated using the mechanisms we have discussed.
- a soma of length 7 which contains 7 pairs simple and second messenger ports (G_i and SM_i, respectively). The simple ports G_i are where sodium and potassium maximum conductance can be modified by direct means. The second messenger ports SM_i accept trigger T_0 and generate proteins which alter the functionality of the cell by accessing the genome.
- Three nuclear membrane gate blocks, NG_i where second messenger proteins T_1 can dock to initiate the transcription of a protein used in altering cell function.

- The protein block is shown differentiating into the gate proteins that are sent to the sites G_i. Although not shown, there could also be proteins that alter Ca^{++} injection currents

The axon hillock voltage is then input into the Hodgkin–Huxley equations to generate an action potential. The voltage at $z = 0$ would have the form

$$\hat{v}_m(0, \tau) = B_0\, e^{-\tau} + \sum_{n=1}^{\infty} B_n\, \cos(7\beta_n) e^{-(1+\beta_n^2)\tau}$$

Our challenge then is two fold:

- Find a quick and efficient way to approximate the axon-hillock voltage
- Given the axon-hillock voltage, find a quick and efficient way to determine the associated BFV.

References

M. Adams, G. Swanson, TINS neurotoxins supplement. Trends Neurosci. Supplement:S1–S36 (1996)

J. Gerhart, M. Kirschner, *Cells, Embryos and Evolution: Towards a Cellular and Developmental Understanding of Phenotypic Variation and Evolutionary Adaptability* (Blackwell Science, Malden, 1997)

A. Harvey, Presynaptic effects of toxins. Int. Rev. Neurobiol. **32**, 201–239 (1990)

P. Kaul, P. Daftari, Marine pharmacology: bioactive molecules from the sea. Annu. Rev. Pharmacol. Toxicol. **26**, 117–142 (1986)

J. Peterson, *Calculus for Cognitive Scientists: Partial Differential Equation Models*. Springer Series on Cognitive Science and Technology (Springer Science+Business Media Singapore Pte Ltd., Singapore, 2015 in press)

J. Peterson, T. Khan, Abstract action potential models for toxin recognition. J. Theor. Med. **6**(4), 199–234 (2006)

G. Schiavo, M. Matteoli, C. Montecucco, Neurotoxins affecting neuroexocytosis. Physiol. Rev. **80**(2), 717–766 (2000)

S. Stahl, *Essential Psychopharmacology: Neuroscientific Basis and Practical Applications*, 2nd edn. (Cambridge University Press, Cambridge, 2000)

G. Strichartz, T. Rando, G. Wang, An integrated view of the molecular toxicology of sodium channel gating in excitable cells. Annu. Rev. Neurosci. **10**, 237–267 (1987)

C. Wu, T. Narahashi, Mechanism of action of novel marine neurotoxins on ion channels. Annu. Rev. Pharmacol. Toxicol. **28**, 141–161 (1988)

Part IV
Models of Emotion and Cognition

Chapter 10
Emotional Models

We begin by looking at some key assumptions made in typical high level or top-down emotional modeling. A representative of this approach is found in the work of Sloman (1997a, b, 1998, 1999a, b) which can be summarized by the following two statements. First, there are information processing architectures existing in the human brain that mediate internal and external behavior. Secondly, there is a correspondence between the high level functionality obtained from artificial and explicitly designed architectures and naturally evolved architectures despite low level implementation differences. The first assumption is part of many of today's cognitive science programs while the second assumption is more problematic. One of the reasons we are developing a software model of cognition is that we feel it will be helpful in both elucidating our understanding of a software cognitive process and in validating this second assumption.

It is clear this desired correspondence will depend on the right kind of abstraction of messy hardware detail. For example, we could look at an abstract wiring diagram for a portion of the cerebral cortex (highly stylized, of course) as discussed in chapter neural-structure and see any attempt to implement high level outlines of software architectures that might subserve cognition or emotional modeling will be inherently highly interconnected. It is also clear that the concept of hierarchical organization is not very rigid in wetware as we have discussed at length in Chap. 14. Thus implementing these architectures in software will require the ability to operate in asynchronous ways using modules whose computations are not parallelizable across multiple nodes.

Instead, these computational objects communicate outputs and assimilate inputs in some sort of a collective fashion. This has led us to propose the use of asynchronous software tools as a means of coordinating many different such objects operating in an asynchronous fashion on a heterogeneous computer network. The broad plan of an approach to this kind of emotional modeling is given in Fig. 10.1: We can then design characters to move autonomously in a three dimensional virtual world and generate a real-time model of its emotions from both interactions with both its

© Springer Science+Business Media Singapore 2016
J.K. Peterson, *BioInformation Processing*, Cognitive Science and Technology,
DOI 10.1007/978-981-287-871-7_10

Fig. 10.1 The emotionally enabled avatar

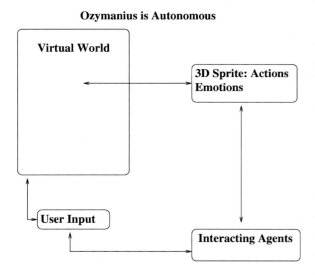

environment and other characters. Such *avatars* will be controlled by an interacting collection of software agents. Our discussions in previous chapters have been giving us the background and tools to understand how to begin the process of building such characters, although there is much yet to do. To understand the simulation environment appropriate for emotional modeling, let's back up and look at the study of emotions in general. We begin with Sloman's approach which is not based on the underlying neurobiology at all. Sloman assumes the study of emotions can be divided into three broad categories. These areas can be

- Semantics Based: we analyze the use of language to uncover implicit assumptions underlying emotion
- Phenomena Based: we assume emotions are a well-specified category; we try to correlate measurable things (physiological changes, neuronal firing rates etc.) with emotional states.
- Design Based: we take an engineering stance and try to build a system exhibiting phenomena that are understood as emotional attributes. We also try to find possible mechanisms to generate the emotional outputs we seek.

We do not want to approach the study of emotions using any of these paradigms; instead, we want emotional attributes to be a consequence of the full brain model with communication modulated by neurotransmitters and hormones. For example, in Rolls (2005), there is an excellent treatment of the biological approach. In fact, emotions probably emerge from interactions between neural modules as is described in Scherer (2009). Another good overview is in by Ono et al. (2003) and that of Damasio and Carvalho (2013). The goal of Chaps. 18 and 19 which are coming up is to find

Fig. 10.2 The three tower model

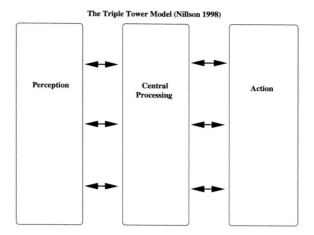

Fig. 10.3 The three layer model

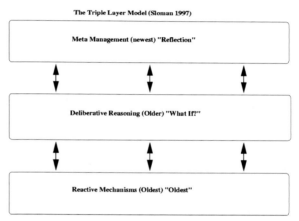

tools to allow us to efficiently code these interactions. We show some simple full brain architectures in Chap. 20 developed with these tools to give a taste of what we can do with these tools and we show how all of the discussion in this text culminates in an approach to a cognitive dysfunction model in Chap. 21. However, for the moment, we think it is important to look at the broad outlines of these other approaches. Consider the following software architecture inspired by Sloman's computational emotional models. A simplistic towered approach divides cognition into three separate areas as shown in Fig. 10.2: This model separates cognition into the three areas: perception, central processing mechanisms and action methods. We don't try to understand how these blocks can be implemented at this point. This diagram comes from the work of Nilsson (2001). Sloman then introduced the simple layered model (Fig. 10.3) which is organized around reasoning concepts. A hybrid model

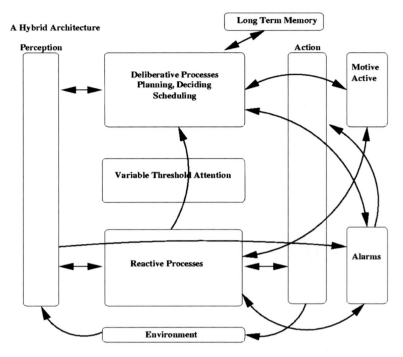

Fig. 10.4   A hybrid combined emotional model architectures

can then be introduced that essentially uses both the towers and layers together to give us more interacting functional subgroups. In Fig. 10.4, the indicated feedback and feedforward paths help us to visualize how the many functional subroutines will be combined. Finally, a full meta-management layer is added as shown in Fig. 10.5.

The meta model gives a simplistic summary of adult human information processing in a modular design. Many researchers believe that such modular designs are essential for defeating the combinatorial explosion that arises in the search for solutions to complex problems. Hence, in the Sloman hybrid model, human information processing uses three distinctly different but **simultaneously** active architectural layers: Reactive, Deliberative and Meta management; plus support modules for Motive Generation, Global Alarm Mechanisms and Long Term Associative Memory Storage. All of these layers can be implemented using asynchronously interacting agents. Note how far we are from the underlying neurobiology and the approaches which use networks of computational nodes such as neurons.

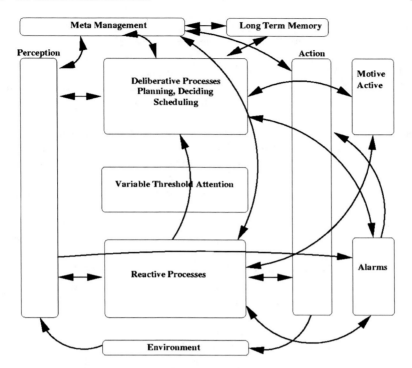

Fig. 10.5 Adding meta management to the hybrid model

10.1 The Sloman Emotional Model

Sloman then classifies emotions in the following way. The Reactive Layer and the Global Alarm process provide outputs that can be classified as **Primary Emotions**. These would be reactions such as *being startled*, *being frozen with terror* and *being sexually aroused*. The Deliberative Layer provides **Secondary Emotions**. This include things such as *apprehension* and *relief*. Sloman argues that such responses, require a "what-if" reasoning ability. Finally, the Meta Management Layer allows for **Tertiary Emotions**. These include complicated and complex responses such as *control of thought and emotion, loss of emotion, infatuations and humiliations* and *thrilled anticipation*. These are crucial to the absorption of culture, mathematics and so forth. Now the crucial point is that all layers and the alarm system operate concurrently (i.e. asynchronously) and none is in total control. Sloman offers a compelling example: we can think of **longing** as a **tertiary emotion**. Longing for your mother requires at least that you know you have a mother, you know she is not present, you understand the possibility of being with her and you find her absence unpleasant. All of these things require that you possess and manipulate information. However, these conditions are not *sufficient* as these conditions could lead you to *regret* her absence but not *long* for her. Note we can consider **regret** as an *attitude* and **longing** as an emotion. So we clearly need more: one possibility is that *longing* has the additional quality that you can't easily put thoughts of her out of your mind. This indicates a **partial loss of control of attention** which is an extraordinarily interesting way to view this phenomenon of longing. This implies that to be able to lose control of this attention means that we can sometimes control it also. There must be some information processing mechanism which can control which information is processed but which is **not always in control**. Partly losing control of thought processes is a perturbation of our normal state.

We can infer that one way to think about **tertiary emotions** is that they are **perturbant states** in our system. Other examples are extreme grief (you can't think of anything else) and extreme infatuation (your thoughts are always drawn to that person). Our task then is to make sure that our interacting society of computational agents can exhibit these sorts of behaviors.

We could then build our emotional models using interacting collections of computational agents on a heterogeneous computer network. Some nodes of the cluster compute emotional attributes and some nodes will drive the movement and expressions of a three dimensional character or avatar that moves autonomously in a virtual world. Silas the dog has already been implemented in this fashion by Blumberg at the MIT Media School as early as 1997 (Blumberg 1997).

However, this model is not neurobiologically based and as such is not suitable for our needs. We want to add biological plausibility to our models.

10.2   PsychoPhysiological Data

In a sequence of seminal papers, (Lang et al. 1998; Codispotti et al. 2001; Bradley and Lang 2000) and (Cuthbert et al. 1996), it has been shown that people respond to emotionally tagged or affective images in a semi-quantitative manner. Human volunteers were shown various images and their physiological responses were recorded in two ways. One was a skin galvanic response and the other a fMRI parameter. Typical results are plotted in Fig. 10.6. In this database of images, extreme images always generated a large plus or minus response while neutral images such as those of an infant generated null or near origin results.

If we followed the original intent and spirit of the Affective Image research, we would like to develop data for each of nine primary emotional states as indicated in Fig. 10.6. These would be emotional states that correspond to the following nine locations on the two dimensional grid:

Fig. 10.6 Human response
to emotionally charged
picture data

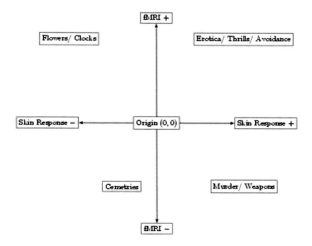

2D Coordinates	Physiological responses	Image type
(High, High)	High galvanic and high fMRI response	Thrills
(High, Null)	High galvanic and flat fMRI response	
(High, Low)	High galvanic and low fMRI response	Murder
(Null, High)	Flat galvanic and high fMRI response	
(Null, Null)	Flat galvanic and flat fMRI response	
(Null, Low)	Flat galvanic and low fMRI response	
(Low, High)	Low galvanic and high fMRI response	Flowers
(Low, Null)	Low galvanic and flat fMRI response	
(Low, Low)	Low galvanic and low fMRI responses	Cemeteries

Clearly, the emotional tags associated with the images in the affective image database are not cleanly separated into primary emotions such as anger, sadness and happiness. However, we can infer that the center (Null, Null) state is associated with images that have no emotional tag. Also, the images do cleanly map to distinct 2D locations on the grid when the emotional contents of the images differ. Hence, we will assume that if a database of images separated into states of anger, sadness, happiness and neutrality were presented to human subjects, we would see a similar separation of response. Our hypothetical response would be captured in the emotion triangle seen in Fig. 10.7. Indeed, in both the musical and painting compositional domain, we will therefore design special matrices called Würfelspiel matrices for the four positions marked in Figs. 10.6 and 10.7. Such matrices were used in the 18th century so that fragments of music could be rapidly prototyped by using a matrix of possibilities. We will develop such matrices for music and painting in Chaps. 11 and 12, respectively. Note, we can also use other types of emotional labels and use a similar triangle to design other sorts of symbolic mapping data. We will be using the emotionally labeled music and painting data to train cortical models in later chapters.

Fig. 10.7 Emotionally
charged compositional data
design

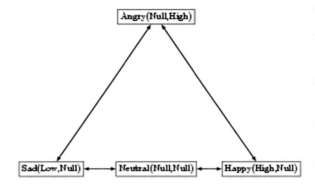

References

B. Blumberg, *Old Tricks, New Dogs: Ethology and Interactive Creatures*, PhD thesis, MIT, The Program in Media Arts and Sciences, School of Architecture and Planning (1997)

M. Bradley, P. Lang, Affective reactions to acoustic stimuli. Psychophysiology **37**, 204–215 (2000)

T. Ono, G. Matsumoto, R. Llinás, A. Berthoz, R. Norgren, H. Nishijo, R. Tamura (eds.), *Cognition and Emotion in the Brain* (Elsevier, Amsterdam, 2003)

M. Codispotti, M. Bradley, P. Lang, Affective reactions to briefly presented pictures. Psychophysiology **38**, 474–478 (2001)

B. Cuthbert, M. Bradley, P. Lang, Probing picture perception: activation and emotion. Psychophysiology **33**, 103–111 (1996)

A. Damasio, G. Carvalho, The nature of feelings: evolutionary and neurobiological origins. Nat. Rev.: Neurosci. **14**, 143–152 (2013)

P. Lang, M. Bradley, J. Fitzimmons, B. Cuthbert, J. Scott, B. Moulder, V. Nangia, Emotional arousal and activation of the visual cortex: an fMRI analysis. Psychophysiology **35**, 199–210 (1998)

N. Nilsson, *Teleo-Reactive Programs and the Triple-Tower Architecture. Technical Report* (Robotics Laboratory, Department of Computer Science, Stanford University, Stanford, 2001)

E. Rolls, *Emotion Explained* (Oxford University Press, Oxford, 2005)

K. Scherer, Emotions are emergent processes: they require a dynamic computational architecture. Philos. Trans. R. Soc. B **364**, 3459–3474 (2009)

A. Sloman, Designing human-like minds (1997a), http://www.cs.bham.ac.uk/research/cogaf. Presented at the European Conference on Artificial Life

A. Sloman, What sort of control system is able to have a personality, in *Creating Personalities for Synthetic Actors: Towards Autonomous Personality Agents*, ed. by R. Trappl, P. Petta (Springer, Heidelberg, 1997b), pp. 166–208

A. Sloman, Architectures and tools for human-like agents (1998). Presented at the European Conference on Cognitive Modeling

A. Sloman, Architectural Requirements for Human-like Agents Both Natural and Artificial (What Sorts of Machines Can Love?), in *Human Cognition and Social Agent Technology*, vol. 19, Advances in Consciousness Research Series, ed. by K. Dautenhahn (John Benjamin Publishing, Amsterdam, 1999a)

A. Sloman, What Sort of Architecture Is Required for a Human-like Agent? in *Foundations of Rational Analysis*, ed. by M. Wooldridge, A. Rao (Kluwer, Boston, 1999b)

Chapter 11
Generation of Music Data: J. Peterson and L. Dzuris

We design musical data using a grammatical approach to musical composition. This is, of course, simplistic, but we want a mechanism for quickly and repeatedly generating large volumes of data that can be used to train the auditory cortex of the cognitive model.

11.1 A Musical Grammar

In the literature, there are many attempts to model musical compositional designs. Several theorists discuss music in terms of large chunks or sections and overall function. One example is found in Caplin (1988), *Classical Form: A Theory of Formal Functions For The Instrumental Music of Haydn, Mozart and Beethoven*, there is a focus on larger musical structures such as those found in the symphony movements of Haydn, Mozart, and Beethoven. Most of the discussion is not relevant to the design of musical fragments of the type we desire, though there is mention of the idea of a musical sentence. He states that the sentence is usually eight measures in length, but that the fundamental melodic material is presented in the opening two measures. Caplin admits that within the *basic idea* contained in these first measures, there may be several distinct *motives*. Since motives are in and of themselves identifiable morsels of music, we can logically compose shorter sentences than those Caplin is analyzing.

Approaching the matter of function from the opposite direction, small units building up to larger ones, we have Davie (1953), *Musical Structure and Design*, who compares musical sounds to words in terms of clauses, sentences, and paragraphs leading to larger structures of form. He refers to musical cadences in terms of various punctuation marks. Depending on the type used, impressions of rest, incompleteness and surprise might be conveyed. He suggests the use of musical cadences which are perfect and plagal for completeness (by which Caplin means V-I with soprano on scale degree 1 or IV-I); imperfect for incompleteness (V-I with soprano on scale degree 3 or 5) and interrupted for surprise (deceptive cadences are V or V7-VI).

© Springer Science+Business Media Singapore 2016
J.K. Peterson, *BioInformation Processing*, Cognitive Science and Technology,
DOI 10.1007/978-981-287-871-7_11

Davie goes on to examine ways to extend phrases. He gives examples of grammatical clauses followed by phrases lengthened by adding adjectives, etc. The musical examples are extended by sequential repetition, additional measures of the sequential repetition, stretching the cadence and repetition of the stretched cadence.

These musical techniques do mimic the end result of adding adjectives in grammar. The sentence becomes longer. We see how the comparison breaks down when we see that an adjective further distinguishes one noun from any other noun. Repetition of a musical fragment is merely a restatement. Sequential repetition can be compared to restatement of a sentence by a different person (different starting pitch).

In defining a musical sentence, Davie believes it is the combination of two or more phrases that are necessary for balance. In other words, there is a question phrase and a response phrase. All possible three measure outcomes from our matrix will be complete musical sentences. The opening fragment functions as a question phrase and the closing fragment functions as the response. We have added a connecting phrase between the two. This insertion extends our musical line, providing a smooth transition between question/answer. In terms of speech, the exchange flows more naturally by sounding less abrupt.

Some, such as Berry (1966), *Form in Music: An Examination Of Traditional Techniques Of Musical Structure and Their Application in Historical and Contemporary Styles*, believe music to be essentially abstract. However, he does believe it has the ability to "*impart a sense, a mood, impression of states or qualities*". Here, too, cadences are considered *musical equivalents of punctuation*. Where Davie focuses on scale degree and harmony (real or implied) to determine degrees of finality, Berry goes a step further. He points out the effect cadences have on the rhythmic motion of a line. A cadence does one of two things, either interrupts rhythmic motion or conveys closure by stopping it all together. Berry's explanation of smaller structural units also supports our three measure phrases as musical sentences. He states that a musical phrase (Berry interchanges the terms phrase and statement) may be compared to a clause that contains at least a subject and a predicate (question/answer). He goes on to say that it has a distinct beginning, a clear course of direction and an ending (cadence).

Though not a common form of analysis, relating nouns and verbs to tonic and dominants has been discussed by theorists. An example from Cope (1991), *Computers and Musical Style*, is that V7-I equals a simple verb-object motion, with the tonic coming as a consequence to the dominant motion. This is similar to the idea of the object of a sentence being a consequence of the verb's action. Cope goes on to say that the ensemble of musical pitches and chords typically relates in terms of major and minor keys. Also, these phenomena function in relation to the key name pitch and chord which are called "tonic" by theorists. In C major, the ensemble of notes would be CDEFGABC. The tonic note is C and the tonic triad (three-note chord) is CEG. There is a detailed section on parsing, which is a technique for diagramming relationships between sentence parts in language. A sentence is broken into two basic parts, a noun phrase and a verb phrase. From there, each can be broken into smaller pieces (article plus noun/adverb plus verb). Cope shows a parse of the C-major scale that shows the basis for the melodic movement used in our own examples. First, he presents the sentence as CDEFGABC. Next, he identifies noun equivalents, C (tonic) and F (subdominant). The G (dominant) functions as a verb, leaving the articles and their modifiers: A (Prolongation), E (Adverb), D (Prolongation) and B.

Logical motions in music are systematically described by Cope in the following way. SPEAC equal to Statement, Preparatic, Extension, Antecedents and Consequent. The desired line of succession then looks like this: S followed by PEA; P by SAC; E by SPAC; A by EC; and C by SPEA Cope believes that a "generated hierarchy" of notes is necessary to produce music that makes sense. He reasons that a random substitution of a word type that would produce a nonsense sentence (noun–verb–noun could equal horse grabs sky), would have the same effect in music. We agree with Cope that tonality itself is a hierarchy where certain notes or scale degrees have specific functions. Additionally, a set of generalities for tonal melodies is given. Cope based this list on examples from composers across generations. These are basic composition rules taught in your average music theory classes: use mostly stepwise motion; follow melodic skips with stepwise motion in the opposite direction; use one or more notes per beat agreeing with harmony; and usually begin and end on tonic or dominant chord members. Cope summarizes by saying

> With proper selection of elemental representations (i.e.,nouns are tonics) and careful coding of appropriate local semantics, the same program that produces sentences can produce music.

Kohs (1973). *Musical Form: Studies in Analysis and Synthesis* stresses that the human mind will organize music into sections based on repetition and contrast and that distinctions exist between functional and decorative (nonharmonic) tones. Then, in making his own connection between speech and music, Koh brings in another factor, **emotion**. We will study the connections of emotion to music more completely in Sect. 11.4.1. Kohs connects to emotional states as follows:

> In prehistory, vocal melody was probably developed as a form of emotionally inflected speech. Melody has been associated with words since the earliest times, and wordless vocalization has always been a rather rare phenomenon. Thus it is not surprising that some of the characteristics of melody are derived from speech, and that some of the melodic forms are related to the forms of prose and poetry.

Poetry can be analyzed in terms of meter, its pattern of accented and unaccented syllables. Musical meter is analyzed the same way; in terms of strong and weak beats. Kohs compares nonmetrical music to the less structured rhythm that is characteristic to prose. Again, we have a writer stress that the key to a sentence or musical phrase is that both words and musical elements are put together based on the function each will carry out. Kohs is explicit about music having

> its own special kind of grammar and syntax. Successive tones may be grouped to form a musical phrase having a sense of completion and unity similar to that found in a verbal sentence. Some tones, like the verbal subject and predicate, are essential to the musical

structure; others are decorative. Suspensions, appogiaturas, neighboring tones and similar decorations cannot stand alone without resolution any more than an adjective may stand without a noun or pronoun. Musical phrases may be simple or complex; a short musical idea may be expanded by a variety of means, such as parenthetical insertions, or extensions at the beginning or the end.

He follows with an example comparison of a sentence and a musical phrase, in which both are transformed by the addition of words and tones respectively. The additions to the basic musical line are tried and true techniques taught to all students of composition.

We have gleamed a primitive view of how we wish to abstract structure from music by building on these works. Although the mapping between a grammatical view of music and the actual way a person composes is imperfect, it pervades many of the discussions on composition, orchestration and so forth in the musical literature. We have thus decided that our first attempt at autonomous music creation will be based on a simplistic grammatical approach: we will try to create a collection of short musical fragments which embody or encapsulate a notion of good compositional style.

11.2   The Würfelspiel Approach

We will start by using an 18th century historical idea called The Musicalisches Würfelspiel. In the 1700s, fragments of music could be rapidly prototyped by using a matrix $\mathcal{A}$ of possibilities. We show an abstract version of a typical Musicalisches Würfelspiel matrix in Eq. 11.1. It consists of P rows and three columns. In the first column are placed the opening phrases or nouns; in the third column, are placed the closing phrases or objects; and in the second column, are placed the transitional phrases or verbs. Each phrase consisted of L notes and the composer's duty was to make sure that any opening, transitional and closing (or noun, verb and object) was both viable and pleasing for the musical style that the composer was attempting to achieve.

$$\mathcal{A} = \begin{bmatrix} \text{Opening 0} & \text{Transition 0} & \text{Closing 0} \\ \text{Opening 1} & \text{Transition 1} & \text{Closing 1} \\ \vdots & \vdots & \vdots \\ \text{Opening P-1} & \text{Transition P-1} & \text{Closing P-1} \end{bmatrix} \tag{11.1}$$

Thus, a musical stream could be formed by concatenating these fragments together: picking the ith Opening, the jth Transition and the kth Closing phrases would form a musical sentence. Since we would get a different musical sentence for each choice of the indices i, j and k (where each index can take on the values 0 to $P - 1$), we can label the sentences that are constructed by using the subscript i, j, k as follows:

$$S_{i,j,k} = \text{Opening i} + \text{Transition j} + \text{Closing k}$$

Note that there are P^3 possible musical sentences that can be formed in this manner. If each opening, transition and closing fragment is four beats long, we can build P^3 different twelve beat sentences.

It takes musical talent to create such a Musicalisches Würfelspiel array, but once created, it can be used in the process of learning fundamental principles of the music compositional process. We will eventually use musical fragments which are tagged with a specific emotional color to build a model which can assemble musical fragments which have a specific emotional content. However, we will start with a proof of concept using a Musicalisches Würfelspiel matrix with emotionally neutral examples.

11.3 Neutral Music Data Design

We will start with simple compositional patterns and ideas; hence, we will be using only quarter and half notes permitted in any of the phrases. Further, we do not want the musical fragments to be too long, so for now each fragment consists of four beats in 4/4 time. We will begin opening phrases and end closing or cadence phrases on tonic C, approaching or leaving by step or tonic chord leap. Finally, the middle phrases are centered around a third or a fifth. Using these guidelines, we created a Musicalisches Würfelspiel array using opening phrases as nouns, middle phrases as verbs and closing or cadence phrases as objects.

11.3.1 Neutral Musical Alphabet Design

We chose the noun phrases as shown in Fig. 11.1.

Now the last note in each of four opening phrases must be able to be played right before any of the first notes in a middle phrase. Correct combinations are not random choices and so the musical composer's skill is captured to some extent in the choices that are made for the middle phrases. Our design alphabet can be encoded as $\mathcal{H}_1 = \{c, d, e, f, g, a, b, C\}$, and $\mathcal{H}_2 = \{c^2, d^2, e^2, f^2, g^2, a^2, b^2, C^2\}$, where C being the C above middle C. All the letters in $\mathcal{H}_1$ label quarter notes from the middle C octave and the notes in the $\mathcal{H}_2$ alphabet denote half notes. Our alphabet is

Fig. 11.1 Musical opening phrases

Fig. 11.2 Musical middle phrases

thus $\{\mathcal{H}_1, \mathcal{H}_2\}$ which has cardinality 16. Within this alphabet, the second opening, $cedf$, can be written as the matrix

$$n_1 = \begin{bmatrix} 1 & 0 & 0 & 0 & 0 & 0 & 0 & 0 & 0 & 0 & 0 & 0 & 0 & 0 & 0 & 0 \\ 0 & 0 & 1 & 0 & 0 & 0 & 0 & 0 & 0 & 0 & 0 & 0 & 0 & 0 & 0 & 0 \\ 0 & 1 & 0 & 0 & 0 & 0 & 0 & 0 & 0 & 0 & 0 & 0 & 0 & 0 & 0 & 0 \\ 0 & 0 & 0 & 1 & 0 & 0 & 0 & 0 & 0 & 0 & 0 & 0 & 0 & 0 & 0 & 0 \end{bmatrix}$$

There are similar matrices for each of the other opening phrases. Each opening phrase is thus a 4×16 matrix that has very special property: a row can only have one 1. A given middle phrase will have a similar structure and only some of the possible middle phrase matrices we could use will be acceptable. Our middle phrase design choices for emotionally flat music are shown in Fig. 11.2.

In essence, when we design these two sets of matrices, we provide samples for a hypothetical mapping from the set of 4×16 opening matrices to the set of 4×16 middle matrices. These samples provide enough information for us to approximate the opening to middle transition function using a blend of neurally inspired architectures and function approximation techniques. Finally, the cadence or closing phrases, as shown in Fig. 11.3, were designed so that they would sound pleasing when attached to any of the possible opening—middle phrase combinations. Again, our design gives us a set of appropriate 4×16 matrices which encode valid closing phrases. These closing phrases need to be coupled to the end notes of any middle phrase. The closing data we design then gives us enough samples to approximate the middle to ending transformation.

In addition, our opening data gives us four examples of starting notes for neutral musical twelve note sequences. Now there are nine possible start notes for each opening phrase and the fact that we do not choose some of them is important. Also,

Fig. 11.3 Musical closing phrases

Fig. 11.4 The neutral music matrix

each four note sequence in any of the three phrases, opening, middle and closing, is order dependent. Given a note in any phrase, the selection of the next note that follows is *not* random. The actual note sequence that appears in each phrase also gives sample data that constrains the phrase to phrase transformations. We can use this information to effectively approximate our mappings using excitation/inhibition neurally inspired architectures. Roughly speaking, if a given subset of notes are good choices to follow another note, then the notes not selected to follow should be actively inhibited while the acceptable notes should be actively encouraged or enhanced in their activity. The complete Musicalisches Würfelspiel matrix, as seen in Fig. 11.4, thus consists of four rows and three columns that will provide a total of 64 distinct musical fragments that are intended to model neutral musical sentence design.

Currently, these fragments are rather short since they are just four beats (i.e. $L = 4$) for each grammatical element. This is enough to help with the prototype development of this stage. However, when we generate the additional Musicalisches Würfelspiel matrices that will correspond to the other emotional shadings, we may find it necessary to use longer note sequences (i.e. increase L) in order to capture the desired emotional colorings. If that turns out to be true, we can easily return to this neutral case and redesign the neutral Musicalisches Würfelspiel matrix to use longer note sequences.

11.3.2 The Generated Musical Phrases

From our 4×3 Musicalisches Würfelspiel matrix, we can generate 64 musical selections of twelve beats each. In Fig. 11.5, we show the selections generated using opening one, all the possible middle phrases and the first cadence phrase. The intent here is

Fig. 11.5 Neutral sequences generated using the first opening phrase, all the middle phrases and the first ending phrase. The first column of the figure provides a label of the form xyz where x indicates the opening used; y, the middle phrase used; and z, the ending phrase. Thus, 131 is the fragment built from the first opening, the third middle and the first ending

that all 64 selections we so generate will be devoid of emotional content. You should try playing these pieces on the piano and compare them to our later selections that are purported to have happy, angry and sad overtones that are displayed as detailed in Sects. 11.4.3, 11.4.4 and 11.4.5.

The fragments shown in Fig. 11.5 show only a few of the possibilities. For example, the selections for opening two, verb one and all the cadence phrases are displayed in Fig. 11.6. Again, we invite you to play the pieces. The important point is that

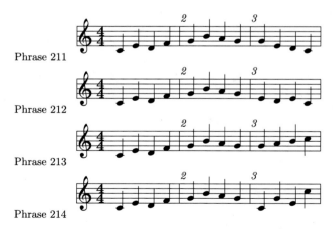

Fig. 11.6 Neutral sequences generated using the second opening phrase, the first middle phrase and all the ending phrases

all 64 pieces we can so generate using the Würfelspiel approach are **equally** valid emotionally neutral choices. Hence, the Musicalisches Würfelspiel matrix we have created *captures* the essence of the solution to emotionally neutral music compositional design problems.

11.4 Emotional Musical Data Design

In earlier sections, we have discussed how we would design a Musicalisches Würfelspiel matrix which contained simple emotionally neutral musical compositions. We now extend our construction process to include Musicalisches Würfelspiel matrices that are sad, angry and happy. Our review of previous work has led us to the following observations.

11.4.1 Emotion and Music

Schellenberg et al. (2000), *Perceiving Emotion in Melody: Interactive Effects of Pitch and Rhythm* specifically addresses music and emotion. The researchers were trying to decipher what effects two specific musical elements (pitch and rhythm) had upon the perception of listeners. Before manipulating elements, they had to establish a set of melodies that "unequivocally expressed one of three emotions: happy, sad, or scary". The authors decided that each of the three emotions: happy, sad, scary, is considered to be a "basic" emotion (see e.g., Ekman and Davidson 1994). Further, based on a review of literature cited in their paper (Hevner 1935, 1936, 1937; Kratus 1993; Slodoba 1991; Terwogt and van Grinsven 1991; Thompson and Robitaille 1992), there is a consensus that "adults from a common culture generally show broad agreement when associating such emotions with particular pieces of music. Other research shows that young children have similar associations (Cunningham and Sterling 1988; Dolgin and Adelson 1990; Giomo 1993; Kastner and Crowder 1990; Kratus 1993; Terwogt and van Grinsven 1991).

Various attributes were used to describe melodies in the each of the three emotion categories. In *happy* melodies, there should be fast tempi (Hevner 1936; Rigg 1940; Scherer and Oshinsky 1977; Wedin 1972); major modes (Crowder 1984, 1985; Geraldi and Gerken 1995; Gregori et al. 1996; Kastner and Crowder 1990; Kratus 1993; Scherer and Oshinsky 1977; Wedin 1972) and staccato articulation (Juslin 1997). On the other hand, in *sad* music, the use of lower pitches (Hevner1936,Crowder1985,Wedin1972) along with minor modes (Crowder 1984, 1985; Geraldi and Gerken 1995; Gregori et al. 1996; Kastner and Crowder 1990; Kratus 1993; Scherer and Oshinsky 1977; Wedin 1972) and legato articulation (Juslin 1997), are necessary to elicit the proper tone of sadness. Finally, in *scary* music, there should be a broad pitch range. Any instrument can produce staccato, legato or broad range sound. However, in the study above, listeners associated the staccato articulation produced by a guitar to a happy state; the legato articulation of a violin to a sad state; and the broad pitch range of an organ to a scared state.

Meyer (1956), *Emotion and Meaning in Music*, examines "meaning" in music. His arguments are based on what he labels an absolute expressionist viewpoint. This specifically means "expressive emotional meanings arise in response to music and that these exist without reference to the extra musical world of concepts, actions, and human emotional states". Expectation is a concept that Meyer considers a product that comes from natural mental processes of perception. The process involves instinctive grouping and organizing of information coming in through the senses. Applying this logic to music, Meyer states that music elicits varying responses from a listener by manipulating the expected. For example, musical progression that moves in an irregular way throughout elicits a feeling of suspense or ambiguity for listener. Why? Meyer claims the listener would begin to doubt the relevance of his own expectations. This may be true for a trained musician, but we are not so sure that the average listener is aware of having certain expectations and therefore does not consciously go through phases of doubt. Meyer's second argument makes more sense to us. It is the opposite notion that if the music is so uniform or repetitive, then the music itself has ambiguity in that it seems static, going nowhere.

Meyer links our experience of modulation (shifts in tonal center) and key changes to departures in a narrative line in a novel. These complications of the plot function like our extensions in musical phrases and is simply understood by the listener. Different musical styles are complex systems of probability. This idea seems to tie into what Cope did in later years (see Cope 1991, 2001). Cope entered many examples of Chopin and then used a complex system of computed probabilities to form Chopin-like pieces. We think that to know what expectation to have, based on probability, and to understand additive elements in writing and music is due to having encountered them before. Thus, we feel this is a human experience rather than some intrinsic response.

In a later section, Meyer states that an expectation must "have the status of an instinctive mental and motor response, a felt urgency, before its meaning can truly be comprehended". He does not go on to say how that line is crossed, but suggests that it is the deviation of the pure tone, exact intonation, perfect harmony, rigid rhythm, etc. which conveys emotion.

As we have seen documented in multiple sources, there is an association between minor mode and the emotional state of sadness. Meyer adds the emotional state of suffering as well. He reasons that these types of emotion are a product of the unstable character of the mode itself. The unstable character refers to the fact that the minor mode is presented in different versions: melodic minor, natural minor, and harmonic minor. Because it is possible and likely to have a combination within one piece, it is very chromatic. Chromaticism de-emphasizes tonal centering. Since tonal centering is the basis of Western musical language, chromaticism seems unstable due to its unpredictability.

Balkwill and Thompson (1999), *A Cross-Cultural Investigation of Emotion in Music: Psychophysical and Cultural Clues* attempts to answer the following question:

Can people identify the intended emotion in music from an unfamiliar tonal system? If they can, is their sensitivity to intended emotion associated with perceived changes in psychophysical dimensions of music [defined later as any property of sound that can be perceived independent of musical experience, knowledge, or enculturation]?

We were very interested in the results to see if our ideas for emotionally tagged music would have characteristics that could be readily identified without cultural constraints. In this small case study, four specific emotions (joy, sadness, anger, and peacefulness) were presented using ragas of India. In this Hindustani system, there is a specific raga or collection of notes for nine individual moods. Participants were asked to rate tempo, rhythmic complexity, melodic complexity, and pitch range in addition to the four emotions.

Balkwill and Thompson give us some preliminary expectations based on other studies. Tempo is most consistently associated with emotional content. A slow pace equates to sadness and, a faster one; joy. Also, simpler melodies with few variations in melodic contour and more repetition are associated with positive and peaceful emotions. Complex melodies with more melodic contour and less repetition are associated with negative anger and sadness. Timbre plays a role as well. Fear and sadness were reported more when expressed by a violin or the human voice. Finally, timpani was associated with anger.

An interesting note they mention is that a narrower pitch range (reduced melodic contour) may be processed as one auditory stream, therefore *easy* to process which may cause positive emotional ratings. This may be linked to Meyer's idea of an instinctive mental or motor response.

The conclusions were that given music that was not culture-specific to the listeners, they were forced to rely on other, psychophysical, cues to perceive emotional content. As predicted, tempo was a strong cue used to successfully identify the ragas intended for joy, sadness, and anger. It did not work with peacefulness. Also, ratings made by expert and non-expert listeners were pretty equal in identifying joy and sadness. The only significant predictor of peacefulness in this study seemed to be timbre. A flute was highly rated as peaceful.

What happens when two real performers are put up against a computer generated performance of a piece is exactly the focus of Clarke and Windsor (2000), *Real and Simulated Expression: A Listening Study*. There is no solid conclusion in this paper, other than the matter will need further investigation. They do state that in this study, the simulated performance treated tempo and dynamics as elements that were correlated based on principles of energy and motion. There were minute differences in the way human performers treated repeated notes, both rhythmically and dynamically. Hence, each performance was perceived in different ways by the listeners.

Basic emotions are defined in Juslin (1997), *Emotional Communication in Music Performance: A Functionalist Perspective and Some Data*, by the following attributes. They have distinct functions that contribute to individual survival. They are found in all cultures and are experienced as unique feeling states. Further, they appear early in the course of human development and are associated with distinct autonomic patterns of physiological cues. Further, he states that most researchers agree on at least four basic emotions: happiness, sadness, anger and fear. In this small study, three guitarists were asked play the same melody five different ways. One was to be without expression. This would correspond to our (null, null) or *neutral* fragments. Two aspects examined were whether or not emotions could be communicated to the listeners, and how the performers' intentions affected expressive cues in the performance (the psychophysical cues studied in Balkwill and Thompson 1999). Like other studies in this commentary, they found that expressions of happiness, sadness, and anger were readily identified by listeners. The fourth emotional state, in this case fear, was a little elusive. Gender and training did not significantly effect ones ability to identify intended emotion. The study suggests that each emotion has certain characteristics as detailed below (the authors point out that in other instruments it is typical to use staccato articulation when expressing anger, but for electric guitarists, they uniformly revert to legato articulation for expressing anger):

	Loud	Quiet	Fast	Slow	Staccato	Legato
Anger	x		x			x
Sadness		x		x		x
Happiness	x		x		x	
Fear		x		x	x	

In Juslin and Madison (1999), *The Role of Timing Patterns in Recognition of Emotional Expression from Musical Performance*, we quote from the abstract:

> We gradually removed different acoustic cues (tempo, dynamics, timing, articulation) from piano performances rendered with various intended expressions (anger, sadness, happiness, fear) to see how such manipulation would effect a listener's ability to decode emotional expression. The results show that (a) removing the timing patterns yielded a significant decrease in listeners' decoding accuracy, (b) timing patterns were by themselves capable of communicating some emotions with accuracy better than chance, (c) timing patterns were less effective in communicating emotions than were tempo and dynamics.

The authors acknowledge the nature of their study as preliminary and in need of further extended study. At any rate, a few hypotheses are put forth. The first is that long and short note durations may be played differently depending on the intended emotion. They found that expressions of happiness were played in shorter note values and patterns in the expressions of sadness were played in longer notes. Secondly, anger and happiness were associated with staccato articulation. This is the first instance we encountered of further distinction between the staccato articulation representing anger and the staccato articulation representing happiness. It was found that the anger expressions used uniform staccato patterns where the happiness ones were more variable depending on the positions within the phrase. It is suggested that more study of this phenomenon needs to be done, as it may be a key component to decoding happiness.

11.4.2 Emotional Music Data Design

The underlying goal in building each matrix was to remain as basic as possible. We decided to work within a monophonic texture, meaning melody line only. Note values were restricted to quarter notes and half notes in quadruple meter. Quarter rests were also allowed, but used sparingly. All four matrices (neutral, happy, sad, angry) are structurally similar. Each consists of three columns with four fragment choices that are one measure in length.

Any fragment from column one from any of the matrices is designed to function as an opening phrase. We define an opening phrase as one that clearly establishes a tonal center. In western tonal music, the tonal focal point can be narrowed to a single tone/pitch that is known as the tonic. We have made C our tonic note in all cases. In all but one case, the opening fragment also established the mode as either major or minor. The exception is made in the angry matrix, where ambiguity is desirable. All fragments in column two of any of the matrices are designed to function as a transition phrase. As the label implies, these transition phrases serve as connectors between a choice from column one and a choice from column three. It is in these middle phrases that movement away from the tonic is made or continued. This movement is necessary for forward progress of a melody. Therefore, each transition phrase is now highlighting a secondary pitch, one other than the tonic note established by the opening phrase. To close our melodic lines, an ending phrase is chosen. Any fragment from column three of any of the matrices will function in the same manner. We designed each to move back to the tonic note in such a way as to produce a quality of closure to our melodic lines. This was done by approaching tonic in the most basic of ways. Using stepwise motion up or down to the tonic logically ends the melodic journey by bringing you back to the home pitch. An alternative is to return to tonic via melodic skip from the third or fifth scale degree. In the key of C, the tonic (C) is scale degree 1, D 2, etc. So, a melodic skip from the third of fifth scale degree means a skip from an E or a G note. Together with the tonic note, third and fifth scale degrees make up a tonic chord (harmony). By using either the third or the fifth to lead back to tonic, we again produce a sense of closure by reinstating tonic as the final destination of our brief melodic journey.

We used the following guidelines to design our Würfelspiel matrices of different emotional slants. When we produced *emotion-deprived* or *neutral* fragments, individual characteristics documented by researchers as being contributing factors of basic emotion in music were neutralized to the best of our abilities. Some of the contributing factors, such as mode, are also essential in a plausible melody, and could not be removed. The use of major mode is the default with a tempo of 45 beats, slow to the point that the individual notes outweigh the overall sense of a melody. The rational comes from reading. The telltale sign of a new reader is the slow pace during which equal emphasis is placed on every word. A beginner will produce an emotion-deprived reading. The same result is our goal here. Further, we use even rhythms and exact note durations with the melody played with a basic computer generated sound.

Likewise, fragments we intended to emotionally tag as *happy* had individual characteristics researched by others incorporated into the design. Each characteristic was chosen based on a general consensus by other researchers and authors as being a contributing factor in representing *happy* in music. Again, this entails choosing a major mode, a very quick tempo (250 beats) and the use of staccato. If we wish, we can use quarter rests. We could choose to present the melodies using a flute, as this particular instrument has often been linked to happiness in the literature.

Once more, individual characteristics that have been researched by others were incorporated into the design of our fragments tagged by *sad*. Each characteristic was chosen based on a general consensus by other researchers and authors as being a contributing factor in representing *sad* in music. There is a use of minor mode with a slow tempo (70 beats). Also, we use slurs and legato and the bass clef to put us in a lower register. We choose to present these melodies using a violin, as stringed instruments are particularly lined to sadness in the literature. Finally, there is some use of chromaticism.

To emotionally tag the fragments as *angry*, individual characteristics that have been researched by others were incorporated into the design. Each characteristic was chosen based on a general consensus by other researchers and authors as being a contributing factor in representing "angry" in music. We use a minor mode, a moderate tempo (180 beats) faster than used for the sad melodies, slightly slower than the tempo used for the happy melodies and increased variation of articulation (slurs, accents). Further, there is the incorporation of larger leaps. We choose to present these melodies using a trumpet, as brass and percussion are often linked to anger in the literature. There are also more repeated notes and the use of an ambiguous fragment where the mode is not clearly established in opening phrase.

11.4.3 **Happy** *Musical Data*

Following the outline above, we have designed the *Happy* musical data as shown in Fig. 11.7.

From our 4 × 3 Musicalisches Würfelspiel matrix, we can generate 64 musical selections of twelve beats each. In Fig. 11.8, we show the selections generated using opening one, all the possible middle phrases and the first cadence phrase.

Then we show all the selections for opening two, verb one and all the closings in Fig. 11.9. This still only shows eight of the sixty four possible pieces, of course. We invite you to sit at the piano and see how they sound. You should hear how that they are distinctly happy. Our interpretation of the *core* meaning of a *happy* musical fragment is based on ideas from two separate disciplines. First, our reading of the relevant literature has given us guidance into the choices for notes, tempo and playing style as outlined above; and second, the psychophysiological studies of Lang et al. (1998) as outlined in many papers has given us a pseudo-quantitative measure of the affective content on an emotionally charged image. Musical studies have shown

Fig. 11.7 The happy music matrix

Fig. 11.8 Happy sequences generated using opening one, all middle phrases and the first closing

that if a composer deliberately attempts to convey a given emotional content in their music, queries of their audience show that the desired emotional flags have been set.

11.4.4 Sad *Musical Data*

Now, we move toward the design of musical data that is intended to be sad. In Fig. 11.10, you can see the musical data that we designed to have an overall tone of sadness. To emotionally tag these fragments as "sad," individual characteristics such

Phrase 211

Phrase 212

Phrase 213

Phrase 214

Fig. 11.9 Happy sequences using opening two, middle phrase one and all the closings

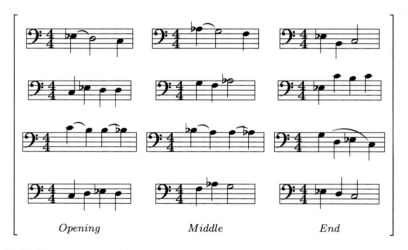

Opening   *Middle*   *End*

Fig. 11.10 The sad music matrix

as the use of minor mode, a slow tempo, the use of slurs and legato, and the use of bass clef to put us in a lower register and so forth were incorporated into the design.

From our 4 × 3 Musicalisches Würfelspiel matrix, we can generate 64 musical selections of twelve beats each. In Fig. 11.11, we show the selections generated using opening one, all the possible middle phrases and the first cadence phrase.

Then we show all the selections for all the openings, the first middle phrase and the second ending in Fig. 11.12. This still only shows eight of the sixty four possible pieces, of course. We invite you to sit at the piano and see how they sound. You should hear a sense of sadness in each.

Phrase 111

Phrase 121

Phrase 131

Phrase 141

Fig. 11.11 Sad sequences using opening one, all the middle phrases and the first closing

Phrase 112

Phrase 212

Phrase 312

Phrase 412

Fig. 11.12 Sad sequences using all the openings, the first middle phrases and the second closing

11.4.5 Angry *Musical Data*

To emotionally tag these fragments as "angry," individual characteristics that have been researched by others were incorporated into the design. Each characteristic was chosen based on a general consensus by other researchers and authors as being a contributing factor in representing "angry" in music. These include the use of minor mode, a moderate tempo (faster than used for the sad melodies (slightly slower than the tempo used for the happy melodies), increased variation of articulation (slurs, accents), incorporation of larger leaps, melody played by a trumpet, more repeated notes and ambiguous fragment (mode not clearly established in opening phrase) (Fig. 11.13).

Fig. 11.13 The angry music matrix

Fig. 11.14 Angry sequences generated using the first opening, all the middle phrases and the first closing

From the angry 4 × 3 Musicalisches Würfelspiel matrix, as usual, we can generate 64 musical selections of twelve beats each. In Fig. 11.14, we show the selections generated using opening one, all the possible middle phrases and the first cadence phrase.

Then we show all the selections for opening two, the fourth middle and all the endings in Fig. 11.15. This still only shows eight of the sixty four possible pieces, of course. We invite you to sit at the piano and see how they sound. You should be able to hear a sense of anger in each selection.

Fig. 11.15 Angry sequences generated using opening two, the fourth middle phrase and all the closings

11.4.6 Emotional Musical Alphabet Selection

As you have seen in Sects. 11.4.3, 11.4.4 and 11.4.5, the emotionally tagged musical data uses a richer set of notes and *articulation* attached to the notes to construct grammatical objects. We can think of the added articulation as punctuation marks. *Slurs* (one note and multiple note), *staccato* and *marcato* accents are attached to various notes in our examples to add emotional quality. Our design alphabet can be encoded as $\mathcal{H} = \{c, d, e, f, g, a, b, r\}$ where each note in this alphabet is now thought of as a *musical object* with a set of defining characteristics. Here r is rest. For our purposes, the attributes of a note are choices from a small set of possibilities from the list $\mathcal{A} = \{p, b, s, a\}$. The index p indicates what pitch we are using for the note: -1, denotes the first octave of pitches below middle C; 0, the pitches of the middle C octave and 1, the first octave of pitches above middle C. The letter b tells us how many beats the note is held. The length of the slur is given by the value of s and a denotes the type of articulation used on the note. We choose to treat slurs as entities which are separate from the other accent markings for clarity. For these examples, we have slurs that range from zero to three in length, so permissible values of s are taken from the set $\{0, 1, 2, 3\}$. This could easily be extended to longer slurs. The beat value b is either one of two as only quarter and half notes are used. There are many possible articulations. An expanded list, for marks either above or below a note for effect, might include *neutral*, no punctuation ($a = 0$); *pizzicato*, a dot ($a = 1$); *marcato* or *sforzando*, a > ($a = 2$); *staccato* or *portato*, a −, ($a = 3$); *strong pizzicato*, an apostrophe ($a = 4$); and *sforzato*, a^ ($a = 5$). A given note n is thus a collection which can be denoted by $n_{p,b,s,a}$ where the attributes take on any of there allowable values. A few examples will help sort out this out. The symbol $d_{1,2,2,1}$ is the half note d above middle d with a pizzicato articulation which is the start of a two note slur which ends on the second note that follows this middle d. The rest

Fig. 11.16 Some angry phrases. **a** Middle phrase. **b** Opening phrase

does not have pitch, articulation or slurring. Thus, we set the value of pitch, slurring and articulation to 0 and use the notation $r_{0,1,0,0}$ or $r_{0,2,0,0}$ to indicate a quarter or half rest, respectively. Our alphabet is thus $\{\mathcal{H}\}$ which has cardinality 8. Each letter has a finite set of associated attributes and each opening, middle or closing phrase is thus a sequence of 4 musical entities.

Within this alphabet, an angry middle phrase such as shown in Fig. 11.16a, can be encoded as $\{e_{0,1,0,2}, \; d_{0,1,0,2}, \; c_{0,2,0,0}\}$. This would be then written as the matrix in Eq. 11.2

$$
n_1 =
\begin{bmatrix}
0 & 0 & \{0, 1, 0, 2\} & 0 & 0 & 0 & 0 & 0 \\
0 & \{0, 1, 0, 2\} & 0 & 0 & 0 & 0 & 0 & 0 \\
\{0, 2, 0, 0\} & 1 & 0 & 0 & 0 & 0 & 0 & 0
\end{bmatrix}
\tag{11.2}
$$

If we had a fragment with a slur such as shown in Fig. 11.16b, this would be encoded as $\{c_{0,1,1,0}, \; d_{0,1,0,0}, \; d_{0,1,0,2}, \; c_{0,1,0,0}\}$. In matrix form, we have Eq. 11.3

$$
n_1 =
\begin{bmatrix}
\{0, 1, 1, 0\} & 0 & 0 & 0 & 0 & 0 & 0 & 0 \\
0 & \{0, 1, 0, 0\} & 0 & 0 & 0 & 0 & 0 & 0 \\
0 & \{0, 1, 0, 0\} & 0 & 0 & 0 & 0 & 0 & 0 \\
\{0, 1, 0, 0\} & 0 & 0 & 0 & 0 & 0 & 0 & 0
\end{bmatrix}
\tag{11.3}
$$

These matrices indicate which musical object is used in a sequence. The four opening phrases in a Würfelspiel music matrix can thus be encoded into matrices that are 2×8 (both notes in the phrase are half notes) to 4×8 (all notes are quarter notes). Each of these matrices has the special property that a row can only have one nonzero entry. A given middle phrase will have a similar structure, making only some of the possible middle phrase matrices acceptable.

Note that encoding music in this way generates a compact data representation. However, we need to model the data so that each possible musical entry is encoded in a unique way. The first seven entities in $\mathcal{H}$ come in a total of 144 distinct states: three pitches, two beats, four slur lengths and six articulations. The rest comes in only two states. Hence, a distinct alphabet here has cardinality $7 \times 144 + 2$ or 1010. The size of this alphabet precludes showing an example as we did with the compact representation, but the matrices that encode these musical samples still possess the property that each row has a single 1. Of course, with an alphabet this large in size, we typically do not use a standard matrix representation; instead, we use sparse matrix or linked list techniques. The data representation we used for the musically neutral data has a much lower cardinality because the neutral data is substantially simpler.

References

L. Balkwill, W. Thompson, A cross-cultural investigation of emotion in music: Psychophysical and cultural clues. Music Percept. **17**(1), 43–64 (1999)

W. Berry, *Form in music: An examination of traditional techniques of musical structure and their application in historical and contemporary styles* (Prentice-Hall, Upper Saddle River, 1966)

W. Caplin, *Classical Form: A Theory of Formal Functions for the Instrumental Music of Haydn* (Oxford University Press, Mozart, 1998)

E. Clarke, W. Windsor, Real and simulated expression: A listening study. Music Percept. **17**(3), 277–313 (2000)

D. Cope, *Computers and Musical Style* (A-R Editions, Middleton, 1991)

D. Cope, *Virtual Music: Computer Synthesis of Musical Style* (MIT press, Cambridge, 2001)

R. Crowder, Perception of the major/minor distinction: I. Historical and theoretical foundations. Psychomusicology **4**, 3–12 (1984)

R. Crowder, Perception of the major/minor distinction: III, Hedonic, musical and affective discriminations. Bull. Psychon. Soc. **23**, 314–316 (1985)

J. Cunningham, R. Sterling, Developmental changes in the understanding of affective meaning in music. Motiv. Emot. **12**, 399–413 (1988)

C. Davie, *Musical Structure and Design* (Dover Publications, New York, 1953)

K. Dolgin, E. Adelson, Age changes in the ability to interpret affect in sung and instrumentally-presented melodies. Psychol. Music **18**, 87–98 (1990)

P. Ekman, R. Davidson (eds.), *The Nature of Emotion: Fundamental Questions* (Oxford University Press, Oxford, 1994)

G. Geraldi, L. Gerken, The development of affective responses to modality and melodic contour. Music Percept. **12**, 279–290 (1995)

A. Gregori, L. Worrall, A. Sarge, The development of emotional responses to music in young children. Motiv. Emot. **20**, 341–348 (1996)

C. Giomo, An experimental study of children's sensitivity to mood in music. Psychol. Music **21**, 141–162 (1993)

K. Hevner, The affective character of the major and minor modes in music. Am. J. Psychol. **47**, 103–118 (1935)

K. Hevner, Experimental Studies of the elements of expression in music. Am. J. Psychol. **48**, 246–268 (1936)

K. Hevner, The affective value of pitch and tempo in music. Am. J. Psychol. **49**, 621–630 (1937)

P. Juslin, Emotional communication in music performance: A functionalist perspective and some data. Music Percept. **14**(4), 383–418 (1997)

P. Juslin, G. Madison, The role of timing patterns in recognition of emotional expression from musical performance. Music Percept. **17**(2), 197–221 (1999)

M. Kastner, R. Crowder, Perception of the major/minor distinction: IV Emotional connections in young children. Music Percept. **8**, 189–202 (1990)

E. Kohs, *Musical Form: Studies in Analysis and Synthesis* (Houghton Mifflin Company, Massachusetts, 1973)

J. Kratus, A developmental study of children's interpretation of emotion in music. Psychol. Music **21**, 3–19 (1993)

L. Meyer, *Emotion and Meaning in Music* (University of Chicago Press, Chicago, 1956)

M. Rigg, The effect of tonality and register upon musical mood. J. Musicol. **2**, 49–61 (1940)

G. Schellenberg, A. Krysciak, J. Campbell, Perceiving emotion in melody: Interactive effects of pitch and rhythm. Music Percept. **18**(2), 155–171 (2000)

K. Scherer, J. Oshinksky, Cue utilization in emotion attribution from auditory stimuli. Motiv. Emot. **1**, 331–346 (1977)

J. Slodoba, Music structure and emotional response. Psychol. Music **19**, 110–120 (1991)

M. Terwogt, F. van Grinsven, Musical expression of moodstates. Psychol. Music **19**, 99–109 (1991)

W. Thompson, B. Robitaille, Can composers expression emotion through music? Exp. Stud. Arts
   10, 79–89 (1992)
L. Wedin, Multidimensional scaling of emotional expression in music. Swed. J. Musicol. **54**, 1–17
   (1972)

Chapter 12
Generation of Painting Data: J. Peterson, L. Dzuris and Q. Peterson

Our simple painting model will be based on a Würfelspiel matrix similar to what we used in the above emotionally neutral music compositions. Since our ultimate goal is to create cognitive models, it is instructive to look at the notion of creativity from the *cortical* point of view. The advent of portions of cortex specialized to fusing sensory information is probably linked in profound ways to the beginnings of the types of creativity we label as art. Hence, models of creativity, at their base, will involve models of visual cortex data which can then be fed into associative learning modules. We have developed data models for the auditory cortex which are music based in Chap. 11. We need to look at visual cortex data now. We therefore focus on the creation of data sets that arise from paintings as we feel that this will give us the data to train the visual input to the area 37 sensor fusion pathway.

12.1 Developing a Painting Model

Consider the two paintings, Fig. 12.1a, b, which have fairly standard compositional designs. Each was painted starting with the background and then successive layers of detail were added one at a time. As usual, the design elements farthest from the viewer's eye are painted first. The other layers are then assembled in a farthest to nearest order. We note that in the generation of computer graphics for games, a similar technique is used. A complex scene could in principle be parsed into individual polygonal shapes each with its own color. Then, we could draw each such polygon into the image plane which we see on the computer screen. However, this is very inefficient. There are potentially hundreds of thousands of such unique polygons to find and finding them uses computationally expensive intersection algorithms. It is much easier to take advantage of the fact that if we draw a polygon on top of a previous image, the polygon paints over, or occludes, the portion of the image that lies underneath it. Hence, we organize our drawing elements in a tree format with the root node corresponding to distances farthest from the viewer. Then, we draw a scene by simply traversing the tree from the root to the various leafs.

© Springer Science+Business Media Singapore 2016
J.K. Peterson, *BioInformation Processing*, Cognitive Science and Technology,
DOI 10.1007/978-981-287-871-7_12

(a) **(b)**

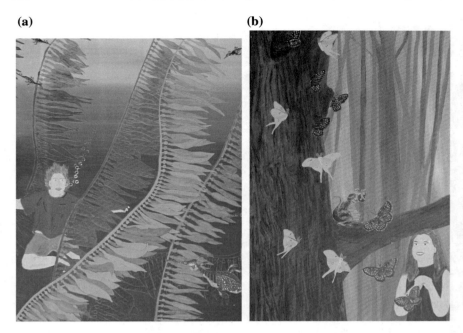

Fig. 12.1 Two simple paintings. **a** This painting uses many separate layers for the kelp. The human figure is in an intermediate layer between kelp layers and the seadragons essentially occupy foreground layers. The background varies from a *dark blue* at the *bottom* to very *light blue*, almost *white*, at the *top*. **b** This painting uses a complicated background layer with many trees. The dragon and the large tree are midground images, while the butterflies and moths are primarily foreground in nature. The girl is midground

The painting seen in Fig. 12.1a started with the background. This used a gradient of blue, ranging from very dark, almost black, at the bottom, to very light, almost white, at the top. There are, of course, many different shades and hues of blue as brushes are used to create interesting blending effects with the various blues that are used. However, we could abstract the background to a simple blue background and capture the basic compositional design element. The many kelp plants are all painted in different planes. The kelp farthest from the viewer are very dark to indicate distance, while the plants closest to the viewer use brighter greens with variegated hues. We note that we could abstract the full detail of the kelp into several intermediate midground layers: perhaps, the farthest midground layer might be one kelp plant that is colored in dark green with the second, closest midground layer, a bright green kelp plant. The human figure is placed between kelp layers, so we can capture this compositional design element by placing a third midground layer between the two midground kelp plant layers. Finally, there are many seadragons in foreground layers at various distances from the viewer. We could simplify this to a single foreground layer with one seadragon painted in a bright red. Hence, the abstract composi tional design of the painting in Fig. 12.1a is as follows:

The abstract seadragons design

Layer	Description
Background one	Blue gradient from dark to light
Midground one	Very dark green kelp plant
Midground two	Human figure
Midground three	Bright green kelp plant
Foreground	Bright red sea dragon

In a similar fashion, we can analyze Fig. 12.1b. The background in this painting is a large collection of softly defined trees. These are deliberately not sharply defined so that their distance from the viewer is emphasized. The midground image is the very large tree that runs from the bottom to the top of the painting. There are then two more midground images: the whimsical dragon figure on the tree branch and the human figure positioned in front of the tree. Finally, there are a large number of Baltimore butterflies and Luna moths which are essentially foreground images. We can abstract this compositional design as follows:

The abstract tree painting design

Layer	Description
Background one	Fuzzy brown trees
Midground one	Large tree in brighter browns
Midground two	Dragon in red
Midground three	Human figure
Foreground	Butterfly in black and luna moth in green

The paintings shown in Fig. 12.1a, b are much more complicated than the simple abstract designs. However, we can capture the essence of the compositional design in these tables. We note that, in principle, a simpler description in terms of one background, one midground and one foreground is also possible. For example, we could redo the abstract designs of Fig. 12.1a, b as follows:

The three element abstract seadragons design

Layer	Description
Background	Blue gradient from dark to light, very dark green kelp plant
Midground	Human figure and bright green kelp plant
Foreground	Bright red sea dragon

The three element abstract tree painting design

Layer	Description
Background	Fuzzy brown trees
Midground	Large tree in brighter browns and dragon in red
Foreground	Human figure, butterfly in black and Luna moth in green

(a)   **(b)**

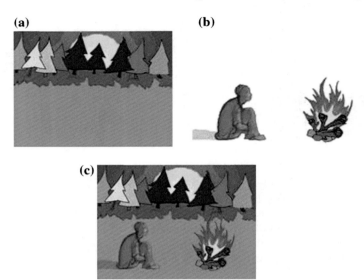

Fig. 12.2 The background and foreground of a simple painting. **a** Background of a *mellow* image. Note this image plane is quite complicated. Clearly, it could be broken up further into midground and background images. **b** Foreground of a *mellow* image which is also very complicated. A simpler painting using only one of the foreground elements would work nicely also. **c** A *mellow* painting

These new designs do not capture as much of the full complexity of the paintings as before, but we believe they do still provide the essential details. Also, we do not believe a two layer approach is as useful. Consider for example a two plane painting as shown in Fig. 12.2a, b. These two planes can then be assembled into a painting as we have described to give what is shown in Fig. 12.2c.

In this design, the foreground is too complex and should be simplified. There is also too much in the background. Some of that should move into a midground image so that the design in more clearly delineated.

We will therefore limit our attention to paintings that can be assembled using a background, midground and foreground plane. The background image is laid down first—just as an artist applies the paint for the background portion first. Then, the midground image is placed on top of the background thereby occluding a portion of the background. Finally, the foreground image is added, occluding even more of the previous layers. We recognize, of course, that a real artist would use more layers and move back and forth between each in a non-hierarchical manner. However, we feel, for the reasons discussed above, that the three layer abstraction of the compositional design process is a reasonable trade off between too few and too many layers. Also, we eventually will be encoding painting images into mathematical forms for computational purposes and hence, there is a great burden on us to design something pleasing using this formalism which at the same time is sufficiently simple to use as data in our cognitive model building process.

Our painting model thus will use a compositional scheme in which a valid painting is constructed by three layers: *background* (BG), *midground* (MG) and *foreground* (FG). A painting is assembled by first displaying the BG, then overlaying the MG which occludes some portions of the BG image and finally adding the FG image. The final FG layer hides any portions of the previous layers that lie underneath it. This simplistic scheme captures in broad detail the physical process of painting. When we start a painting, we know that if we paint the foreground images first, it will be technically difficult and aesthetically displeasing to paint midground and background images after the foreground. A classical example is painting a detailed tree in the foreground and then realizing that we still have to paint the sky. The brush strokes in the paint medium will inevitably show wrong directions if we do this, because we can not perform graceful side to side, long brush strokes since the foreground image is already there. Hence, a painter organizes the compositional design into abstract physical layers—roughly speaking, organized with the background to foreground layers corresponding to how far these elements are away from the viewer's eye. Recall the Würfelspiel matrix used in musical composition had the general form below.

$$
\mathcal{A} = \begin{bmatrix} \text{Opening 0} & \text{Transition 0} & \text{Closing 0} \\ \text{Opening 1} & \text{Transition 1} & \text{Closing 1} \\ \vdots & \vdots & \vdots \\ \text{Opening P-1} & \text{Transition P-1} & \text{Closing P-1} \end{bmatrix}
$$

The building blocks of a typical musical phrase here are an **Opening** followed by a **Transition** and then completed by a **Closing**. We can use this technique to generate painting compositions by letting the **Opening** phrase be the **Background** of a painting; the **Transition**, the **Midground** and the **Closing**, the **Foreground**. Hence, the painting matrix would be organized as in the matrix shown in Eq. 12.1

$$
\mathcal{A} = \begin{bmatrix} \text{Background 0} & \text{Midground 0} & \text{Foreground 0} \\ \text{Background 1} & \text{Midground 1} & \text{Foreground 1} \\ \vdots & \vdots & \vdots \\ \text{Background P-1} & \text{Midground P-1} & \text{Foreground P-1} \end{bmatrix} \tag{12.1}
$$

This painting matrix would then allow us to rapidly assemble P^3 different paintings. Both of these data sets are examples of Würfelspiel data matrices that enable us to efficiently generate large amounts of emotionally labeled data.

(a)   **(b)**   **(c)**   **(d)**

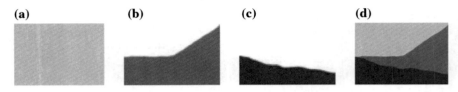

Fig. 12.3 A neutral painting. **a** Background. **b** Midground. **c** Foreground. **d** Assembled painting

Fig. 12.4 The neutral
painting matrix

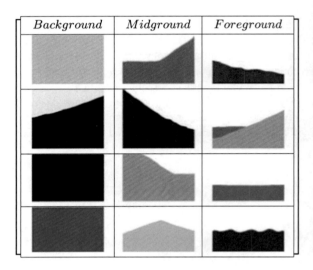

12.2   Neutral Painting Data

With this said, we can begin to design a series of paintings that are assembled
from three pieces: the background, midground and foreground thereby generating a
Künsterisches Würfelspiel matrix approach, an artistic toss of the dice. index-
The design of emotionally labeled painting data!the neutral painting Würfelspiel
matrix!neutral data design: flat and monochromatic colors with sparse lines Our first
Künsterisches Würfelspiel matrix will consist of four backgrounds, midgrounds and
foregrounds assembled into the usual 4 × 3 matrix and we will specifically attempt
to create painting compositions that have a neutral emotional tone. We have thought
carefully about what emotionally neutral paintings assembled in this primitive fashion from backgrounds, midgrounds and foregrounds should look like both as drawn
and as colored elements. We have decided on flat and monochromatic colors with
sparse lines. A typical painting composition design can be seen in Fig. 12.3a, b and c.
We can then assemble the background, middle ground and foreground elements into
a painting as shown in Fig. 12.3d. Note that this painting is quite bland and elicits no
emotional tag. In Fig. 12.4, we see a matrix of four backgrounds, four midgrounds
and four foregrounds which are assembled in a Würfelspiel fashion.

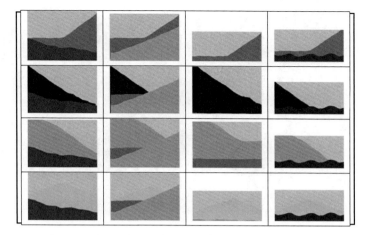

Fig. 12.5 16 Neutral compositions: background 1, midground 1, foregrounds 1–4, background 1, midground 2, foregrounds 1–4, background 1, midground 3, foregrounds 1–4, background 1, midground 4, foregrounds 1–4

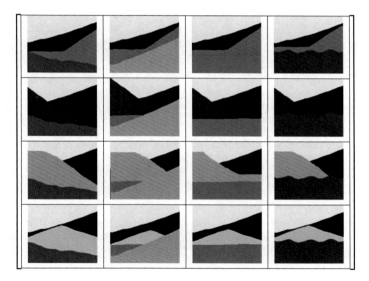

Fig. 12.6 16 Neutral compositions: background 2, midground 1, foregrounds 1–4, background 2, midground 2, foregrounds 1–4, background 2, midground 3, foregrounds 1–4, background 2, midground 4, foregrounds 1–4

12.2.1 The Neutral Künsterisches Würfelspiel Approach

We can use this matrix to assemble 64 individual paintings. We are displaying them in thumbnail form in the Figs. 12.5, 12.6, 12.7 and 12.8.

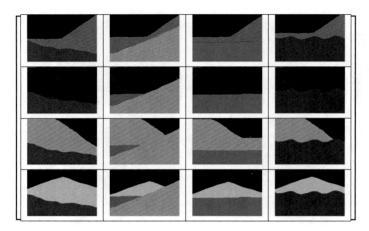

Fig. 12.7 16 Neutral compositions: background 3, midground 1, foregrounds 1–4, background 3, midground 2, foregrounds 1–4, background 3, midground 3, foregrounds 1–4, background 3, midground 4, foregrounds 1–4

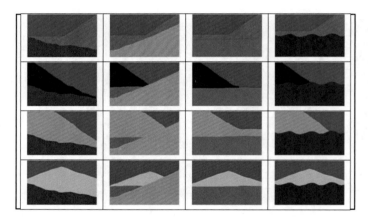

Fig. 12.8 16 Neutral compositions: background 4, midground 1, foregrounds 1–4, background 4, midground 2, foregrounds 1–4, background 4, midground 3, foregrounds 1–4, background 4, midground 4, foregrounds 1–4

12.3   Encoding the Painting Data

Our abstract painting compositions are encoded as the triple $\{b, m, f\}$ where b denotes the background,

m, the midground and f the foreground layer, respectively. Each of these layers is modeled with a collection of graphical objects. Each object in such a list has the following attributes: inside color, c_i; boundary color, c_b; and a boundary curve, $\partial\Omega$, described as an ordered array $\{(x_i, y_i)\}$ of position coordinates.

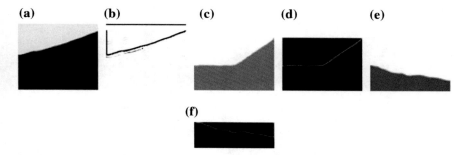

Fig. 12.9 Encoding a neutral painting. **a** The second background image. **b** The background edges. Note there are two edge curves. **c** The first midground image. **d** The midground edges. **e** The first foreground image. **f** The foreground edges

Consider the neutral painting constructed from the second background, first midground and first foreground image. Compute the edges of each image layer using the Linux tool **convert** via the command line **convert -edge 2 -negate <file in> <file out>**. The **-edge 2** command option extracts the edges giving us an image that is all black with the edge in white. The second **-negate** then swaps black and white in the image to give us the edges as black curves.

For expositional convenience, we will assume all of the paintings are 100×100 pixels in size. We can divide this rectangle into a 10×10 grid which we will call the coarse grid. Note that in Fig. 12.9b, the edge is a curve which can be

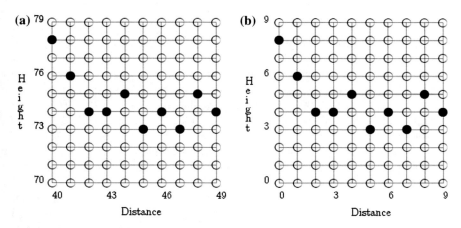

Fig. 12.10 Approximating the edge curve. **a** A 10×10 portion of a typical background image that contains part of an edge curve. The grid from *horizontal* pixels 40–49 and *vertical* pixels 70–79 is shown. The 10×10 block of pixels defines a node in the coarse grid that must be labeled as being part of the edge curve. **b** A typical coarse scale edge curve in which each filled in *circle* represents a 10×10 block in the original image which contains part of the fine scale edge curve

(a)   **(b)**   **(c)**   **(d)**

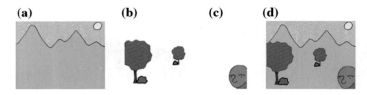

Fig. 12.11 Assembling a happy painting. **a** Happy background. **b** Happy midground. **c** Happy foreground. **d** A happy painting

Fig. 12.12 The happy painting matrix

approximated by a collection of positions $\{(x_i, y_j)\}$, where the index i ranges from 0 to some positive integer N, by drawing the line segments between successive points $p_i(t) = t(x_i, y_i) = (1 - t)(x_{i+1}, y_{i+1})$ for t in $[0, 1]$ for all appropriate indices i. The smaller 10×10 grid allows us to approximate these edge positions as shown in Fig. 12.10a. Each of the 100 large scale boxes that comprise the 10×10 grid that contain a portion of an edge curve are assigned a filled in circle in Fig. 12.10b. We see that the fine scale original edge has been redrawn using a coarser scale. Although there is inevitable loss of definition, the basic shape of the edge is retained.

In Chap. 11, we discussed the problem of generating neutral musical data alphabets. Let's revisit that now.

We can extend those ideas by realizing that in the context of music, part of our alphabet must represent notes of a variety of pitches. The music alphabet included letters which represented the standard seven notes of the octave below middle C, middle C octave and the octave above middle C. These notes could be listed as the matrix of Eq. 12.2 where the subscript $-1, 0$ and 1 on each note indicate the pitches below middle C, of middle C and above middle C, respectively.

$$\mathcal{M} = \begin{bmatrix} c_{-1} & d_{-1} & e_{-1} & f_{-1} & g_{-1} & a_{-1} & b_{-1} \\ c_0 & d_0 & e_0 & f_0 & g_0 & a_0 & b_0 \\ c_1 & d_1 & e_1 & f_1 & g_1 & a_1 & b_1 \end{bmatrix} \tag{12.2}$$

We note that there are more musical octaves that could be used and hence the matrix $\mathcal{M}$, now 3×7, could be as large as 7×7, if we wished to encode all of the notes on a piano keyboard. If we equate the last row of $\mathcal{M}$ with vertical pixel position 0 and the first row with 2, we see there is a nice correspondence between the matrix of positional information we need for the paintings and the matrix of pitch information we use for music. Hence, the manner in which we choose an alphabet for the painting data is actually closely aligned with the way in which we choose the alphabet for musical data despite the seeming differences between them.

We can also approximate the edge from the foreground, Fig. 12.9d, in a similar fashion. These edges define closed figures in the plane under the following conditions: first, if the first and last point of the curve are the same; and second, if the first and last point of the curve are on the boundary of the box. For example in Fig. 12.9c, the top edge hits the boundary of the image box on both the right and left side. Hence, we will interpret this as two simple closed figures in the plane: first, the object formed by including all the portions of the image box below the edge; and second, the complement of the first object formed by the portions of the image box above the edge. In addition, there is a second edge which is not a closed figure which therefore does not have an inside portion.

It is clear the edges will always have an edge color and will possibly have an inside color. We will let E^{pj} denote an edge set. The first superscript, p, takes on the values 0, 1 or 2, and indicates whether the edge set is part of a background, midground or foreground image. The second superscript tells us which edge set in a particular image we are focusing on. For example, Fig. 12.9c has two edge sets. Therefore, this background image can be encoded as the set E^0 defined by

$$\mathcal{E}^0 = \{E^{00}, c_i^{00}, c_b^{00}\}, \{E^{01}, c_b^{01}\}\}$$

since the second edge set does not require an inside color. To complete the description of this background image, we note that the entire image is drawn against a chosen base color. Hence, the background image can be described fully by the set $\mathcal{D}^0$

$$\mathcal{D}^0 = \{\beta^0, \mathcal{E}\}$$

where β^0 represents the base color of the background image. Note that we do not need to add a base color to the midground and foreground images. In general, we have a finite number of edges in each image. Let M^0, M^1 and M^2 denote the number of edges the background, midground and foreground images contain. Further, the ith ordered pair of the jth edge in an image is labeled as (x_{ij}^p, y_{ij}^p) where the value of p indicates whether we are in a background, midground or foreground image.

We will label the cardinality of each edge set by the integers N^{0j}, N^{1j} and N^{2j} with the first superscript having the same meaning as before and the second labeling which edge set we are considering. Thus, we can let E_i^{1j} represent the ith ordered pair in the jth edge in the midground. Further, we let inside and edge colors be denoted by c_i^{1j} and c_e^{1j}, respectively. Also, we will assume that the paintings use a small number of colors, here 8. We can encode a painting into the matrix shown in Eq. 12.3.

$$
\begin{array}{|c|ccc|}
\hline
\text{bg} & \{E^{00}, N^{00}, \beta^0, c_i^{00}, c_e^{00}\} & \cdots & \{E^{0M^0}, N^{0M^0}, c_i^{0M^0}, c_e^{0M^0}\} \\
\hline
\text{mg} & \{E^{10}, N^{10}, c_i^{10}, c_e^{10}\} & \cdots & \{E^{1M^1}, N^{1M^1}, c_i^{1M^1}, c_e^{1M^1}\} \\
\hline
\text{fg} & \{E^{20}, N^{20}, c_i^{20}, c_e^{20}\} & \cdots & \{E^{2M^0}, N^{2M^2}, c_i^{2M^2}, c_e^{2M^2}\} \\
\hline
\end{array}
\tag{12.3}
$$

The encoding of any of the background, midground or foreground images will use the alphabet we have described above. There are 100 possible (x_i, y_i) coordinates which can be painted in any of 8 colors. Each edge set in an image is thus a word in our alphabet comprised of a sequence of letters chosen from our alphabet of 100 positional dependent letters with eight possible colors. For example, the background image of Fig. 12.9a, is built from two words. The first word represents the first edge set and it consists of ten letters as the edge set goes all the way across the image. The second word denotes the second edge set and it consists of four letters.

12.4   Emotionally Labeled Painting Data

How can we capture emotional element in data that is sufficiently simple to be encoded efficiently for auditory cortex training? We begin with a short survey of the relevant literature.

12.4.1   Painting and Emotion in the Literature

The artist Schneider (2001), (*Capturing Emotion*), develops a philosophy about painting. Believing he "cannot paint feelings, but can apply paint in a manner that brings about an idea of those feelings," he consciously decides on particular elements to incorporate that are going to best portray the emotion he is seeking. The first important choice to make is the initial tone (color) laid down on a blank canvas. This would be our background layer. Schneider says this choice immediately establishes mood based on whether you choose a cool or warm tone. Next, he discusses edges and values. He places a value of nine for pure black and pure white is a one value. The degree of sharpness in the edges and the range of value used create varying degrees of contrast. Sharp edges and full range use of values gives a harsh effect that would not be good choices for paintings that are supposed to evoke emotions such as peacefulness, etc. Further contrast is established by lighting and shadows. Schneider links emotions such as horror, anger, fear, or surprise with high contrast lighting. It is interesting to note that an instance of spotlighting, which is of course a high contrast setting, is thought by this painter to evoke loneliness. He explains that the harshness of the contrast conveys isolation. Feelings such as affection, contentment, peace and introspection are all likely to be painted with subdued contrast. Schneider ends the article with some generalities about color. Pure colors suggest intensity or agitation; grays suggest peace, harmony, or reflection; and combinations of the elements addressed above will alter effects of the painting. The choices made will either reinforce or detract from the goal emotion. It is logical then to conclude that using soft muted colors to evoke peacefulness or contentment would be significantly negated by inserting sharp lines. The result would probably be one of indifference, like our neutral painting examples.

Within Pickford (1972), *Psychology and Visual Aesthetics*, the section titled "Feelings and Emotions Expressed by Lines and Shapes," discusses three research experiments done with lines. The first was by H. Lundholm circa 1921 which focused on distinguishing beautiful from ugly. In it, participants were asked to draw lines they thought should fall into each category. The "beautiful" lines tended to be unified in direction or movement, had continuity, lacked angles and intersections, and were symmetrical. The "ugly" lines were typically drawn with many irregularities and angles. Other research, done by Poffenberger and Barrows from 1924, gave 500 subjects 18 lines, both curved and angular, to look at and then assign an adjective from a given list. Although there were vast differences in the opinions of the 500 participants, some generalizations could be gathered from the data. Curves were labeled as sad, quiet, and lazy. Angles were labeled as hard, powerful, and agitating. Direction of the lines further differentiated the responses: horizontal lines were labeled quiet; downward sloping curves were sad or gentle; rising lines were merry or agitating; and downward sloping lines were sad, weak, or lazy. In the last research cited, Hevner from 1935, more lines were studied as well as different versions of the same portrait. This one also incorporated the colors red and blue. Again, curves were assigned qualities such as serene, graceful, and tender, while angles were rough, robust, vigorous and dignified. For the colors, red was happy and exciting, while blue was labeled serene, sad, or dignified. In a later chapter on "Associations and Attitudes to Colour", a variety of research experiments are cited. Consistent in all of them are the following: red is linked to excitement, cheer, defiance, and power; black is linked to sad, despondent, distressed, and power; blue and green are linked to calm, secure, tender; and yellow is linked with cheerfulness and playfulness.

In his overview of expression, (Hogbin 2000, *Appearance and Reality*), states that the visual elements of light and color evoke emotion depending on certain qualities or force of marking. As examples of this, he lists the taut curve, languid movement, and short staccato repetition. A viewer will be drawn into the intentions of the artist if correct essential characteristics are chosen by the painter. Hogbin suggests four emotional responses: desire (for pleasure), need (for survival), rejection (for danger), or wonderment (for the unknown). Hogbin then discusses light, color, marks, and line. Intensity of light will have an influence on mood. He gives several examples:

flashes of light will warn; glowing light will comfort; shadows from a spotlight may frighten; morning light may bring hope. Colors can be subjective, with much meaning linked to culture and association. Consider the colors blue and red. Blue may be associated with peace and calmness, but also sadness. Red might evoke feelings of anger and violence, but also love. More generalized statements can be made about colors and combinations. Complementary colors, those opposite one another on a color wheel, cause a *visual vibration* when placed together. This creates a more dynamic and lively feel. Also, greater intensity of color tends to imply emotional excitement. Use of colors near each other on the color wheel will quiet the energy of a painting. Hogbin divides marks into three types: non-objective, abstraction, or representational depending on the intent of the artist. Emotion enters as the viewer attaches meaning to a mark and does one of three things: empathizes, recoils, or is curious. Finally, lines can suggest activity or passivity as it moves from wiggly to a curve, and ending straight.

Useful sources of information on how artists address the issue of emotion in their work are included in the collection of writings by artists about art Goldwater (1974), *Artists on Art*. Though the turn toward expressing emotion in art is often considered a 19th century change, this collection includes source references that show that artists far earlier than that were considering how to evoke emotion within their paintings. Leon Battista Alberti's *Kleinere Künsttheoretische Schriften*, Vienna, 1877, passim. (Collated with De pictura, Basel, 1540.) (Sch. p. 111): this 15th century artist, 1404–1472, wrote, "A narrative picture will move the feelings of the beholder when the men painted therein manifest clearly their own emotions....emotions are revealed by the movements of the body". Giovanni Paolo Lomazzo's *Trattata dell'arte della pittura, scultura*, ed architettura, Milan, 1585, (Sch. p. 359): within a 16th century artist's (1538–1600) definition of painting we read, "Painting is an art which...even shows visibly to our eyes many feelings and emotions of the mind". Nicolas Poussin's *Correspondance de Nicolas Poussin*, ed. Ch. Jouanny, Paris, 1911, Letters 147, 156.- G.P. Bellori, Le vite de'pittori, scultori, ed architetti moderni, Rome, 1672, pp. 460–462: here, from a 17th century (1594–1665) artist, we find mention of emotions via discussion on form and color. "The form of each thing is distinguished by the thing's function or purpose. Some things produce laughter, others terror; these are their forms," and "Colors in painting are as allurements for persuading the eyes, as the sweetness of meter is to poetry".

Charles Le Brun's *Conference Upon Expression*, tr. John Smith, London, 1701, passim. (Paris, 1667) (Sch. p. 555): the artist, whose dates are 1619–1690, gives us some specific instruction about how to portray different expressions with the face and body language within paintings after stressing the importance of expression overall.

It is a necessary Ingredient in all the parts of Painting, and without it no picture can be perfect...Expression is also a part which marks the Motions of the Soul, and renders visible the Effects of Passion. Horror can be portrayed by the following suggestions, "the Eyebrow will be still more frowning; the Eye-ball instead of being in the middle of the Eye, will be drawn down to the under lid; the Mouth will be open, but closer in the middle than at the corners, which ought to be drawn back, and by this action, wrinkles in the Cheeks; the Colour of the Visage will be pale; and the Lips and Eyes something livid; this Action has some resemblance of Terror", and If to Joy succeed Laughter, this Motion is expressed by the Eyebrow raised about the middle, and drawn down next the Nose, the Ees almost shut; the Mouth shall appear open, and shew the Teeth; the corners of the Mouth being drawn back and raised up, will make a wrinkle in the Cheeks, which will appear puffed up, and almost hiding the Eyes; the face will be Red, the Nostrils open; the Eyes may seem Wet, or drop some Tears, which being very different from those of Sorrow, make no alteration in the Face; but very much when excited by Grief.

In the Pierre-Paul Prud'hon quote from Charles Clément's *Prud'hon, sa vie, ses oeuvres, et sa correpondance*, Paris, 1872, pp. 127, 178–80, written in 18th century Rome (1787), Prud'hon states,

> ...in general, there is too much concern with how a picture is made, and not enough with what puts life and soul into the subject represented...no one remembers the principal aim of those sublime masters who wished to make an impression on the soul, who marked strongly each figures character and, by combining it with the proper emotion, produce an effect of life and truth that strikes and moves the spectator....

From Gustave Coquiot's *Seurat*, Paris, ca. 1924, pp. 232–33, we find that on August 28, 1890, Seurat made the following comments about aesthetics. "Gaiety of tone is given by the dominance of light; of color, by the dominance of warm colors; of line by the dominance of lines above the horizontal. Calm of tone is given by the equivalence of light and dark; of color, by an equivalence of warm and cold; and of line, by horizontals. Sadness of tone is given by the dominance of dark; of color, by the dominance of cold colors; and by line, by downward directions".

Expression beyond that portrayed by a human subject in a painting is referenced in letters Matisse wrote in 1908, from *The Museum of Modern Art, Henri Matisse*, ed. Alfred H. Barr, New York, 1931, pp. 29–36. (La Grand Revue, Paris, Dec. 25, 1908). Matisse says

> What I am, above all, is expression...Expression, to my way of thinking, does not consist of the passion mirrored upon a human face or betrayed by a violent gesture. The whole arrangement of my picture is expressive. The place occupied by figures or objects, the empty spaces around them, the arranging in a decorative manner the various elements at the painter's disposal for the expression of his feelings, In a picture every part will be visible and will play the role conferred upon it, be it principal or secondary. All that is not useful in the picture is detrimental...

Gino Severini's *Du Cubisme au Classicisme*, Paris, 1921. and *Ragionamenti sulle arti figurative*, Milan, 1936. discusses certain rules in art. Music is specifically named as a similarly composed creation.

> An art which does not obey fixed and inviolable laws is to true art what a noise is to a musical sound. To paint without being acquainted with these fixed and very severe laws is tantamount to composing a symphony without knowing harmonic relations and the rules of counterpoint. Music is but a living application of mathematics. In painting, as in every constructive art, the problem is posed in the same manner. To the painter, numbers become magnitudes and color tones; to the musician, notes and sound tones.

Now our task is to take these ideas and use them to create useful emotionally labeled painting data.

12.4.2   The Emotional Künsterisches Würfelspiel Approach

As usual, we will think of paintings as assembled from three pieces: the background, midground and foreground. We will develop Künsterisches Würfelspiel matrices that correspond to sad and happy emotional states.

12.4.2.1   The Happy Würfelspiel Painting Matrix

A typical happy painting (BG, MG, FG) would be constructed to give the overall impression of a happy emotional state. An example is shown in Fig. 12.11a, b and c. We can assemble the background, middle ground and foreground elements into a painting as shown in Fig. 12.11d.

Now consider the array we get from four kinds of happy background, midground and foreground images. In Fig. 12.12, we see the resulting matrix. The first column of images are the backgrounds, the middle column, the midgrounds and finally, the last column, the foregrounds. Note that this matrix has a definite emotional state associated with it as all the images are somewhat happy—definitely not emotionally neutral images.

12.4.2.2   The Happy Paintings

We can use this matrix to assemble 64 individual paintings. We are displaying them in thumbnail form in the Figs. 12.13, 12.14, 12.15 and 12.16.

Fig. 12.13   16 Happy compositions: background 1, midground 1, foregrounds 1–4, background 1, midground 2, foregrounds 1–4, background 1, midground 3, foregrounds 1–4, background 1, midground 4, foregrounds 1–4

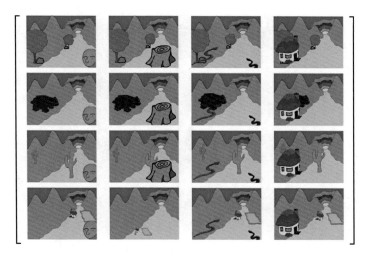

Fig. 12.14 16 Happy compositions: background 2, midground 1, foregrounds 1–4, background 2, midground 2, foregrounds 1–4, background 2, midground 3, foregrounds 1–4, background 2, midground 4, foregrounds 1–4

Fig. 12.15 16 Happy compositions: background 3, midground 1, foregrounds 1–4, background 3, midground 2, foregrounds 1–4, background 3, midground 3, foregrounds 1–4, background 3, midground 4, foregrounds 1–4

12.4.2.3 The Sad Würfelspiel Painting Matrix

As usual, a typical sad painting would be constructed from a given background, midground and foreground image which we have designed to give the overall impression of a sad emotional state. A typical painting composition design of this sort can

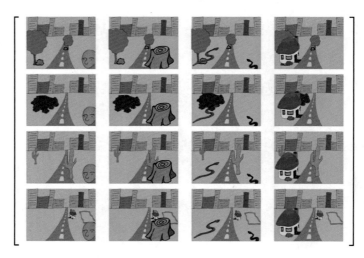

Fig. 12.16 16 Happy compositions: background 4, midground 1, foregrounds 1–4, background 4, midground 2, foregrounds 1–4, background 4, midground 3, foregrounds 1–4, background 4, midground 4, foregrounds 1–4

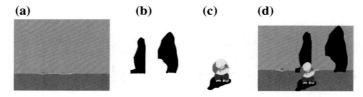

Fig. 12.17 Assembling a sad painting. **a** Sad background. **b** Sad midground. **c** Sad foreground. **d** A sad painting

be seen in Fig. 12.17a, b and c. We can assemble the background, middle ground and foreground elements into a painting as shown in Fig. 12.17d.

Now consider the array we get from four kinds of sad background, midground and foreground images. In Fig. 12.18, we see the resulting matrix. The first column of images are the backgrounds, the middle column, the midgrounds and finally, the last column, the foregrounds. Note that this matrix has a definite emotional state associated with it as all the images are somewhat happy—definitely not emotionally neutral images!

12.4.2.4   The Sad Paintings

We can use this matrix to assemble 64 individual paintings. We are displaying them in thumbnail form in the Figs. 12.19, 12.20, 12.21 and 12.22.

Fig. 12.18 The sad painting matrix

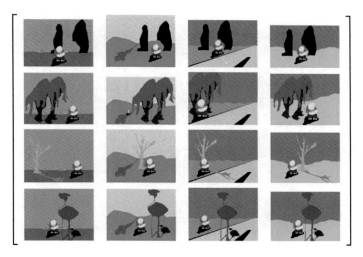

Fig. 12.19 16 Sad compositions: background 1, midground 1, foregrounds 1–4, background 1, midground 2, foregrounds 1–4, background 1, midground 3, foregrounds 1–4, background 1, midground 4, foregrounds 1–4

We now have enough data from both music and paintings to begin the process of training cortex in a full brain model.

Fig. 12.20 16 Sad compositions: background 2, midground 1, foregrounds 1–4, background 2, midground 2, foregrounds 1–4, background 2, midground 3, foregrounds 1–4, background 2, midground 4, foregrounds 1–4

Fig. 12.21 16 Sad compositions: background 3, midground 1, foregrounds 1–4, background 3, midground 2, foregrounds 1–4, background 3, midground 3, foregrounds 1–4, background 3, midground 4, foregrounds 1–4

Fig. 12.22 16 Sad compositions: background 4, midground 1, foregrounds 1–4, background 4, midground 2, foregrounds 1–4, background 4, midground 3, foregrounds 1–4, background 4, midground 4, foregrounds 1–4

References

R. Goldwater, M. Treves (eds.), *Artists on Art* (Pantheon Books, New York, 1974)

S. Hogbin, *Appearance and Reality* (Cambium Press, Bethel, 2000)

R. Pickford, *Psychology and Visual Aesthetics* (Hutchinson Educational LTD, London, 1972)

W. Schneider, Capturing emotion. Am. Artist **65**(703), 30–35 (2001)

Chapter 13
Modeling Compositional Design

We believe that one of the hardest problems to overcome in our attempt to develop software models of cognition is that of *validation*. How do we know that our theoretical abstracts of cellular signaling are giving rise to network architectures of interacting objects that are reasonable? Since we are attempting to create outputs that are good approximations of real cognitive states such as emotional attributes, it is not clear at all how to measure our success. Traditional graphs of ordinal data, tables of generated events versus measured events couched in the language of mathematics and the data structures of computer science, while valid, do not help us to see that the models are correct. To address this problem, we decided that in addition to developing cognitive models that are measured in traditional ways, we would also develop models that can be assessed by experts in other fields for validity. There are two such models we are attempting to build: one is a model of music composition in which short stanzas of music are generated autonomously that are emotionally colored or tagged. The second is a model of painting composition in which primitive scenes comprised of background, foreground and primary visual elements are generated autonomously with emotional attributes.

The cognitive models are based on simplified versions of biological information processing whose salient elements are sensory and associative cortex, the limbic system and pathways for the three key neurotransmitters serotonin, norepinephrine and dopamine. We will discuss some of the details of these computational modules in the sections to come. To validate these models, we have begun by constructing musical and painting data that will serve as samples of the cognitive output states we wish to see from Area 37 of the temporal cortex. The associative cortex devoted to assigning higher level meaning to musical phrases is trained with musical data built using an 18th century approach called a Würfelspiel matrix and a simple grammar based code. We have constructed examples of neutral, sad, happy and angry twelve beat musical sentences which are to be the desired output of this portion of area 37

© Springer Science+Business Media Singapore 2016   227
J.K. Peterson, *BioInformation Processing*, Cognitive Science and Technology,
DOI 10.1007/978-981-287-871-7_13

which functions as a polymodal fusion device in Chap. 11. In effect, we are using this data, to constrain the area 17 to area 37 pathways. Similarly, the associative cortex associated to assigning higher level meaning to a class of painting compositions built from assembling the three layers, background, midground and foreground, is trained from similar Würfelspiel matrices devoted to emotionally labeled painting compositions which were developed in Chap. 12. The resulting cognitive model then develops outputs in four emotional qualia for the separate types of inputs: auditory (music) and visual (painting).

13.1   The Cognitive Dysfunction Model Review

The first step that we need to take in building our models is to use the musical and painting data to constrain or *train* our model of the associative cortex. In general, our model takes this specialized sensory input and generates a high level output as shown in Fig. 13.1. In order to perform this training, we need to develop an abstract model of information processing in the brain. We have chosen the abstraction presented in Fig. 5.3. There is much biological detail missing, of course. We are focusing on the associative cortex, the limbic system and a subset of neurotransmitter generation pathways that begin in various portions of the midbrain. The architecture of the limbic system that we will use is shown in Fig. 5.2 and the information processing pathways we will focus on are shown in Fig. 13.2. Our cortex model itself is shown in Fig. 5.1. The musical data provides the kind of associated output that might come from area 37 of the temporal cortex. The low level inputs that start the creation of a music phrase correspond to the auditory sensory inputs into area 41 of the parietal cortex which are then processed through areas 5, 7 and 42 before being sent to the further associative level processing in the temporal cortex. The painting data then

Fig. 13.1 The sensory to output pathways

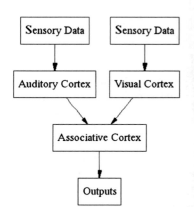

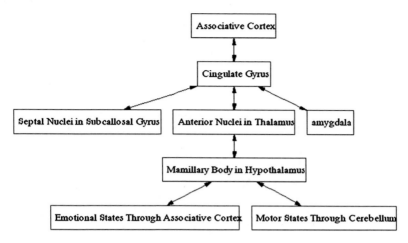

Fig. 13.2 A generic model of the limbic system

provides a similar kind of associated input into area 37 from the occipital cortex. Inputs that create the paintings correspond to the visual sensory inputs into area 17 of the occipital cortex which are then further processed by area 18 and 19 before being sent to the temporal cortex for additional higher level processing. We fully understand how simplified this view of information processing actually is, but we believe it captures some of the principle features. In Fig. 13.3, we show how we will use the musical and painting data to constrain the outputs of the associative cortex and limbic system. The data we discuss here will allow us to build the model shown in Fig. 13.4. In this model, we are essentially testing to see if the we generate the kinds of high level meta outputs we expect. Hence, this is part of our initial validation phase. However, a much more interesting model is obtained by inversion as shown in Fig. 13.5. Here, we use fMRI and skin conductance inputs and starting elements for music and painting data that have not been used in the training phase to generate new compositional designs in these various modalities.

In Fig. 13.3, we indicate that the outputs of the limbic system of our model are sent to the meta level pathways we use for music, painting and so forth and also to the cerebellum for eventual output to motor command pathways.

The emotional tagged data sets contain examples of equally valid solutions to compositional design processes. We have discussed musical data that is neutral and emotionally tagged (Chap. 11), neutral and neutral paintings (Chap. 12). In each of these chapters, we discuss the generation of 64 examples of solutions to compositional design tasks in different emotional modalities that are equally acceptable according to some measure. So, can we *learn* from this kind of data how the experts that designed these data samples did their job? Essentially, buried in these data sets are important *clues* about what makes a great design. How do we begin to understand the underlying compositional design principles?

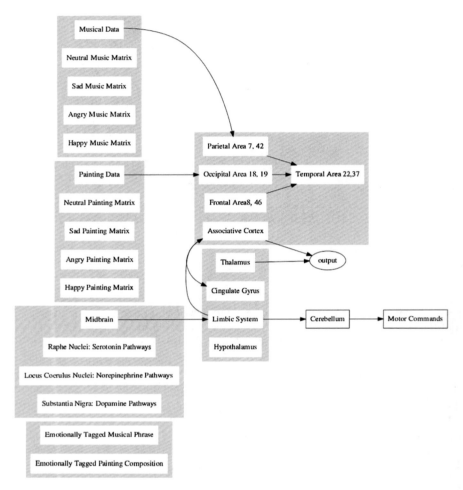

Fig. 13.3 The musical and painting data is used to constrain the outputs of the associative cortex and limbic system

Each data set that is encoded into a Würfelspiel matrix, whether using music or art, therefore contains crucial information about equally valid examples of data in different emotional modalities. From this data, we can build mappings that tell us which sequences of choices are valid and which are not from the perspective of the expert who has designed the examples. In a very real sense, a cognitive model built from this data is automatically partially validated from a psychological point of view.

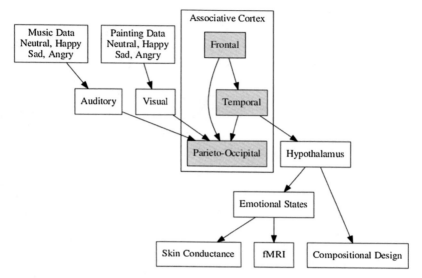

Fig. 13.4 Music or painting initialization generates fMRI and skin conductance outputs and a full musical or painting design

13.2   Connectionist Based Compositional Design

Let's now look at how we might develop a standard connectionist approach to building models that **understand** our data. We will move to an explicit neurobiological model later. Each data set that is encoded into a Würfelspiel matrix, whether using music, art or something else, contains crucial information about equally valid design solutions in different modalities. From this data, we want to build mappings that tell us which sequences of choices are valid, and which are not, from the perspective of the expert who has designed the examples. In a very real sense, a cognitive model built from this data is automatically partially validated from a psychological point of view. For example, the neutral and emotionally tagged music and painting data sets contain examples of equally valid solutions to a compositional design process in both neutral and emotionally labeled flavors. Each set of data contains 64 examples of solutions to compositional design tasks that are equally acceptable according to some measure.

13.2.1   Preprocessing

Each data set has an associated alphabet with which we can express noun, verb and object units. For our purposes, let's say that each of the nouns, verbs or objects is a finite list of actions from an alphabet of R symbols, where the meaning of the symbols is, of course, highly dependent on the context of our example. In a simple

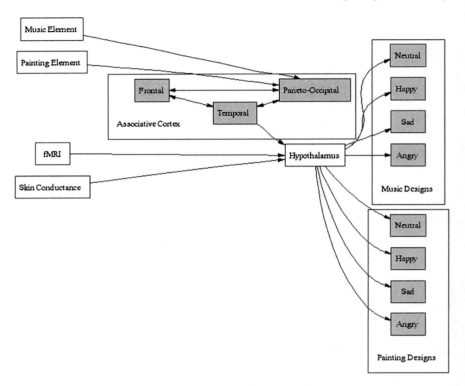

Fig. 13.5 fMRI and skin conductance inputs plus music or painting initialization generate a full musical or painting design in a given emotional modality

music example, the list of actions might be a sequence of four beats and the alphabet could be the choices from the C major scale. Thus, we will assume each of the P nouns consists of a list of length L from an alphabet of R symbols. The raw inputs we see as the noun vector n are thus normally processed into a specialized vector for use in algorithms that model the compositional process. For example, in the generation of a painting, we use our painting medium to create the background which is a process using pigment on a physical canvas. This is the raw input n and each element of a noun n is a letter from the alphabet. Let the letters in the alphabet be denoted by a_0 through a_{R-1}. Then we can associate the noun with a vector of L components, $n[0], \ldots, n[L-1]$ where component $n[j]$ is a letter we will represent by a_j^n. The letter a_j^n is then encoded into a vector of length R all of whose entries are 0 except for the entry in slot j. In general, we can choose to encode n into a form more amenable to computation and data processing in many ways. In our work, we have encoded the raw data into an abstract grammar, thereby generating the feature vector N. In general, the input nouns n are preprocessed to create output noun states denoted by N. In a similar fashion, we would preprocess verb and object inputs to create verb and object output states denoted by V and O, respectively. The preprocessing is carried out by mappings f_n, f_v and f_o, respectively. For example, the mapping from raw data to the feature vector form for nouns is represented by Eq. 13.1:

$$n = \begin{bmatrix} a_0^n \\ \vdots \\ a_{L-1}^n \end{bmatrix} \xrightarrow{f_n} N = \begin{bmatrix} N_0 \\ N_1 \\ \vdots \\ N_{L-1} \end{bmatrix}$$

Our association of a noun n with the feature vector N is thus but one example of the mapping f_n. In a similar fashion, we would preprocess verb and object inputs to create verb and object output states denoted by V and O, respectively. The preprocessing is carried out by mappings f_v and f_o, respectively.

Now, a primitive object in our purported compositional grammar has length L. We can denote such an input noun object as n_i and a corresponding output noun object as N_i. Our mapping problem is thus to determine the rule behind the mapping from the noun feature vectors N to the verb feature vectors V, g_{NV}, and from the verb feature vectors V to the object feature vectors O, g_{VO}. We can express this mathematically by Eq. 13.1.

$$\begin{bmatrix} N_0 \\ N_1 \\ \vdots \\ N_{L-1} \end{bmatrix} \xrightarrow{g_{NV}} \begin{bmatrix} V_0 \\ V_1 \\ \vdots \\ V_{L-1} \end{bmatrix} \text{ and } \begin{bmatrix} V_0 \\ V_1 \\ \vdots \\ V_{L-1} \end{bmatrix} \xrightarrow{g_{VO}} \begin{bmatrix} O_0 \\ O_1 \\ \vdots \\ O_{L-1} \end{bmatrix} \quad (13.1)$$

We can combine these processing steps into the diagram shown in Fig. 13.6.

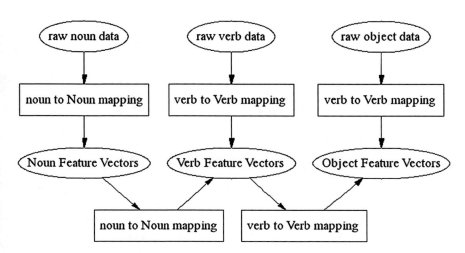

Fig. 13.6 Raw sentence to feature vector processing

13.2.2   Noun to Verb Processing

The data we are given is order dependent. For example, if we are given a noun, n_i of form $\{a_{i0}^n, \ldots, a_{i,L-1}^n\}$, then we attend to the letters of this noun sequentially as $a_{i0}^n \to a_{i1}^n \to \cdots \to a_{i,L-1}^n$. We are given that each noun n_i is associated with a set of possible verbs, $\{v_j\}$ equal to $\{a_{j0}^v, \ldots, a_{j,L-1}^v\}$ for $0 \le j < P - 1$ and one task is thus to understand the noun to verb mapping. However, another task is to understand how to generate the original noun sequence. Why are some noun sequences useful or pleasing in this context and others are not? To generate a noun sequence n equal to $\{a_0^n, \ldots, a_{L-1}^n\}$ means we choose a random start letter a_0^n and then from that preferred sequences are generated while non-interesting words are biased against. Hence, we think of a mapping, the Noun Generating Map or NGM as accepting an input, a_0^n and generating a preferred second note a_1^n. Then a_1^n is used as an input to generate a preferred third note, a_2^n and so on until the full string of letters is finished. To model this mapping, we start by using the information about useful noun strings we have. Given letter a_{i0}^n, we know that a_{i1}^n is preferred.

We embed the original data into an analog vector by converting each letter a_{i0}^n of the noun, which is a 0 or a 1 in our initial encoding, into a real number ξ_{i0}^n. To set the value of the real number ξ_{i0}^n, we choose a thresholding tolerance, ϵ, in the interval $(0, 0.25)$ and choose a real number y randomly from the interval $[-0.5\epsilon, 0.5\epsilon]$ and then set the value of ξ_{i0} to be $\epsilon \pm y$. Therefore, the value of ξ_{i0}^n lies in the interval $[0.5\epsilon, 1.5\epsilon]$. For example, if ϵ was chosen to be 0.20, then for all indices in the binary encoding of the letter a_{i0}^n that are 0, we would randomly choose y from $[-0.1, 0.1]$ generating ξ_{i0}^n values that lie in $[0.10, 0.30]$. We will call the number ϵ our analog threshold. The entry in the binary encoded letter that corresponds to a 1 will be randomly chosen from $[1 - 1.5\epsilon, 1 - 0.5\epsilon]$. Hence, for $\epsilon = 0.2$, entry with a 1 will be assigned a real number in the interval $[0.7, 0.9]$. Consequently, the raw binary noun, verb and object data is mapped into a new analog representation in which each entry is a real number chosen as above.

Since we know that only certain letters should follow a_{i0}^n, only certain analog states ξ_{i1}^n are permissible given a start state of ξ_{i0}^n. We infer from this that there is an unknown mapping h^{01} which maps the analog encoding of letter 0 to the analog encoding of letter 1, $h^{01}(\xi_{i0}^n) = \xi_{i1}^n$. This mapping has special characteristics: our data tells us that only certain letters that can follow letter 0. Each acceptable second letter is a vector in R dimensional space whose components are analog zero except the one the corresponds to the second letter. That component is an analog one. We have at most P examples of acceptable second letters. This means we have $R - P$ second letters that are not acceptable. In other words, the preferred output for a given noun is a R row matrix formed from the acceptable verbs that has at most P columns.

We can do this for all of the nouns in our data set. Hence, we will have at most P first letter choices and each of these will have at most $R - P$ unacceptable second letters. Let T and T' denote the set of all acceptable and unacceptable outputs respectively. Following a traditional machine learning approach, we could model the mapping h^{01} as a chained feed forward network, with feedforward and feedback connections between artificial neurons. This mapping takes the first letter of a noun and outputs a set of acceptable second letters. Training is done by matching input to output using excitation and inhibition (i.e., using a Hebbian approach). We know which elements in the analog output vector should be close to one and which should be close to zero for the analog input.

We initialize all of the tunable parameters to be small positive if the connection from component k in the input to component j on the output is between two analog ones. All other connections are initialized to small negative numbers. For each first letter we have data for, we do the following: pick the initial first letter in our data set and compute the relevant output for the first associated second letter. Increase the connection weights on any path between a high input and a high output and decrease the connection weight on any other paths. Cycle to the next second letter and redo until all possibilities are exhausted. We thus continue this process until every input generates an output with a high component value in the location that corresponds to the index for the second letter. At this point, we say we have trained our nonlinear mapping h^{01} so that first letters in our data noun sequences are biased to connect to their corresponding second letters. A second letter is then chosen randomly from the set of acceptable second letters via an additional input line which is a sense is a coarse model of *creativity*.

If we let the set of all the generated weights be the matrix W^{01}, we note this is an $R \times PR$ size matrix. We can develop a similar mapping for the second to third letter, h^{12} with weights W^{12}, the third to fourth letter, h^{23} with weights W^{23}, and finally, the mapping from letter $L - 1$ to letter L, $h^{L-2,L-1}$ with weights $W^{L-2,L-1}$.

The procedure for creating a valid noun sequence can now be given. Choose a valid starting letter for a noun, a_{i0}^{n} and we map it to its analog form, ξ_{i0}^{n}. Then, applying the first to second letter map, we find an acceptable second letter by the computation $h^{01}(\xi_{i0}^{n}) = \{\xi_{i1}^{n}\}$. This second letter can be used as the input into the next map, generating an acceptable third letter. Hence, the composite map, $h^{12}h^{01}$ takes a valid first letter and creates the three letter analog sequence defined by Eq. 13.2:

$$\begin{bmatrix} \xi_{i0}^{n} \\ \xi_{i1}^{n} \\ \xi_{i2}^{n} \end{bmatrix} \in \begin{bmatrix} \xi_{i0}^{n} \\ h^{01}(\xi_{i0}^{n}) \\ h^{12}(h^{01}(\xi_{i0}^{n})) \end{bmatrix} \tag{13.2}$$

The analog sequences are then mapped into three letter sequences by assigning an analog value to either a one or zero using a threshold tolerance τ. This means we map a component whose value is above τ to 1 and one whose value is below τ to 0. This can of course generate invalid sequences as we are only supposed to have a single 1 assigned from any analog sequence. We do have to make sure that our developed map does not allow this. For example, for $\tau = .6$, the vector

$$\begin{bmatrix} 0.83 \\ 0.55 \\ 0.35 \end{bmatrix} \xrightarrow{\tau=0.6} \begin{bmatrix} 1 \\ 0 \\ 0 \end{bmatrix}$$

generates a valid noun, but if $\tau = .5$, an invalid binary sequence is generated.

$$\begin{bmatrix} 0.83 \\ 0.55 \\ 0.35 \end{bmatrix} \xrightarrow{\tau=0.5} \begin{bmatrix} 1 \\ 1 \\ 0 \end{bmatrix}$$

which we would not know how to interpret as part of a noun. Nevertheless, despite these obvious caveats, the procedure above learns how to generate all acceptable three letter nouns given an initial start letter. There will be at most P^2 possible second and third letters in this set. Since this set of possibilities will grow rapidly, after generating the letter two set of possibilities, we randomly choose one of the columns of the letter two matrix as the second letter choice and apply the h^{12} mapping to that letter. We then randomly choose one of the columns of the letter three matrix as the third letter choice.

We then extend this procedure to the generation of all P letters with concatenation $h^{P-2,P-1}\cdots h^{12}h^{01}$ which we will denote by the symbol H^n, where the superscript indicates this is the mapping we will use for noun sequences. The mapping H^n is the Noun Generating Map or NGM that we seek. It generates a set of P^{P-1} letter two to letter P sequences. By making a random column choice at each letter, we generate one random P letter noun sequence for each initial letter we use. We can do something similar for the verb and object data generating the Verb and Object Generating Maps H^v and H^o, respectively. These three mappings are the noun, verb and object generator mappings we were seeking. Then we need to connect nouns to verbs and verbs to objects. The mapping from N to V is where the real processing lies. The Würfelspiel matrix training data approach tells us that it is permissible for certain nouns to be linked to certain verbs. While we could memorize a look-up table based on this data, that is not what we wish to do. We want to determine underlying rules behind these associations as emergent behavior in a complex system of interacting agents. Thus, each output noun N_i in the the collection of P nouns $\{N_i : 0 \leq i < P\}$ should activate any of the P verbs $\{V_j : 0 \leq j < P\}$ via the map g_{NV}. Further, each verb V_j in $\{V_j : 0 \leq j < P\}$ should activate the output nouns $\{O_i : 0 \leq i < P\}$ by the action of the map g_{VO}. To build the mappings g_{NV} and g_{VO}, we will eventually use a more sophisticated model which is graph based with feedback which is another discussion entirely.

13.2.3   Sentence Construction

There are then two ways to create a valid emotionally labeled music or painting composition. The first does not use the mappings g_{NV} and g_{VO}. For a painting, a random choice of background letter is chosen to begin the sentence selection process. This input generates a valid background image. Then, a randomly chosen starting letter for the midground is then used to generate a valid midground image. Finally, a random start letter for the foreground generates the last foreground image from which the painting is constructed. The output of the cognitive module is thus a short sentence, that is painting, of the type we have discussed. The second method is more interesting. The randomly generated noun N generates a valid verb $g_{NV}(N)$ and the valid object is generated by the concatenation $g_{VO}(g_{NV}(N))$. Thus, the composite map $g_{VO}\,g_{NV}$ provides a sentence generator.

Once we can generate sentences, we note that we can move to the generation of streams consisting of sentences concatenated to other sentences after we create an Object to Noun mapping. This is done in a way that is similar to what we have done before using a Würfelspiel array approach. Other possibilities then come to mind; for example, in this painting context, such transitions can create arbitrarily long visual streams we can call animations. The analog of key changes in music is then perhaps visual scene changes. Since key changes are logical transitions, we can create arbitrarily long musical streams punctuated by appropriate key changes by using the Würfelspiel array approach to model which key changes between given key signatures are pleasing between two musical streams. Note, this construction process has all the problems of a traditional connectionist approach. The g_{NV}, g_{VO} and g_{NO} maps depend on our data and would have to be rebuilt when new data is used. We believe it will be much better to replace this connectionist approach with a biologically based simple brain model.

13.3   Neurobiologically Based Compositional Design

We can see that in principle, there is a lot that can be accomplished by building the maps we have discussed above using any architectures we desire. However, we would like our connectionist architecture choices to be based on what we know of neurobiology in either humans or other creatures with interesting neural systems such as spiders, honeybees and cephalopods.

13.3.1   Recalling Data Generation

In the 1700s, fragments of music could be rapidly prototyped by using a matrix of possibilities called a Würfelspiel matrix. The historical examples were constructed as 10×10 matrices where each entry was a 16 beat sequence. The musician would toss a die to determine which choice to use from each column. Hence, a musical Würfelspiel matrix could be used to rapidly prototype 1000 160 beat samples of a chosen musical type. In essence, a Würfelspiel music matrix captures in an abstract and succinct form the genius of the artist. We show an abstract version of a typical Musikalisches Würfelspiel matrix, $\mathcal{M}$, in Eq. 13.3. It consists of P rows and three columns. In the first column are placed the opening (O) phrases or nouns; in the third column, are placed the closing phrases (C) or objects; and in the second column, are placed the transitional phrases (T) or verbs. Each phrase consisted of L beats and the composer's duty was to make sure that any opening, transitional and closing (or noun, verb and object) was both viable and pleasing for the musical style that the composer was attempting to achieve. Thus, a musical stream could be formed by concatenating these fragments together: picking the ith Opening (O_i) , the jth Transition (T_j) and the kth Closing (C_k) phrases would form a musical sentence. Note that there are P^3 possible musical sentences that can be formed in this manner. If each opening, transition and closing fragment is four beats long, we can build P^3 different twelve beat sentences.

$$
\mathcal{M} = \begin{bmatrix} O_0 & T_0 & C_0 \\ O_1 & T_1 & C_1 \\ \vdots & \vdots & \vdots \\ O_{P-1} & T_{P-1} & C_{P-1} \end{bmatrix} \quad \mathcal{P} = \begin{bmatrix} B_0 & M_0 & F_0 \\ B_1 & M_1 & F_1 \\ \vdots & \vdots & \vdots \\ B_{P-1} & M_{P-1} & F_{P-1} \end{bmatrix} \tag{13.3}
$$

Further, a simple painting can also be considered as a different sort of triple; it consists of a background (B), a midground (M) painted on top of the background and finally, a foreground (F) layer which further occludes additional portions of the combined background and midground layers. Hence, paintings could be organized as a matrix $\mathcal{P}$ as shown in Eq. 13.3 also. Thus, a painting could be formed by concatenating these images together: picking the ith background (B_i), the jth midground (M_j) and the kth foreground (F_k) phrases would form a painting $B_i M_j F_k$. Note that there are again P^3 possible paintings that can be formed. A sample music matrix is shown in Fig. 11.13 and a painting matrix in Fig. 12.12.

Clearly, each data set encoded into a Würfelspiel matrix therefore contains crucial information about equally valid examples of data that are solutions to a certain design problem. The existence of this data means that mappings exist that tell us which sequences of choices are valid and which are not from the perspective of the expert who has designed the examples. Each data set therefore has an associated alphabet with which we can express noun, verb and object units. For our purposes, let's say that each of the nouns, verbs or objects is a finite list of actions from an alphabet of R symbols, where the meaning of the symbols is of course highly dependent on the context of our example. In a simple music example, the list of actions might be a sequence of four beats and the alphabet could be the choices from the C major scale. Thus, we will assume each of the P nouns consists of a list of length L from an alphabet of R symbols. Since the general alphabet has R symbols, each element of a noun can be thought of as a vector of size R whose components are all 0 except for a single 1. Let the letters in the alphabet be denoted by a_0 through a_{R-1}. Then a noun has components 0 to $L-1$ where component $n[j]$ is the letter a_j^n. The letter a_j^n is then encoded into a vector of length R all of whose entries are 0 except for the entry in slot j.

Fig. 13.7 A three cortical column model

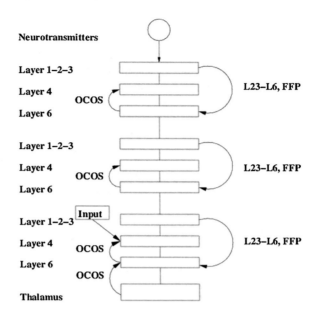

13.3.2 Training the Isocortex Model

Consider a typical multiple column portion of our cortex model, say Area 17, 18 and 19. We could think of it as shown abstractly in Fig. 13.7. In this picture, we show a stack of three cortical columns. At the top of the figure, we show the neurotransmitters input from specialized midbrain areas. In our model, we are limiting our attention to serotonin from the Raphe nuclei, norepinephrine from the Locus Coeruleus and dopamine from the dopamine producing neurons. At the bottom, we indicate that information flow through the cortical columns the thalamic control circuitry.

The situation is of course much more complicated and you must remember that the arrows designated as OCOS and FFP actually refer to abstractions of specialized groupings of neurons as shown in earlier figures as shown in Sect. 5.4.1. Let's assume that we have obtained auditory and visual cortex data (for example, the musical and painting data matrices $\mathcal{M}$ and $\mathcal{P}$ previously discussed) which are encoded into feature vectors of our design using an abstract grammar. In our own physiology, raw sensory input is also coded into progressively more abstract versions of the actual signals. Hence, we view our chosen algorithms for computing the feature vector of the given data as the analog of that process. These feature vectors are then used as the inputs into layer four of the bottom cortical column corresponding to each sensory modality. The powerful OCOS and FFP circuits then serve as devices that do feature grouping. The outputs of associative cortex areas such as area 20, 21, 22 and 37 of the temporal cortex would then be constrained to match the known outputs due to our data.

To build the information processing models we are discussing thus requires that we understand the input to output maps that are based on the Würfelspiel data generated for our given sensory modalities. As discussed above, this data will be used to train the auditory and visual cortex in preparation for the training of the associative cortex. The trained model of the associative cortex provides use with a neurobiologically motivated model of sensor fusion. To understand this process better, we consider the details of how a model of sentence construction for a given abstract grammar works. This show quite clearly the perceptual grouping tasks that must be accomplished with the OCOS and FFP training algorithms applied to the cortical columns of Fig. 5.24. Recall, the raw data n, v and o is encoded into feature vectors N, V and O, respectively. The preprocessing is carried out by mappings f_n, f_v and f_o, respectively. Now, a primitive object in our compositional grammar has length L. Our mapping problem is thus to determine the rule behind the mapping from the noun feature vectors N to the verb feature vectors V, g_{NV}, and from the verb feature vectors V to the object feature vectors O, g_{VO}. We can express this mathematically by Eq. 13.1. To generate a noun sequence n_i equal to $\{a_{i0}^n, \ldots, a_{i,L-1}^n\}$ means we choose a random start letter and then from that preferred sequences are generated while non-interesting words are biased against. Hence, we think of a mapping, the Noun Generating Map or **NGM** as accepting an letter input, a_0^n and generating a preferred noun. Note if we start with a_0^n, we then need to generate a preferred second letter, a_1^n. Then a_1^n is used as an input to generate a preferred third note, a_2^n and so on until the full string of letters is finished. The verb and object data provide us with samples of a verb and object generating map, **VGM** and **OGM** also. To model this mapping, we start by using the information about useful noun strings we have.

Since we know which letters are preferred as the next letter given a starting choice, we know that we have at most P first letter choices and each of these will have at most $R - P$ unacceptable second letters. Let T and T' denote the set of all acceptable and unacceptable outputs respectively. As we noted earlier, we could model this mapping as a chained feed forward network, with feedforward and feedbackward connections between artificial neurons using standard training techniques. This mapping would take the first letter of a noun and output a set of acceptable second letters. However, we would need to build such a mapping for each of the letter to letter transitions and this is certainly computationally messy and even if it is workable for the nouns, it is still not at all like the circuitry we see in Fig. 5.24. Hence, we will instead choose to think of this mapping as a cortical column circuit whose outputs are produced by OCOS and FFP interactions as described in Grossberg (2003). That is, the noun inputs a_{ij}^n go into layer 4 of the bottom cortical column of Fig. 5.24 and the OCOS and FFP circuits then allow us to adjust the strengths of connections between the neurons in the column so that we obtain the groupings of letters that are preferred in the nouns of our sample. For convenience of exposition, we note the cortical column for nouns can be labeled using the three six layer cortical groups, G_{n0}, G_{n1} and G_{n2}. Further, let G_{ni}^ℓ denote the ℓth in layer G_{ni}. Then, we have the training sequence

$$G_{n0}^4 \rightarrow G_{n0}^2 \rightarrow G_{n1}^4 \rightarrow G_{n1}^2 \rightarrow G_{n2}^4 \rightarrow G_{n2}^2$$

where each arrow denotes perceptual grouping training using OCOS and FFP laminar circuits. The output from layer 2 of the top cortical group of the noun column thus consists of the preferred letter orders for nouns. We handle the verb and object data for groups G_{vi} and G_{oj} in a similar way using the training sequences

$$G_{v0}^4 \rightarrow G_{v0}^2 \rightarrow G_{v1}^4 \rightarrow G_{v1}^2 \rightarrow G_{v2}^4 \rightarrow G_{v2}^2$$
$$G_{n0}^4 \rightarrow G_{o0}^2 \rightarrow G_{o1}^4 \rightarrow G_{o1}^2 \rightarrow G_{o2}^4 \rightarrow G_{o2}^2.$$

To generate the preferred Noun to Verb and Verb to Object mappings, we use two more cortical columns with the groupings G_{NVi} and G_{VOj}. We then use the training sequences

$$G_{n2}^2, G_{v2}^2 \rightarrow G_{NV0}^4 \rightarrow G_{NV0}^2 \rightarrow G_{NV1}^4 \rightarrow G_{NV1}^2 \rightarrow G_{NV2}^4 \rightarrow G_{NV2}^2$$
$$G_{V2}^2, G_{o2}^2 \rightarrow G_{VO0}^4 \rightarrow G_{VO0}^2 \rightarrow G_{VO1}^4 \rightarrow G_{VO1}^2 \rightarrow G_{VO2}^4 \rightarrow G_{VO2}^2$$

The output from Noun–Verb cortical column is a preferred verb for a given noun and the output from the Verb–Object cortical column is a preferred object for a given verb. We thus have five cortical column stacks whose bottom column handles input sentence parts and outputs a grammatically correct following part of the growing sentence. At this point, we have the noun, verb and object generating maps, NGM, VGM and OGM, that we have sought, as well as instantiations of Noun to Verb and Verb to Object mappings.

13.3.3 Sensor Fusion in Area 37

A given set of sensory data encoded into a Würfelspiel matrix using a rudimentary grammar based on an abstraction of the signal can be used to imprint a five cortical column model of isocortex. Let's assume we have data from two sensory modalities, A and B. We posit an information processing architecture which consists of the four mutually interacting agents: isocortex imprinted by sensory data of type A, CA; isocortex imprinted by sensory data of type B, CB; an isocortex Area 37 module, A37, to be imprinted by the expected outcome due to the fusing of information from CA and CB; and an isocortex limbic module, LM, whose purpose is to provide modulatory influence to the inputs of the other three modules. Using notation similar to what we have used previously, the training sequence, suppressing internal details, is shown in Eq. (13.5).

$$\text{Sensory A input} \rightarrow G_{CA0}^4 \rightarrow G_{CA2}^2$$
$$\text{Sensory B input} \rightarrow G_{CB0}^4 \rightarrow G_{CB2}^2$$
$$G_{CA2}^2, G_{CB2}^2 \rightarrow G_{A37,0}^4 \rightarrow G_{A37,2}^2 \qquad (13.4)$$
$$G_{CA2}^2, G_{CB2}^2, G_{A37,2}^2 \rightarrow G_{LM0}^4 \rightarrow G_{LM2}^2$$
$$G_{LM2}^2 \rightarrow G_{CA0}^4, G_{CB0}^4, G_{A37,0}^4$$

13.4 Integration of the Models

We can thus construct mappings that encode the compositional designs from our painting and music data. This suggests some basic principles: if we use the superscripts α and β to denote emotional modality and data type, respectively, we can label the mappings in the form $\{g_{NV}^{\alpha\beta}, g_{VO}^{\alpha\beta}\}$. We let $\alpha = 0, 1, 2, 3$ denote neutral, happy, sad and angry and $\beta = 0, 1$ indicate the choice of music and painting designs, respectively. We thus have a collection of mappings

$\{g_{NV}^{00}, g_{VO}^{00}\}$	$\{g_{NV}^{01}, g_{VO}^{01}\}$	$\{g_{NV}^{02}, g_{VO}^{02}\}$	$\{g_{NV}^{03}, g_{VO}^{03}\}$	Music
$\{g_{NV}^{10}, g_{VO}^{10}\}$	$\{g_{NV}^{11}, g_{VO}^{11}\}$	$\{g_{NV}^{12}, g_{VO}^{12}\}$	$\{g_{NV}^{13}, g_{VO}^{13}\}$	Painting

The high level cortex processing is shown in Figs. 13.8 and 13.9. A simplified version of the auditory and visual cortex to associative cortex processing is illustrated. Only some of the interconnections are shown. Each cortex type is modeled as four layers which are associated with the specific time frames of milliseconds (10^{-3} s), seconds, days (10^3 s) and weeks (10^6 s). Each time frame connects feedforward and feedback to all time frames with longer time constants (i.e. the millisecond time frame outputs project to the second, day and week time frames). The associative cortex processing is assumed to be output via the four time frames of the temporal cortex. We show very few of these connections as the diagram quickly becomes difficult to read. However, the frontal, parieto-occipital and temporal cortex modules are all fully interconnected. The portions of these diagrams labeled as frames of specific time

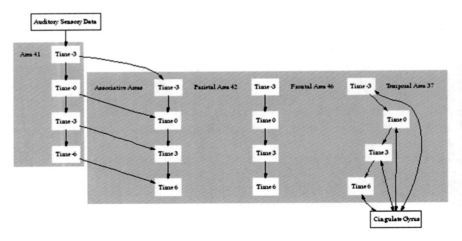

Fig. 13.8 A simplified version of the auditory cortex to associative cortex processing. Only some of the interconnections are shown. Each cortex type is modeled as four layers which are associated with specific time frames. The different processing time frames are labeled -3, 0, 3, and 6 for millisecond, seconds, days and weeks. The temporal cortex time frames shown connect to limbic system represented by the cingulate gyrus

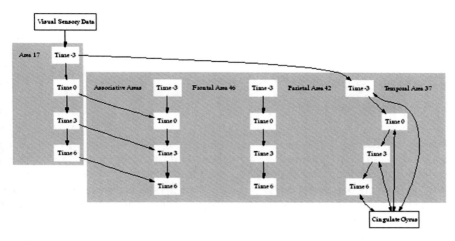

Fig. 13.9 A simplified version of the visual cortex to associative cortex processing. Only some of the interconnections are shown. Each cortex type is modeled as four layers which are associated with specific time frames. The different processing time frames are labeled −3, 0, 3, and 6 for millisecond, seconds, days and weeks. The temporal cortex time frames shown connect to limbic system represented by the cingulate gyrus

constants are functionally equivalent to the different levels of cortical processing discussed in the isocortex models. Specifically, the millisecond time frame is the cortical structure closest to raw sensory input, the second time frame is the cortex module the first one passes information to and so forth.

A few of these connections are shown in Figs. 13.8 and 13.9, in which we only see connections illustrated between the different time frames of the parieto-occipital and temporal lobes. There are simple one line double arrows drawn for the frontal to temporal and frontal to parieto-occipital to indicate the other connections. For each emotional modality, our musical and painting Würfelspiel data provides 64 equally valid data points. Consider the 64 *sad* data points for music. We know as humans that this data is *sad*. Each such *sad* data point provides auditory cortex training data and 64 examples of the *sad* emotional attribute. The painting data gives us 64 examples of visual cortex training data as well as 64 additional *sad* emotional attributes. Hence, we have 128 examples of *sadness* split between the sensory pathways of hearing and vision. In this way, we build 128 examples of each emotional attribute from music and painting split equally between hearing and vision. We know from studies of neural processing, that the front of the auditory and visual cortex is closely aligned to sensory data and as you progress into more interior layers of cortex, neural ensembles begin to respond to progressively higher and more abstract patterns. For example, in the auditory cortex, initially the nerve cells respond to simple phonemes of perhaps 20 mS duration and higher levels are responsive to words, then sentences and so forth. We can make similar analogies to processing in the visual cortex. Outputs from primary sensory cortex are fed into higher level associative cortex where more abstract processing is performed. Hence, our data provides a validating pathway for two types of primary sensory cortex as well as a primitive model of higher level associative cortex.

13.5 Depression Models

We can use the developed models to model forms of depression following the *monoamine—neurokinin* and *BDNF* hypothesis of dysfunction. In the *monoamine* hypothesis, we assume that depression is due to a deficiency of the monoamine neurotransmitters norepinephrine *NE* and serotonin *5HT*. Since we are dealing with monoamine NTs, all of the machinery needed to construct new NT on demand, break it down into components for reuse and transport it back from the cleft into the pre-synapse via re-uptake pumps is available in the pre-synaptic cell. The enzyme used to break down the monoamine is called *monoamine oxidase* or *MAO*. In a malfunctioning system, there may be too little monoamine NT, too few receptors for it on the post-synapse side, improper breakdown of the NT and/or improper re-uptake. If MAO inhibitors are given, the breakdown is prevented, effectively increasing monoamine concentration. If *tricyclic* compounds are given, the re-uptake pump is blocked, also effectively increasing the monoamine concentration.

Another hypothesis for the etiology of depression is that there is a malfunction in one or more second messenger systems which initiate intracellular transcription factors that control critical gene regulation. A candidate for this type of malfunction are the pathways that initiate activation of *brain derived neurotrophic factor* or *BDNF*. As discussed in Stahl (2000), BDNF is critical to the normal health of brain neurons but under stress, the transcription of the protein from the gene for BDNF is somehow decreased leading to possible cell death of neurons. The administering of MAO inhibitors and tricyclics then reverse this by effectively stimulating the production of BDNF.

Finally, we know that serotonin and a peptide called substance P of a class of peptides known as neurokinins are co-transmitters. There is some evidence that if an antagonist to substance P is administered, it functions as an antidepressant. The amygdala has both monoamine and substance P receptors and hence modulation of the monoamine pre-synaptic terminal by substance P is a real possibility.

Our simple pharmacological model will allow us to model the critical portions of these depression hypotheses in our abstract cells which can then be linked together into larger computational ensembles.

The full cognitive model is built as an interacting collection of the intelligent software agents encompassing auditory, visual, frontal, limbic cortex agents, serotonin and noradrenergic pathway agents, environmental input agents for music, painting and generic compositional design agents and a emotional amygdala agent. The connections between the agents can be modified by pharmacological agents at the critical time scales of $1-10$ ms (immediate), $10-1000$ ms (fast onset) and

10,000–10,000,000 ms (slow onset) via alteration of abstract cell feature vectors, abstract fiber outputs, and abstract ensemble outputs via Hebbian interactions and phase locking.

13.6 Integration into a Virtual World

The final step is to integrate the model into a three dimensional (3D) virtual world. This is done by using the motor and thalamus outputs of our model to influence the movement and texture mappings of a 3D character. Characters or avatars are generated in such worlds using computer code which constructs their body as a 3D mesh of polygons given a 3D position. Then 2D matrices of a texture are applied to portions of the 3D mesh by warping the texture to fit the 3D geometry of the character. This is called the texture mapping phase and it is only then that the character takes on a realistic appearance. In our model, the fMRI and skin conductance scalars output from the thalamus serve to locate a position in 2D emotional space as determined by psychophysiological experiments (see Chap. 10) which may be a more complex emotional state than that of sadness, happiness or anger. We can assign different texture mappings to different emotional vectors and thereby influence the physical appearance of the avatar based on emotional input. In addition, we can assign different categories of motor movements to emotional vectors in order to have distinct motor patterns that are associated with emotional states. There is much work to be done to work out all the details of these mappings, of course. Some of this detail is shown in Fig. 13.10. Finally, we note that this simple model will be used as the foundation upon which we will assemble our models of cognitive dysfunction. The diagram of Fig. 13.3 clearly shows that there are neurotransmitter inputs into the limbic system and cortex which can influence the output that reaches the avatar. Indeed, they influence the outputs we call meta level in music and painting compositional design. There are established connections between neurotransmitter levels in cortex associated with cognitive dysfunctions. Also, there is a body of literature on the musical and painting outputs of cognitively disturbed patients which can therefore be associated with the neurotransmitter levels. This enables us to develop another set of training data that is similar in spirit to what we have already done with the musical and painting data. Once our models are constrained by this new data, we have models that generate outputs of music and painting in a variety of emotional modalities. The alterations in neurotransmitter levels can be obtained by specific lesions in our software models, thereby allowing us to do controlled lesion studies which we believe will be of therapeutic use in the study and treatment of cognitive dysfunction. The resulting cognitive models are then attached to the software that generates the attributes of a character that moves autonomously in a 3D virtual world Our high level graphical interfaces thus realize our models as 3D artificial avatars in virtual worlds and hence can interact with both other avatars and the user of the software. We will use these tools to develop models of cognitive dysfunction such as depression for possible use in diagnostic/therapy efforts in collaboration with

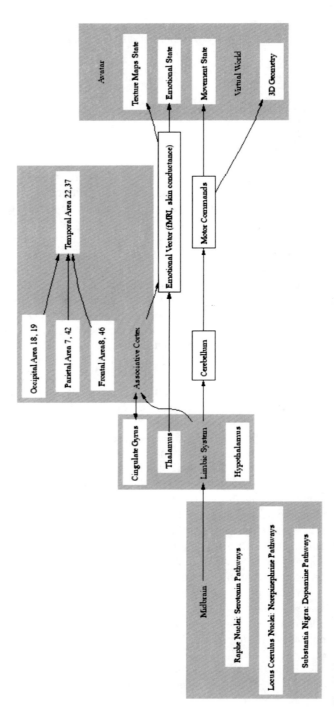

Fig. 13.10 Simplified connections to the avatar in a 3D virtual world

cognitive scientists. We can envision that in the future, we can develop interfaces allowing signals obtained from multielectrode electrophysiological probes to drive the avatars as well.

13.7   Lesion Studies

Using our proposed tools, we can thus build autonomous art, music and general design agents that are coupled to avatars that *live* in appropriate 3D worlds that are not pastiche based but instead are based on emerging structure. This approach thus has significant broad scientific impact. In addition, the utilization of music and design professionals as our first round of clinicians to help shape the data sets used to shape our cognitive models is innovative and will provide an infrastructure for future involvement of such personnel in the even more important planned lesion studies. If our software model of human cognition is capable of producing valid emotional responses to ordinary stimuli from music and art, then it will provide a powerful tool to study cognitive dysfunction. If such a model could be built, it would be possible to introduce errors in many ways into the architecture. The output from the model could then be assessed and we would in principle have an experimentally viable mechanism for lesion studies of human cognition which is repeatable and fine grained. A few such experimental protocols could be

- Alter how messages are sent or received between abstract neurons by changes in the pharmacological inputs they receive (enhancement or blocking of neurotransmitters via drugs at some stage in the agonist spectrum),
- Alter the physical connection strategies between the fibers in the ensembles to model physiological wiring issues between cognitive submodules,
- Alter how messages are sent or received between ensembles to model high level effects of neurotransmitter or other signaling molecule problems,
- Since each software agent could itself be a collection of interacting finer-grained agents, we could introduce intracellular or extracellular signaling errors of many types inside a given agent to study communication breakdown issues.

We believe that the capability for lesion studies that are software based could be of great value. If we can build reasonable software models of various aspects of human cognition, then we can also alter those models in ways dictated to us by experimental evidence and can provide insight into aspects of cognitive dysfunction. For example, we can couple our models to mid level data (functional MRI and gross anatomical structure) and low level data (neurochemical and neurotransmitter changes and microstructure deviations). The resulting models will be therefore faithful to data encompassing multiple scales and will be an important first step toward abstract models that are useful in artificial (and hence controllable) lesion studies of human cognition. Further, by altering the cognitive model in a particular lesion study, the functioning autonomous music and painting composer can generate musical fragments and drawings/paintings that can be compared to the existing literature on the artistic output of dysfunctional individuals for further validation (Leo 1983).

The next step is then to tie the development of our avatars to information obtained from mouse, eye and body movements of cognitively impaired individuals as they play a video style game based on the 3D virtual world previously developed. This enables us to build a real-time model of the emotions and other cognitive functions of a patient from their own interactions. We believe this could be of great value as a diagnostic and therapeutic tool, although we have a lot of work to do to get to that point.

There is a large and growing gap between the scientists who obtain laboratory data and the scientists who are attempting to use the data to obtain models of higher order processes. We believe that communication between these disparate research endeavors can be enhanced by providing a tool which makes it easier to integrate low level detailed biological knowledge into high level meta modeling efforts. In many conversations with either side of this divide, there is a fundamental lack of appreciation for the other's work. Hence, a bridge must be built which enables each side to more fully appreciate the important results of the other. We believe our proposed models in the way we suggest is a useful step in this direction.

13.8   The Complete Cognitive Model

The model we have generated so far creates valid emotionally labeled musical compositions and paintings. Some design principles about this construction process can now be gleaned from this work for the generation of musical and painting compositional designs. In general, our model takes specialized sensory input and generates a high level output as shown in Fig. 13.1. We can design algorithms to train our cortical tissue models using laminar cortical processing via on-surround excitation/off-surround inhibition and Hebbian based connection strengthening. Our discussion so far shows us that the full computational model can be described by the diagram of Fig. 13.4. As discussed for emotional data in the context of music, given a random starting note say from column one of a Würfelspiel music matrix of given emotional modality, our model will generate an entire valid musical composition. Note that this output should actually be interpreted as two separate pieces of information: one, as a musical composition and two, as an emotional state. Our data thus provides training for the correct output of a model of emotions as well as a total of 128 happy, sad and angry input/output training samples. We can thus generate valid compositional designs that have a specific emotional tag for both the auditory, visual and associative cortex pathways.

We will start with an emotional model which outputs two parameters (loosely based on the psychophysiological data experiments). The first is actually a skin conductance parameter and the second is a complicated computed value that arises from certain fMRI measurements. The interesting thing about these values is that in experiments with human subjects, when people saw pictures with emotional contents such as "sad", "angry" and so forth, the two parameters mentioned above determined a x-y coordinate system in which different emotional attributes were placed experimentally in very different parts of this plane. For example, "sad" images might go to quadrant 2 and "angry" images might be mapped to quadrant 3. This is an over simplification of course, but the idea that images of different emotional attributes would be separate in the plane is powerful. Our 128 examples of each emotional attribute thus give us 128 data points which should all be mapped to the same decision region of this two dimensional plane. Thus, we have data that unambiguously gives us a desired two dimensional output for our emotional model. At this point, we have a model that given an abstract auditory and visual input stream from the data will generate given emotional attributes. We can then turn the system around if you like, by noting that each data set corresponds to a certain decision region in "emotional space". Further, recognize that we have a coupled model of sensory processing and emotional computation. We have an auditory and visual agent which given input from a known emotion decision region, generates a musical and artistic stream of a given emotional attribute. We model this as a three software agent construction: auditory, visual and emotional. Each of these agents accepts inputs from the others. We have enough data to develop a first pass at a model which given a two dimensional emotional input and a visual and auditory random start, will generate an emotional tagged auditory and visual stream. This model is shown in Fig. 13.5 for music and painting. We can then *validate* this model easily by just letting anyone listen or look at our output and tell us if it is good. Hence, we are validating the whole model instead of just the equivalent of a local patch of hippocampal tissue in a slice. Indeed, it should also be possible to include validation in the learning algorithm.

13.9 Virtual World Constructions

A cognitive model which is implementable within an environment of distributed computation would be able to give three dimensional characters in a virtual world cognitive/emotional states so that more realistic "human" interactions are obtainable in complex social, political and military simulations. Such a cognitive model, of course, entails reasonable abstractions of biological detail. We believe that the single most important item in the development of such a model is the training data and so we explain as carefully as possible how we organize our training data into easily generated Würfelspiel matrices for the auditory, visual and temporal cortex portions of the cognitive model. The trained model is then capable of outputs which can be interpreted as emotional states which can be tied to various attributes of a character so as to enable believable response.

Consider again, the abstraction presented in Fig. 5.3. Musical data has provided the kind of associated output that might come from area 37 of the temporal cortex. The low level inputs that start the creation of a music phrase correspond to the auditory sensory inputs into area 41 of the parietal cortex which are then processed through areas 5, 7 and 42 before being sent to the further associative level processing in the temporal cortex. The painting data has then provided a similar kind of associated input into area 37 from the occipital cortex. Inputs that create the paintings correspond to the visual sensory inputs into area 17 of the occipital cortex which are then further processed by area 18 and 19 before being sent to the temporal cortex for additional higher level processing. The musical and painting data inputs correspond to specific fMRI and skin conductance outputs in addition to an encoded compositional design. Our data thus allows us to build the model shown in Fig. 13.4.

At this point we have a simple model of the thalamic outputs as shown in Fig. 5.3. These thalamic outputs can then be integrated into a larger model which can be part of the character behavior modules in a three dimensional (3D) virtual world. In effect, this is done by using the motor and thalamus outputs of the model to influence the movement and texture mappings of a 3D character. Characters or avatars are generated in such worlds using computer code which constructs their body as a 3D mesh of polygons given a 3D position. Then 2D matrices of a texture are applied to portions of the 3D mesh by warping the texture to fit the 3D geometry of the character. In our model, the fMRI and skin conductance scalars output from the thalamus thus serve to locate a position in 2D emotional space as determined by psychophysiological experiments which may be a more complex emotional state than that of sadness, happiness or anger. We can assign different texture mappings to different emotional vectors and thereby influence the physical appearance of the avatar based on emotional input. In addition, we can assign different categories of motor movements to emotional vectors in order to have distinct motor patterns that are associated with emotional states. Some of this detail is shown in Fig. 13.10.

We will return to these ideas in Chap. 21 where we outline a much better method for the training of graph based neural models. But first, we need to discuss in much detail how we would build mappings from inputs to outputs for the complicated neural models we have been talking about. Before we get into these details, we will talk a bit about network models of neurons in general in Chap. 14 and start going through some training ideas in Chap. 15.

References

J. Di Leo, *Interpreting Children's Drawings* (Burnner/Mazel Publishers, Routledge, 1983)

S. Grossberg, How does the cerebral cortex work? development, learning, attention and 3D vision by laminar circuits of visual cortex. (Boston University Technical Report CAS/CS TR-2003-005, 2003)

I. Peretz, R. Zatorre, Brain organization for music processing. Annu. Rev. Psychol. **56**, 89–114 (2005). (Annual Reviews)

S. Stahl, *Essential Psychopharmacology: Neuroscientific Basis and Practical Applications*, 2nd edn. (Cambridge University Press, Cambridge, 2000)

L. Stewart, K. von Kriegstein, J. Warren, T. Griffiths, Music and the brain: disorders of musical listening. Brain **129**, 2533–2553 (2006)

Chapter 14
Networks of Excitable Neurons

We know that inputs into the dendritic tree of an excitable nerve cell can be separated into first and second messenger classes. The first messenger group consists of Hodgkin–Huxley voltage dependent ion gates and the second messengers includes molecules which bind to the dendrite through some sort of interface and then trigger a series of secondary events inside the cytoplasm of the cell body. We will primarily be interested in second messengers which are neurotransmitters. We need to have a more detailed discussion of the full dendrite—soma—axon system so we can know how to connect one excitable nerve cell to another and build networks. We will have a more abstract look at second messenger systems later, but for now we will be focused on a class of neurotransmitters that include dopamine.

14.1 The Basic Neurotransmitters

Let's look at the class of catecholamine neurotransmitters in the nervous system. We will focus on only a few types: **DA**, dopamine; **NE**, norepinephrine; and **E**, epinephrine. These neurotransmitters can be lumped together into the category called **CA**s because they all share a common core biochemical structure, the catechol group. These neurotransmitters are very important and how they interact in a neural system determines many behaviors. An important review of serotonin effects is given in Roberts (2011) and in Dayan and Huys (2009) and that of catecholamines in general can be found in Arnsten (2011). A discussion of dopamine's effects is given in Friston et al. (2014). We only touch on basic things here. NE and E have very similar structure, consisting of a benzene ring and a certain type of side chain. A benzene ring can be denoted by a cyclic structure of six carbons as is seen in Fig. 14.1. The carbons are numbered one through six, starting from the eastern-most one on the ring and moving counterclockwise. Replace the hydrogens on the third and fourth carbon with a hydroxyl group as seen in Fig. 14.1. The structure is now called a **catechol** group. We can also replace the hydrogen on carbon one by an ethane molecule (C_2H_6) (see Fig. 14.2): note one hydrogen is removed from ethane so that it can attach to

© Springer Science+Business Media Singapore 2016
J.K. Peterson, *BioInformation Processing*, Cognitive Science and Technology,
DOI 10.1007/978-981-287-871-7_14

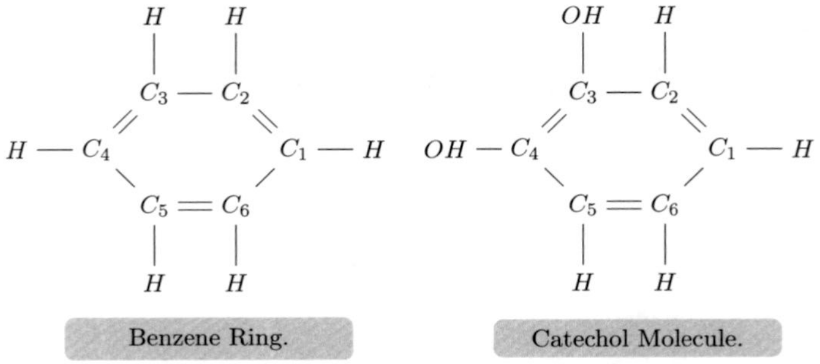

Benzene Ring.

Catechol Molecule.

Fig. 14.1 Benzene and Catechol molecule

Fig. 14.2 Catechol molecule
with ethyl chain on C_1

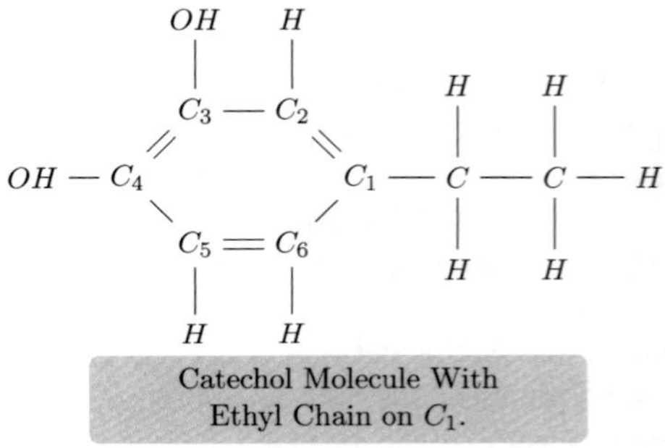

Catechol Molecule With
Ethyl Chain on C_1.

Fig. 14.3 Dopamine

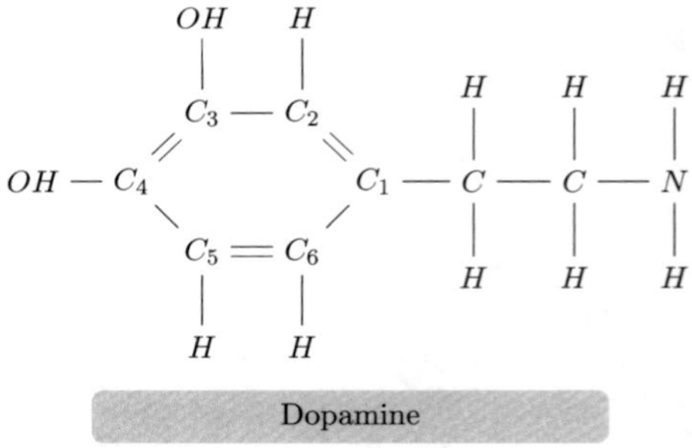

Dopamine

the benzene ring. We can then add an ammonia molecule NH_3 (**amine** group) to the outermost carbon in the ethyl side chain, the amine losing one hydrogen in order to bond and thus appearing on the chain as NH_2: This is the 3, 4-dihydroxy derivative of the **phenylethylamine** molecule which is the neurotransmitter **Dopamine** (DA) (see Fig. 14.3). Further derivatives of this then give us **Norepinephrine** (NE) and **Epinephrine** (E). If the beta carbon on the side chain replaces one hydrogen with a hydroxyl group by hydrolysis we obtain NE (see Fig. 14.4). If the hydrogen on the second carbon of the ethyl side chain by a methyl ion, we get E (see Fig. 14.5). Note that all of these molecules are *monoamines*—"one amine" constructions and

Fig. 14.4 Norepinephrine

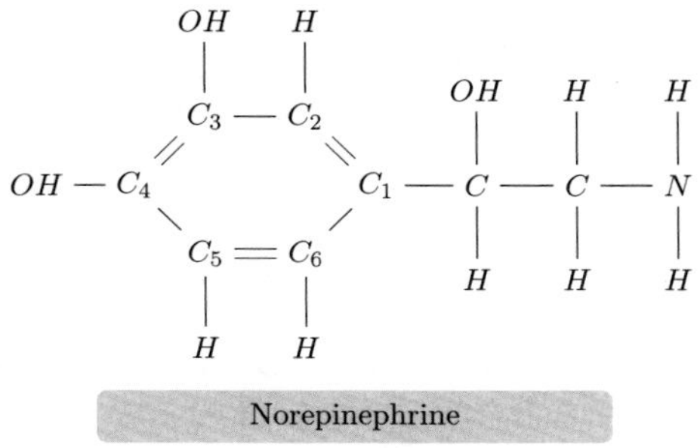

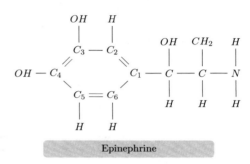

Norepinephrine

Fig. 14.5 Epinephrine

Epinephrine

all are derivatives of the catechol group. Without going into detail, DA, NE and E are constructed from the nutrient soup inside our cells via the following pathway (see Fig. 14.6):

1. Tyrosine hydroxylase (adds a OH group) converts Tyrosine into L-DOPA. This is called the rate limiting substance in CA synthesis because if it is not present, the reaction will not continue. So there must be a mechanism to request from our genome appropriate levels of Tyrosine synthesis when needed.
2. L-DOPA, formed by hydroxylation of tyrosine, is then converted into DA by DOPA decarboxylase (take away a carboxyl group).

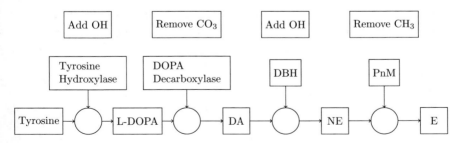

Fig. 14.6 The catecholamine synthesis pathway

3. DA is then converted into NE by DBH, dopamine-beta-hydroxylase (add a hydroxyl group to the beta carbon).
4. NE is then converted into E by phenylethanolamine-N-methyltransferase (PnM) (add a methyl group to the amine group on the alpha carbon).

We can see this construction pathway diagrammatically in Fig. 14.6. All of the enzymes above, *Tyrosine hydroxylase*, *DOPA decarboxylase*, *dopamine-beta-hydroxylase* and *phenylethanolamine-N-methyltransferase* are regulated in complex ways. Further, all three enzymes are critical to the proper functioning of CA bio synthesis.

14.2 Modeling Issues

Consider the following perceptive quotation (Black 1991):

> More generally, regulation of transmitter synthesis shares important commonalities in neurons that differ functionally, anatomically, and embryologically. This point is worthy of emphasis. Increased impulse activity stimulates transmitter synthesis in dopaminergic nigral neurons that regulate coordination of motor function, noradrenergic locus coerulus neurons that may play a critical role in attention and arousal, and adrenergic adrenomedullary cells central to the stress response. Simply stated, the common biochemical and genomic organization of these diverse populations determines how environmental, epigenetic information, through altered impulse activity, is translated into neural information. **In this prototypical example, cellular biochemical organization, not behavioral modality, is a key determinant of how external stimuli are converted into neural language. In this domain, modes of information storage are biochemically specific, not modality specific, indicating that synaptic systems subserving entirely different behavioral and cognitive functions may share common modes of information processing**

Hence, in our search for a core or parent object whose instantiation is useful in constructing software architectures capable of subserving learning and memory function, the evidence presented above suggests we focus on *prototypical neurotransmitters*. It is clear a model of dendritic–axonal interaction where dendrite and axon objects interact via some sort of intermediary agent. These agents need to look at both dendritic and axonal specific information. Both the dendrite and the axon should have some dependency on neurotransmitter objects whose construction and subsequent reabsorption for further reuse should follow modulatory pathways that mimic in principle the realities of CA transmitter synthesis. Further, there is a need for *Object Recognition Sites*, as dendritic objects need receptor objects that interact with the neurotransmitter objects. There should also be a finite number of types of neurotransmitter objects and possibly different finite number of receptor object types. The number and variety of these sites grow or shrink in accordance with learning laws.

How should we model CA release, CA termination, CA identification via CA specific receptors and synaptic plasticity? CA is released through a large variety of mechanisms, each of which is modulated in varying degrees by many other agents. Most importantly, this release needs Ca^{++} which is mediated by second messenger action such hormones and intraneural cAMP in what can be a rather global way. Hence, release of CA itself can modulate subsequent CA release. Further, there are more global mechanisms which interact with the CA release and production cycle via **non-transmitter mechanisms**. Note Black (1991, p. 34)

> ...angiotensin II receptors on the [presynaptic] membrane also modulate norepinephrine release. Angiotensin is a potent vasoconstrictor, derived from circulating angiotensinogen by the action of renin, an enzyme secreted by the kidney...the principle is startling: the kidney can communicate with sympathetic neurons through **nonsynaptic mechanisms**...Circulating hormone regulates transmitter release at the synapse. Synaptic communication, then, may be modulated by nonsynaptic mechanisms, and distant structures may talk to receptive neurons. Consequently aspects of **communication with the nervous system are freed from hard-wiring constraints**.

Clearly, dendritic–axonal object interaction should be mediated via pathways of both local scope (using neurotransmitter objects) and global scope, using additional objects which could be modeled after hormones.

We also know CA termination is critical to CA function. CA substance is deactivated by reabsorption into the presynaptic structure itself. Any agent that would interfere with CA uptake through the presynaptic structure allows CA neurotransmitter to remain available for use too long. This has a profound effect on the functioning of the nervous system. A typical drug agent that does this is cocaine. It follows that our software system should have efficient collection and recycling schemes for the neurotransmitter objects. For example, dopamine levels are controlled by three separate mechanisms.

- Dopamine packets are released into the synaptic cleft and bind to receptors on the dendrite. So, the number of receptors per unit area of dendritic membrane provides a control of dopamine concentration in the cleft.
- Dopamine is broken down by enzymes in the cleft all the time which also control dopamine concentration.
- Dopamine is pumped back into the presynaptic bulb for reuse providing another mechanism.

Now, CA receptors can be classed into two broad types: the α- and β-receptors. Each can be inhibitory and/or excitatory in nature. These structures can be located on both presynaptic and postsynaptic sites. The α class is located on presynaptic tissue and when activated, inhibits further CA release; thereby functioning as an autofeedback loop—these receptors have been called *autoreceptors*. It follows that receptor objects should be usable in both dendritic and axonal objects.

The interaction between the presynaptic and postsynaptic neurons is always changing. This mutability is called their synaptic plasticity. It is mediated by both local and global pathways. In Black (1991, pp. 38–39), we see the Post Synaptic Density or PSD structure mediates this plasticity. The PSD contains many components that can be altered via second messenger triggers which gives rise to synaptic plasticity. The appropriate software agent connecting dendritic and axonal objects is therefore an object of type say PSD—a bridging object whose function and structure is alterable via objects of type hormone, neurotransmitters and perhaps others through some sort of software mechanism.

14.3   Software Implementation Thoughts

We now discuss some of the issues that are involved in the implementation of a software environment that will facilitate rapid prototyping of biologically motivated models of cognition. Using ideas from a language such as C++, python or others, would lead us to neural architectures as derived classes from a virtual class of neural objects. The computational strategies that support functions such as training would also be classes derived from a core primitive class of computational engines. We wish to allow classes of architectures which are general enough to allow each computational element in an architecture to itself be a neural network of some type, a statistical model and so forth. This also implies that we must develop more sophisticated models of connections between computational elements (synaptic links between classical lumped sum neuronal models are one such type of connection). In addition, we wish to have the capability in our modeling environment to adapt abstract principles of neuronal processing that are more closely inspired by current neurobiological knowledge. Consider this comment Black (1991, pp. xii–xiii):

> Briefly, certain types of molecules in the nervous system occupy a unique functional niche. These molecules subserve multiple functions simultaneously. It is well recognized that biochemical transformations are the substance of cellular function and mediate cell interactions in all tissues and organ systems. Particular subsets of molecules simultaneously serve as intermediates in cellular biochemistry, function as intercellular signals, and respond in characteristic modes to environmental stimuli. These critical molecules actually incorporate environmental information into the cell and nervous system. Consequently these molecules simultaneously function as biochemical intermediates and as symbols representing specific environmental conditions....

> The principle of multiple function implies that there is no clear distinction among the processes of cellular metabolism, intercellular communication, and symbolic function in the nervous system. Representation of information and communication are part of the functioning fabric of the nervous system. This view also serves to collapse a number of other traditional categorical distinctions. For example, the brain can no longer be regarded as the hardware underlying the separate software of the mind. Scrutiny will indicate that these categories are ill framed and that hardware and software are one in the nervous system.

From this first quote of Black, you can see the enormous challenge we face. We can infer from the above that to be able to have the properties of plasticity and response to environmental change, certain software objects must be allowed to subserve the dual roles of communication and architecture modification. Black refers to this as the "principle of polyfunction" (Black 1991, p. 3).

> Shorn of all detail, the software–hardware dichotomy is artificial. As we shall see, software and hardware are one and the same in the nervous system. To the degree that these terms have any meaning in the nervous system, software changes the hardware upon which the software is based. For example, experience changes the structure of neurons, changes the signals that neurons send, changes the circuitry of the brain, and thereby changes output and any analogue of neural software.

This is an interesting comment. How are we to design a software architecture system which essentially is self-modifying? Architectural modification in the nervous system involves the environmental input (or some other impulse) triggering increased production of some important protein, enzyme etc. for the purpose of constructing synapses, growing dendrites and so forth. However, these mechanisms are already in place and they are activated when needed. So, the architecture is modified, but **no new mechanisms are constructed**. Hence, low-level software objects that implement these mechanisms and which are accessed via some sort of request for service might be a reasonable way to implement this capability.

Next, note Black (1991, pp. 8–9)

> Increasing evidence indicates that ongoing function, that is, communication itself, alters the structure of the nervous system. In turn altered structure changes ongoing function, which continues to alter structure. The essential unity of structure and function is a major theme... In summary, a neurotransmitter molecule, which is known to convey millisecond-to-millisecond excitatory information, also regulates circuit architecture by stimulating synaptic growth. In this system, then, signal communication, growth, altered architecture, altered neural function, and memory are causally interrelated; there is no easy divide between hardware and software. The rules of function are the rules of architecture, and function governs architecture, which governs function

It seems clear from the above that our underlying software architecture must be event-based in some fashion. If we think of architectures consisting of loosely coupled computational modules, each module is a kind of input–output mapping which we can call an object of type IOMAP. Models with computational nodes linked by edges are called connectionist models. Hence, classical connectionist modeling asserts that the information content of the architecture is stored in the weight values that are attached to the links and these weight values are altered due to environmental influence via some sort of updating mechanism. It is now clear the architecture itself must be self-modifying. Now it is fairly easy to write add and delete type functions into connectionist schemes (just think of the links and weights implemented as doubly linked lists), but what is clear from the above that much more is needed. We need mutable synaptic efficacy, the ability to add synapses and alter the function of neuron bodies themselves. Consider the dual roles played by neural processing elements (Black 1991, pp. 13–14):

> [Weak reductionism] involves identification of scientifically tractable structural elements that process information in the nervous system....Any set of elements is relevant only insofar as it processes information and simultaneously participates in ongoing neural function; these dual roles require the neural context.

What structural elements may be usefully examined? It may be helpful to outline appropriate general characteristics.... First, neural elements of interest must change with environment. That is, environmental stimuli must, in some sense, regulate the function of these particular units such that the units actually serve to represent conditions of the real world. **The potential units, or elements of interest, thereby function as symbols representing external or internal reality. The symbols, then, are actual physical structures that constitute neural language representing the real world**. Second, the symbols must govern the function of the nervous system such that the representation itself constitutes a change in neural state. **Consequently symbols do not serve as indifferent repositories of information but govern ongoing function of the nervous system**. Symbols in the nervous system simultaneously dictate the rules of operation of the system. In other words, the symbols are central to the architecture of the system; architecture confers the properties that determine behavior of the system. **The syntax of symbol operation is the syntax of neural function**.

We now have a potential blueprint for our software architectural design: we need software objects that

1. Change with environmental conditions.
2. Their state is either a direct or indirect representation of external environmental conditions.
3. Their change in state corresponds to a change in the underlying architecture.

Information processing also involves a combinatorial strategies Ira Black, (Black 1991, p. 17):

> Two related strategies are employed by the nervous system in the manipulation of the transmitter molecular symbols. First, individual neurons use multiple transmitter signal types at any given time. Second, each transmitter type may respond independently of others to environmental stimuli.... The neuron appears to use a **combinatorial strategy**, a simple and elegant process used repeatedly in nature in a variety of guises. A series of distinct elements, of relatively restricted number, are used in a wide variety of combinations and permutations. For example, if a given neuron uses four different transmitters, and each can exist in only three different concentrations (based on environmental influences) independently, the neuron can exist in 3^4 discrete transmitter states....However, this example appears to vastly underestimate combinatorial potential of the neuron in reality.

Each IOMAP module must therefore be capable of a certain number of active states. This capability cannot be achieved with a simple scalar parameter for the information processing of a neuron. How are we to implement such a combinatorial possibility? How should we imbue our software objects with the equivalent of activation via a variety of neurotransmitter pathways of varying temporal signatures?

To implement an architecture where computational modules communicate—an agent based architecture, we will turn to functional programming languages such as **Erlang**, **Haskell** and so forth which do not use shared memory. These languages can easily use thousands of cores and so they are very promising. Also, **Erlang** allows us to write code that can continue to run even if some of its processes are lost which will allow us to implement lesion studies very nicely.

14.4 How Would We Code Synapse Interaction?

Let's put together all of the insight we have gathered from our discussions. We know that

1. The principle of multiple function implies that there is no clear distinction among the processes of cellular metabolism, intercellular communication and symbolic function in the nervous system. **Hence our software infrastructure should possess the same blurring of responsibility**.
2. As mentioned before, software and hardware are the same in the nervous system. This implies that the basic building blocks of our software system should be able to alter the software architecture itself. We could implement part of this capability using the notion of *dynamically bound* computational objects and strategies. using the C++ language. We could also do prototyping in *Octave*, *python* and others. If we use the compiled C++ language, we will be forced to deal with the inadequacies of the C++ language's implementation of objects whose type is bound at run-time. As we mentioned above, we will not use C++ at this time and instead focus on **Erlang**.
3. Another basic principle from Black (1991) in the context of ongoing function is that the very fact of communication, alters nervous system structure. This implies that the structure of our software architecture be mutable in response to the external input environment.

The evidence for these organizing principles can be summarized as follows:

1. We know that dendritic-axon coincident electrical activation strengthens synaptic efficacy through what are called Hebbian mechanisms. There is substantial evidence of this: in the rat hippocampus, long term potentiation (LTP) leads to structural modification of the synapse via the glutamate neurotransmitter. Clearly, **neurotransmitters which are known to convey millisecond to millisecond excitatory information also regulate circuit architecture via synaptic structural change**.
2. Macroscopic behavioral change is mirrored by distinct structural change at the synaptic level. These points have been verified experimentally by the many excellent investigations into the nervous system of the the sea snail *Aplysia Californica*. In this animal, long term habituation and long term sensitization behaviors are associated with definite structural changes in the number of synaptic vesicles, the number of synaptic sites and so forth. In addition, these microscopic changes in structure have the same temporal signature that the behavioral changes exhibit; i.e. these changes last as long or as short as the corresponding behavior.

Neurotransmitters give the nervous system a capability to effect change on an extraordinary range of time scales; consider the following simplified version of neurotransmitter interaction. We abstract out of the wealth of low-level biological detail an abstract version of neurotransmitter pathways: start with an initial change in the external environment ΔE_0 coming into a neuronal ensemble. This elicits a corresponding change in impulse ΔI_0 from the ensemble which in turn triggers a change in neurotransmitter level, ΔNT_0. This change in neurotransmitter is further modulated by the breakdown rate and reabsorption rate for the neurotransmitter. For simplicity, assume this is modeled by the constant factor R. Thus the net change in neurotransmitter level is given by $\Delta NT_0 - R$. Consider the time series corresponding to an initial environmental change of ΔE_0 which is never repeated. We let time be modeled in discrete ticks labeled by $\{t_0, t_1, t_2, \ldots\}$ and let the subscript of each variable indicate its value at that time tick. We obtain the series:

$$\Delta E_0 \rightarrow \Delta I_0 \rightarrow (\Delta NT_0 - R)$$
$$\rightarrow \Delta I_1 \rightarrow (\Delta NT_1 - R)$$
$$\rightarrow \Delta I_2 \rightarrow (\Delta NT_2 - R)$$
$$\vdots$$
$$\rightarrow \Delta I_p \rightarrow (\Delta NT_p - R)$$

where the process terminates at step p because $\Delta NT_p - R \approx 0$. We see that the interplay between creation of neurotransmitter and its reabsorption rate allows for the effect of an environmental signal to have a temporal signature that lasts beyond one time tick.

Our model must then clearly allow for nonlocal interaction and long term temporal response to a given signal. From the above discussion, we also see that the basic response of change in environment implies change in neurotransmitter level ($\Delta E \rightarrow \Delta NT$) does not depend on what the signal type is or which neurotransmitter pathway is activated. Evidence for this includes the following neurotransmitter-signal pairs which all modulate response to a change in signal via the change in transmitter level:

1. **Dopamine** releasing neurons regulating **motor function**.
2. **Norepinephrine** releasing neurons regulating **attention** and **arousal**.
3. **Adrenaline** releasing neurons regulating **stress response**.

Here, we see clearly that cellular biochemical organization, not the type of environmental signal or behavior, is the key to how external stimuli are converted into neural language. We seek to construct a core group of software objects that have similar function. Their organization and their methods of interaction will determine how external stimuli are converted into our *neural object* language.

14.4.1 The Catecholamine Abstraction

We begin by drawing from the wealth of information above, a simplified version which will serve as the building block for our abstraction of these ideas into a functional software architecture. Consider Fig. 14.7. In it, we try to suggest the units from which we need to build the abstract neuronal object. Each dendrite object contains a number of receptor objects, while an axon object contains neurotransmitter objects.

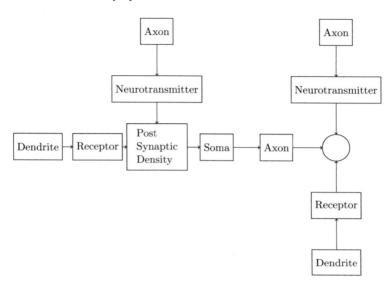

Fig. 14.7 Abstract neuronal process

A dendrite–axon object pair interact via the PSD object which plays a role that is similar to its biological function; however, we will also see that we will be able to build extremely rich mathematical numerical processing objects as well using this paradigm.

The output of the PSD object is sent to the computational body of the neuronal object, the soma object. The details of how the soma will process the inputs it receives will be discussed later. For now, we just note that some sort of information processing takes place within this object. The output of the soma is collected into an axon object containing the previously mentioned neurotransmitter objects. In this simplified illustration shown in Fig. 14.7, we do not try to indicate that there could be many different types of neurotransmitters. We indicate the interaction pathways in Fig. 14.7 on the dendritic and axonal side of the soma. We provide further detail of the PSD in the close up view given by Fig. 14.8. Here, we assume that the dendrite object contains five types of neurotransmitters and the axon uses the associated receptor objects for these neurotransmitters. The result of the dendritic–axonal interaction is then computed by the PSD object via some as yet unspecified means. Now, how should we handle the intricacies of the neurotransmitter–receptor interactions? We do not want to model **all** of the relevant details. For our purposes, we will concentrate on a few salient characteristics:

1. The probability of neurotransmitter efficacy will be denoted by the scalar parameter p. The variable p models neurotransmitter efficacy in a lumped parameter manner. The amount of transmitter produced via the pathways that access the genome, the probability of interaction with receptors and so forth are combined into one number. More neurotransmitter is modeled by increasing p; increased

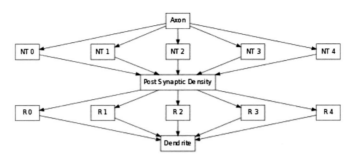

Fig. 14.8   Dendrite–axonal interaction pathway

numbers of receptors or increased efficacy of receptors can also be handled by increasing p. We can be even more specific. Let $\mathcal{N}$ denote the amount of neurotransmitter units available (a unit is essentially a synaptic vesicle) and $\mathcal{R}$ the number of receptor sites in the PSD. Let ρ be the probability that a neurotransmitter finds a receptor. The amount of transmitter available for binding is then $\rho \mathcal{N}$; further, let $\xi(\mathcal{R})$ be the probability that a receptor site is able to bind with the transmitter. This is of course dependent on the number of receptor sites. Thus the activity level p of the neurotransmitter is given by

$$p = \rho \mathcal{N} \, \xi(\mathcal{R})$$

This gives a transmitter efficacy model that is a function of amount of transmitter $\mathcal{N}$, number of receptors $\mathcal{R}$ and probabilities of finding and binding. Hence, our efficacy model p can be written $p(\rho, \xi, \mathcal{N}, \mathcal{R})$ to explicitly indicate these dependencies.

2. Each neurotransmitter will have its own reabsorption rate denoted by q.
3. Each neurotransmitter has its own intrinsic time interval of action: we will not model this explicitly at this time. Instead, we can mimic this effect by controlling the (p, q) interactions for a given neurotransmitter.
4. A neurotransmitter has an associated locality that sets the scope of its interaction with dendrites and so forth. For example, a neurotransmitter might only have a local interaction via the PSD structure between one axon and one dendrite. On the other hand, the neurotransmitter might permeate the intracellular fluid surrounding the dendrite and axonal trees of our artificial neural system. Hence, in software, we might want a given neurotransmitter to effect PSD structures on other dendrites. We might also want other axons contributing this neurotransmitter to contribute to the net dendrite–PSD–axon interaction. These effects will be modeled with a locality object, say LOCAL. The equations that we will define for various sorts of dendritic–axonal interactions (Eqs. 14.1–14.4) are all written at this point in terms of sets of active neuronal indices for our artificial neural system.

14.4.2   PSD Computation

Let's denote the PSD computation by the symbol •. We need to model how to obtain the value of a prototypical interaction, I. The simplest case is that of a *One Axon–One Dendrite Interaction*. We will let the term $\mathcal{A}_j^i$ denote a collection of things organized using standard vector notion. We have

$$\mathcal{A}_j^i = \begin{cases} \mathcal{A}_j^i[0] = v & \text{value attached to the axon.} \\ \mathcal{A}_j^i[1] = p_\sigma & \text{the efficacy for an associated neurotransmitter } \sigma \\ \mathcal{A}_j^i[2] = q_\sigma & \text{the reabsorption rate for an associated neurotransmitter } \sigma \end{cases}$$

We use a similar notation of the dendrite, $\mathcal{D}_j^i$. We allow for more fields in the dendrite than the value field, so we show this as a dotted entry.

$$\mathcal{D}_j^i = \begin{cases} \mathcal{D}_j^i[0] = v & \text{value attached to the dendrite.} \\ \vdots \end{cases}$$

Thus, the value of the interaction is given by

$$I = \left(\mathcal{A}_j^i[1] - \mathcal{A}_j^i[2] \right) \mathcal{A}_j^i[0] \bullet \mathcal{D}_\ell^j[0] \tag{14.1}$$

where $\mathcal{A}_j^i$ is the jth axon connection from neuron i. In this case, it connects to the ℓth dendrite of neuron j, $\mathcal{D}_\ell^j$. Another situation occurs when the total efficacy and reabsorption rate for a given axon are determined by efficacies and reabsorption rates from a pool of surrounding axons also. However, there is still just one dendrite involved. We can call this a *One Axon (Local Domain)–One Dendrite* interaction whose interaction value is now given by

$$I = \sum_{k \in \mathcal{L}^i(\sigma)} \left(\mathcal{A}_j^k[1] - \mathcal{A}_j^k[2] \right) \mathcal{A}_j^i[0] \bullet \mathcal{D}_\ell^j[0] \tag{14.2}$$

where the new symbol $\mathcal{L}^i(\sigma)$ denotes the **axonal locality** of the neurotransmitter σ in the ith neuron's axon. It is easy to extend to a situation where in addition to the pool of local axons that contribute to the efficacy and reabsorption rate of the axon, we now add the possibility of local effects on the dendritic values. This is the *One Axon (Local Domain)–K Dendrites* case and is the most complicated situation. The interaction value is now

$$I = \sum_{m \in \mathcal{M}^j(\sigma)} \sum_{k \in \mathcal{L}^i(\sigma)} \left(\mathcal{A}_j^k[1] - \mathcal{A}_j^k[2] \right) \mathcal{A}_j^i[0] \bullet \mathcal{D}_\ell^m[0] \tag{14.3}$$

where the new symbol $\mathcal{M}^j(\sigma)$ denotes the **dendritic locality** of the neurotransmitter σ in the ith neuron's axon. Even this simple model thus allows for long term temporal effects of a given neurotransmitter by using appropriate values of p and q. Indeed, this model exhibits the capability for imbuing our architectures with great plasticity. For example, we can model the effects of $T + 1$ several different transmitters $\{\sigma_0, \sigma_1, \ldots, \sigma_T\}$ at one PSD junction by using the notation $(\mathcal{A}_j^k)^t$ and $(\mathcal{D}_\ell^m)^t$ to indicate which neurotransmitter we are looking at and then summing the interactions as follows:

$$ I = \sum_{t=0}^{T} \sum_{m \in \mathcal{M}^j(\sigma_t)} \sum_{k \in \mathcal{L}^i(\sigma_t)} \left((\mathcal{A}_j^k)^t[1] - (\mathcal{A}_j^k)^t[2] \right) (\mathcal{A}_j^i)^t[0] \bullet (\mathcal{D}_\ell^m)^t[0] \quad (14.4) $$

Let's stop and reflect on all of this. These complicated chains of calculation can be avoided by moving to graphs of computational nodes which communicate with each other by passing messages. The local and more global portions of these computations are simply folded into a given node acting on its own input queue of messages using protocols we set up for the processing. This is what we will be doing with chains of neural objects.

14.5   Networks of Neural Objects

We will discuss what is called a chained network in Chap. 17, but it is easy enough to jump in a bit early to see how our neural computions could be handled. We can connect a standard feedforward architecture of neurons as a chain as shown in Fig. 14.9: Note we can also easily add self feedback loops and general feedback pathways as shown in Fig. 14.10. Note that in Figs. 14.9 and 14.10, the interaction between dendritic and axonal objects has not been made explicit. The usual simplistic approximation to this interaction transforms a summed collection of weighted inputs via a saturating transfer function with bounded output range. The dendritic and axonal interaction is modeled by a scalar W_{ij} as described above. It *measures* the *intensity* of the interaction between the input neuron i and the output neuron j as a simple scalar weight. Since we have developed the ballstick neural processing model, we will eventually try to do better than this.

14.5.1   Chained Architecture Details

Let's look at the Chained Feed Forward Network (CFFN) in more detail. In this network model, all the information is propagated forward. Hence, Fig. 14.10 is not of this type because it has a feedback connection as well as a self loop and so is not strictly feedforward. The CFFN is quite well-known and our first exposure to it occurred in the classic paper of Werbos (1987). Consider the function

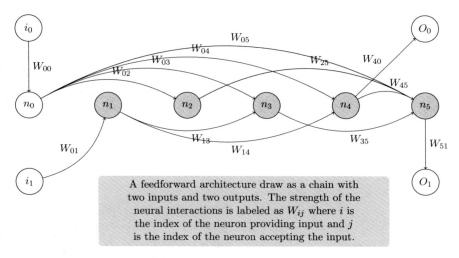

Fig. 14.9 A feedforward architecture shown as a chain of processing elements

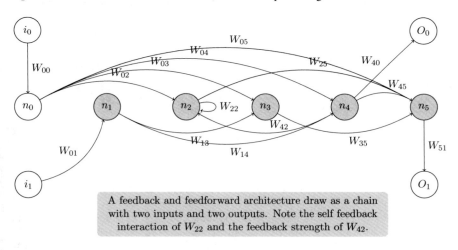

Fig. 14.10 A chained neural architecture with self feedback loops and general feedback

$H : \Re^{n_I} \to \Re^{n_O}$ that has a very special nonlinear structure which consists of a chain of computational elements, generally referred to as *neurons* since a very simple model of neural processing models the action potential spike as a sigmoid function which transitions rapidly from a binary 0 (no spike) to 1 (spike). This sigmoid is called a transfer function and since it can not exceed 1 as 1 is a horizontal asymptote, it is called a *saturating* transfer function also. This model is known as a lumped sum model of post-synaptic potential; now that we know about the ball—stick model of neural processing, we can see the lumped sum model is indeed simplistic. Each *neuron* thus processes a summed collection of weighted inputs via a saturating trans-

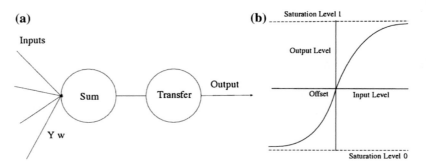

Fig. 14.11 Nodal computations. **a** Schematic. **b** Processing

fer function with bounded output range (i.e. [0, 1]). The neurons whose outputs connect to a given target or **postsynaptic** neuron are called **presynaptic** neurons. Each presynaptic neuron has an output Y which is modified by the synaptic weight $W_{pre,post}$ connecting the presynaptic neuron to the postsynaptic neuron. This gives a contribution $W_{pre,post}$ Y to the input of the postsynaptic neuron. A typical saturating transfer function model is shown in the Fig. 14.11a, b. Figure 14.11a shows a postsynaptic neuron with four weighted inputs which are summed and fed into the transfer function which then processes the input into a bounded scalar output. Figure 14.11b illustrates more details of the processing. A typical transfer function could be modeled as $\sigma(x, o, g) = 0.5\left(1.0 + \tanh\left(\frac{x-o}{\phi(g)}\right)\right)$ with the usual transfer function derivative given by

$$\frac{\partial \sigma}{\partial x}(x, o, g) = \frac{\phi'(g)}{2\phi(g)} sech^2\left(\frac{x-o}{\phi(g)}\right)$$

where o denotes the offset indicated in the drawing and $\phi(g)$ is a function controlling slope. The slope controlling parameter is usually called the gain of the transfer function as it represents amplification of the incoming signal. Note that at the offset point, $\frac{\partial \sigma}{\partial x} = \frac{\phi'(g)}{2\phi(g)}$; hence, g effectively controls the slope of the most sensitive region of the transfer functions domain. The function $\phi(g)$ is for convenience only; it is awkward to allow the denominator of the transfer function model to be zero and to change sign. A typical function to control the range of the gain parameter might be $\phi(g) = g_m + \frac{g_M - g_m}{2}(1.0 + \tanh(g))$ where g_m and g_M denote the lower and upper saturation values of the gain parameter's range. The chained model then consists of a string of N neurons, labeled from 0 to $N - 1$. Some of these neurons can accept external input and some have their outputs compared to external targets. We let

$$\mathcal{U} = \text{ indices } i \text{ where neuron } i \text{ is an input}$$
$$= \{u_0, \ldots, u_{n_I - 1}\} \tag{14.5}$$
$$\mathcal{V} = \text{ indices } i \text{ where neuron } i \text{ is an output}$$
$$= \{v_0, \ldots, v_{n_O - 1}\} \tag{14.6}$$

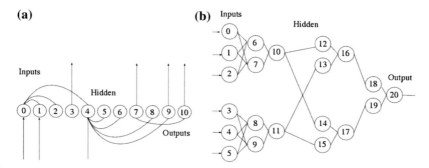

Fig. 14.12 Local networks. **a** Chained feedforward. **b** Local expert

We will let n_I and n_O denote the size of $\mathcal{U}$ and $\mathcal{V}$ respectively. The remaining neurons in the chain which have no external role are sometimes called hidden neurons with dimension n_H. Note in a chain, it is also possible for an input neuron to be an output neuron; hence $\mathcal{U}$ and $\mathcal{V}$ need not be disjoint sets. The chain is thus divided by function into three possibly overlapping types of processing elements: n_I input neurons, n_O output neurons and n_H internal or hidden neurons. In Fig. 14.12a, we see a prototypical chain of eleven neurons. For clarity only a few synaptic links from pre to post neurons are shown. We see three input neurons (neurons 0, 1 and 4) and four output neurons (neurons 3, 7, 9 and 10). Note input neuron 0 feeds its output forward to input neuron 1 in addition to feeding forward to other postsynaptic neurons. The set of postsynaptic neurons for neuron 0 can be denoted by the symbol $\mathcal{F}(0)$ which here is the set $\mathcal{F}(0) = \{1, 2, 4\}$. Similarly, we see $\mathcal{F}(4) = \{5, 6, 8, 9\}$.

We will let the set of postsynaptic neurons for neuron i be denoted by $\mathcal{F}(i)$, the set of **forward links** for neuron i. Note also that each neuron can be viewed as a postsynaptic neuron with a set of presynaptic neurons feeding into it: thus, each neuron i has associated with it a set of backward links which will be denoted by $\mathcal{B}(i)$. In our example, $\mathcal{B}(0) = \{\}$ and $\mathcal{B}(4) = \{0\}$, where in general, the backward link sets will be much richer in connections than these simple examples indicate. For example, the chained architecture could easily instantiate what is called a *local expert model* as shown in Fig. 14.12b. The weight of the synaptic link connecting the presynaptic neuron i to the postsynaptic neuron j is denoted by $W_{i \to j}$. For a feedforward architecture, we will have $j > i$, however, as you can see in Fig. 14.10, this is not true in more general chain architectures. The input of a typical postsynaptic neuron therefore requires summing over the backward link set of the postsynaptic neuron in the following way:

$$y^{post} = x + \sum_{pre \in \mathcal{B}(post)} W_{pre \to post} Y^{pre}$$

Fig. 14.13 Postsynaptic
output calculation

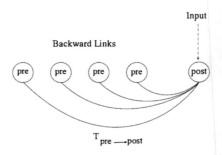

where the term x is the external input term which is only used if the post neuron is
an input neuron. This is illustrated in Fig. 14.13. We will use the following notation
(some of which has been previously defined) to describe the various elements of the
chained architecture.

x^i External input to the ith input neuron
y^i Summed input to the ith neuron
o^i Offset of the ith neuron
g^i Gain of the ith neuron
σ^i Transfer function of the ith neuron
Y^i Output of the ith neuron
$T_{i \rightarrow j}$ Synaptic efficacy of the link between neurons i and j
$\mathcal{F}(i)$ Forward link set for neuron i
$\mathcal{B}(i)$ Backward link set for neuron i

The chain FFN then processes an arbitrary input vector $x \in R^{n_I}$ via an iterative
process as shown below.

$$
\begin{aligned}
&\text{for}(i = 0; i < N; i++) \{ \\
&\quad \text{if } (i \in \mathcal{U}) \\
&\qquad y^i = x^i + \sum_{j \in \mathcal{B}(i)} T_{j \rightarrow i} Y^j \\
&\quad \text{else} \\
&\qquad y^i = \sum_{j \in \mathcal{B}(i)} T_{j \rightarrow i} Y^j \\
&\quad Y^i = \sigma^i(y^i, o^i, g^i) \\
&\}
\end{aligned}
$$

The output of the CFFN is therefore a vector in $\Re^{n_O}$ defined by $H(x) = \{Y^i \mid i \in \mathcal{V}\}$;
that is,

$$
H(x) = \begin{bmatrix} Y^{v_0} \\ \vdots \\ Y^{v_{n_O-1}} \end{bmatrix}
$$

and we see that $H : \Re^{n_I} \to \Re^{n_o}$ is a highly nonlinear function that is built out of chains of nonlinearities. The parameters that control the value of $H(x)$ are the link values, the offsets and the gains for each neuron. Note computations are performed on each nodes input queue!

14.5.1.1 The Traditional Structure of the Chain Model

The CFFN model consists of a string of N neurons. Since each neuron is a general input/ output processing module, we can think of these neurons as type IOMAP. The neurons are labeled from 0 to $N - 1$. A dendrite of any neuron can accept external input and an axon of any neuron can have an external tap for comparison to target information. In Fig. 14.10, we see a chain of six neurons. We see two neurons that accept external input (neurons 0 and 1) and two neurons having external taps (neurons 4 and 5). The set of postsynaptic neurons for a given neuron i is denoted by the symbols $\mathcal{F}(i)$; the letter $\mathcal{F}$ denotes the *feedforward links*, of course. Note the self and feedback connection in the set $\mathcal{F}(2)$. Also each neuron can be viewed as a postsynaptic neuron with a set of presynaptic neurons feeding into it: thus, each neuron i has associated with it a set of backward links which will be denoted by $\mathcal{B}(i)$. In our example, these are the sets given below

$$
\begin{array}{llll}
\mathcal{F}(0) & \{1, 2, 3, 4, 5\} & \mathcal{B}(0) & \{\} \\
\mathcal{F}(1) & \{3, 4\} & \mathcal{B}(1) & \{\} \\
\mathcal{F}(2) & \{2, 5\} & \mathcal{B}(2) & \{0, 2, 4\} \\
\mathcal{F}(3) & \{5\} & \mathcal{B}(3) & \{0, 1\} \\
\mathcal{F}(4) & \{5\} & \mathcal{B}(4) & \{0, 1\} \\
\mathcal{F}(5) & \{\} & \mathcal{B}(5) & \{0, 2, 3, 4\}
\end{array}
$$

where in general, the backward link sets can be much richer in connections than this simple example indicates. This model thus uses the computational scheme

$$
Y^{post} = \sigma\left(y^{post}, o, g\right) = \sigma\left(x + \sum_{pre \in \mathcal{B}(post)} W_{pre \to post} Y^{pre}, o, g\right)
$$

which can be rewritten

$$
a^{post} = S\left(x + \sum_{pre \in \mathcal{B}(post)} a^{pre} \bullet d^{post}, p\right)
$$

where a indicates the value of a given axon, d, the value of a given dendrite and S, the soma computational engine, which depends on a vector of parameters, p. In the simple sigmoidal transfer function used in this chain, the parameters are the gain, offset and the minimum and maximum shaping parameters for the sigmoid; hence, the parameter vector p would have values $p[0] = o, p[1] = g, p[2] = g_m$ and $p[3] = g_M$.

Fig. 14.14 A chained neural
architecture with PSD
computational sites shown

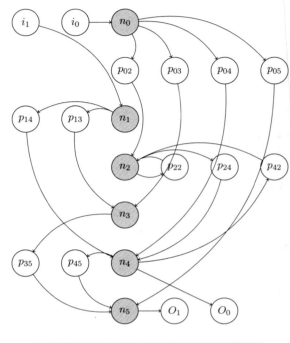

A chained neural architecture with self
feedbackloops and general feedback.

Finally, we let the symbol $a^{pre} \bullet d^{post}$ indicate the computation performed by the PSD
between the pre and post neuronal elements. Note that the PSD $\bullet$ operation in the
chain model therefore corresponds to a simple multiplication of the pre-axon value
d and the post-dendrite value d.

14.5.2   Modeling Neurotransmitter Interactions

The use of neurotransmitters in our axon models and receptors in our dendrite models
requires more discussion. The original Figs. 14.9 and 14.10 did not show the PSD
computational structures, while the larger illustration Fig. 14.14 does. This gives
some insight into our new modeling process, but it will be much more illuminating
to look at careful representations of this six neuron system in the case of two receptors
on all dendrites and two neurotransmitters on all possible axons. Note that this is just
for convenience and the number of neurotransmitters and receptors does not need to
be constant in this fashion. Further, the number of neurotransmitters and receptors
at each dendrite and axon need not match. Nevertheless, the illustrations presented

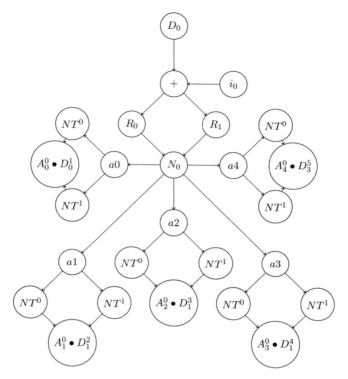

Fig. 14.15 The dendritic tree details for Neuron N_0 showing two neurotransmitters per pre axon connection to the dendrite

in Fig. 14.15 give a taste of the plasticity of this approach. We will use a common notation for these illustrations given by

D_0, D_1: The dendritic objects associated with a neuron.

I: The external input to a dendrite.

R_0, R_1: The receptor objects associated with a dendrite.

NT^0, NT^1: The neurotransmitter objects associated with an axon.

$N_0, N_1, N_2, N_3, N_4, N_5$: The six neurons in our example.

$A_0, A_1, \ldots$: The axons associated with a neuron.

$A_j^i \bullet D_q^p$: The PSD object handling interaction between the jth axon of neuron i and the qth dendrite of neuron p.

E_O: The external tap on an axon.

With all this said, the extended pictures of neuron 0 gives a very detailed look at the network previously shown as Fig. 14.14. We could also specify axonal and dendritic locality sets (although we do not illustrate such things in these pictures for reason of excessive clutter), but we don't really have to as long as we think of all of this calculation as taking place in a given nodes input queue asynchronously. But we could setup this sort of careful linking as follows. We might specify the axonal locality set for the axon 2 of neuron 1 for neurotransmitter t to consist of two axons: axon 1 of neuron 1 and axon 2 of neuron 1 (itself). This information would then be stored in an appropriate linking description, but for convenience of exposition, we list it here in set form. The name $AL_t(i)$ denotes the *axon link set* for neuron i for neurotransmitter t. So, this example would give $AL(A_2^1)_t = \{A_1^1, A_2^1\}$. In a similar way, we could define the dendritic locality set, DL for dendrite 1 of neuron 2 for receptor r in set form (using the standard A_j^i and D_q^p notation): $DL(D_1^1)_r = \{D_0^2, D_1^2\}$.

Look carefully at neuron 0. Figure 14.15 shows its full structure. It is the first neuron in the chain and the computational capability of the neuron and its dendritic and axonal tree is highly nonlinear due to the dependence on the two neurotransmitters and two receptors present at each of the five PSD calculations. There are also time delay possibilities that are controlled by gate opening and closings for the receptors. The other neurons would have similar diagrams and we will not show them here.

14.6   Neuron Calculations

The discussion above shows us how to define various types of neurons that all have a similar structure. A neuron type is typically called a *class*. We will go back to our ball stick model to organize our thoughts. A neuron class must compute the axon hillock voltage, of course. This computation explicitly depends on

- L, the length of the cable,
- λ_c, the electrotonic length,
- The value of $\rho = \frac{G_D}{G_S}$, the ratio of the dendritic conductance to soma conductance,
- The solution to the eigenvalue equation

$$\tan(\alpha L) = -\frac{\tanh(L)}{\rho L}(\alpha L),$$

- Q, the number of eigenfunctions to use in our expansions,
- The vectors c_j we obtain for voltage impulses at location j on the dendrite. There are versions of this data vector for each choice of j; hence, there are L data vectors.
- The M matrix which is only computed once.
- The solution to the matrix system $MA = c_j$ for each j.

Thus, we can define a neuron class for each possible choice of this parameter set. The solutions of the matrix equation and the eigenvalue problem need only be computed once for each neuron class. Now for a given neuron of the a class, consider the model of Fig. 14.16 in which the details of the soma cable are hidden within the circle that is drawn to represent the cell body. Each neuron's soma is actually a Rall cable of length L_S with electrotonic distance L_{SE} and has an attached dendrite with length L_D with electrotonic distance L_{DE}. The axon hillock is identified with the $z = 0$ position

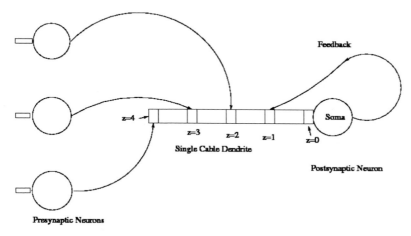

Fig. 14.16 A network model

on the soma cable. In the figure, the dendritic and soma cables of the presynaptic neurons are not shown in detail and, for convenience, the cable length is shown as $L_D = 4$. We can now write down in organized form how computations will take place in this network.

- At $\tau = 0$, calculate the voltage, v, at the axon hillock via

$$v_D(0) = \sum_{n=0}^{Q_D} \sum_{j=0}^{L_D} \sum_{i=0}^{N_j^{D,0}} V_{0,D,i}^j \, \hat{A}_n^j \, \cos[\alpha_n(L_D - j)]$$

$$v_S(0) = \sum_{n=0}^{Q_S} \sum_{k=0}^{L_S} \sum_{i=0}^{N_k^{S,0}} V_{0,S,i}^k \, \hat{B}_n^k \, \cos[\beta_n(L_S - k)]$$

$$v(0) = v_D(0) + v_S(0)$$

where $N_j^{D,0}$ and $N_k^{S,0}$ denote the number of pulses arriving at node j and node k of the dendrite and soma, respectively, at time $\tau = 0$. We note the scalars such as $V_{0,D,i}^j$ are actually the attenuated values $V_{0,D,i}^j \, e^{-(1+alpha_n^2)\epsilon}$ where ϵ is a small amount of time—say .05 time constants. The voltage at the axon hillock is then used to initialize the Hodgkin–Huxley action potential response as we discussed in Peterson (2015). If an axon potential is generated, it can be sent back as a negative impulse to the dendrites to model the refractory period of the neuron.
- Advance the time clock one unit.
- At $\tau = 1$, at the axon hillock, the voltages applied at $\tau = 0$ have now decayed to

$$v_D(0, 1) = \sum_{n=0}^{Q_D} \sum_{j=0}^{L_D} \sum_{i=0}^{N_j^{D,0}} V_{0,D,i}^j \, \hat{A}_n^j \, \cos[\alpha_n(L_D - j)] \, e^{-(1+\alpha_n^2)}$$

$$v_S(0, 1) = \sum_{n=0}^{Q_S} \sum_{k=0}^{L_S} \sum_{i=0}^{N_k^{S,0}} V_{0,S,i}^k \, \hat{B}_n^k \, \cos[\beta_n(L_E - k)] \, e^{-(1+\beta_n^2)}$$

The new pulses that have arrived at $\tau = 1$ are given by

$$v_D(1) = \sum_{n=0}^{Q_D} \sum_{j=0}^{L_D} \sum_{i=0}^{Q_j^{D,1}} V_{1,D,i}^j \, \hat{A}_n^j \, \cos[\alpha_n(L_D - j)]$$

$$v_S(1) = \sum_{n=0}^{Q_S} \sum_{k=0}^{L_S} \sum_{i=0}^{Q_k^{S,1}} V_{1,S,i}^k \, \hat{B}_n^k \, \cos[\beta_n(L_E - k)]$$

$$z(0, 1) = v_D(1) + v_S(1)$$

and the total synaptic potential arriving at the axon hillock at time $\tau = 1$ is thus

$$v(1) = (v_D(1) + v_D(0, 1)) + (v_S(1) + v_S(0, 1)) \tag{14.7}$$

Again, the voltage at the axon hillock is then used to initialize the Hodgkin–Huxley action potential response.

- At time $\tau = 2$, we would find

$$v_D(0, 2) = \sum_{n=0}^{Q_D} \sum_{j=0}^{L_D} \sum_{i=0}^{Q_j^{D,0}} V_{0,D,i}^j \, \hat{A}_n^j \, \cos[\alpha_n(L_D - j] \, e^{-2(1+\alpha_n^2)}$$

$$v_S(0, 2) = \sum_{n=0}^{Q_S} \sum_{k=0}^{L_S} \sum_{i=0}^{Q_k^{S,0}} V_{0,S,i}^k \, \hat{B}_n^k \, \cos[\beta_n(L_E - k] \, e^{-2(1+\beta_n^2)}$$

$$v_D(1, 1) = \sum_{n=0}^{Q_D} \sum_{j=0}^{L_D} \sum_{i=0}^{Q_j^{D,1}} V_{1,D,i}^j \, \hat{A}_n^j \, \cos[\alpha_n(L_D - j] \, e^{-(1+\alpha_n^2)}$$

$$v_S(1, 1) = \sum_{n=0}^{Q_S} \sum_{k=0}^{L_S} \sum_{i=0}^{Q_k^{S,1}} V_{1,S,i}^k \, \hat{B}_n^k \, \cos[\beta_n(L_E - k] \, e^{-(1+\beta_n^2)}$$

The axon hillock voltage at $\tau = 2$ would then be

$$v(2) = (v_D(2) + v_D(1, 1) + v_D(0, 2)) + (v_S(2) + v_S(1, 1) + v_S(0, 2)) \tag{14.8}$$

where the notation $v_D(0, i)$ means the input voltage at time 0 after i time units of attenuation. Hence, $v_S(1, 1)$ means the voltage that came in at time 1 1 time unit later.

- Continuing in this fashion, at time $\tau = T$, the total synaptic potential seen at the axon hillock is

$$v(T) = \left(v_D(T) + \sum_{j=0}^{T} v_D(T - j, j) \right) + \left(v_S(T) + \sum_{j=0}^{T} v_S(T - j, j) \right) \quad (14.9)$$

and an axon potential is generated with possible refractory feedback as before.

The voltage at the axon hillock is then used to initialize the Hodgkin–Huxley action potential response as we discussed in Peterson (2015).

It should be clear you that this is a very complex process even for a single neuron. It is quite complicated to understand how to model a network of such computational modules efficiently. We have spent a lot of time in this text detailing how to test the models using snippets of MatLab code and it has been very instructive. However, the interpretative nature of the MatLab integrated development environment begins to hamper us if we want to build larger interactions. From the discussions in this section, we can also see that differential equation models that are based on continuous time rather than discrete time have difficulties too. In order to build even simple software architecture implementations for networks of neurons, it is therefore clear we must develop an abstract framework for the implementation of neural processing models. To do this, we have developed abstractions of the action potential we call *biological feature vectors* or *BFVs* which have been discussed already. For example, at each time tick in the computational chain above, we can compute the BFV that associated with the change in membrane voltage. We can then trigger an action potential if the axon hillock voltage exceeds a threshold voltage Ψ. Thus if $V(T) > \Psi$, the the output of the cell, Y, is a BFV, ξ. Thus,

$$Y = \begin{cases} 0 & \text{if } V(T) < \Psi, \\ \xi & \text{otherwise} \end{cases}, \quad (14.10)$$

For these calculations, we need to decide on an active time period. For example, this active period here could be the five time ticks t_0 to t_4. Hence, any incoming signals into the cell prior to time t_4 are ignored. Once, the time period is over, the cell can possibly generate a new BFV.

Now step back again. All of this intense timing with its need to pay attention to locality sets and so forth, can be removed by simply focusing on the chain architecture and the backward and forward link sets of the graph of nodes. It is easy to implement in a truly asynchronous way and since computation is on whatever is in a nodes input queue, much of this messy notation is unnecessary! We will explore this approach in later chapters.

References

A. Arnsten, Catecholamine influences on dorsolateral prefrontal cortical networks. Biol. Psychiatry **69**, e89–e99 (2011)

I. Black, *Information in the Brain: A Molecular Perspective* (A Bradford Book, MIT Press, Cambridge, 1991)

P. Dayan, Q. Huys, Serotonin in affective control. Annu. Rev. Neurosci. **32**, 95–126 (2009). (Annual Reviews)

K. Friston, P. Schwartenbeck, T. FitzGerald, M. Moutoussis, T. Behrens, R. Dolan, The anatomy of choice: dopamine and decision-making. Philos. Trans. R. Soc. B **369**(20130481), 1–12 (2014)

J. Peterson, *Calculus for Cognitive Scientists: Partial Differential Equation Models*. Springer Series on Cognitive Science and Technology (Springer Science+Business Media Singapore Pte Ltd., Singapore, 2015 in press)

A. Roberts, The importance of serotonin for orbitofrontal function. Biol. Psychiatry **69**, 1185–1191 (2011)

P. Werbos, Learning how the world works, in *Proceedings of the 1987 IEEE International Conference on Systems, Man and Cybernetics*, vol. 1. IEEE (1987), pp. 302–310

Chapter 15
Training the Model

As we have seen in the previous chapters, it is going to be difficult to model the information processing that we feel is present in neural computation. To get started, let's take what we have sketched about our networks being graphs of computational nodes and see if we can figure out a way to make such a graph map a given set of inputs into a desired set of objects. We are going to use what are called **Directed Graphs** as the edges between nodes are directional; we do not assume information flows both ways between two nodes. A specified edge will tells us information goes from the input or **pre** node to the output or **post** node as we have shown in our earlier discussions on chained objects. We will remain focused on how we can do this within MatLab although it will quickly become clear we eventually want to move to other computer languages. So to illustrate how we can *train* or imprint clusters of computational objects, let's consider a typical *On-Center, Off-Surround* cortical circuit and the associated *Folded Feedback Pathway* which connects two stacked cortical blocks. These circuits have been taken from Raizada and Grossberg (2003). We show their respective directed graph (DG) representations in Fig. 5.22a, b. Associated with any such DG structure, there is a notion of flow given by the Laplacian of the graph. It is easiest to show how this is done for the OCOS and FFP DG. It will then be evident how to extend to the DG's associated with other computational modules in the simple brain model. Although not shown in the implementation source code, there are methods for the Laplacian calculations and the resulting flows as well.

15.1 The OCOS DAG

Let's redo the OCOS graph as we did in Chap. 18. In Fig. 5.22a, there are 8 neural nodes and 9 edges. We relabel the node numbers seen in the figure to match our usual feedforward orientation. The OCOS DG is a graph G which is made up of the vertices V and edges E given by

© Springer Science+Business Media Singapore 2016
J.K. Peterson, *BioInformation Processing*, Cognitive Science and Technology,
DOI 10.1007/978-981-287-871-7_15

$$V = \{N_1, N_2, N_3, N_4, N_5, N_6, N_7, N_8\}$$
$$E = \{E_1, E_2, E_3, E_4, E_5, E_6, E_7, E_8, E_9\}$$

where you can see the nodes (we identify N1 in the graph model with N8 in the figure and so forth). We will use the identifications

$$\begin{bmatrix} E_1 = & \text{edge from node } N_1 \text{ to } N_2 \; E_2 = & \text{edge from node } N_2 \text{ to } N_3 \\ E_3 = & \text{edge from node } N_2 \text{ to } N_4 \; E_4 = & \text{edge from node } N_2 \text{ to } N_5 \\ E_5 = & \text{edge from node } N_3 \text{ to } N_6 \; E_6 = & \text{edge from node } N_4 \text{ to } N_7 \\ E_7 = & \text{edge from node } N_5 \text{ to } N_8 \; E_8 = & \text{edge from node } N_2 \text{ to } N_7 \\ E_9 = & \text{edge from node } N_1 \text{ to } N_7 \end{bmatrix}$$

We then denote the OCOS DG by $G(V, E)$. Recall, its *incident* matrix is then the matrix K whose K_{ne} entry is defined to be $+1$ if there is an edge going out of node n; -1 if there is an edge going into node n and 0 otherwise. For the OCOS DG, K is the 8×9 matrix given by

$$K = \begin{bmatrix}
1 & 0 & 0 & 0 & 0 & 0 & 0 & 0 & 1 \\
-1 & 1 & 1 & 1 & 0 & 0 & 0 & 1 & 0 \\
0 & -1 & 0 & 0 & 1 & 0 & 0 & 0 & 0 \\
0 & 0 & -1 & 0 & 0 & 1 & 0 & 0 & 0 \\
0 & 0 & 0 & -1 & 0 & 0 & 1 & 0 & 0 \\
0 & 0 & 0 & 0 & -1 & 0 & 0 & 0 & 0 \\
0 & 0 & 0 & 0 & 0 & -1 & 0 & -1 & -1 \\
0 & 0 & 0 & 0 & 0 & 0 & -1 & 0 & 0
\end{bmatrix}$$

Now assume that there is a vector function f which assigns a scalar value to each node. Then, the *gradient* of f on the DG is defined to be $\nabla f = K^T f$ given by Eq. 15.1.

$$\nabla f = \begin{bmatrix}
f(N1) - f(N2) \\
f(N2) - f(N3) \\
f(N2) - f(N4) \\
f(N2) - f(N5) \\
f(N3) - f(N6) \\
f(N4) - f(N7) \\
f(N5) - f(N8) \\
f(N2) - f(N7) \\
f(N1) - f(N7)
\end{bmatrix} = \begin{bmatrix}
f_1 - f_2 \\
f_2 - f_3 \\
f_2 - f_4 \\
f_2 - f_5 \\
f_3 - f_6 \\
f_4 - f_7 \\
f_5 - f_8 \\
f_1 - f_7 \\
f_2 - f_7
\end{bmatrix} \tag{15.1}$$

where we identify $f(N1) \equiv f_1$ and so forth. It is easy to see that this is a measure of *flow* through the graph. We then define the Laplacian $\nabla^2 f = KK^T f$ for this DG as shown in Eq. 15.2.

$$\nabla^2 f = \begin{bmatrix} 2f_1 - f_2 - f_7 \\ -f_1 + 5f_2 - f_3 - f_4 - f_5 - f_7 \\ -f_2 + 2f_3 - f_6 \\ -f_2 + 2f_4 - f_7 \\ -f_2 + 2f_5 - f_8 \\ -f_3 + f_6 \\ -f_1 - f_2 - f_4 + 3f_7 \\ -f_5 + f_8 \end{bmatrix} \tag{15.2}$$

Since KK^T is symmetric, the eigenvalues of this Laplacian are positive and there is nice eigenvector structure. The standard cable equation for the voltage v across a membrane is

$$\lambda^2 \nabla^2 v - \tau \frac{\partial v}{\partial t} - v = -r \lambda^2 k$$

where the constant λ is the space constant, τ is the time constant and r is a geometry independent constant. The variable k is an input source. We will assume the flow of information through our OCOS DG is given by the graph based partial differential equation

$$\nabla^2_{OCOS} f - \frac{\tau_{OCOS}}{\lambda^2_{OCOS}} \frac{\partial f}{\partial t} - \frac{1}{\lambda^2_{OCOS}} f = -rk.$$

Now, we relabel the fraction $\tau_{OCOS}/\lambda^2_{OCOS}$ by μ_1 and the constant $1/\lambda^2_{OCOS}$ by μ_2 where we drop the *OCOS* label as it is understood from context. Each computational DG will have constants μ_1 and μ_2 associated with it and if it is important to distinguish them, we can always add the labellings at that time. This equation gives us the Laplacian graph based information flow model. This gives, using a finite difference for the $\frac{\partial f}{\partial t}$ term, Eq. 15.3, where we define the finite difference $\Delta f_n(t)$ as $f_n(t+1) - f_n(t)$.

$$\begin{bmatrix} 2f_1 - f_2 - f_7 \\ -f_1 + 5f_2 - f_3 - f_4 - f_5 - f_7 \\ -f_2 + 2f_3 - f_6 \\ -f_2 + 2f_4 - f_7 \\ -f_2 + 2f_5 - f_8 \\ -f_3 + f_6 \\ -f_1 - f_2 - f_4 + 3f_7 \\ -f_5 + f_8 \end{bmatrix} - \mu_1 \begin{bmatrix} \Delta f_1(t) \\ \Delta f_2(t) \\ \Delta f_3(t) \\ \Delta f_4(t) \\ \Delta f_5(t) \\ \Delta f_6(t) \\ \Delta f_7(t) \\ \Delta f_8(t) \end{bmatrix} = \mu_2 \begin{bmatrix} f_1(t) \\ f_2(t) \\ f_3(t) \\ f_4(t) \\ f_5(t) \\ f_6(t) \\ f_7(t) \\ f_8(t) \end{bmatrix} - r \begin{bmatrix} k_1(t) \\ k_2(t) \\ k_3(t) \\ k_4(t) \\ k_5(t) \\ k_6(t) \\ k_7(t) \\ k_8(t) \end{bmatrix}$$

$$\tag{15.3}$$

Then, if we assume that we want the node values $f(N1)$, $f(N2)$ and $f(N3)$ to be clamped to A, B and C respectively for a given input k, we find after substitution into Eq. 15.3, we obtain Eq. 15.4.

$$
\begin{bmatrix}
2A - B - f_7 \\
-A + 5B - C - f_4 - f_5 - f_7 \\
-B + 2C - f_6 \\
-B + 2f_4 - f_7 \\
-B + 2f_5 - f_8 \\
-C + f_6 \\
-A - B - f_4 + 3f_7 \\
-f_5 + f_8
\end{bmatrix}
- \mu_1
\begin{bmatrix}
0 \\
0 \\
0 \\
\Delta f_4(t) \\
\Delta f_5(t) \\
\Delta f_6(t) \\
\Delta f_7(t) \\
\Delta f_8(t)
\end{bmatrix}
= \mu_2
\begin{bmatrix}
A \\
B \\
C \\
f_4(t) \\
f_5(t) \\
f_6(t) \\
f_7(t) \\
f_8(t)
\end{bmatrix}
- r
\begin{bmatrix}
k_1(t) \\
k_2(t) \\
k_3(t) \\
k_4(t) \\
k_5(t) \\
k_6(t) \\
k_7(t) \\
k_8(t)
\end{bmatrix}
$$

$$(15.4)$$

This gives us an iterative equation to solve for f_4 through f_8 for each input k. We adjust parameters in the OCOS DG using Hebbian ideas. We can perform a similar Laplacian flow and Hebbian update strategy on any DG. In particular, you can work out the details of this process applied to the DG of Fig. 5.22b and the *Layer Six–Four Connections to Layer Two–Three* in Fig. 5.23a and the *Combined OCOS/ FFP model* of Fig. 5.24.

15.1.1   Some MatLab Comments

For the OCOS DG we developed in Chap. 18, we found

Listing 15.1: The OCOS Laplacian

```
L = laplacian(OCOS)
L =
    2    -1     0     0     0     0    -1     0
   -1     5    -1    -1    -1     0    -1     0
    0    -1     2     0     0    -1     0     0
    0    -1     0     2     0     0    -1     0
    0    -1     0     0     2     0     0    -1
    0    -1     0     0     0     1     0     0
   -1    -1     0    -1     0     0     3     0
    0     0     0     0    -1     0     0     1
f = [0.2; 3.0; 1.0; -3.4; 4.6; 7.8; 1.2; 0.4];
```

and so for a given function vector

Listing 15.2: Sample Node function vector

```
f = [0.2; 3.0; 1.0; -3.4; 4.6; 7.8; 1.2; 0.4];
```

we can easily compute ∇f and $\nabla^2 f$.

Listing 15.3: Computing the graph gradient and Laplacian

```
gradf = K'*f
gradf =
    -2.8000
     2.0000
     6.4000
    -1.6000
    -6.8000
    -4.6000
     4.2000
     1.8000
    -1.0000
laplacianf = K*K'*f
laplacianf =
    -3.8000
    11.4000
    -8.8000
   -11.0000
     5.8000
     6.8000
     3.8000
    -4.2000
```

15.2 Homework

Exercise 15.2.1 *Develop the Laplacian Training method equations for the* The Folded Feedback Pathway DG *as shown in Fig. 5.22b.*

Exercise 15.2.2 *Develop the Laplacian Training method equations for the* Layer Six–Four Connections to Layer Two–Three *in Fig. 5.23a.*

Exercise 15.2.3 *Develop the Laplacian Training method equations for the* Combined OCOS/FFP *model of Fig. 5.24.*

Exercise 15.2.4 *Develop the Laplacian Training method equations for the general* OCOS/FFP Cortical Model *as shown in Fig. 5.24. Compare the eigenvalue and eigenvector structure of the component* Combined OCOS/FFP *models and* Layer Six–Four Connections to Layer Two–Three *DAGs that comprise this model to the eigenvalues and eigenvectors of the whole model.*

15.3 Final Comments

In this chapter, we begin the development of a software architecture that would be capable of subserving autonomous decision making. It is clear an interesting small brain architecture would have the following components.

- At least two primary sensory cortex modules. When we combine the OCOS, FFP and 2/3 building blocks into a cortical stack such as shown in Fig. 5.24, this will require about 30 computational nodes to implement. If we adjoin cortical stacks into a $n \times m$ grid, we have a model of sensory cortex that requires $30nm$ nodes. Let's look at a few specific architectures.

1. a 3×3 grid implying 270 nodes per sensory cortex model.
2. a 3×2 grid implying 180 nodes per sensory cortex model.
3. a 3×1 grid implying 90 nodes per sensory cortex model.
4. a 2×1 grid implying 60 nodes per sensory cortex model.
5. a 1×1 grid implying 30 nodes per sensory cortex model.

Now each of these cortex stack models have modulatory connections to a thalamus, midbrain and cerebellum components.

- A cortical stack model for motor cortex which will provide our output commands.
- A cortical stack model for associative cortex which will give us the sensor fusion we need to drive the motor cortex.
- Simple thalamus, midbrain and cerebellum models which we will discuss in later books.

The small brain model we build then has the fixed directed graph architecture we obtain by building the graph using various neural circuit components such as the OCOS and FFP. There are similar circuit modules we can use to build the thalamus, midbrain and cerebellum models. For example, there are a number of books that discuss theoretical models of brain function carefully. We have used these extensively in our development work. In the subsequent volumes of this multi-volume work we will be discussing this in detail, but for now, we want to mention this source material.

- General theories of brain evolution give many clues to help us design small brain models. A great reference is Striedter's book (Striedter 2004).
- Another source of information on the development of the nervous system in general is the four volume series by Kaas and Bullock (2007a, b), Kaas and Krubitzer (2007) and Kaas and Preuss (2007).
- A general theory of how the thalamus might work is given by Sherman and Guillery (2006).
- The brain in a higher level organism such as ourselves is fundamentally asymmetric which has profound software and wetware architectural ramifications. This is discussed in Hellige (1993).
- There are vast modulatory effects due to hormone modulation which has a big effect on the plasticity or rewiring of our architectures. There is a good introduction to this material in Garcia-Segura (2009).
- The principles of frontal lobe function are discussed in the Stuss and Knight edited volume (Stuss and Knight 2002).

In Table 15.1, we show various computational node counts for the small brain models we are proposing.

We know that *C. Elegans* has approximately 300 neurons and 9000 connections so the last two proposed architectures are similar in size. Many small brain animals function quite well without long term memory storage, so we don't include

Table 15.1 Computational node count for small brain models

Sense I	Sense II	Associative	Motor	Thalamus	Mid brain	Cerebellum	Total
(270) 3 × 3	(270) 3 × 3	(270) 3 × 3	(270) 3 × 3	100	100	100	1380
(180) 3 × 2	(180) 3 × 2	(180) 3 × 2	(180) 3 × 2	100	100	100	1020
(90) 3 × 1	(90) 3 × 1	(90) 3 × 1	(90) 3 × 1	100	100	100	660
(60) 2 × 3	(60) 2 × 1	(60) 2 × 3	(60) 2 × 3	100	100	100	540
(30) 1 × 1	(30) 1 × 1	(30) 1 × 1	(30) 1 × 3	100	100	100	320
(30) 1 × 1	(30) 1 × 1	(30) 1 × 1	(30) 1 × 3	30	30	30	210

a memory module at this time. We could test the ability of an architecture like this to be autonomous within a virtual world. In this book, we program in the interpreted language MatLab, but for models of this size we will need to recode in a language such as **C** or **C++**. It is easy to see that organizing our graph architectures into objects would be very useful. Hence, subsequent volumes, we introduce object oriented programming so that we can rewrite our ideas into that kind of code. Our design criteria at this stage are to see what kind of autonomy we can achieve in our small brain models of 300 neurons or less. The power requirements for such an architecture are reasonable for on board loading into an autonomous robot such as those fielded for Mars exploration. In addition, we want our architectures to support second messenger activity through a subset of neurotransmitter modulation using the thalamus–midbrain–cerebellum loop. Hence, the small brain model will be useful for software based lesion studies for depression and schizophrenia. To make it easier to see how our models are performing, the first step builds an avatar which is a character in a 3D virtual world such as we can build using Crystal Space, (Tybergheim 2003). The avatar then interacts with the virtual world using commands generated in our motor cortex module due to associative cortex processing of environmental inputs. A reasonable visualization would require a 15 frame per second (fps) rate; hence, the upper bound on computation time in our model is 1/15 s which is well within our computational capabilities if we abstract from the bioinformation processing details discussed in this volume simplified computational strategies.

We can assume the edge and node structure of our model is static—certainly, plasticity is possible but we will ignore that for now. We have a model of how to process information flow through the graph using the graph Laplacian based cable equation. It is then clear that we have to decide how to process information flowing on the edges and in the nodes themselves. Hence, each graph model will have associated edge and node functions which can capture as little or as much of the underlying biology as we want. The desired 15 fps rate will help us decide on our approximation strategies. In the chapters to follow, we will finish this book with an abstraction of second messenger systems, Chap. 6, and an approximate way to model the Ca^{++} current injections into the cytosol which form the background of a second messenger trigger event, Chap. 7.

References

L. Garcia-Segura, *Hormones and Brain Plasticity* (Oxford University Press, Oxford, 2009)

J. Hellige, *Hemispheric Asymmetry: What's Right and What's Left* (Hardvard University Press, Cambridge, 1993)

J. Kaas, T. Bullock (eds.), Evolution of Nervous Systems: A Comprehensive Reference Editor J. Kaas (Volume 1: Theories, Development, Invertebrates) (Academic Press Elsevier, Amsterdam, 2007a)

J. Kaas, T. Bullock (eds.), Evolution of Nervous Systems: A Comprehensive Reference Editor J. Kaas (Volume 2: Non-Mammalian Vertebrates) (Academic Press Elsevier, Amsterdam, 2007b)

J. Kaas, L. Krubitzer (eds.), Evolution of Nervous Systems: A Comprehensive Reference Editor J. Kaas (Volume 3: Mammals) (Academic Press Elsevier, Amsterdam, 2007)

J. Kaas, T. Preuss, (eds.), Evolution of Nervous Systems: A Comprehensive Reference Editor J. Kaas (Volume 4: Primates) (Academic Press Elsevier, Amsterdam, 2007)

R. Raizada, S. Grossberg, Towards a theory of the laminar architecture of cerebral cortex: Computational clues from the visual system. Cereb. Cortex **13**, 100–113 (2003)

S. Murray Sherman, R. Guillery, *Exploring The Thalamus and Its Role in Cortical Function* (The MIT Press, Cambridge, 2006)

G. Striedter, *Principles of Brain Evolution* (Sinauer Associates, Sunderland, 2004)

D. Stuss, R. Knight (eds.), *Principles of Frontal Lobe Function* (Oxford University Press, Oxford, 2002)

J. Tybergheim Crystal Space 3D Game Development Kit. (Open Source, 2003), http://www.sourceforge.net

Part V
Simple Abstract Neurons

Chapter 16
Matrix Feed Forward Networks

We now begin to explore strategies for updating the graph models based on experience. Let's begin with a very simple graph model called the Matrix Feedforward Networks or MFFNs model. Instead of using graph based code, we do everything using matrices and vectors. This graph model is based on information moving forward through the network—hence, the adjective *feed forward*.

16.1 Introduction

Assuming we have a collection of **inputs** we want our model to map to desired **outputs**, each input is fed forward through the model to an output. The discrepancy between the output the model gives and the desired output is the **error**. Since the model has no feedback in it, we can calculate (albeit with some pain!) the partial derivatives of the resulting error with respect to all the free parameters in the model and use gradient descent to try to minimize the collection of errors we obtain from using all the inputs. The gradient descent technique used here is historically called **backpropagation** in the literature. Later, we will rewrite this architecture using more general notation which is called the *chained feed forward architecture* which itself is simply a special case of our general directed graph architectures for brain modeling. However, it is very instructive to look at the MFFN as it shows us in detail the difficulties we will have mapping inputs to outputs in a general graph. We consider a function

$$F : R^{n_0} \to R^{n_{M+1}} \tag{16.1}$$

that has a special nonlinear structure. The nonlinearities in the FFN are contained in the neurons which are typically modeled by sigmoid functions,

$$\sigma(y) = 0.5\left(1 + \tanh(y)\right), \tag{16.2}$$

© Springer Science+Business Media Singapore 2016
J.K. Peterson, *BioInformation Processing*, Cognitive Science and Technology,
DOI 10.1007/978-981-287-871-7_16

typically evaluated at

$$y = \frac{x - o}{g},\tag{16.3}$$

where x is the input, o is the offset and g is the gain, respectively, of each neuron. We can also write the transfer functions in a more general fashion as follows:

$$\sigma(x, o, g) = 0.5\left(1.0 + \tanh\left(\frac{x - o}{g}\right)\right)$$

The feed forward network consists of $M + 1$ layers of neurons connected together with connection coefficients. For each i, $0 \le i \le M + 1 - 1$, the ith layer of n_i neurons is *connected* to the $i + 1$st layer of n_{i+1} neurons by a *connection* matrix, T^i. Schematically, we will use the following notation for an MFFN:

$$\begin{matrix} T^0 & & & T^M & \\ n_0 \to n_1 \to & \dots & \to n_M & \to & n_{M+1} \end{matrix}\tag{16.4}$$

We can make this notation even more succinct by simply using

$$n_0 \to n_1 \to \dots \to n_M \to n_{M+1};\tag{16.5}$$

or even more simply, we can just refer to a $n_0 n_1 \cdots n_{M+1}$ MFFN. we will use either one of these notations from this point on. The feed forward network processes an arbitrary $I \in R^{n_0}$ in the following way. First, for $0 \le \ell \le M + 1$ and $1 \le i \le n_\ell$ and layer ℓ, we define some additional terms to help us organize our work. We let

σ_i^ℓ denote the *transfer* function, $\quad Y_i^\ell$ denote the output,
y_i^ℓ denote the input, $\quad O_i^\ell$ denote the offset,

g_i^ℓ denote the gain for the ith neuron and X_i^ℓ denote$\left(\frac{y_i^\ell - O_i^\ell}{g_i^\ell}\right)$. $\tag{16.6}$

Then, given $x \in R^{n_0}$, for $0 \le i \le M$, we have the inputs are processed by the zeroth layer as follows.

$$y_i^0 = x_i\tag{16.7}$$
$$X_i^0 = (y_i^0 - O_i^0)/g_i^0\tag{16.8}$$
$$Y_i^0 = \sigma_i^0(X_i^0)\tag{16.9}$$

For the next layer, more interesting things happen.

$$y_i^1 = \sum_{j=1}^{n_1-1} T_{ji}^0 Y_j^0 \tag{16.10}$$

$$X_i^1 = (y_i^1 - O_i^1)/g_i^1 \tag{16.11}$$

$$Y_i^1 = \sigma_i^1(X_i^1) \tag{16.12}$$

We continue in this way. At the layer $\ell + 1$, we have

$$y_i^{\ell+1} = \sum_{j=1}^{n_\ell-1} T_{ji}^\ell Y_j^\ell \tag{16.13}$$

$$X_i^{\ell+1} = (y_i^{\ell+1} - O_i^{\ell+1})/g_i^{\ell+1} \tag{16.14}$$

$$Y_i^{\ell+1} = \sigma_i^{\ell+1}(X_i^{\ell+1}) \tag{16.15}$$

So the output from the MFFN at layer $M + 1$ is given by

$$y_i^{M+1} = \sum_{j=1}^{n_{M+1}-1} T_{ji}^M Y_j^M \tag{16.16}$$

$$X_i^{M+1} = (y_i^{M+1} - O_i^{M+1})/g_i^{M+1} \tag{16.17}$$

$$Y_i^{M+1} = \sigma_i^{M+1}(X_i^{M+1}) \tag{16.18}$$

With this notation, for any $I \in R^{n_0}$, we know how to calculate the output from the MFFN. It is clearly a process that flows forward from the input layer and it is very nonlinear in general even though it is built out of relatively simple layers of nonlinearities. The parameters that control the value of $F(I)$ are the $n_i n_{i+1}$ coefficients of $T^i, 0 \le i \le M$; the offsets $O^i, 0 \le i \le M + 1$; and the gains $g^i, 0 \le i \le M + 1$. This gives the number of parameters,

$$N = \sum_{i=0}^{M} n_i n_{i+1} + 2 \sum_{i=0}^{M+1} n_i. \tag{16.19}$$

Now let

$$\mathcal{I} = \left\{ I_\alpha \in R^{n_0} : 0 \le \alpha \le S - 1 \right\}$$

and

$$\mathcal{D} = \left\{ D_\alpha \in R^{n_{M+1}} : 0 \le \alpha \le S - 1 \right\}$$

be two given sets of data of size $S > 0$. The set $\mathcal{I}$ is referred to as the set of *exemplars* and the set $\mathcal{D}$ is the set of outputs that are associated with exemplars. Together, the sets $\mathcal{I}$ and $\mathcal{D}$ comprise what is known as the *training* set. The **training problem** is to choose the N network parameters, $T^0, \ldots, T^M, O^0, \ldots, O^{M+1}, g^0, \ldots, g^{M+1}$, such that we minimize,

$$E = 0.5 \sum_{\alpha=0}^{S-1} \| F(I_\alpha) - D_\alpha \|_2^2, \tag{16.20}$$

$$= 0.5 \sum_{\alpha=0}^{S-1} \sum_{i=0}^{n_{M+1}-1} \left(Y_{\alpha i}^{M+1} - D_{\alpha i} \right)^2,$$

where the subscript notation α indicates that the terms defined by (16.6) correspond to the αth exemplar in the sets $\mathcal{I}$ and $\mathcal{D}$.

16.2 Minimizing the MFFN Energy

We will use the notation established in Sect. 16.1. If all the partial derivatives of the feed forward energy function (16.20) with respect to the network parameters were known, we could minimize (16.20) using a standard gradient descent scheme. This is known as the MFFN Training Problem. Relabeling the N MFFN variables temporarily as $w_i, 0 \le i \le N - 1$, the optimization problem we face is

$$\min_{w \in R^N} \quad E\left(\mathcal{I}, \mathcal{D}, w \right) \tag{16.21}$$

where we now explicitly indicate that the value of E depends on $\mathcal{I}$, $\mathcal{D}$ and the parameters w. In component form, the equations for gradient descent are then

$$w_i^{new} = w_i^{old} - \lambda \frac{\partial E}{\partial w_i}\bigg|_{w^{old}} \tag{16.22}$$

where λ is a scalar parameter that must be chosen efficiently in order for the gradient descent method to work. The above technique is the heart of the *back propagation* technique which is currently used in many cases to solve the MFFN training problem. The MFFN back propagation equations will now be derived. For further information, see the discussions in Rumelhart and McClelland (1986).

16.3 Partial Calculations for the MFFN

We need an algorithm to find the gradient of the MFFN error function so that we can apply gradient descent.

16.3.1 The Last Hidden Layer

First, we'll look at the weight update equations for the weights between the last hidden layer and the output layer. Using the chain rule, we have

$$\frac{\partial E}{\partial T_{pq}^M} = \sum_{\beta=0}^{S-1} \left(\frac{\partial E}{\partial X_{\beta q}^{M+1}} \right) \left(\frac{\partial X_{\beta q}^{M+1}}{\partial T_{pq}^M} \right) \tag{16.23}$$

Now let $e_{\alpha j}^{M+1} = Y_{\alpha j}^{M+1} - D_{\alpha j}$. Using our rules for computing the output of a node, we then obtain

$$\frac{\partial E}{\partial X_{\beta q}^{M+1}} = \sum_{\alpha=0}^{S-1} \sum_{j=0}^{n_{M+1}-1} e_{\alpha j}^{M+1} \left(\frac{\partial Y_{\alpha j}^{M+1}}{\partial X_{\beta q}^{M+1}} \right),$$

$$= \sum_{\alpha=0}^{S-1} \sum_{j=0}^{n_{M+1}-1} e_{\alpha j}^{M+1} \left(\sigma_j^{M+1} \right)' \left(X_{\alpha j}^{M+1} \right) \left(\frac{\partial X_{\alpha j}^{M+1}}{\partial X_{\beta q}^{M+1}} \right),$$

$$= \sum_{\alpha=0}^{S-1} \sum_{j=0}^{n_{M+1}-1} e_{\alpha j}^{M+1} \left(\sigma_j^{M+1} \right)' \left(X_{\alpha j}^{M+1} \right) \delta_\beta^\alpha \delta_q^j,$$

$$= e_{\beta q}^{M+1} \left(\sigma_q^{M+1} \right)' \left(X_{\beta q}^{M+1} \right). \tag{16.24}$$

where δ_β^α is the standard *kronecker delta function*. Letting $\xi_{\beta q}^{M+1} = \left(\frac{\partial E}{\partial X_{\beta q}^{M+1}} \right)$ we have,

$$\xi_{\beta q}^{M+1} = e_{\beta q}^{M+1} \left(\sigma_q^{M+1} \right)' \left(X_{\beta q}^{M+1} \right),$$

$$\frac{\partial E}{\partial T_{pq}^M} = \sum_{\beta=0}^{S-1} \xi_{\beta q}^{M+1} \left(\frac{\partial X_{\beta q}^{M+1}}{\partial T_{pq}^M} \right),$$

$$X_{\beta q}^{M+1} = \frac{\sum_{l=0}^{n_M-1} T_{lq}^M Y_{\beta l}^M - O_q^{M+1}}{g_q^{M+1}}. \tag{16.25}$$

Since the offset and gain terms are independent of the choice of the input set index β, the subscripts β are unnecessary on those terms and are not shown. Taking the indicated partial, we have

$$\frac{\partial X_{\beta q}^{M+1}}{\partial T_{pq}^{M}} = \frac{\sum_{l=0}^{n_M-1} \left(\frac{\partial T_{lq}^{M}}{\partial T_{pq}^{M}}\right) Y_{\beta l}^{M}}{g_q^{M+1}},$$

$$= \frac{Y_{\beta p}^{M}}{g_q^{M+1}}. \tag{16.26}$$

Using (16.26), we can then rewrite (16.25) as

$$\xi_{\beta q}^{M+1} = e_{\beta q}^{M+1} \left(\sigma_q^{M+1}\right)' \left(X_{\beta q}^{M+1}\right),$$

$$\frac{\partial E}{\partial T_{pq}^{M}} = \frac{\sum_{\beta=0}^{S-1} \xi_{\beta q}^{M+1} Y_{\beta p}^{M}}{g_q^{M+1}}. \tag{16.27}$$

The gain and offset terms can also be updated. The appropriate equations follow below.

$$\frac{\partial E}{\partial O_q^{M+1}} = \sum_{\beta=0}^{S-1} \left(\frac{\partial E}{\partial X_{\beta q}^{M+1}}\right) \left(\frac{\partial X_{\beta q}^{M+1}}{\partial O_q^{M+1}}\right)$$

$$= -\frac{\sum_{\beta=0}^{S-1} \xi_{\beta q}^{M+1}}{g_q^{M+1}}$$

$$\frac{\partial E}{\partial g_q^{M+1}} = \sum_{\beta=0}^{S-1} \left(\frac{\partial E}{\partial X_{\beta q}^{M+1}}\right) \left(\frac{\partial X_{\beta q}^{M+1}}{\partial g_q^{M+1}}\right)$$

$$= -\frac{\sum_{\beta=0}^{S-1} \xi_{\beta q}^{M+1} (y_{\beta q}^{M+1} - O_q^{M+1})}{(g_q^{M+1})^2} \tag{16.28}$$

The back propagation equations for the last hidden layer to the output layer are thus given by Eqs. (16.27) and (16.28).

16.3.2   The Remaining Hidden Layers

We now derive the back propagation equations for the remaining layers. Consider

$$\frac{\partial E}{\partial T_{pq}^{M-1}} = \sum_{\beta=0}^{S-1} \left(\frac{\partial E}{\partial X_{\beta q}^{M}}\right) \left(\frac{\partial X_{\beta q}^{M}}{\partial T_{pq}^{M-1}}\right) \tag{16.29}$$

Defining $\xi_{\beta q}^M = \frac{\partial E}{\partial X_{\beta q}^M}$, we obtain

$$\xi_{\beta q}^M = \sum_{\alpha=0}^{S-1} \sum_{j=0}^{n_{M+1}-1} e_{\alpha j}^{M+1} \left(\frac{\partial Y_{\alpha j}^{M+1}}{\partial X_{\beta q}^M} \right)$$

$$= \sum_{\alpha=0}^{S-1} \sum_{j=0}^{n_{M+1}-1} e_{\alpha j}^{M+1} \left(\sigma_j^{M+1} \right)' \left(X_{\alpha j}^{M+1} \right) \left(\frac{\partial X_{\alpha j}^{M+1}}{\partial X_{\beta q}^M} \right) \qquad (16.30)$$

we then calculate

$$\frac{\partial X_{\alpha j}^{M+1}}{\partial X_{\beta q}^M} = \frac{\left(\sum_{l=0}^{n_M-1} T_{lj}^M \left(\frac{\partial Y_{\alpha l}^M}{\partial X_{\beta q}^M} \right) \right)}{g_j^{M+1}},$$

$$= \frac{\left(\sum_{l=0}^{n_M-1} T_{lj}^M \left(\sigma_l^M \right)' (X_{\alpha l}^M) \left(\frac{\partial X_{\alpha l}^M}{\partial X_{\beta q}^M} \right) \right)}{g_j^{M+1}},$$

$$= \frac{\left(\sum_{l=0}^{n_M-1} T_{lj}^M \left(\sigma_l^M \right)' (X_{\alpha l}^M) \delta_\beta^\alpha \delta_q^l \right)}{g_j^{M+1}},$$

$$= \frac{T_{qj}^M \left(\sigma_q^M \right)' (X_{\beta q}^M) \delta_\alpha^\beta}{g_j^{M+1}}, \qquad (16.31)$$

Hence,

$$\xi_{\beta q}^M = \sum_{\alpha=0}^{S-1} \sum_{j=0}^{n_{M+1}-1} e_{\alpha j}^{M+1} \left(\sigma_j^{M+1} \right)' \left(X_{\alpha j}^{M+1} \right) \left(\frac{\partial X_{\alpha j}^{M+1}}{\partial X_{\beta q}^M} \right)$$

$$= \sum_{\alpha=0}^{S-1} \sum_{j=0}^{n_{M+1}-1} e_{\alpha j}^{M+1} \left(\sigma_j^{M+1} \right)' \left(X_{\alpha j}^{M+1} \right) \frac{T_{qj}^M \left(\sigma_q^M \right)' (X_{\beta q}^M) \delta_\alpha^\beta}{g_j^{M+1}}$$

$$= \sum_{j=0}^{n_{M+1}-1} e_{\beta j}^{M+1} \left(\sigma_j^{M+1} \right)' \left(X_{\beta j}^{M+1} \right) \frac{T_{qj}^M \left(\sigma_q^M \right)' (X_{\beta q}^M)}{g_j^{M+1}} \qquad (16.32)$$

From (16.28), it the follows that

$$
\xi_{\beta q}^{M} = \left(\sum_{j=0}^{n_{M+1}-1} \frac{\xi_{\beta j}^{M+1} T_{qj}^{M}}{g_{j}^{M+1}} \right) \left(\sigma_{q}^{M} \right)' (X_{\beta q}^{M}) \tag{16.33}
$$

Equation (16.33) defines a *recursive* method of computing the $\xi_{\beta q}^{M}$ in terms of the previous *layer* $\xi_{\beta j}^{M+1}$. The calculations for the next *layer* of partial derivatives is similar:

$$
\frac{\partial X_{\beta q}^{M}}{\partial T_{pq}^{M-1}} = \frac{\left(\frac{\partial y_{\beta q}^{M}}{\partial T_{pq}^{M-1}} \right)}{g_{q}^{M}},
$$

$$
= \frac{\sum_{l=0}^{n_{M-1}-1} \left(\frac{\partial T_{lq}^{M-1}}{\partial T_{pq}^{M-1}} \right) Y_{\beta l}^{M-1}}{g_{q}^{M}},
$$

$$
= \frac{Y_{\beta p}^{M-1}}{g_{q}^{M}}. \tag{16.34}
$$

Substituting (16.34) into (16.29), we find

$$
\xi_{\beta q}^{M} = \sum_{j=0}^{n_{M+1}-1} \frac{\xi_{\beta j}^{M+1} T_{qj}^{M}}{g_{j}^{M+1}} (\sigma_{q}^{M})' (X_{\beta q}^{M})
$$

$$
\frac{\partial E}{\partial T_{pq}^{M-1}} = \frac{\sum_{\beta=0}^{S-1} \xi_{\beta q}^{M} Y_{\beta p}^{M-1}}{g_{q}^{M}} \tag{16.35}
$$

The update equations for the gain and offset terms are given below.

$$
\frac{\partial E}{\partial O_{q}^{M}} = \sum_{\beta=0}^{S-1} \left(\frac{\partial E}{\partial X_{\beta q}^{M}} \right) \left(\frac{\partial X_{\beta q}^{M}}{\partial O_{q}^{M}} \right)
$$

$$
= - \frac{\sum_{\beta=0}^{S-1} \xi_{\beta q}^{M}}{g_{q}^{M}}
$$

$$
\frac{\partial E}{\partial g_{q}^{M}} = \sum_{\beta=0}^{S-1} \left(\frac{\partial E}{\partial X_{\beta q}^{M}} \right) \left(\frac{\partial X_{\beta q}^{M}}{\partial g_{q}^{M}} \right)
$$

$$
= - \frac{\sum_{\beta=0}^{S-1} \xi_{\beta q}^{M} (y_{\beta q}^{M} - O_{q}^{M})}{(g_{q}^{M})^{2}} \tag{16.36}
$$

16.4 The Full Backpropagation Equations for the MFFN

The update equations for the next back propagation step are thus given by Eqs. (16.35) and (16.36). Equations (16.27), (16.28), (16.35) and (16.36) give the back propagation equations for the last two layers. The back propagation equations between any two interior layers, including the equations governing the back propagation between the first hidden layer and the input layer, can now be derived using induction. We obtain for all layer indices $l, 0 \leq l \leq M$:

$$\xi_{\beta q}^{M+1} = e_{\beta q}^{M+1} \left(\sigma_q^{M+1} \right)' \left(X_{\beta q}^{M+1} \right),$$

$$\xi_{\beta q}^{M-l} = \sum_{j=0}^{n_{M-l+1}-1} \frac{\xi_{\beta j}^{M-l+1} T_{qj}^{M-l}}{g_j^{M-l+1}} \left(\sigma_q^{M-1} \right)' (X_{\beta q}^{M-l}),$$

$$\frac{\partial E}{\partial T_{pq}^{M-l}} = \frac{\sum_{\beta=0}^{S-1} \xi_{\beta q}^{M-l+1} Y_{\beta p}^{M-l}}{g_q^{M-l+1}},$$

$$\frac{\partial E}{\partial O_q^{M-l+1}} = -\frac{\sum_{\beta=0}^{S-1} \xi_{\beta q}^{M-l+1}}{g_q^{M-l+1}},$$

$$\frac{\partial E}{\partial g_q^{M-l+1}} = -\frac{\sum_{\beta=0}^{S-1} \xi_{\beta q}^{M-l+1} (y_{\beta q}^{M-l+1} - O_q^{M-l+1})}{(g_q^{M-l+1})^2}. \tag{16.37}$$

Note that at $l = M$, we need to evaluate $X_{\beta q}^0$. This evaluation will depend on the choice of transfer function that is used for the input layer. It is actually common to use a linear transfer function for the input layer with a zero offset and unit gain. In other words, for a linear transfer function, $O_{\beta q}^0 = 0.0$, $g_{\beta q}^0 = 1.0$, $Y_{\beta q}^0 = y_{\beta q}^0$ and $\left(\sigma_q^0 \right)' (X_{\beta q}^0) = 1.0$. For this case, $l = M$, we have

$$\xi_{\beta q}^1 = \sum_{j=0}^{n_2-1} \frac{\xi_{\beta j}^2 T_{qj}^1}{g_j^2} \left(\sigma_q^1 \right)' (X_{\beta q}^1),$$

$$\frac{\partial E}{\partial T_{pq}^0} = \frac{\sum_{\beta=0}^{S-1} \xi_{\beta q}^1 Y_{\beta p}^0}{g_q^1},$$

$$\frac{\partial E}{\partial O_q^1} = -\frac{\sum_{\beta=0}^{S-1} \xi_{\beta q}^1}{g_q^1}$$

$$\frac{\partial E}{\partial g_q^1} = -\frac{\sum_{\beta=0}^{S-1} \xi_{\beta q}^1 (y_{\beta q}^1 - O_q^1)}{(g_q^1)^2}. \tag{16.38}$$

There are no connection coefficients before T^0 and so to derive the update equations for the gains and offsets for the input layer, we need to step back a bit. From our derivations, we know

$$\frac{\partial E}{\partial O_q^0} = \sum_{\beta=0}^{S-1} \xi_{\beta q}^0 \frac{\partial X_{\beta q}^0}{\partial O_q^0}$$

$$= -\frac{\sum_{\beta=0}^{S-1} \xi_{\beta q}^0}{g_g^0} \tag{16.39}$$

$$\frac{\partial E}{\partial g_q^0} = \sum_{\beta=0}^{S-1} \xi_{\beta q}^0 \frac{\partial X_{\beta q}^0}{\partial g_q^0}$$

$$= -\frac{\sum_{\beta=0}^{S-1} \xi_{\beta q}^0 \, (y_q^0 - O_q^0)}{(g_g^0)^2}. \tag{16.40}$$

Hence, if we want to have parameters in the input layer tuned, we can do so with Eqs. (16.39) and (16.40). Equations (16.37), (16.39) and (16.40) gives the complete set of back propagation equations for this type of neural network architecture. If the parameters of the MFFN's input layer are not mutable, we only need to use Eq. (16.37), of course.

16.5   A Three Layer Example

Let's go through all of the forbidding equations above for a three layer network which uses the same sigmoidal transfer function for all neurons in the hidden and output layer and uses some other common transfer function for the input layer. We assume all the transfer functions are of type σ. In the notation we used before, we have $M = 1$ so that the output layer is $M + 1 = 2$ and the input layer is $M - 1 = 0$. Let's also assume we have n_0 neurons in the input layer, n_1 neurons in the hidden layer and n_2 neurons in the output layer. Our equations become:

16.5.1   The Output Layer

$$X_{\beta q}^2 = \frac{y_{\beta q}^2 - O_q^2}{g_q^2}$$

$$e_{\beta q}^2 = Y_{\beta q}^2 - D_{\beta q}$$

$$\xi_{\beta q}^2 = e_{\beta q}^2 \sigma'\left(X_{\beta q}^2\right),$$

$$\frac{\partial E}{\partial T_{pq}^1} = \frac{\sum_{\beta=0}^{S-1} \xi_{\beta q}^2 Y_{\beta p}^1}{g_q^2},$$

$$\frac{\partial E}{\partial O_q^2} = -\frac{\sum_{\beta=0}^{S-1} \xi_{\beta q}^2}{g_q^2}$$

$$\frac{\partial E}{\partial g_q^2} = -\frac{\sum_{\beta=0}^{S-1} \xi_{\beta q}^2 (y_{\beta q}^2 - O_q^2)}{(g_q^2)^2} \qquad (16.41)$$

16.5.2 The Hidden Layer

$$X_{\beta q}^1 = \frac{y_{\beta q}^1 - O_q^1}{g_q^1}$$

$$\xi_{\beta q}^1 = \left(\sum_{j=0}^{n_2-1} \frac{\xi_{\beta j}^2 T_{qj}^1}{g_j^2}\right) \left(\sigma_q^1\right)' (X_{\beta q}^1)$$

$$\frac{\partial E}{\partial T_{pq}^0} = \frac{\sum_{\beta=0}^{S-1} \xi_{\beta q}^1 Y_{\beta p}^0}{g_q^1}$$

$$\frac{\partial E}{\partial O_q^1} = -\frac{\sum_{\beta=0}^{S-1} \xi_{\beta q}^1}{g_q^1}$$

$$\frac{\partial E}{\partial g_q^1} = -\frac{\sum_{\beta=0}^{S-1} \xi_{\beta q}^1 (y_{\beta q}^1 - O_q^1)}{(g_q^1)^2} \qquad (16.42)$$

16.5.3 The Input Layer

$$\xi_{\beta q}^0 = \sum_{j=0}^{n_1-1} \frac{\xi_{\beta j}^1 T_{qj}^0}{g_j^1} \left(\sigma_q^0\right)' (X_{\beta q}^0),$$

$$\frac{\partial E}{\partial O_q^0} = -\frac{\sum_{\beta=0}^{S-1} \xi_{\beta q}^0}{g_q^0},$$

$$\frac{\partial E}{\partial g_q^0} = -\frac{\sum_{\beta=0}^{S-1} \xi_{\beta q}^0 (x_{\beta q} - O_q^0)}{(g_q^0)^2}. \qquad (16.43)$$

where $x_{\beta q}$ is the input into the qth input neuron for sample β.

16.6   A MatLab Beginning

Let's get started at coding a MatLab implementation of the MFFN. We will not be clever yet and instead code up a quick and dirty example for a specific 2-3-1 MFFN. Let's begin with some initializations and the code for calculating the summed squared error over the data set. We will assume each piece of data is sent through the 2-3-1 MFFN in a function **myffneval** which will be written later. Using the evaluation function and the initializations, we can then calculate the summed squared error over the data. Consider the following code for the 2-3-1 initialization part. The function **MyFFN** assumes there is some data stored in **Data** which consists of triples of the form (x, y, z). The built in function **rand** is very easy to use. The command **rand(2,3)** returns random numbers in $(0, 1)$ for all the entries of a 2×3 matrix and so the line **-1+2*rand(2,3)** sets up a random matrix with entries in $(-1, 1)$. We use this approach to set up a random matrix for our T^0 and T^1 matrices which will be called **T1** and **T2** in MatLab because the numbering in MatLab starts at 1 instead of the 0 we use in our derivations. We also set up random vectors of the offsets and gains which we call **O** and **G**.

   We also set up our nodal processing functions. We need 6 of them here, so we set them up in a cell data structure **nodefunction{i}** as follows;

Listing 16.1: Initializing the nodefunctions for the 2-3-1 MFFN

```
  sigmainput = @(x,o,g) (x-o)/g;
  sigma = @(x,o,g) 0.5*( 1 + tanh( (x-o)/g ) );
  % assign the node functions
  nodefunc{1} = sigmainput;
5 nodefunc{2} = sigmainput;
  nodefunc{3} = sigma;
  nodefunc{4} = sigma;
  nodefunc{5} = sigma;
  nodefunc{6} = sigma;
```

This makes it easy for us to refer to nodal processing functions by their node number. Next, we set up the loop over the data triples which enables us to calculate the summed squared error. We store the outputs of the nodes as **Y**, the raw errors as **e = Y - Z** and the squared errors as **e^2**. We then use this information to sum the squared errors over the data set and compute the error function **sumE**.

Listing 16.2: The error loop for the 2-3-1 MFFN

```
% setup the MFFN evaluation loop for this data
Y = [];
e = [];
E = [];
for i = 1:N
    Y(i) = myffneval(X1(i),X2(i),Z(i),T1,T2,Off,Gain,nodefunc);
    e(i) = Y(i) - Z(i);
    E(i) = e(i)^2;
end
sumE = 0.5*sum(E);
end
```

All of this is put together in the function **MyFFN** below.

Listing 16.3: Initializing and finding the error for a 2-3-1 MFFN

```
function [sumE,e,E,Y,Z] = MyFFN(Data)
%
% Sample FFN code for a specific 2-3-1 MFFN
%
% Data is triples (X1,X2,Z)
%
N = length(Data);
X1 = Data(:,1);
X2 = Data(:,2);
Z = Data(:,3);
%
% Set up weight matrices
% T1 and T2 are randomly initialized in (-1,1)
T1 = -1 + 2*rand(2,3);
T2 = -1 + 2*rand(3,1);
%
% Setup offsets and gains
% Off is 6x1 and Gain is 6x1. Initialize randomly
Off = rand(6,1);
Gain = rand(6,1);
% Setup the processing functions
sigmainput = @(x,o,g) (x-o)/g;
sigma = @(x,o,g) 0.5*( 1 + tanh( (x-o)/g ) );
% assign the node functions
nodefunc{1} = sigmainput;
nodefunc{2} = sigmainput;
nodefunc{3} = sigma;
nodefunc{4} = sigma;
nodefunc{5} = sigma;
nodefunc{6} = sigma;
%
% setup the MFFN evaluation loop for this data
Y = [];
e = [];
E = [];
for i = 1:N
    Y(i) = myffneval(X1(i),X2(i),Z(i),T1,T2,Off,Gain,nodefunc);
    e(i) = Y(i) - Z(i);
    E(i) = e(i)^2;
end
sumE = 0.5*sum(E);
end
```

To finish we need to write the 2-3-1 evaluation code. This is given in **myffneval**. This code follows our algorithm closely and you should look at it carefully to make sure you see that.

Listing 16.4: The 2-3-1 MFFN evaluation

```
    function [Y6] = myffneval(x1,x2,z,T1,T2,Off,Gain,nodefunc)
    %
    % x1,x2,z is the data triple from Data
    % T1 is the 2x3 weight matrix between layer 1 (input layer)
 5  %      and layer 2 (middle layer)
    % T2 is the 3x1 weight matrix between layer 2 (middle layer)
    %      and layer 3 (output layer)
    % Off is the offset vector
    % Gain in the gain vector
10  % nodefunc is a cell of function handles used for
    %      the node processing
    %
    % input layer processing
    y1 = x1;  Y1 = nodefunc{1}(y1,Off(1),Gain(1));
15  y2 = x2;  Y2 = nodefunc{2}(y2,Off(2),Gain(2));
    % middle layer processing
    y3 = T1(1,1)*Y1 + T1(2,1)*Y2;  Y3 = nodefunc{3}(y3,Off(3),Gain(3));
    y4 = T1(1,2)*Y1 + T1(2,2)*Y2;  Y4 = nodefunc{4}(y4,Off(4),Gain(4));
    y5 = T1(1,3)*Y1 + T1(2,3)*Y2;  Y5 = nodefunc{5}(y5,Off(5),Gain(5));
20  % output layer processing
    y6 = T2(1,1)*Y3+T2(2,1)*Y4+T2(3,1)*Y5;  Y6 = nodefunc{6}(y6,Off(6),
       Gain(6));

    end
```

We can test this on some simple data. There are two inputs x_1 and x_2 and if $2x_1/x_2 > 1$, the target value is 1 and otherwise, it is zero. We setup some data and run our codes as follows:

Listing 16.5: Testing Our 2-3-1 Code

```
    X1 = [0.3;0.7;1.4;2.7];
 2  X2 = [0.1;1.3;0.6;7.2];
    C  = [X1,X2,(2*X1./X2)]
    C =

         0.3000        0.1000       6.0000
         0.7000        1.3000       1.0769
 7       1.4000        0.6000       4.6667
         2.7000        7.2000       0.7500
    % It is easy to set the targets now: the first three are 1
    % and the last one is 0.
    Target = [1;1;1;0];
12  Data = [X1,X2,Target]
    Data =

         0.3000        0.1000       1.0000
         0.7000        1.3000       1.0000
         1.4000        0.6000       1.0000
17       2.7000        7.2000          0
    % now find the summed squared error
     [sumE,e,E,Y,Z] = MyFFN(Data);
    % The Y(6) node values are returned for each piece
    % of data and stored in Y
22  Y
    Y =

         0.1128        0.0539       0.0449       0.0291
    % The targets are
    Z
27  Z =
         1
         1
         1
         0
```

```
32 The squared errors are
   E
   E =
        0.7872      0.8951      0.9123      0.0008
   % and the error is
37 sumE
   sumE =
        1.2977
```

16.7   MatLab Implementations

Let's write our first attempt at implementing a standard MFFN in MatLab. We are
going to be quite general in this implementation; the previous 2-3-1 example was just
a warmup. From studying our code there, we can see more clearly how the places
where we should write more generic code. But make no mistake; there are always
tradeoffs to any software implementation decision we make.

We will break our implementation into fairly standard pieces: an initialization
function, an evaluation function, an update function and finally a training function.

16.7.1   Initialization

Listing 16.6: Initialization code mffninit.m

```
1 function [G,O,T] = mffninit(GL,GH,OL,OH,TL,TH,LayerSizes)
  %
  %    This is a P layer MFFN where P is the number
  %    of layers
  %    number of nodes in the input layer 0 = LayerSizes(1)
6 %    number of nodes in the middle layer 1 = LayerSizes(2)
  %    ...
  %    number of nodes in the output layer = LayerSizes(P)
  %
  % GL is the lower bound for the gains
11 % GH is the upper bound for the gains
  % OL is the lower bound for the offsets
  % OH is the upper bound for the offsets
  % TL is the lower bound for the edge weights
  % TH is the upper bound for the edge weights
16 T = {};
  O = {};
  G = {};

  [P,m] = size(LayerSizes);
21 % initialize parameters
  for p=1:P-1
      rows = LayerSizes(p);
      cols = LayerSizes(p+1);
      T{p} = TL+(TH-TL)*rand(rows,cols);
26    O{p} = OL+(OH-OL)*rand(1,rows);
      G{p} = GL+(GH-GL)*rand(1,rows);
  end
  rows = LayerSizes(P);
  O{P} = OL+(OH-OL)*rand(1,rows);
31 G{P} = GL+(GH-GL)*rand(1,rows);

  end
```

16.7.2   Evaluation

Listing 16.7: Evaluation: mffneval2.m

```
     function [y,Y,RE,E] = mffneval2(Input,Target,nodefunction,Gain,Offset
         ,T,LayerSizes)
 2   %
     %   This is a P layer MFFN where P is the number
     %   of layers
     %   number of nodes in the input layer 0 = LayerSizes(1)
     %   number of nodes in the middle layer 1 = LayerSizes(2)
 7   %   ...
     %   number of nodes in the output layer = LayerSizes(P)
     %
     %  sigma is the node transfer function
     %  Input is input matrix of S rows of input data
12   %       S x LayerSizes(1)
     %  target is desired target value matrix of S rows of target data
     %       S xLayerSizes(P)
     %  nodefunction is the node transfer function cell data
     %  nodefunctionprime is the node transfer function derivative cell
         data
17   sigma = nodefunction{1};
     sigmainput = nodefunction{2};
     sigmaoutput = nodefunction{3};

     %  Extract number of layers
22   %  so
     %  output layer = P
     %  input layer = 1
     [P,m] = size(LayerSizes);

27   %  Assume InputSize = TargetSize!!
     %
     %  Extract the number of training samples S
     [S,InputSize] = size(Input);

32   %  set up storage for all MFFN values
     y = {};
     Y = {};
     for p = 1:P
         cols = LayerSizes(p);
37       y{p} = zeros(S,cols);
         Y{p} = zeros(S,cols);
     end

     %  set up storage for both raw error RE and local squared error LE
42   cols = LayerSizes(P);
     LE = zeros(S,cols);
     RE = zeros(S,cols);
     Output = zeros(S,cols);

47   energy = 0;
     for s=1:S
         %  loop on training examplars
         for p = 1:P
             if p ==1
52               y{p}(s,:) = Input(s,:);
                 for i=1:LayerSizes(1)
                     input = y{p}(s,i);
                     offset = Offset{p}(i);
                     gain = Gain{p}(i);
```

```
57              Y{p}(s,i) = sigmainput(input,offset,gain);
            end
         else
            for i=1:LayerSizes(p)
               y{p}(s,i) = dot(T{p-1}(:,i),Y{p-1}(s,:));
62             input = y{p}(s,i);
               offset = Offset{p}(i);
               gain = Gain{p}(i);
               if p ==P
                  Y{p}(s,i) = sigmaoutput(input,offset,gain);
67             else
                  Y{p}(s,i) = sigma(input,offset,gain);
               end
            end
         end
72   end % p loop
     Output(s,:) = Y{P}(s,:);
     RE(s,:) = Output(s,:) - Target(s,:);
     LE(s,:) = RE(s,:).^2;
     energy = energy + sum(LE(s,:));
77 end % s loop

   E = 0.5*energy;

   end
```

16.7.3   Updating

Now let's consider the update process. We will follow the update algorithms we have presented. Our node evaluation functions have the form of the σ listed below for various values of L and H where we assume $L < H$, of course.

$$\sigma(y, o, g) = \frac{1}{2}\left((H + L) + (H - L)\tanh\left(\frac{y - o}{g}\right)\right)$$

The three partials are

$$\frac{\partial\sigma}{\partial y} = \frac{H - L}{2g}\left(\operatorname{sech}\left(\frac{y - o}{g}\right)\right)^2$$

$$\frac{\partial\sigma}{\partial o} = -\frac{H - L}{2g}\left(\operatorname{sech}\left(\frac{y - o}{g}\right)\right)^2$$

$$= -\frac{\partial\sigma}{\partial y}$$

$$\frac{\partial \sigma}{\partial o} = -\frac{H-L}{2g^2}\left(\text{sech}\left(\frac{y-o}{g}\right)\right)^2$$
$$= -\frac{1}{g}\frac{\partial \sigma}{\partial y}.$$

In the code for our updates, we will assume we have coded all of our σ' functions as if the argument to the node function was simply u instead of the ratio $\frac{y-o}{g}$. Hence, we need the following derivative for all our node functions.

$$\sigma'(u) = \frac{H-L}{2}\,l(\text{sech}(u))^2.$$

In code we would have

Listing 16.8: Node function initialization

```
      OL = -1.4;
      OH = 1.4;
      GL = 0.8;
 4    GH = 1.2;
      TL = -2.1;
      TH = 2.1;
      nodefunction = {};
      nodefunctionprime = {};
 9    sigma = @(y,offset,gain) 0.5*(1 + tanh( (y - offset)/gain) );
      nodefunction{1} = sigma;
      sigmainput = @(y,offset,gain) (y - offset)/gain;
      nodefunction{2} = sigmainput;
      SL = -.5;
 14   SH = 1.5;
      % ouput nodes range from SL to SH
      sigmaoutput = @(y,offset,gain) 0.5*( (SH+SL) + (SH-SL)* tanh( (y -
          offset)/gain) );
      nodefunction{3} = sigmaoutput;
      % these derivatives are really partial sigma/ partial y
 19   % so all have a 1/gain built in
      sigmaprime = @(y,offset,gain) (1/(2))* sech((y - offset)/gain).^2;
      nodefunctionprime{1} = sigmaprime;
      sigmainputprime = @(y,offset,gain) 1;
      nodefunctionprime{2} = sigmainputprime;
 24   sigmaoutputprime = @(y,offset,gain) ( (SH-SL)/(2) )* sech((y - offset
          )/gain).^2;
      nodefunctionprime{3} = sigmaoutputprime;
```

Note **nodefunction{1}** is the main node processing function, **nodefunction{2}** is the input node function and **nodefunction{3}** is the output node function.

You see this used in the update code as follows

Listing 16.9: Node function assignment in the update code

```
% main sigma
sigma = nodefunction{1};
% input sigma
sigmainput = nodefunction{2};
5 % output sigma
sigmaoutput = nodefunction{3};
% main sigma prime
sigmaprime = nodefunctionprime{1};
% input sigma prime
10 sigmainputprime = nodefunctionprime{2};
% output sigma prime
sigmaoutputprime = nodefunctionprime{3};
```

Listing 16.10: Updating: mffnupdate.m

```
function [Gain, Offset, T, length] = mffnupdate(Input, Target,
    nodefunction, nodefunctionprime, Gain, Offset, T, LayerSizes, y, Y, lambda
    )
%
3 % We train a P layer MFFN to math input to targets where P is the
    number
% of layers
% number of nodes in the input layer 0 = LayerSizes(1)
% number of nodes in the middle layer 1 = LayerSizes(2)
% ...
8 % number of nodes in the output layer = LayerSizes(P)
%
% Input is input matrix of S rows of input data
%      S x LayerSizes(1)
% target is desired target value matrix of S rows of target data
13 %      S xLayerSizes(P)
%
% Gain is the MFFN gain parameters
% Offset is the MFFN offset parameters
% T is the MFFN edge coefficients
18 % y is the node input data
% Y is the node output data
%
% nodefunction is the node transfer function cell data
% nodefunctionprime is the node transfer function derivative cell
    data
23 sigma = nodefunction{1};
sigmainput = nodefunction{2};
sigmaoutput = nodefunction{3};
sigmaprime = nodefunctionprime{1};
sigmainputprime = nodefunctionprime{2};
28 sigmaoutputprime = nodefunctionprime{3};
%
% turn gain updates on or off: DOGAIN = 0
% turns them off
DOGAIN = 0;
```

```
33
    % Extract number of layers
    % so
    % output layer = P
    % input layer = 1
38  [P,m] = size(LayerSizes);

    %
    % Extract the number of training samples S
    [S,InputSize] = size(Input);
43
    % setup storage for generalized errors
    xi = {};
    for p=P:-1:1
      cols = LayerSizes(p);
48    xi{p} = zeros(S,cols);
    end

    % setup storage for partial derivatives
    DT = {};
53  DO = {};
    DG = {};
    for p=1:P-1
      rows = LayerSizes(p);
      cols  = LayerSizes(p+1);
58    DT{p} = zeros(rows,cols);
      DO{p} = zeros(1,rows);
      DG{p} = zeros(1,rows);
    end
    rows = LayerSizes(P);
63  DO{P} = zeros(1,rows);
    DG{P} = zeros(1,rows);

    % find output layer generalized error
    for s=1:S
68    for i = 1:LayerSizes(P)
        error = Y{P}(s,i) - Target(s,i);
        xi{P}(s,i) = error*sigmaoutputprime(y{P}(s,i),Offset{P}(i),Gain{
            P}(i));
      end
    end
73
    for p=P-1:-1:1
      for s = 1:S
        % find xi{p}(s,:)
        rows = LayerSizes(p);
78      cols = LayerSizes(p+1);
        for q = 1:rows
          for j=1:cols
            if p == 1
              f = sigmainputprime(y{p}(s,q),Offset{p}(q),Gain{p}(q));
83          else
              f = sigmaprime(y{p}(s,q),Offset{p}(q),Gain{p}(q));
            end
            xi{p}(s,q) =  xi{p}(s,q) +  xi{p+1}(s,j)*T{p}(q,j)*f/Gain{
                p+1}(j);
          end % inner j node loop
```

```
88      end % outer q node loop
        % find DT{p}, DO{p+1}, DG{p+1}
        for u = 1:rows
          for v = 1:cols
            DT{p}(u,v) = DT{p}(u,v) + xi{p+1}(s,v)*Y{p}(s,u)/Gain{p
              +1}(v);
93        end%v
        end%u
        for v=1:cols
          DO{p+1}(v) = DO{p+1}(v) - xi{p+1}(s,v)/Gain{p+1}(v);
          %DG{p+1}(v) = DG{p+1}(v) -  xi{p+1}(s,v)*(y{p+1}(s,v) -
              Offset{p+1}(v))/(Gain{p+1}(v)^2);
98        end% v loop
      end % s loop
    end % layer loop
    % now do input layer
    for s = 1:S
103     % find DO{1}, DG{1}
        cols = LayerSizes(1);
        for v = 1:cols
          DO{1}(v) = DO{1}(v) - xi{1}(s,v)/Gain{1}(v);
          %DG{1}(v) = DG{1}(v) -  xi{1}(s,v)*(y{1}(s,v) - Offset{1}(v))
              /(Gain{1}(v)^2);
108     end % v loop
    end % s loop
    % find the norm of the gradient
    squarelength = 0.0;
    for p = 1:P-1
113     squarelength = squarelength + (norm(DT{p},'fro'))^2;
        if DOGAIN == 1
          squarelength = squarelength + (norm(DG{p},'fro'))^2;
        end
        squarelength = squarelength + (norm(DO{p},'fro'))^2;
118 end
    if DOGAIN == 1
        squarelength = squarelength + (norm(DG{P},'fro'))^2 + (norm(DO{P},
          'fro'))^2;
    else
        squarelength = squarelength + (norm(DO{P},'fro'))^2;
123 end
    length = sqrt(squarelength);
    Normgrad = length;
    NormGradMFFN = Normgrad;
    lengthtol = 0.05;
128 if length < lengthtol
      scale = lambda;
    else
      scale = lambda/length;
    end
133 fscale = scale;
    % now do training
    for p = 1:P-1
        rows = LayerSizes(p);
        cols = LayerSizes(p+1);
```

```
138      for  u = 1:rows
            for  v = 1:cols
               T{p}(u,v) = T{p}(u,v) - scale*DT{p}(u,v);
            end% v loop
         end% u loop
143      for  u = 1:rows
               Offset{p}(u) = Offset{p}(u) - scale*DO{p}(u);
               %Gain{p}(u) = Gain{p}(u) - scale*DG{p}(u);
            end%u loop
      end% p loop
148   rows = LayerSizes(P);
      for  u = 1:rows
         Offset{P}(u) = Offset{P}(u) - scale*DO{P}(u);
         %Gain{P}(u) = Gain{P}(u) - scale*DG{P}(u);
      end
153
      end
```

16.7.4   Training

Listing 16.11: Training: mffntrain.m

```
1  function  [G,O,T,Energy] = mffntrain(Input,Target,nodefunction,
      nodefunctionprime,G,O,T,LayerSizes,y,Y,lambda,NumIters)
   %
   %    We train a P layer MFFN to math input to targets where P is the
        number
   %    of layers
   %    number of nodes in the input layer 0 = LayerSizes(1)
6  %    number of nodes in the middle layer 1 = LayerSizes(2)
   %    ...
   %    number of nodes in the output layer = LayerSizes(P)
   %
   %  Input is input matrix of S rows of input data
11 %        S x LayerSizes(1)
   %  target is desired target value matrix of S rows of target data
   %        S xLayerSizes(P)
   %
   %  Gain is the MFFN gain parameters
16 %  Offset is the MFFN offset parameters
   %  T is the MFFN edge coefficients
   %  y is the node input data
   %  Y is the node output data
   %
21 Energy = [];
   [y,Y,RE,E] = mffneval2(Input,Target,nodefunction,G,O,T,LayerSizes);
   Energy = [Energy E];
   for  t = 1:NumIters
      [G,O,T] = mffnupdate(Input,Target,nodefunction,nodefunctionprime,G,
         O,T,LayerSizes,y,Y,lambda);
26    [y,Y,RE,E] = mffneval2(Input,Target,nodefunction,G,O,T,LayerSizes);
      Energy = [Energy E];
   end
   plot(Energy);
   end
```

16.8   Sample Training Sessions

16.8.1   Approximating a Step Function

Now we can test the code. We use 31 exemplars to train a 1-5-1 MFFN with the gain updates turned off to approximate a simple step function. So we have $10 + 7 = 17$ tunable parameters and 31 constraints. We have to setup inputs and outputs. Here we want every input less than 1 to be the target 0 and every input bigger than 1 to be the target 1. The traditional if conditional in Matlab is not suited for vectors such as we create below for **x**. Hence we use the function **iff** below to do the testing. The builtin Matlab function **nargchk** is used with **condition** set to be the test **x < 1** and the code line **iff(x<1,1,0)** is expanded into **error(nargchk(x<1,1,0,31))** as the size of **x** is 31. When the test is true, the return value is 1 and otherwise it is 0. This enables us to use conditional testing on a vector.

Listing 16.12: A vector conditional test function

```
function  result  =  iff(condition, trueResult, falseResult)
    error(nargchk(3,3,nargin));
    if condition
        result  =  trueResult;
5   else
        result  =  falseResult;
    end
end
```

Listing 16.13: Testing the Code

```
    X = linspace(0,3,31);
2  Input = X';
    U = arrayfun(@(x) iff(x<1,1,0),X);
    Target = U';
    LayerSizes = [1;5;1];
    OL = -1.4;
7  OH = 1.4;
    GL = 0.8;
    GH = 1.2;
    TL = -2.1;
    TH = 2.1;
12 [G,O,T] = mffninit(GL,GH,OL,OH,TL,TH,LayerSizes);
    nodefunction = {};
    nodefunctionprime = {};
    sigma = @(y,offset,gain) 0.5*(1 + tanh( (y - offset)/gain) );
    nodefunction{1} = sigma;
17 sigmainput = @(y,offset,gain) (y - offset)/gain;
    nodefunction{2} = sigmainput;
    SL = -.5;
    SH = 1.5;
    % ouput nodes range from SL to SH
22 sigmaoutput = @(y,offset,gain) 0.5*( (SH+SL) + (SH-SL)* tanh( (y -
        offset)/gain) );
    nodefunction{3} = sigmaoutput;
    % these derivatives are really partial sigma/ partial y
    % so all have a 1/gain built in
    sigmaprime = @(y,offset,gain) (1/(2))* sech((y - offset)/gain).^2;
```

```
27 nodefunctionprime{1} = sigmaprime;
   sigmainputprime = @(y,offset,gain) 1;
   nodefunctionprime{2} = sigmainputprime;
   sigmaoutputprime = @(y,offset,gain) ( (SH-SL)/(2) )* sech((y - offset
       )/gain).^2;
   nodefunctionprime{3} = sigmaoutputprime;
32 %
   [y,Y,RE,E] = mffneval2(Input,Target,nodefunction,G,O,T,LayerSizes);
   NumIters = 10;
   lambda = .005;
   [G,O,T,Energy] = mffntrain(Input,Target,nodefunction,
       nodefunctionprime,...
37 G,O,T,LayerSizes,y,Y,lambda,NumIters);
   NumIters = 10;
   lambda = .005;
   [G,O,T,Energy] = mffntrain(Input,Target,nodefunction,
       nodefunctionprime,...
   G,O,T,LayerSizes,y,Y,lambda,NumIters);
42 ...
   [G,O,T,Energy] = mffntrain(Input,Target,nodefunction,
       nodefunctionprime,...
   G,O,T,LayerSizes,y,Y,lambda,200);
   [G,O,T,Energy] = mffntrain(Input,Target,nodefunction,
       nodefunctionprime,...
   G,O,T,LayerSizes,y,Y,lambda,1000);
47 [G,O,T,Energy] = mffntrain(Input,Target,nodefunction,
       nodefunctionprime,...
   G,O,T,LayerSizes,y,Y,0.006,1000);
```

After all this training, we still haven't quite got it. What we do have is an approximation to this step function shown in Fig. 16.1. To generate this graph, we evaluate the FFN at new points and plot along with the graph of the real function.

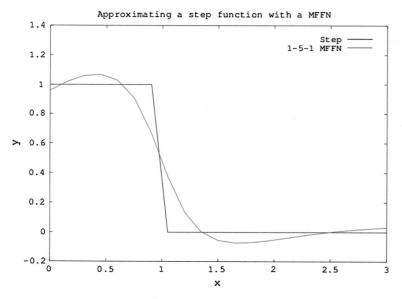

Fig. 16.1 Approximating a step function using a 1-5-1 MFFN

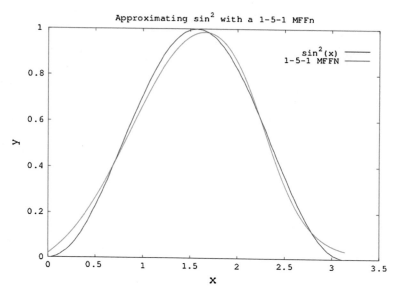

Fig. 16.2 Approximating a $\sin^2$ using a 1-5-1 MFFN

Listing 16.14: Generating the plot that test our approximation

```
  X = linspace(0,3,21);
2 Input = X';
  U = arrayfun(@(x) iff(x<1,1,0),X);
  Target = U';
  [y,Y,RE,E] = mffneval2(Input,Target,nodefunction,G,O,T,LayerSizes);
  Z = Y{3};
7 plot(X,U,X,Z);
```

16.8.2 Approximating sin²

We are going to approximate $\sin^2(x)$ on the interval $[0, \pi]$. We use 21 examplars
with a 1-5-1 MFFN which is 17 parameters that can be adjusted and 21 constraints.

Listing 16.15: Approximating $\sin^2$

```
  X = linspace(0,3,21);
  Input = X';
3 U = sin(X).^2;
  Target = U';
  LayerSizes = [1;5;1];
  OL = -1.4;
  OH = 1.4;
8 GL = 0.8;
  GH = 1.2;
  TL = -2.1;
  TH = 2.1;
  [G,O,T] = mffninit(GL,GH,OL,OH,TL,TH,LayerSizes);
```

```
13 nodefunction = {};
   nodefunctionprime = {};
   sigma = @(y,offset,gain) 0.5*(1 + tanh( (y - offset)/gain) );
   nodefunction{1} = sigma;
   sigmainput = @(y,offset,gain) (y - offset)/gain;
18 nodefunction{2} = sigmainput;
   SL = -.5;
   SH = 1.5;
   % ouput nodes range from SL to SH
   sigmaoutput = @(y,offset,gain) 0.5*( (SH+SL) + (SH-SL)* tanh( (y -
       offset)/gain) );
23 nodefunction{3} = sigmaoutput;

   sigmaprime = @(y,offset,gain) (1/(2))* sech((y - offset)/gain).^2;
   nodefunctionprime{1} = sigmaprime;
   sigmainputprime = @(y,offset,gain) 1;
28 nodefunctionprime{2} = sigmainputprime;
   sigmaoutputprime = @(y,offset,gain) ( (SH-SL)/(2) )* sech((y - offset
       )/gain).^2;
   nodefunctionprime{3} = sigmaoutputprime;
   %
   [y,Y,RE,E] = mffneval2(Input,Target,nodefunction,G,O,T,LayerSizes);
33 [G,O,T,Energy] = mffntrain(Input,Target,nodefunction,
       nodefunctionprime,G,O,T,LayerSizes,y,Y,0.005,100);
   [G,O,T,Energy] = mffntrain(Input,Target,nodefunction,
       nodefunctionprime,G,O,T,LayerSizes,y,Y,lambda,NumIters);
   ...
   [G,O,T,Energy] = mffntrain(Input,Target,nodefunction,
       nodefunctionprime,G,O,T,LayerSizes,y,Y,lambda,50);
   [G,O,T,Energy] = mffntrain(Input,Target,nodefunction,
       nodefunctionprime,G,O,T,LayerSizes,y,Y,lambda,50);
38 [G,O,T,Energy] = mffntrain(Input,Target,nodefunction,
       nodefunctionprime,G,O,T,LayerSizes,y,Y,lambda,50);

   X = linspace(0,3.14,101);
   Input = X';
   U = sin(X).^2;
43 Target = U';
   [y,Y,RE,E] = mffneval2(Input,Target,nodefunction,G,O,T,LayerSizes);
   Z = Y{3};
   plot(X,U,X,Z);
```

After all this training, we are doing pretty well. The approximation is shown in Fig. 16.2 and it is clearly phase shifted a bit.

16.8.3   Approximating sin² Again: Linear Outputs

The choice of linear output node functions does not do as well. Even after significant training, the approximation is not as good. The approximation is shown in Fig. 16.3.

Listing 16.16: Approximating sin^2 Again: Linear Outputs

```
  X = linspace(0,3.14,31);
  Input = X';
  U = sin(X).^2;
4 Target = U';
  LayerSizes = [1;4;1];
  OL = -1.4;
  OH = 1.4;
  GL = 0.8;
9 GH = 1.2;
  TL = -2.1;
  TH = 2.1;
  [G,O,T] = mffninit(GL,GH,OL,OH,TL,TH,LayerSizes);
  nodefunction = {};
```

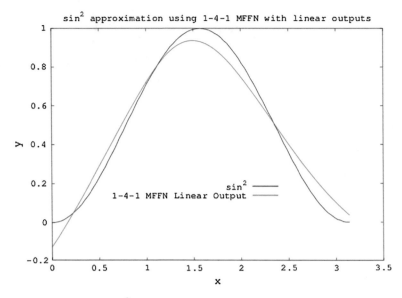

Fig. 16.3 Approximating a sin² using a 1-5-1 MFFN with linear outputs

```
14 nodefunctionprime = {};
   sigma = @(y, offset, gain)  0.5*(1 + tanh( (y - offset)/gain) );
   nodefunction{1} = sigma;
   sigmainput = @(y, offset, gain)  (y - offset)/gain;
   nodefunction{2} = sigmainput;
19 sigmaoutput = @(y, offset, gain)  (y - offset)/gain;
   nodefunction{3} = sigmaoutput;

   sigmaprime = @(y, offset, gain)  (1/(2))* sech((y - offset)/gain).^2;
   nodefunctionprime{1} = sigmaprime;
24 sigmainputprime = @(y, offset, gain)  1;
   nodefunctionprime{2} = sigmainputprime;
   sigmaoutputprime = @(y, offset, gain)  1;
   nodefunctionprime{3} = sigmaoutputprime;
   %
29 [y, Y, RE, E] = mffneval2(Input, Target, nodefunction, G, O, T, LayerSizes);
   NumIters = 10;
   lambda = .0005;
   [G, O, T, Energy] = mffntrain(Input, Target, nodefunction,
       nodefunctionprime, G, O, T, LayerSizes, y, Y, lambda, NumIters);
   [G, O, T, Energy] = mffntrain(Input, Target, nodefunction,
       nodefunctionprime, G, O, T, LayerSizes, y, Y, lambda, 100);
34 [G, O, T, Energy] = mffntrain(Input, Target, nodefunction,
       nodefunctionprime, G, O, T, LayerSizes, y, Y, lambda, 100);
   [G, O, T, Energy] = mffntrain(Input, Target, nodefunction,
       nodefunctionprime, G, O, T, LayerSizes, y, Y, lambda, 400);
   [G, O, T, Energy] = mffntrain(Input, Target, nodefunction,
       nodefunctionprime, G, O, T, LayerSizes, y, Y, lambda, 400);
   [G, O, T, Energy] = mffntrain(Input, Target, nodefunction,
       nodefunctionprime, G, O, T, LayerSizes, y, Y, lambda, 1000);
   [G, O, T, Energy] = mffntrain(Input, Target, nodefunction,
       nodefunctionprime, G, O, T, LayerSizes, y, Y, lambda, 2000);
39 .
   more iterations
   .
   X = linspace(0, 3.14, 101);
   Input = X';
```

```
44 U = sin (X) .^2;
   Target = U';
   [y,Y,RE,E] = mffneval2 (Input ,Target ,nodefunction ,G,O,T,LayerSizes);
   Z = Y{3};
   plot (X,U,X,Z);
```

Reference

D. Rumelhart, J. McClelland, *Parallel Distributed Processing: Explorations in the Microstructure of Cognition* (MIT Press, Cambridge, 1986)

Chapter 17
Chained Feed Forward Architectures

We now introduce a version of the feed forward architecture known as the chained feedforward network or CFFN and discuss the backpropagation method of training in this context. Here, we will implement the FFN as a chain of computational elements which will provide a very general and flexible backbone for future discussion and expansion. The CFFN is quite well-known and our first exposure to it occurred in the classic papers of Werbos (1987a, b, 1988, 1990a, b). We will apply the methods derived in Sect. 17.3 below to calculate the needed partial derivatives.

17.1 Introduction

We consider a function $H: \Re^{n_I} \rightarrow \Re^{n_O}$ that has a very special nonlinear structure. This structure consists of a string or chain of computational elements, generally referred to as neurons in deference to a somewhat tenuous link to a lumped sum model of post-synaptic potential. Each neuron processes a summed collection of weighted inputs via a saturating transfer function with bounded output range. The neurons whose outputs connect to a given target or **postsynaptic** neuron are called **presynaptic** neurons. Each presynaptic neuron has an output Y which is modified by the synaptic weight $T_{pre,post}$ connecting the presynaptic neuron to the postsynaptic neuron. This gives a contribution $T_{pre,post} Y$ to the input of the post-synaptic neuron. A typical saturating transfer model was shown in the Fig. 14.11a in Chap. 14. Figure 14.11a shows a postsynaptic neuron with four weighted inputs which are summed and fed into the transfer function which then processes the input into a bounded scalar output. Figure 14.11b in Chap. 14 illustrates more details a typical sigmoid function processing node. As discussed in Chap. 14, a typical transfer function could be modeled as $\sigma(x, o, g) = 0.5\left(1.0 + \tanh\left(\frac{x-o}{\phi(g)}\right)\right)$ with the usual transfer function derivative given by $\frac{\partial \sigma}{\partial x}(x, o, g) = \frac{\phi'(g)}{2\phi(g)} sech^2\left(\frac{x-o}{\phi(g)}\right)$ where o denotes the offset indicated in the drawing and $\phi(g)$ is a function controlling slope

© Springer Science+Business Media Singapore 2016
J.K. Peterson, *BioInformation Processing*, Cognitive Science and Technology,
DOI 10.1007/978-981-287-871-7_17

which is usually called the gain of the transfer function. The function $\phi(g)$ is for convenience only; it is awkward to allow the denominator of the transfer function model to be zero and to change sign. A typical function we use to control the range of the gain parameter is $\phi(g) = g_m + \frac{g_M - g_m}{2}\left(1.0 + \tanh(g)\right)$ where g_m and g_M denote the lower and upper saturation values of the gain parameter's range. The CFFN model consists of a string of N neurons, labeled from 0 to $N - 1$. Some of these neurons can accept external input and some have their outputs compared to external targets. We let

$$\mathcal{U} = \{i \in \{0, \ldots, N - 1\} \mid \text{neuron i is an input neuron}\}$$
$$= \{u_0, \ldots, u_{n_I - 1}\} \tag{17.1}$$
$$\mathcal{V} = \{i \in \{0, \ldots, N - 1\} \mid \text{neuron i is an output neuron}\}$$
$$= \{v_0, \ldots, v_{n_O - 1}\} \tag{17.2}$$

We will let n_I and n_O denote the cardinality of $\mathcal{U}$ and $\mathcal{V}$ respectively. The remaining neurons in the chain which have no external role will be called hidden neurons with dimension n_H. Note that $n_H + |\mathcal{U}| = N$. Note that it is possible for an input neuron to be an output neuron; hence $\mathcal{U}$ and $\mathcal{V}$ need not be disjoint sets. The chain is thus divided by function into three possibly overlapping types of processing elements: n_I input neurons, n_O output neurons and n_H internal or hidden neurons. In Fig. 14.12a, we see a prototypical chain of eleven neurons. For clarity only a few synaptic links from pre to post neurons are shown. We see three input neurons (neurons 0, 1 and 4) and four output neurons (neurons 3, 7, 9 and 10). Note input neuron 0 feeds its output forward to input neuron 1 in addition to feeding forward to other postsynaptic neurons. The set of postsysnaptic neurons for neuron 0 can be denoted by the symbol $\mathcal{F}(0)$ which here is the set

$$\mathcal{F}(0) = \{1, 2, 4\}$$

Similarly, we see

$$\mathcal{F}(4) = \{5, 6, 8, 9\}$$

We will let the set of postsynaptic neurons for neuron i be denoted by $\mathcal{F}(i)$, the set of **forward links** for neuron i. Note also that each neuron can be viewed as a postsynaptic neuron with a set of presynaptic neurons feeding into it: thus, each neuron i has associated with it a set of backward links which will be denoted by $\mathcal{B}(i)$. In our example,

$$\mathcal{B}(0) = \{\}$$
$$\mathcal{B}(4) = \{0\}$$

where in general, the backward link sets will be much richer in connections than these simple examples indicate. The weight of the synaptic link connecting the presynaptic

Table 17.1 FFN evaluation process

$$
\begin{aligned}
&\text{for}(i = 0;\ i < N;\ i++) \{ \\
&\quad \text{if } (i \in \mathcal{U}) \\
&\qquad y^i = x^i + \sum_{j \in \mathcal{B}(i)} T_{j \to i} Y^j \\
&\quad \text{else} \\
&\qquad y^i = \sum_{j \in \mathcal{B}(i)} T_{j \to i} Y^j \\
&\qquad Y^i = \sigma^i(y^i, o^i, g^i) \\
&\}
\end{aligned}
$$

neuron i to the postsynaptic neuron j (it is assumed $j > i$) will be denoted by $T_{i \to j}$. The input of a typical postsynaptic neuron therefore requires summing over the backward link set of the postsynaptic neuron in the following way:

$$
y^{post} = x + \sum_{pre \in \mathcal{B}(post)} T_{pre \to post} Y^{pre}
$$

where the term x is the external input term which is only used if the post neuron is an input neuron. This is illustrated in Fig. 14.13. We will use the following notation (some of which has been previously defined) to describe the various elements of the chained FFN:

x^i: The external input to the ith input neuron

y^i: The summed input to the ith neuron

o^i: The offset of the ith neuron

g^i: The gain of the ith neuron

σ^i: The transfer function of the ith neuron

Y^i: The output of the ith neuron

$T_{i \to j}$:   The synaptic efficacy of the link between neuron i and neuron j. This link is only defined for $i < j$. However, for convenience, we can set up a connection between any neuron i and neuron j and simply set $T_{i \to j} = 0$ for both infeasible connections (such as $j \leq i$) and connections not realized in the architecture.

$\mathcal{F}(i)$:   The forward link set for the ith neuron

$\mathcal{B}(i)$:   The backward link set for the ith neuron

The chain FFN processes an arbitrary $x \in R^{n_I}$ via an iterative process as shown in Table 17.1.

The output of the CFFN is therefore a vector in $\Re^{n_O}$ defined by

$$
H(x) = \{Y^i \mid i \in \mathcal{V}\} \tag{17.3}
$$

that is,

$$
H(x) = \begin{bmatrix} Y^{v_0} \\ \vdots \\ Y^{v_{n_O - 1}} \end{bmatrix}
$$

and we see that $H: \mathfrak{R}^{n_I} \rightarrow \mathfrak{R}^{n_O}$ is a highly nonlinear function that is built out of chains of feedforward nonlinearities. The parameters that control the value of $H(x)$ are the forward links, the offsets and the gains for each neuron. The cardinality of these parameter sets are given by

$$n_s = \sum_{i=0}^{N-1} |\mathcal{F}(i)|$$
$$n_o = N$$
$$n_g = N$$

where $|\mathcal{F}(i)|$ denotes the size of the forward link set for the ith neuron; n_s denotes the number of synaptic links; n_o, the number of offsets; and n_g, the number of gains. The number of parameters for a CFFN is then written as N_p:

$$N_p = n_s + n_o + n_g \qquad (17.4)$$

Now let $\mathcal{I} = \{x_\alpha \in R^{n_I} : 0 \le \alpha \le S - 1\}$ and $\mathcal{D} = \{D_\alpha \in R^{n_O} : 0 \le \alpha \le S - 1\}$ be two given sets of data of size $S > 0$. The set $\mathcal{I}$ is referred to as the set of *input exemplars* and the set $\mathcal{D}$ is the set of outputs that are associated with exemplars. Together, the sets $\mathcal{I}$ and $\mathcal{D}$ comprise what is known as the *training* set. Also, from now on, we will use the subscript notation α to indicate the dependence of various variables on the αth exemplar in the sets $\mathcal{I}$ and $\mathcal{D}$.

Now, the indices which denote components of a given target D_α and the components of the CFFN output $H(I_\alpha)$ are embarassingly mismatched. We will write the target D_α as:

$$D_\alpha = \begin{bmatrix} D_\alpha^{d(v_0)} \\ \vdots \\ D_\alpha^{d(v_{n_O-1})} \end{bmatrix}$$

Now this notation is quite convoluted; however, we must be clear which component of the target are to be matched with which CFFN output neuron. The mapping $d : V \rightarrow \{0, \dots, n_O - 1\}$ is, of course, the trivial correspondence $d(v_i) = i$. For notational cleanness, we will simply denote $d(v_i)$ by d_i. The CFFN training Problem is then to choose the N_p chain FFN parameters such that we minimize an energy function, E, given by

$$E = 0.5 \sum_{\alpha=0}^{S-1} \sum_{i=0}^{n_O-1} f\left(Y_\alpha^{v_i} - D_\alpha^i\right),$$
$$= 0.5 \sum_{\alpha=0}^{S-1} \sum_{i \in V} f\left(Y_\alpha^i - D_\alpha^{d_i}\right) \qquad (17.5)$$

where f is a nonnegative function of the error term $Y_\alpha^i - D_\alpha^{d_i}$; e.g. using the function $f(x) = x^2$ gives the standard L^2 or least squares energy function.

17.2 Minimizing the CFFN Energy

We will use the notation established in Sect. 17.1. If all the partial derivatives of the energy function (17.5) with respect to the chain FFN parameters were known, we could minimize (17.5) using a standard gradient descent scheme which we call CFFN back propagation. Relabeling the N_p CFFN variables temporarily as w_i, $0 \le i \le N_p - 1$, the optimization problem we face is

$$\min_{w \in R^{N_p}} E(\mathcal{I}, \mathcal{D}, w) \tag{17.6}$$

where we now explicitly indicate that the value of E depends on $\mathcal{I}$, $\mathcal{D}$ and the parameters w. In component form, the equations for gradient descent are then

$$w_i^{new} = w_i^{old} - \lambda \frac{\partial E}{\partial w_i}\bigg|_{w^{old}} \tag{17.7}$$

where λ is a scalar parameter that determines what percentage of the gradient step to actually take. We can write (17.7) more compactly using vector notation:

$$w^{new} = w^{old} - \lambda \nabla E(w^{old}) \tag{17.8}$$

In practice, we typically replace ∇E with the descent vector D obtained by normalizing ∇E by its length:

$$D = \begin{cases} \frac{\nabla E}{\|\nabla E\|} & \|\nabla E\| > 1 \\ \nabla E & \|\nabla E\| \le 1 \end{cases}$$

and use the optimization scheme:

$$w^{new} = w^{old} - \lambda D(w^{old})$$

17.3 Partial Derivative Calculation in Generalized Chains

In any sort of a model that is built from tunable parameter with an associated energy function minimized by choosing the parameters via gradient descent, we must be able to calculate the needed partial derivatives. We now discuss a recursive method for the computation of the partial derivatives general computational chains.

The calculation of the partial derivatives required to implement gradient descent in the context of backpropagation derivation is a bit intimidating and confusing because of its recursive nature. To help explain this, let's look at an abstract example: suppose we have a function E which is a function of the three variables Y^0, Y^1 and Y^2. These variables are assumed to have the following dependencies:

$$E = E(Y^0, Y^1, Y^2)$$
$$Y^2 = f^2(Y^0, Y^1)$$
$$Y^1 = f^1(Y^0)$$

where f^i denotes the functional dependencies. Note that these computational elements are organized in a chain like fashion. Now, consider a particular example of the chained functions Y^2 and Y^1 given below:

$$Y^2 = (3Y^0 + 4Y^0 Y^1)^2$$
$$Y^1 = 2(Y^0)^3$$

The function Y^2 depends on Y^0 in two different ways: the first is a **direct** dependence which we can denote by the usual partial derivative symbols:

$$\frac{\partial Y^2}{\partial Y^0} = 2(3Y^0 + 4Y^0 Y^1) \times (3 + 4Y^1)$$

However, there are additional hidden dependencies; e.g., Y^1 itself depends on Y^0. Using the usual partial derivative notation, the complete of **total** dependence of Y^2 on Y^0 is given by a much more complicated expression:

$$\frac{dY^2}{dY^0} = \frac{\partial Y^2}{\partial Y^0} + \frac{\partial Y^2}{\partial Y^1}\frac{\partial Y^1}{\partial Y^0}$$

where we denoted this **total** dependence using the notation $\frac{dY^2}{dY^0}$ which is the **total derivative** of Y^2 with respect to Y^0. The situation can get more complicated. If the function Y^1 had additional **hidden** dependencies on the variable Y^0, the direct dependency term $\frac{\partial Y^1}{\partial Y^0}$ would be inadequate. In this case, it should be replaced by the total dependency term $\frac{dY^1}{dY^0}$. Thus, to compute the **total derivative** of Y^2 with respect to Y^0, we should use:

$$\frac{dY^2}{dY^0} = \frac{\partial Y^2}{\partial Y^0} + \frac{\partial Y^2}{\partial Y^1}\frac{dY^1}{dY^0}$$
$$= 2(3Y^0 + 4Y^0 Y^1) \times (3 + 4Y^1) + 2(3Y^0 + 4Y^0 Y^1) \times (4Y^0) \times 6(Y^0)^2$$

We therefore must make a distinction between the **total derivative** of a function with respect to a variable u and the **direct derivative** of function with respect to u. We will denote **total derivatives** with the **d** and **direct derivatives** with the **@** notation, respectively. Of course, in some cases these derivatives are the same and so we can use either notation interchangeably.

Let's go back to the original example and add a function E which depends on Y^0, Y^1 and Y^2. We let

$$E = (Y^0)^2 + (Y^1)^4 + (Y^2)^8$$
$$Y^2 = (3Y^0 + 4Y^0Y^1)^2$$
$$Y^1 = 2(Y^0)^3$$

Note that the direct dependence of E on Y^0 is:

$$\frac{\partial E}{\partial Y^0} = 2Y^0$$

However, the **total** dependence of E on Y^0 must be calculated as follows:

$$
\begin{aligned}
\frac{dE}{dY^0} &= \frac{\partial E}{\partial Y^0} + \frac{\partial E}{\partial Y^1}\frac{dY^1}{dY^0} + \frac{\partial E}{\partial Y^2}\frac{dY^2}{dY^0} \\
&= \frac{\partial E}{\partial Y^0} + \frac{\partial E}{\partial Y^1}\frac{\partial Y^1}{\partial Y^0} + \frac{\partial E}{\partial Y^2}\left\{\frac{\partial Y^2}{\partial Y^0} + \frac{\partial Y^2}{\partial Y^1}\frac{\partial Y^1}{\partial Y^0}\right\} \\
&= 2Y^0 + 4(Y^1)^3 \times 6(Y^0)^2 \\
&\quad + 8(Y^2)^7 \times \left\{2(3Y^0 + 4Y^0Y^1) \times \left[(3 + 4Y^1) + (4Y^0) \times 6(Y^0)^2\right]\right\}
\end{aligned} \quad (17.9)
$$

This certainly looks complicated! However, the chained nature of these computations allows us to reorganize this is much easier format: consider the recursive scheme below:

$$\frac{dE}{dY^2} = \frac{\partial E}{\partial Y^2} \tag{17.10}$$

$$\frac{dE}{dY^1} = \frac{\partial E}{\partial Y^1} + \frac{dE}{dY^2}\frac{\partial Y^2}{\partial Y^1} \tag{17.11}$$

$$\frac{dE}{dY^0} = \frac{\partial E}{\partial Y^0} + \frac{dE}{dY^1}\frac{\partial Y^1}{\partial Y^0} + \frac{dE}{dY^2}\frac{\partial Y^2}{\partial Y^0} \tag{17.12}$$

After calculation, this simplifies to:

$$\frac{dE}{dY^2} = 8(Y^2)^7$$

$$\frac{dE}{dY^1} = 4(Y^1)^3 + \frac{dE}{dY^2} \times 2(3Y^0 + 4Y^0Y^1) \times (4Y^0)$$

$$= 4(Y^1)^3 + (8(Y^2)^7) \times 2(3Y^0 + 4Y^0Y^1) \times (4Y^0)$$

$$\frac{dE}{dY^0} = 2Y^0 + \frac{dE}{dY^1} \times 6(Y^0)^2 + \frac{dE}{dY^2} \times 2(3Y^0 + 4Y^0Y^1) \times (3 + 4Y^1)$$

$$= 2Y^0 + \{4(Y^1)^3 + (8(Y^2)^7) \times 2(3Y^0 + 4Y^0Y^1) \times (4Y^0)\} \times 6(Y^0)^2$$
$$+ 8(Y^2)^7 \times 2(3Y^0 + 4Y^0Y^1) \times (3 + 4Y^1)$$

$$= 2Y^0 + 4(Y^1)^3 \times 6(Y^0)^2$$
$$+ 8(Y^2)^7 \times \{2(3Y^0 + 4Y^0Y^1) \times [(3 + 4Y^1) + 6(Y^0)^2 \times (4Y^0)]\}$$

The recursive calculation matches the one that was obtained directly via (17.9). The discussion above motivates a general recursive scheme for computing the rates of change of a given function E with respect to chained variables. Assume the function E depends on the variables Y^0 to Y^N in the following way:

$$E = E(Y^0, Y^1, \dots, Y^N)$$
$$Y^N = f^N(Y^0, Y^1, \dots, Y^{N-1})$$
$$Y^{N-1} = f^{N-1}(Y^0, Y^1, \dots, Y^{N-2})$$
$$\vdots$$
$$Y^i = f^i(Y^0, Y^1, \dots, Y^{i-1})$$
$$\vdots$$
$$Y^1 = f^1(Y^0)$$

We compute the required partial derivatives recursively as follows:

$$\frac{dE}{dY^N} = \frac{\partial E}{\partial Y^N} \tag{17.13}$$

$$\frac{dE}{dY^{N-1}} = \frac{\partial E}{\partial Y^{N-1}} + \frac{dE}{dY^N}\frac{\partial Y^N}{\partial Y^{N-1}} \tag{17.14}$$

$$\frac{dE}{dY^{N-2}} = \frac{\partial E}{\partial Y^{N-2}} + \frac{dE}{dY^N}\frac{\partial Y^N}{\partial Y^{N-2}} + \frac{dE}{dY^{N-1}}\frac{\partial Y^{N-1}}{\partial Y^{N-2}}. \tag{17.15}$$

$$\frac{dE}{dY^i} = \frac{\partial E}{\partial Y^i} + \sum_{j=i+1}^{N} \frac{dE}{dY^j}\frac{\partial Y^j}{\partial Y^i}. \tag{17.16}$$

$$\frac{dE}{dY^0} = \frac{\partial E}{\partial Y^0} + \sum_{j=1}^{N} \frac{dE}{dY^j}\frac{\partial Y^j}{\partial Y^0} \tag{17.17}$$

Note that this can easily be organized into the scheme given in Table 17.2. Also, we use the notation $\frac{\partial Y^j}{\partial Y^i}$ for the more correct form $\frac{\partial f^j}{\partial Y^i}$. Now, let's consider another complication: each function Y^i has an internal structure of functional parameters

Table 17.2 Recursive chained partial calculation

$$\text{for}(i = N; i \geq 0; i--) \{$$
$$\quad \text{if } (i == N)$$
$$\quad\quad \frac{dE}{dY^i} = \frac{\partial E}{\partial Y^i}$$
$$\quad \text{else}$$
$$\quad\quad \frac{dE}{dY^i} = \frac{\partial E}{\partial Y^i} + \sum_{j=i+1}^{N} \frac{dE}{dY^j} \frac{\partial Y^j}{\partial Y^i}$$
$$\}$$

$\boldsymbol{w}_i = \{w_{i0}, \ldots, w_{i,n_i-1}\}$ which are used in addition to the inputs from elements Y^0 to Y^{i-1}. The symbol n_i denotes the number of internal parameters in function Y^i. This internal structure is **not** based on a chained architecture. To be explicit, our previous example can be written in this new context as follows:

$$Y^0 = f^0(w_{00}, \ldots, w_{0,n_0-1})$$
$$\quad = f^0(w_0)$$
$$Y^1 = f^1(Y^0, w_{10}, \ldots, w_{1,n_1-1})$$
$$\quad = f^1(Y^0, w_1)$$
$$Y^2 = f^2(Y^0, Y^1, w_{20}, \ldots, w_{2,n_2-1})$$
$$\quad = f^2(Y^0, Y^1, w_2)$$
$$E = E(Y^0, Y^1, Y^2)$$

To calculate $\frac{\partial E}{\partial w_{ij}}$, we can modify the previous recursive scheme for all appropriate indices i:

$$\frac{dE}{dY^2} = \frac{\partial E}{\partial Y^2}$$
$$\frac{dE}{dw_{2i}} = \frac{dE}{dY^2} \frac{\partial Y^2}{\partial w_{2i}}$$
$$\frac{dE}{dY^1} = \frac{\partial E}{\partial Y^1} + \frac{dE}{dY^2} \frac{\partial Y^2}{\partial Y^1}$$
$$\frac{dE}{dw_{1i}} = \frac{dE}{dY^1} \frac{\partial Y^1}{\partial w_{1i}}$$
$$\frac{dE}{dY^0} = \frac{\partial E}{\partial Y^0} + \frac{dE}{dY^1} \frac{\partial Y^1}{\partial Y^0} + \frac{dE}{dY^2} \frac{\partial Y^2}{\partial Y^0}$$
$$\frac{dE}{dw_{2i}} = \frac{dE}{dY^0} \frac{\partial Y^0}{\partial w_{0i}}$$

We summarize the recursive partial calculations when internal parameters are present in Table 17.3. We can use these recursive techniques to derive the **backpropagation** algorithm for the case of the Chain Feed Forward Network, **CFFN**. Indeed, we can also derive the backpropagation algorithm for many other architectures in a similar way.

Table 17.3 Recursive
chained partial calculation
with internal parameters

$$
\begin{aligned}
&\text{for}(i = N; i \geq 0; i - -)\ \{ \\
&\quad \text{if } (i == N) \\
&\qquad \frac{dE}{dY^i} = \frac{\partial E}{\partial Y^i} \\
&\quad \text{else} \\
&\qquad \frac{dE}{dY^i} = \frac{\partial E}{\partial Y^i} + \sum_{j=i+1}^{N} \frac{dE}{dY^j} \frac{\partial Y^j}{\partial Y^i} \\
&\qquad \text{for}(j = 0; j \leq n_i - 1; j + +) \\
&\qquad\quad \frac{dE}{dw_{ij}} = \frac{dE}{dY^i} \frac{\partial Y^i}{\partial w_{ij}} \\
&\ \}
\end{aligned}
$$

17.4 Partial Calculations for the CFFN

We now derive the fundamental equations which allow us to recursively compute
the needed partial derivatives starting at the last output neuron and sweeping right
to left toward the first input neuron. This recursive computational scheme is known
as backpropagation and in this form it is inherently serial in nature. We will use the
discussion from Sect. 17.3 and the information in Tables 17.2 and 17.3 heavily in
what follows.

We will begin by laying out some notational ground rules:

Neuron Transfer Functions: These will be modeled somewhat generally. Each
neuron i has an associated transfer function σ^i which we will model as a function of three arguments; i.e. $\sigma^i(y^i, o^i, g^i) \equiv \sigma^i$ using our previously established
notations for the input, offset and gain of neuron i. There are then three partial
derivatives of interest:

$\frac{\partial \sigma^i}{\partial y}$:   This is the rate of change of the transfer function with respect to its
first argument. At sample α, we will denote this derivative by $\sigma^i_{0,\alpha} \equiv$
$\sigma^i_0(y^i_\alpha, o^i, g^i)$.

$\frac{\partial \sigma^i}{\partial o}$:   This is the rate of change of the transfer function with respect to its second argument. At sample α, we will denote this derivative by $\sigma^i_{1,\alpha} \equiv$
$\sigma^i_1(y^i_\alpha, o^i, g^i)$.

$\frac{\partial \sigma^i}{\partial g}$:   This is the rate of change of the transfer function with respect to its
third argument. At sample α, we will denote this derivative by $\sigma^i_{2,\alpha} \equiv$
$\sigma^i_2(y^i_\alpha, o^i, g^i)$.

Now, the output of a given transfer function is denoted by

$$
Y^i = \sigma^i(y^i, o^i, g^i)
$$

$$
= \sigma^i \left(\sum_{j \in \mathcal{B}(i)} T_{j \to i}\, Y^j, o^i, g^i \right)
$$

$$
\equiv f^i(Y^0, Y^1, \ldots, Y^{i-1}, T_{0 \to i}, \ldots, T_{i-1 \to i}, o^i, g^i)
$$

where the internal parameters of the function Y^i are the gain and offset of the transfer function as well as the link weights connecting to neuron i. Hence, this particular type of chained architecture can be written in functional form using the internal parameters

$$w_i = \{T_{0 \to i}, \ldots, T_{i-1 \to i}, o^i, g^i\}$$

This internal parameter set has cardinality $i + 2$. Thus, each transfer function has the instantiated form

$$Y^i = \sigma^i(y^i, o^i, g^i)$$
$$\equiv f^i(Y^0, Y^1, \ldots, Y^{i-1}, w_i)$$

Error Notation: The error between the value from output neuron i due to sample α and the desired output component $D_\alpha^{d_i}$ will be denoted by the term e_α^i. For notational purposes, we will define the generalized output error as follows:

$$e_\alpha^i = \begin{cases} Y_\alpha^i - D_\alpha^{d_i}, & i \in V \\ 0, & else \end{cases}$$

Energy Function: The energy function depends explicitly on the outputs of the n_O output neurons in the chain FFN. The partial with respect to the ith output neuron will be denoted $E_i = \frac{\partial E}{\partial Y^i}$ whose value at a given sample α is given by

$$E_{i,\alpha} = \begin{cases} Y_\alpha^i - D_\alpha^{d_i} = e_\alpha^i & i \in V \\ 0, & else \end{cases}$$

Note that $E_{i,\alpha}$ is equivalent to the **direct** partial term $\frac{\partial E}{\partial Y^i}$.

Indicial Notation: Our indices derive from a C based notation. Hence, sums typically run from 0 to an upper bound of the form $M - 1$. To avoid abusive notation, we will use the convention $M' \equiv M - 1$ to indicate that the variable M has been decremented by 1. We will also let θ denote n_O'. Note that we also have a nice identity connecting the forward and backward link sets: for a given node i, we have

$$\sum_{k=i+1}^{N-1} \sum_{i \in B(k)} = \sum_{k \in F(i)} \tag{17.18}$$

This formula is interpreted as follows: since the value i is fixed, the inner sum over $i \in B(k)$ is empty unless i is in the set $B(k)$. We will now apply the notions about recursively organized chained partial calculations to the CFFN. Since we have seen how to interpret the CFFN as a general chained architecture using the notation of Sect. 17.3, we can apply the recursive partial calculations for chains with internal parameters directly to obtain (see Table 17.3) the backpropagation equations for the CFFN.

17.4.1   The $\frac{\partial Y^j}{\partial Y^i}$ Calculation

We begin by finding the critical partial $\frac{\partial Y^j}{\partial Y^i}$ so that we can find the total dependence of E on Y^i.

$$
\begin{aligned}
\frac{\partial Y^j}{\partial Y^i} &= \sigma_0^j(y^j, o^j, g^j) \frac{\partial}{\partial Y^i} \left\{ x^j + \sum_{k \in \mathcal{B}(j)} T_{k \to j} Y^k \right\} \\
&= \sigma_0^j(y^j, o^j, g^j) \sum_{k \in \mathcal{B}(j)} T_{k \to j} \delta_i^k \\
&= \sigma_0^j(y^j, o^j, g^j) \sum_{i \in \mathcal{B}(j)} T_{i \to j}
\end{aligned}
$$

where we interpret x^j as zero in j is not in the input set $\mathcal{U}$. Hence, the general expansion term

$$
\frac{dE}{dY^i} = \frac{\partial E}{\partial Y^i} + \sum_{j=i+1}^{N-1} \frac{dE}{dY^j} \frac{\partial Y^j}{\partial Y^i}
$$

in this context becomes

$$
\begin{aligned}
\frac{dE}{dY^i} &= \frac{\partial E}{\partial Y^i} + \sum_{j=i+1}^{N-1} \sum_{i \in \mathcal{B}(j)} \frac{dE}{dY^j} \sigma_0^j(y^j, o^j, g^j) T_{i \to j} \\
&= \frac{\partial E}{\partial Y^i} + \sum_{j \in \mathcal{F}(i)} \frac{dE}{dY^j} \sigma_0^j(y^j, o^j, g^j) T_{i \to j}
\end{aligned}
$$

where we use the identity (17.18) to simplify the equations above. Of course, the term $\frac{dE}{dY^i}$ is not there unless the energy has a direct dependence on Y^i. This only happens if the node i is an output node.

17.4.2   The Internal Parameter Partial Calculations

The partials with respect to the internal parameters consist of two separate types: the first are partials with respect to the offset and gain variables and the second are partials with respect to linking terms. The appropriate partial calculation from Table 17.3 for a given internal parameter w is given by

$$
\frac{dE}{dw} = \frac{dE}{dY^i} \frac{\partial Y^i}{\partial w}
$$

Offset and Gain Parameter Partials

For $w = 0^i$ and $w = g^i$, we obtain

$$\frac{dE}{do^i} = \frac{dE}{dY^i} \sigma_1^i \text{ and } \frac{dE}{dg^i} = \frac{dE}{dY^i} \sigma_2^i$$

Linking Parameter Partials

For $w = T_{j \to i}$, we obtain

$$\frac{dE}{dT_{j \to i}} = \frac{dE}{dY^i} \sigma_0^i(y^i, o^i, g^i) \frac{\partial}{\partial T_{j \to i}} \left\{ x^i + \sum_{k \in \mathcal{B}(i)} T_{k \to i} Y^k \right\}$$

$$= \frac{dE}{dY^i} \sigma_0^i(y^i, o^i, g^i) \left\{ \sum_{k \in \mathcal{B}(i)} \delta_j^k Y^k \right\}$$

$$= \frac{dE}{dY^i} \sigma_0^i(y^i, o^i, g^i) Y^j$$

These equations are summarized in the Table 17.4. For simplicity, in this table, we assume there is only one output node—$N - 1$. Since we are really trying to compute the gradient for the energy function E defined in terms of the full sample set, the calculations outlined above must be done at each sample. This leads to the full backpropagation equations given in Table 17.5. We observe that the backpropagation algorithm to compute $\frac{dE}{dY_\beta^{pre}}$ at a given presynaptic neuron pre requires summing over the forward link set and a possible external error signal in the following way:

$$\frac{dE}{dY_\beta^{pre}} = e + \sum_{post \in \mathcal{F}(pre)} \sigma_0^{post} T_{pre \to post} \frac{dE}{dY_\beta^{post}}$$

where all variables subscripted with *post* are associated with a postsynaptic neuron. An illustration of this process is shown in Fig. 17.1. This is a very convenient way to abstract this process. Combining our evaluation procedure (Table 17.1) and the

Table 17.4 Backpropagation in the CFFN: one sample

$$
\begin{aligned}
&\text{for}(i = N - 1; i \geq 0; i - -) \ \{ \\
&\quad \text{if } (i == N - 1) \\
&\quad\quad \tfrac{dE}{dY^i} = E_i \\
&\quad \text{else} \\
&\quad\quad \tfrac{dE}{dY^i} = E_i + \sum_{j \in \mathcal{F}(i)} \tfrac{dE}{dY^j} \sigma_0^j(y^j, o^j, g^j)\, T_{i \to j} \\
&\quad \text{for}(j \in \mathcal{F}(i)) \\
&\quad\quad \tfrac{\partial E}{\partial T_{i \to j}} = \tfrac{dE}{dY^j} \sigma_0^j Y^i \\
&\quad \tfrac{\partial E}{\partial o^i} = \tfrac{dE}{dY^i} \sigma_1^i \\
&\quad \tfrac{\partial E}{\partial g^i} = \tfrac{dE}{dY^i} \sigma_2^i \\
&\}
\end{aligned}
$$

Table 17.5 Backpropagation
for CFFN: multiple samples

$$
\begin{aligned}
&\text{for}(\beta = 0;\ \beta < S;\ \beta ++)\ \{ \\
&\quad \text{for}(i = N - 1;\ i \geq 0;\ i --)\ \{ \\
&\qquad \tfrac{dE}{dY^i_\beta} = E_{i,\beta} + \sum_{j \in \mathcal{F}(i)} \tfrac{dE}{dY^j_\beta}\ \sigma^j_{0,\beta}\ T_{i \to j} \\
&\quad \} \\
&\} \\
&\text{for}(i = N - 1;\ i \geq 0;\ i --)\ \{ \\
&\quad \text{for}(j \in \mathcal{F}(i)) \\
&\qquad \tfrac{\partial E}{\partial T_{i \to j}} = \sum_{\beta=0}^{S'} \tfrac{dE}{dY^j_\beta}\ \sigma^j_{0,\beta}\ Y^i_\beta \\
&\qquad \tfrac{\partial E}{\partial o^i} = \sum_{\beta=0}^{S'} \sigma^i_{0,\beta}\ \tfrac{dE}{dY^i_\beta}\ \sigma^i_{1,\beta} \\
&\qquad \tfrac{\partial E}{\partial g^i} = \sum_{\beta=0}^{S'} \sigma^i_{0,\beta}\ \tfrac{dE}{dY^i_\beta}\ \sigma^i_{2,\beta} \\
&\quad \}
\end{aligned}
$$

Fig. 17.1 Presynaptic error
calculation

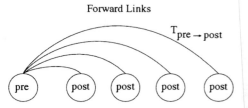

backpropagation procedure (Table 17.5), we see that information about both back-ward (evaluation) and forward (partial derivative calculation) links is required. Also, note that as formulated here, this algorithm is inherently **sequential**! It is not easily implemented in parallel!

We will discuss how to implement the chain training algorithms in MatLab soon.

17.5   Simple MatLab Implementations

Let's look at how to setup code to implement the CFFN ideas. We won't do a proper job yet as there are lots of details that are difficult to code properly. But we will start the process now by looking at how to code the evaluation portion of the of a 2-3-1 CFFN which is the same network as our usual 2-3-1 MFFN we have discussed earlier. We begin with the usual initializations we put into the function **chaininit**. This function sets up a vector of six node numbers which we only need to know how many nodes there are. You'll note we don't really use the individual entries on the variable **N** here. We store the **in** node of an edge and the **out** node as well as the weight value of the edge **w** in a data structure MatLab uses called a **struct**. Our struct is called **e** and we can access the three fields in **e** using the notation **e.in, e,out** and **e.w**. We set the edges for the 2-3-1 network architecture in the cell variable **a** as a collection of row vectors of the form **[in, out]**. We use **e** to construct a cell **E** to hold the edge information we need.

Now if you think about the evaluation step in a CFFN, recall we will have to compute $\sum_{j \in \mathcal{B}(i)} T_{j \to i} Y^j$. The backward set $\mathcal{B}(i)$ consists of a set of nodes. The weight values $T_{j \to i}$ are stored in the cell **E**. Suppose we had a CFFN with 100 nodes and we were looking at the backward set $\mathcal{B}(36)$. The edges in the cell **E** are labeled starting at 1 and with the last edge in the CFFN. Each node listed in $\mathcal{B}(36)$ corresponds to an index j in a $T_{j \to i}$ value we need. However, the value $T_{j \to i}$ is stored as a value **w** in some edge **E{k}**. We need to find the indices k where **E{k}.in = j** which implies a search over the edges for each index j from the backward set. This is expensive! It is a lot better to save us the search by storing edges corresponding to the backward sets in a new cell **BE**. Exactly what is stored in each **BE{k}** depends on how we choose to number our edges so be careful to keep track of your edge numbering scheme as you will need to remember it here.

After we set up **BE**, we initialize the node functions as usual.

Listing 17.1: Setting up a 2-3-1 CFFN

```
function [N,O,G,E,BE,nodefunction] = chaininit()
%
N = [1;2;3;4;5;6];
e = struct();
E = {};
a = {[1,3];[1,4];[1,5];[2,3];[2,4];[2,5];[3,6];[4,6];[5,6]};
W = -1+2*rand(length(a),1);
O = -1+2*rand(length(a),1);
G = 0.6+0.6*rand(length(a),1);
for i = 1:length(a)
    e.in = a{i}(1);
    e.out = a{i}(2);
    e.w = W(i);
    E{i} = e;
end

BE = {};
BE{1} = {};
BE{2} = {};
BE{3} = [E{1};E{4}];
BE{4} = [E{2};E{5}];
BE{5} = [E{3};E{6}];
BE{6} = [E{7};E{8};E{9}];

sigma = @(x,o,g) 0.5*(1+tanh( (x-0)/g ));
sigmainput = @(x,o,g) (x-o)/g;
nodefunction{1} = sigmainput;
nodefunction{2} = sigmainput;
nodefunction{3} = sigma;
nodefunction{4} = sigma;
nodefunction{5} = sigma;
nodefunction{6} = sigma;

end
```

Next, we tackle the evaluation part with the function **chaineval**. Note we handle the input nodes separately and the other nodes that implement the $\sum_{j \in \mathcal{B}(i)} T_{j \to i} Y^j$ calculation use the **BE** cell data. For node **i**, **BE{i}(j)** gives us an edge connecting to **i**. Hence, we can extract the needed weight value easily, without a search, using

BE{i}(j).w. Note we never use the backward sets **B** here as the backward edge sets are more useful.

Listing 17.2: The evaluation loop for a 2-3-1 CFFN

```
   function  [y,Y]  =  chaineval(X,N,O,G,BE,nodefunction)
   %
   %
   for  i  =  1:length(N)
5    if  (i == 1 || i == 2)
        y(i) = X(i);  Y(i) = nodefunction{i}(y(i),O(i),G(i));
     else
        y(i) = 0;
        [row,col] = size(BE{i});
10      for  j = 1:row
           weight = BE{i}(j).w;
           y(i) = y(i) + weight*Y(j);
        end
        Y(i) = nodefunction{i}(y(i),O(i),G(i));
15   end
   end
   end
```

To see this 2-3-1 CFFN in action, we make up some data and push it through the architecture. We find

Listing 17.3: Testing our 2-3-1 CFFN

```
   X = [1.2;2.5];
   [N,O,G,E,BE,nodefunction] = chaininit();
3  [y,Y] = chaineval(X,N,O,G,BE,nodefunction);
   Y
   Y =
   0.2513     2.7088     0.9921     0.9509     0.0138     0.8594
```

References

P. Werbos, Learning how the world works, in *Proceedings of the 1987 IEEE International Confer-ence on Systems, Man and Cybernetics*, vol. 1 (IEEE, 1987a), pp. 302–310

P. Werbos, Building and understanding adaptive systems: A statistical/numerical approach to factory automation and brain research. IEEE Trans. Syst. Man Cybern. **17**, 7–19 (1987b)

P. Werbos, Generalization of backpropagation with application to a recurrent gas market model. Neural Netw. **1**, 339–356 (1988)

P. Werbos, A menu of designs for reinforcement learning over time, in *Neural Networks for Control*, eds. by W. Miller, R. Sutton, P. Werbos (1990a), pp. 67–96

P. Werbos, Backpropagation through time: What it does and how to do it. Proc. IEEE **78**(10), 1550–1560 (1990b)

Part VI
Graph Based Modeling In Matlab

Chapter 18
Graph Models

Our cognitive model will consist of many artificial neurons interacting on multiple time scales. Although it is possible to do detailed modeling of neurobiological systems using GENESIS (Bower and Beeman 1998) and NEURON (Hines 2003), we do not believe that they are useful tools in modeling the kind of information flow between cortical modules that is needed for a cognitive model. Progress in building large scale models that involve many cooperating neurons will certainly involve making suitable abstractions in the information processing that we see in the neuron. Neurons transduce and integrate information on the dendritic side into wave form pulses and there are many models involving filtering and transforms which attempt to "see" into the action potential and find its informational core so to speak. However, all of these methods are hugely computationally expensive and even a simple cognitive model will require ensembles of neurons acting together locally to create global effects. We believe the simple biological feature vector (BFV) discussed here is able to discriminate subtle changes in the action potential wave form.Hence, we will use ensembles of abstract neurons whose outputs are BFVs (11 dimensional say) to create local cortical computationally phase locked groups. The effects of modulatory agents such as neurotransmitters can then be modeled as introducing changes in the BFVs. The kinds of changes one should use for a given neurotransmitters modulatory effect can be estimated from the biophysical and toxin literature. An increase in sodium ion flow, Ca^{++} gated second messenger activity can be handled at a high level as a suitable change in one of the 11 parameters of the BFV. Note this indeed is an implication for realistic neuronal processing. Further, this raises a very interesting question. How much information processing is possible in a large scale interacting neuron population given an abstract form of neuronal output? How can we get insight into the minimal amount of information carried per output wave form that can subserve cognition?

We believe this type of interaction is inherently asynchronous and hence the artificial neurons we use for our model must be treated as agents interacting asynchronously. The discussion in the previous chapters gives us powerful clues for the

© Springer Science+Business Media Singapore 2016

J.K. Peterson, *BioInformation Processing*, Cognitive Science and Technology,

DOI 10.1007/978-981-287-871-7_18

design of a software architecture that may subserve computational intelligence purposes. The components of this design are a finite collection of abstract neurons $\mathcal{N}_i$. A
given neuron $\mathcal{N}_{i_0}$ accepts inputs from a sub population $\mathcal{N}_j \, j \in \mathcal{D}_i$, where $\mathcal{D}_i$ denotes
the input into the *dendritic* system of the artificial neuron. It then generates an output
we will denote by ζ_i which is of the form

$$\zeta_i = \left[t_0^i, V_0^i, t_1^i, V_1^i, t_2^i, V_2^i, t_3^i, V_3^i, g^i, t_4^i, V_4^i \right]'.$$

This output takes the form of the low dimensional BFV we discussed earlier. Hence,
the input $\mathcal{D}_i$ is a sequence

$$\mathcal{D}_i = \left\{ \zeta_j \, j \in \mathcal{D}_i \right\}.$$

where each ζ_j is a BFV structure. Event sequences in our neural model are thus
sequences of the 11 dimensional outputs $[t_0, V_0, t_1, V_1, t_2, V_2, t_3, V_3, g, t_4, V_4]'$ that
are processed by a given neuron agent.

    These neurons need to interact with each other and form temporally and spatially
localized ensembles of activity whose outputs are entrained. For a given computational intelligence model, there is a global system time clock as well. Thus, the
neurons function in a timed environment and so all variables in all inputs and outputs
are also tagged with time indices. The representation at this level is quite messy and
is usually avoided. We can always do computations at each agent simply in terms of
information that has arrived and is to be processed. However, to be precise the output
of a neuron would be

$$\zeta_i(t) = \left[t_0^{it}, V_0^{it}, t_1^{it}, V_1^{it}, t_2^{it}, V_2^{it}, t_3^{it}, V_3^{it}, g^{it}, t_4^{it}, V_4^{it} \right].$$

    The FFP and OCOS cortical circuits can be combined into a multi-column model
(Raizada and Grossberg 2003) as seen in Fig. 18.1. This model can be implemented
using the artificial neurons introduced above. The generic software architecture for
our model is shown in Fig. 18.2. In Fig. 18.1, each cortical stack consists of 3 cortical
columns. If each cortical layer consists of 9 neurons, this allows for 3 OCOS/FFP
pathways inside each column. This is roughly 27 neurons per column. In Fig. 18.3,
a more complete architecture is shown. Here, for convenience, each cortical module
consists of 4 cortical stacks, all of which are interconnected. Hence, the 4 column
structure illustrated in Fig. 18.3 can be implemented with approximately 108 neurons
each. Thus, the five cortical modules (Frontal, Parietal, Occipital, Temporal and
Limbic) illustrated require a total of 540 artificial neurons for their implementation.
In general, since we know we start with isocortex, we can assume the number of
stacks is the same in each module. Further, we can allow more OCOS/FFP circuits
per layer. Letting $N_{OCOS/FFP}$ denote the number of OCOS/FFP circuits per layer,
we need $3N_{OCOS/FFP}$ neurons per stack. Thus, if we use N_S cortical stacks for each
cortical module, we will use $3N_{OCOS/FFP}N_S$ neurons for each of the five cortical
modules (Frontal, Parietal, Occipital, Temporal and Limbic). This leads to a total of
$15N_{OCOS/FFP}N_S$ for the simulation of the cortical model.

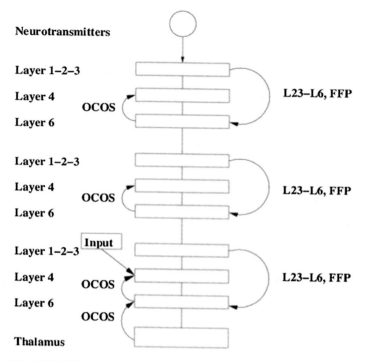

Fig. 18.1   The OCOS/FFP cortical model

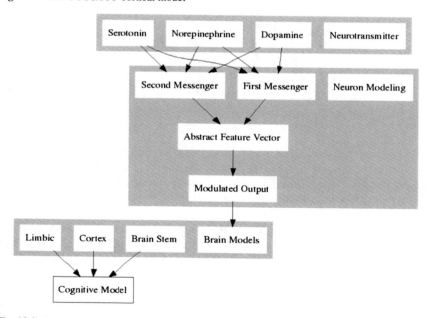

Fig. 18.2   Basic cognitive software architecture design

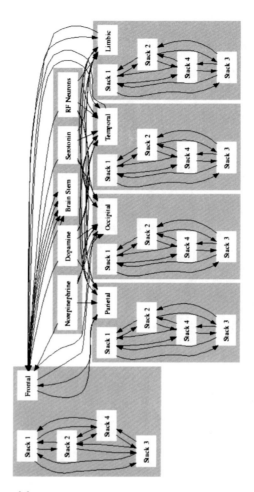

Fig. 18.3 Cognitive model components

We also need a collection of monoamine producing and RF modulatory neurons. We know the number of monoamine producing neurons is small relative to the total brain population. Hence, for convenience, we will choose one neuron of each type for each cortical module in the simulation. This gives 5 neurons of each monoamine type to modulate each cortical model. We also require one *RF* neuron for each of the 5 cortical elements as well as an *RF* neuron to modulate each of the 3 monoamine types. Thus a general simulation requires $15N_{OCOS/FFP}N_S$ neurons to implement the cortex and 25 neurons to implement monoamine modulation of cortex and 5 neurons to modulate the monoamine producing neurons via RF cells in the brain stem.

This model also does not use short or long term memory storage. These elements can clearly be added, but studies of the small scale nervous systems of the spider *portia fimbriata* (Harland and Jackson 2004), the honeybee (Zhang and Srinivasan 2004) and the praying mantis (Kral and Prete 2004) indicate that certain elements of cognitive function have probably emerged in these animals despite the small number of neurons in their brains. Estimates of the neuron population of *portia fimbriata* are in the 400,000 range. *Portia* does not have much long term memory so our choice to ignore long term memory needs seems reasonable for a first model of cognition. Since our artificial neurons use specific data structures to store local agent information, we do use short term memory which is probably similar to the use of short term memory in the spider, honeybee and praying mantis for route planning and decision choices.

Using our simple model of monoamine and RF modulation, 400,000 neurons in our model would imply $400,000/15 = 26,667$ neurons for the cortex models. Thus we have the constraint $N_{OCOS/FFP}N_S = 26,667$. If we use 200 OCOS/FFP circuits per layer, this implies 133 cortical stacks per cortical module. Even a cursory examination of the cortex of an animal shows that such choice will mimic real cortical structure nicely. Hence, a model at this level of cortical complexity has a reasonable probability of subserving cognitive function such as decision making. One way to implement this software architecture is to use *non-blocking system calls* that handle all connections between agents in one thread. A function is called whenever a connection between agents is ready for processing. Hence, this type of programming is known as event-drive or callback based in addition to being called *asynchronous*. Each of our artificial neurons will function in a networked environment and must accept inputs from other agents and generate outputs or decisions. A computational element can be forced to wait and therefore be blocked if it is waiting on the result of data from other elements which it must have to complete its own calculations. Since a neurons accepts inputs from other neurons, it may sometimes perform a computation that takes enough time to keep other neurons from processing. We will choose not to use this type of architecture and instead implement as much code as possible in a language such as **Erlang** in which all objects are immutable and hence the blocking of calls and the sharing of memory is not an issue.

In order to build even simple brain models, we thus need to write code that implements graphs that connect our processing nodes. We will start by building simple graph tools in MatLab/Octave using object oriented techniques as it is a programming language readily available and easy to learn. This will enable us to construct the underlying architecture of connections for our model and then we can do a proof of concept for the edge and nodal processing in MatLab as well. Eventually, we will discuss how to do this processing in other languages but that will be done in detail in later volumes. We will now start on the objects of real interest to us: graphs. We wish to build neural models and to do that we need a way to implement directed graphs where the nodes are neurons and the edges are the synaptic connections between neurons.

18.1 Building Global Graph Objects One

The implementation of the class oriented code for the an object is done in a special directory which must begin with the **@** symbol. Following the Octave documentation, we would set up a subdirectory called **@CoolObject** to use to define our coolobject class. The code that makes an object for us in the **CoolObject** class is called a **constructor**. It has this general look:

Listing 18.1: CoolObject

```
   function p = CoolObject(a)
   %
   % default constructor
   if (nargin==0)
5    p.c = [0];
     p = class(p,'CoolObject');
   elseif (nargin==1)
     if(strcmp(class(a),'CoolObject'))
        p = a(:);
10   elseif (isvector(a) && isreal(a))
        p.c = a(:) ';
        p = class(p,'CoolObject');
     else
        error('Not a valid argument to build a CoolObject');
15   end
   else
     error('too many arguments to the constructor.')
   end
   end
```

The test **nargin == 0** checks to see how many arguments are being used to create our object. If there are no arguments, the line in Octave would be **p = CoolObject();** and we set the *field element* **c** of **p** to be a vector with **0** in it. Note that we are *redefining* how Octave interprets the statement **p.c** here as normally the dot **.** has its own intrinsic meaning in Octave. In general, as usual in object oriented programming, a class object, **p** can have multiple fields associated with it and we generally must overload the meaning of various built in Octave commands to access these fields. We have already mentioned we must overload the meaning of the dot **"."** and we generally also overload the meaning the evaluation parenthesis so that the line **p()** is interpreted in a special way for our class object. If there is one argument to our object, **nargin == 1**, we then use that argument to fill in the field **.cc**. In this code, we also test to see that the argument we use is correct for our object and leave an error if it is not. We also check to see that the number of arguments does not exceed one. In this example, we assume we want our argument, if present, to be either another **CoolObject** or a real valued vector. Note without redefining the meaning of the dot **.** and the parentheses **()** for a **CoolObject p**, the lines **p.c** and **p(x)** would not be allowed. To make these work we have to overload the **subsref** function **Octave** uses to parse these statements. Once this is done, we can then use the lines **p.c** and **p(x)** as we wish. To create a **CoolObject** class, we set up a directory called **@CoolObject**. Note the name of the directory and the name of class are almost the same. The directory however must start with an **@** symbol. Inside this directory, you find the code needed to build our objects.

Listing 18.2: Inside the CoolObject directory

```
[petersj@inkapod ]$ ls @CoolObject
CoolObject.m    subsref.m
```

The code that builds a **CoolObject** object is called the *constructor* and is in the file
CoolObject.m. The name of this file, **CoolObject.m** without the extension **.m**
must match the name of the class subfolder **@CoolObject** without the **@**. Finally,
the overloaded **subsref** function is in **subsref.m**. Our code listed here is not
functional and instead is an outline of what we have to do.

Listing 18.3: Outline of the subsref function

```
function out = subsref(p,s)
%
%
    if (isempty (s))
5       error ('CoolObject: missing index');
    end
    switch s.type
      case '()'
        x = s.subs{1};
10      % call code to determine output of p()
        out = 2;
      case '.'
        fld = s.subs;
        if (strcmp(fld,'c'))
15        % access the field element c
          out = p.c;
        else
          error('invalid property for CoolObject object');
        end
20      otherwise
          error('invalid subscript type for CoolObject');
    end
end
```

If we add the line **s** at the top of this function, we would see what the input **s** is for
the various cases. Using the dot, **.** gives us the following **type** and **subs** elements.

Listing 18.4: The subs elements of p.c

```
  b = p.c
  s =
    type = '.'
    subs =' c'
5
  ans = whatever the action on c was
```

Hence the input **s** has **type** **.** and **subs c**. Then in the next one

Listing 18.5: The p() elements

```
  z = p(x)
  s =
    type = '()'
    subs = {[ whatever x was]}
5
  ans = whatever () did to x
```

For example, having added **s** at the top of **subsref**, we could run this example:

Listing 18.6: CoolObject Example

```
   a = 'jim is cool';
   b = [1;2;3];
   C = CoolObject(a);
   C.c
 5 s =
       type: '.'
       subs: 'c'
   ans =
   jim is cool
10 D = CoolObject(b);
   D.c
   s =
       type: '.'
       subs: 'c'
15 ans =
        1      2      3
```

This is probably not very clear at this point, so let's build a graph class in MatLab (or Octave!). A graph model or graph object will consist of edge and vertices objects glued together in the right way, so we have to build three classes to make this happen. We need to do this because in order to build neural models we need a way to implement directed graphs where the nodes are neurons and the edges are the synaptic connections between neurons.

Make a directory **GraphsGlobal** and inside it make the usual class directories **@edges**, **@vertices** and **@graphs**. We call this directory **GraphsGlobal** because here all of our nodes and edges will be globally labeled. This means as we glue subgraphs together, their original node and edge numbering will be lost as the new graph is built. Now we will eventually want to keep track of the subgraph information which we will do using an addressing scheme. However, that is another story. For now, we will stick to global numbering. In the directory, **@vertices**, we set up a class to handle vertices.

18.1.1   Vertices One

We now write the constructor code and the code for the overloaded indexing. The constructor is as follows:

Listing 18.7: Vertices class

```
function V = vertices(a)
%
% create nodes from the vector a
%
5 s = struct();
  n = length(a);
  for i=1:n
    if isvector(a)
      s.node = a(i);
10  end
    if iscell(a)
      s.node = a{i};
    end
    V.v(i) = s;
15 end
  V = class(V,'vertices');
  end
```

We allow the vertices object to be constructed from a node list which is a vector (**isvector(a)**) or cell data (**iscell(a)**). The overloaded indexing is in **subsref.m**.

Listing 18.8: The vertices subsref overloading

```
function out = subsref(V,s)
%
% give access to private elements
% of p
5 %
  switch s.type
    case "."
      fld = s.subs;
      if(strcmp(fld,"v"))
10      n = length(V.v);
        out = [V.v(1).node];
        for i=2:n
          out = [out V.v(i).node];
        end
15    else
        error("invalid property for vertices object");
      end
    otherwise
      error("@vertices/subsref: invalid subscript type for vertices")
        ;
20  end
  end
```

We can then have an session like these. First, use a column vector to construct the vertices object.

Listing 18.9: Simple vertices session: column vector input

```
   b = [1;2;3;4;5];
   N2 = vertices(b);
   N2.v
4  ans =

        1   2   3   4   5
```

We can also use a row vector or cell data.

Listing 18.10: Simple vertices session: row vector or cell input

```
   b = [1,2,3,4,5];
   N2 = vertices(b);
   N2.v
4  ans =
        1   2   3   4   5
   b = {1,2,3,4,5};
   N2 = vertices(b);
   N2.v
9  ans =
        1   2   3   4   5
```

18.1.2   Edges One

In the subdirectory **@edges**, we then write the code both the constructor and for
overloading the **subsref** command so we can access field elements of an edge
object. Here is the edge constructor code. It takes a vector of ordered pairs and sets
up a collection of edge structures with fields **in** and **out**.

Listing 18.11: Edges

```
    function E = edges(a)
    %
    % create edges from the vector of pairs a
    % of the form [in;out]
5   %
    s = struct();
    n = length(a);
    for i=1:n
        s.in = a{i}(1);
10      s.out = a{i}(2);
        E.e(i)  = s;
    end
    E = class(E,'edges');
    end
```

and the overloaded indexing operators are in **subsref.m**.

Listing 18.12: The edge subsref overloading

```
   function  out  =  subsref(E,s)
   %
   % give  access  to  private  elements
   % of  edges
 5 %
     switch  s.type
       case  ".”
         fld  =  s.subs;
         if(strcmp(fld ,"e"))
10         n  =  length(E.e);
           out  =  [E.e(1).in;E.e(1).out];
           for  i=2:n
             out  =  [out  [E.e(i).in;E.e(i).out]];
           end
15       else
           error("invalid  property  for  edges  object");
         end
       otherwise
         error(" @edges/subsref:  invalid  subscript  type  for  edges");
20   end
   end
```

Then in use, we would have a session like this.

Listing 18.13: Simple edges session: column vectors

```
   a  =  {[1;2];[3;5];[3;7]};
   E2  =  edges(a);
   E2.e
 4 ans  =

       1    3    3
       2    5    7
```

We can also use row vectors in the constructor. Hence, this is fine also.

Listing 18.14: Simple edges session: row vectors

```
   a  =  {[1 ,2];[3 ,5];[3 ,7]};
   E2  =  edges(a);
 3 E2.e
   ans  =

       1    3    3
       2    5    7
```

Now that we can construct a vertices and edges object, we can build a graph class.

18.1.3   A First Graph Class

Next, in the directory **@graphs**, we write the constructor and indexing code.

Listing 18.15: The graphs class

```
function g = graphs(V,E)
%
% constructor for graph g
%
% V = vertices object
% E = edges object
%
g.v = V;
g.e = E;
g = class(g,'graphs');
end
```

Here is the indexing code.

Listing 18.16: The graph subsref overloading

```
function out = subsref(g,s)
  switch s.type
    case '.'
      fld = s.subs;
      if(strcmp(fld,'v'))
        Nodes = g.v;
        out = Nodes.v;
      elseif (strcmp(fld,'e'))
        Edges = g.e;
        out = Edges.e;
      else
        error('invalid property for vertices object');
      end
    otherwise
      error('@vertices/subsref: invalid subscript type for vertices')
      ;
  end
end
```

Notice the pair of statements we use as access via **g.v.v** is not possible. We do this pair of steps:

Listing 18.17: Access via g.v.v is impossible

```
Nodes = g.v;
out = Nodes.v;
```

This is necessary because the line **out = g.v.v** does not parse correctly. So we do it in two steps. The same comment holds for the way we handle the edges. We then build a graph in this session.

Listing 18.18: Simple graph session

```
    G2 = graphs(N2,E2);
    G2.v
    ans =

  5    1    2    3    4    5
    G2.e
    ans =

       1    3    3
 10    2    5    7
```

Next, we add some methods such as compute the **incidence matrix**.

18.2 Adding Class Methods First Pass

We will now add methods to each of our three classes. So far, we haven't found a way to force **Octave** to update its knowledge of the new code as we write it. Note, we have to save our work in sessions and manually restart and reload code each time we change the class codes. This is not a problem in MatLab, but you should be aware of this if you use Octave.

18.2.1 Adding Edge Methods

We add code to add an edge to an existing **edge** object. This is in the file **@edge/add.m**. Here is the code which specifically needs to have both the **to** and **from** nodes as arguments.

Listing 18.19: Adding an edge to an existing edge

```
    function W = add(E,u,v)
    %
    % edge is added to edge list
    %
  5 n = length(E.e);
    out = {[E.e(1).in;E.e(1).out]};
    for i=2:n
      temp = {[E.e(i).in;E.e(i).out]};
      out = [out,temp];
 10 end
    temp = {[u;v]};
    out = [out, temp];
    W = edges(out);
    end
```

The arguments **(u,v)** are the **in node** and **out node** values of the new edge. The syntax for adding a method to a class is that the first argument of the function must be the name we give the edge object in the edge constructor. We take the existing **E.e** data and construct a new data container by concatenation. When we finish the loop, we have copied all the old data over. We then concatenate the new edge data.

Then, we have the data object we can use to create a new edge object. We do that in the line **W = edges(out);** and return **W**. The reason we use this design is that we are not allowed to concatenate the incoming new data to the end of the old edge list. Hence, we must do a complete rebuild. In a language that supports object oriented programming ideas better, this would not have been necessary.

18.2.2   Adding Vertices Methods

We add code to add a node to an existing **vertices** object. This is in the file **@vertices/add.m**. Here is the code:

Listing 18.20: Adding a node to existing vertices object

```
   function W = add(V,j)
   %
   % node is added to vertex list
   %
 5 n = length(V.v);
   out = V.v(1).node;
   for i=2:n
     out = [out V.v(i).node];
   end
10 out = [out j];
   W = vertices(out);
   end
```

We take the existing **V.v** data and again construct a new data container by concatenation. When we finish the loop, we have copied all the old data over. We then concatenate the new vertex node to the data. At this point, we have the data object we can use to create a new vertex object. We do that in the line **W = vertices(out);** and return **W**. The same comments we made in the edge addition apply here. The language MatLab or Octave forces us to do the global rebuild.

18.2.3   Adding Graph Methods

We now write code to add an edge, in the file **addedge.m** and add a node, in the file **addnode.m**. First, look at the adding an edge code. This builds on the add code in the vertices class.

Listing 18.21: Adding an edge in a global graph

```
   function W = addedge(g,i,j)
   %
   % adjoin an edge to an existing graph
   %
 5 E = add(g.e,i,j);
   W = graphs(g.v,E);
   end
```

We now add a node using the add node method for the vertices class.

Listing 18.22: Adding a node in a global graph

```
   function W = addnode(g,j)
   %
 3 % adjoin a node to an existing graph
   %
   V = add(g.v,j);
   W = graphs(V,g.e);
   end
```

Once this code is built, we can add methods for calculating the incidence matrix for the graph and its corresponding Laplacian. To find the incidence matrix, we use a simple loop.

Listing 18.23: Simple incidence matrix for global graph

```
    function K = incidence(g)
    %
  3 % g is a graph having
    % vertices g.v
    % edges g.e
    %
    % Get the edge object from g
  8 E = g.e;
    % get the edges from the edge object
    e2 = E.e;
    % get the vertices object from g
    V = g.v;
 13 % get the vertices from the vertices object
    v2 = V.v;
    % find out how many vertices and edges there are
    [row,sizeV] = size(v2);
    [row,sizeE] = size(e2);
 18 K = zeros(sizeV,sizeE);
    %
    % setup incidence matrix
    %
    for i = 1:sizeV
 23    for j = 1:sizeE
          if e2(1,j)== i
             K(i,j) = 1;
          else if e2(2,j) == i
             K(i,j) = -1;
 28       end
       end
    end

    end
```

The Laplacian of the graph is then easy to find (recall we discussed the Laplacian for a graph in Chap. 15!).

Listing 18.24: Laplacian for global graph

```
function L = laplacian(g)
%
3 % g is a graph having
% vertices g.v
% edges g.e
%
K = incidence(g);
8 L = K*K';
end
```

18.2.4   Using the Methods

Here is a session where we build the **OCOS** and **FFP** networks and find their Lapla-
cians with eigenvalue structure. We add some annotation here and there. First, we
build an OCOS graph directly. This OCOS circuit includes the thalamus connection
which we will later remove as the thalamus will be implemented separately. We
define the node and edge vectors using integers for each node. We will label the
nodes in ascending order now so the resulting graph is feedforward.

Listing 18.25: Define the OCOS nodes and edges data

```
1 % define the OCOS nodes
V = [1;2;3;4;5;6;7;8];
% define the OCOS edges
E = {[1;2],[2;3],[2;4],[2;5],[3;6],[4;7],[5;8],[2;7],[1;7]};
% construct the OCOS edge object
```

We then construct edge and vertices objects.

Listing 18.26: Construct the OCOS edges and vertices object

```
e = edges(E);
% construct the OCOS vertices object
v = vertices(V);
```

Next, we construct the graph object and check to see if our overloaded **subsref.m**
code is working correctly.

Listing 18.27: Construct the OCOS graph

```
% construct the OCOS graph
OCOS=graphs(v,e);
% verify the OCOS edge list
OCOS.v
5 ans =
    1    2    3    4    5    6    7    8
% verify the OCOS edges list
OCOS.e
ans =
10      1       2       2       2       3       4       5       2       1
        2       3       4       5       6       7       8       7       7
```

We then explicitly add nodes and edges to build an FFP graph. Note for a small graph this is not very expensive, but the way nodes and edges are added is actually very inefficient as we have to rebuild the node and edge lists from scratch with each add.

Listing 18.28: Create FFP by adding nodes and edges to OCOS

```
% add the FFP nodes 9, 10 and 11
FFP = addnode(OCOS,9);
FFP = addnode(FFP,10);
4 FFP = addnode(FFP,11);
% add the FFP edges
FFP = addedge(FFP,11,2);
FFP = addedge(FFP,10,11);
FFP = addedge(FFP,9,10);
```

Now let's verify we added the nodes and edges.

Listing 18.29: Verify the added nodes and edges

```
% verify the FFP node list
2 FFP.v
  ans =
     1    2    3    4    5    6    7    8    9   10   11
% verify the FFP edge list
  FFP.e
7 ans =
     1    2    2    2    3    4    5    2    1   11   10    9
     2    3    4    5    6    7    8    7    7    2   11   10
```

Next, let's find incidence matrices.

Listing 18.30: Get Incidence Matrix

```
1 % find the OCOS incidence matrix
  K = incidence(OCOS)
  K =
       1    0    0    0    0    0    0    0    1
      -1    1    1    1    0    0    0    1    0
6      0   -1    0    0    1    0    0    0    0
       0    0   -1    0    0    1    0    0    0
       0    0    0   -1    0    0    1    0    0
       0    0    0    0   -1    0    0    0    0
       0    0    0    0    0   -1    0   -1   -1
11     0    0    0    0    0    0   -1    0    0
  % find the FFP incidence matrix
  K2 = incidence(FFP)
  K2 =
       1    0    0    0    0    0    0    0    1    0    0
            0
```

16	-1	1	1	1	0	0	0	1	0	-1	0
											0
	0	-1	0	0	1	0	0	0	0	0	0
											0
	0	0	-1	0	0	1	0	0	0	0	0
											0
	0	0	0	-1	0	0	1	0	0	0	0
											0
	0	0	0	0	-1	0	0	0	0	0	0
											0
21	0	0	0	0	0	-1	0	-1	-1	0	0
											0
	0	0	0	0	0	0	-1	0	0	0	0
											0
	0	0	0	0	0	0	0	0	0	0	0
											1
	0	0	0	0	0	0	0	0	0	0	1
											-1
	0	0	0	0	0	0	0	0	0	1	-1
											0

We can then find the Laplacian of the graphs.

Listing 18.31: Get Laplacian

```
% find the OCOS Laplacian
L = laplacian(OCOS)
L =

    2   -1    0    0    0    0   -1    0
   -1    5   -1   -1   -1    0   -1    0
    0   -1    2    0    0   -1    0    0
    0   -1    0    2    0    0   -1    0
    0   -1    0    0    2    0    0   -1
    0    0   -1    0    0    1    0    0
   -1   -1    0   -1    0    0    3    0
    0    0    0    0   -1    0    0    1
% find the FFP Laplacian
L2 = laplacian(FFP)
L2 =

    2   -1    0    0    0    0   -1    0    0    0    0
   -1    6   -1   -1   -1    0   -1    0    0    0   -1
    0   -1    2    0    0   -1    0    0    0    0    0
    0   -1    0    2    0    0   -1    0    0    0    0
    0   -1    0    0    2    0    0   -1    0    0    0
    0    0   -1    0    0    1    0    0    0    0    0
   -1   -1    0   -1    0    0    3    0    0    0    0
    0    0    0    0   -1    0    0    1    0    0    0
    0    0    0    0    0    0    0    0    1   -1    0
    0    0    0    0    0    0    0    0   -1    2   -1
    0   -1    0    0    0    0    0    0    0   -1    2
```

Once we have these matrices, we can use standard tools within **Octave** to find their corresponding eigenvalues and eigenvectors which we might want to do now and then.

Listing 18.32: Get OCOS and FFP Laplacian eigenvalue and eigenvectors

```
% get the eigenvalue and eigenvectors for
% the OCOS Laplacian
[eigvecocos, eigvalocos]=eig(L);
% extract the OCOS eigenvalues
5 vals = diag(eigvalocos)
  vals =
        -0.0000
         0.3820
         0.5607
10       2.0000
         2.3389
         2.6180
         4.0000
         6.1004
15 % get the eigenvalue and eigenvectors for
% the FFP Laplacian
[eigvelap, eigvallap]=eig(L2);
% extract the FFP eigenvalues
  vals2 = diag(eigvallap)
20 vals2 =
        -0.0000
         0.2338
         0.3820
         0.5786
25       1.4984
         2.0000
         2.3897
         2.6180
         3.1986
30       4.0000
         7.1010
```

18.2.5 Adding a Graph to an Existing Graph

If we want to add a second graph to an existing graph, we need to write an **addgraph** method for our graph class. Here is one way to do it. It is in the file **addgraph.m** in the **@graph** directory. Our design for this code is to specify the new edge by simply in the form of cell data

Listing 18.33: Links data

```
{[to;from],...};
```

where we have to remember that the nodes **to** and **from** need to be given using the node numbering from the new larger graph we just constructed. That is hard to do so later we will use a vector address strategy for building graphs from subgraphs, but that is for a later time. Here is the annotated code.

Listing 18.34: Adding a graph to an existing graph

```
function W = addgraph(g,H,links)
%
% adjoin a graph H using a list of edges
% to connect the two graphs which is given in a cell data
5 % structure, links
%
% get edge object of H
eh = H.e;
% get the edges of eh
10 ehedges = eh.e;
% get vertices object of H
nh = H.v;
% get the nodes of nh
nhnodes = nh.v;
15 % get edge object of g
eg = g.e;
% get the edges of eg
egedges = eg.e;
% get vertices object of g
20 ng = g.v;
% get the nodes of ng
ngnodes = ng.v;

% copy nodes of g to temporary a
25 a = ngnodes;

% get sizes of all nodes and edges
[n,nodesizeg] = size(ngnodes);
[n,nodesizeh] = size(nhnodes);
30 [n,edgesizeg] = size(egedges);
[n,edgesizeh] = size(ehedges);

% create a copy of the nodes of H
% and add the number of nodes of g
35 % to each entry
for i = 1:nodesizeh
   b(i) = nhnodes(i) + nodesizeg;
end

40 % create the new adjoined node list
v = [a, b];
% create the new vertices object
V = vertices(v);

45 % copy the edges of g into c
% this is data in vector form
c = egedges;
% add the number of nodes of g
% to each in and out value in the
50 % edges of H
d = ehedges + nodesizeg;
% now create the { [i;j],[u;v] }
% type data structure we need for
% the edges constructor
55 % initialize the data e
e = {egedges(:,1)};
% add more entries until the edges
% of g are used up
for i = 2:edgesizeg
60    e{i} = c(:,i);
end
% now add the edges of H
% making sure we set the counter
% to start after the number of edges
65 % in g
for j=1:edgesizeh
   e{j+edgesizeg} = d(:,j);
end

70 % now e is in the proper format
% and we can construct the new edge object
E = edges(e);
W = graphs(V,E);
sizelinks = length(links);
75 for i=1:sizelinks
   from = links{i}(1,1);
   to   = links{i}(2,1);
   W = addedge(W,from,to);
end
```

It is now straightforward to build more complicated graphs. Consider the following session. Note we still have to manually setup the links between the two graphs using the new global node number; that is irritating but we will work on that later when we do address graphs.

Listing 18.35: Building a graph from other graphs

```
 1 V =  [1;2;3;4;5;6;7;8];
   E =  {[1;2],[2;3],[2;4],[2;5],[3;6],[4;7],[5;8],[2;7],[1;7]};
   e =  edges(E);
   v =  vertices(V);
   OCOS=graphs(v,e);
 6 OCOS2 = graphs(v,e);
   % adjoin the second OCOS graph by adding an edge from
   % node 7 of the original OCOS graph to node 1 of the new OCOS2 graph
   W = addgraph(OCOS,OCOS2,{[7;9]});
   W.e
11 ans =
      Columns 1 through 16
        1       2       2       2       3       4       5       2       1       9      10
               10      10      11      12      13
        2       3       4       5       6       7       8       7       7      10      11
               12      13      14      15      16
      Columns 17 through 19
16     10       9       7
       15      15       9
   W.v
   ans =
      Columns 1 through 14:
21     1       2       3       4       5       6       7       8       9      10      11      12      13
               14
      Columns 15 and 16:
       15      16
```

Notice how awkward this code is. The design here removes all the old node and edge information from the incoming graph and renumbers and relabels everything to fit into one new graph. It also assumes there is an edge used to *glue* the graphs together. If we want to build a small brain model, we will want to assemble many modules. Each module will have its own organization and we will want to remember that. However, to use the Laplacian of the small brain model, we will need to have an incidence matrix written in terms of global node numbers. Hence, we need a way to organize nodes in terms of their module structure. This requires an address rather than a node number. We will explore this in the next section where we build a new graph class based on these new requirements.

18.2.6   Drawing Graphs

We need a way to visualize our graphs. We will use visualization software package called **Graphviz** (North and Ganser 2010) which includes the tool **dot** which draws directed graphs. The documentation for **dot** tells us how to write a file which the **dot** command can use to generate the drawing. Consider the file **incToDot.m** which takes the incidence matrix of a graph, **inc** and uses it to write the needed file. This file can have any name we want but it uses the extension **.dot**. Assume the file name we generate is **MyFile.dot**. Once the file is generated, outside of **MatLab** we then run the **dot** command to generate the **.pdf** file **MyFile.pdf** as follows:

Listing 18.36: Using dot to create a graph visualization

```
dot -Tpdf -oMyFile.pdf MyFile.dot
```

We choose the *pdf* format as it is a vector graphic and hence can be scaled up without loss of resolution as far as we want. This is very important in very large graphs. Here is the code.

Listing 18.37: The incidence to dot code: incToDot

```
function incToDot(inc, width, height, aspect, filename)

    [N,E] = size(inc);
4   [fid, msg] = fopen(filename, 'w');
    fprintf(fid, 'digraph G {\n');
    fprintf(fid, 'size=\"%f,%f\";\n', width, height);
    fprintf(fid, 'rankdir=TB;\n');
    fprintf(fid, 'ratio=%f;\n', aspect);
9   for i = 1:N
        fprintf(fid, '%d [shape=\"rectangle\" fontsize=20 fontname=\"Times
            new Roman\" label=\"%d\"];\n', i, i);
    end
    fprintf(fid,'\n');
    for j = 1:E
14      a = 0;
        b = 0;
        for i = 1:N
            if inc(i,j) == 1
                a = i;
19          elseif inc(i,j) == -1
                b = i;
            end
        end
        if (a ~= 0) && (b ~= 0)
24          fprintf(fid, '%d->%d;\n', a, b);
        end
    end
    fprintf(fid, '\n}');
    fclose(fid);
29  end
```

Let's use this on the OCOS circuit. First, we generate the **OCOS.dot** file.

Listing 18.38: Generate the OCOS dot file

```
incToDot(KOCOS,6,6,1.0,'ocos.dot');
```

The dot file has the following look. Note the graph is organized simply. Each node is identified with an integer and then in brackets there is a string which specifies details about its shape and its label. Here we use the same label as the identifier. The edges

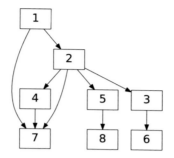

Fig. 18.4 OCOS graph

are as simple as possible, specified with the string **4->1** and so forth. You'll need to look at the **dot** documentation to see more details about how to do this.

Listing 18.39: A generated dot file

```
digraph G {
size ="6.000000,6.000000";
rankdir=TB;
ratio =1.000000;
1 [shape="rectangle" fontsize=20 fontname="Times new Roman" label
  ="1"];
2 [shape="rectangle" fontsize=20 fontname="Times new Roman" label
  ="2"];
3 [shape="rectangle" fontsize=20 fontname="Times new Roman" label
  ="3"];
4 [shape="rectangle" fontsize=20 fontname="Times new Roman" label
  ="4"];
5 [shape="rectangle" fontsize=20 fontname="Times new Roman" label
  ="5"];
6 [shape="rectangle" fontsize=20 fontname="Times new Roman" label
  ="6"];
7 [shape="rectangle" fontsize=20 fontname="Times new Roman" label
  ="7"];
8 [shape="rectangle" fontsize=20 fontname="Times new Roman" label
  ="8"];

1->2;
2->3;
2->4;
2->5;
3->6;
4->7;
5->8;
2->7;
1->7;
}
```

Then, in a separate terminal window, we generate the **OCOS.pdf** file which is shown in Fig. 18.4.

Listing 18.40: Generate OCOS graphic as pdf file

```
dot -Tpdf -oOCOS.pdf OCOS.dot
```

18.2.7   *Evaluation and Update Strategies*

At any node i in our graph, the nodes which interact with it via an edge contact are
in the backward set for node i, $\mathcal{B}(i)$. Hence, the input to node i is the sum

$$\sum_{j\in\mathcal{B}(i)} E_{j\to i}\, Y(j)$$

where $E_{j\to i}$ is the value we assign the edge between j and i in our graph and $Y(j)$ is
the output we assign to node j. We also know the nodes that node i sends its output
signal to are in its forward set $\mathcal{F}(i)$. Both of these sets are readily found by looking
at the incidence matrix for the graph. We can do this in the code **BFsets** which is a
new graph method placed in the **@graphs** directory. In it, we find the backward and
forward sets for each neuron in both the six dimensional address form and the global
node form. In a given row of the incidence matrix, the positive 1's tells us the edges
corresponding to the forward links and the negative 1's give us the backward edges.
So we look at each row of the incidence matrix and calculate the needed nodes for
the backward and forward sets.

Listing 18.41: Finding backward and forward sets: BFsets0

```
    function [BackGlobal, ForwardGlobal, BackEdgeGlobal] = BFsets0(g,Kg)
    %
    % g is the graph
    % Kg is the incidence matrix of the graph
  5 % g.e is the edge object of the graph
    %

    % get edges
    Eg = g.e;
 10 % get edge array
    E = Eg.e;

    [Kgrows, Kgcols] = size(Kg);

 15 BackGlobal = {};
    ForwardGlobal = {};
    BackEdgeGlobal = {};
    for i = 1:Kgrows
      BackGlobal{i} = [];
 20   ForwardGlobal{i} = [];
      BackEdgeGlobal{i} = [];
      for j = 1:Kgcols
        d = E(:,j);
        u = d(1);
 25     v = d(2);
        if Kg(i,j) == 1
          ForwardGlobal{i} = [ForwardGlobal{i},v];
        else if Kg(i,j) == -1
          BackGlobal{i} = [BackGlobal{i},u];
 30       BackEdgeGlobal{i} = [BackEdgeGlobal{i},j];
        end
      end
    end

 35 end
```

The simplest evaluation strategy then is to let each node have an output value determined by a simple sigmoid function such as given in the code **sigmoid.m**.

Listing 18.42: The sigmoid function

```
function y = sigmoid(x)
%
% x is the input
%
5 y = 0.5*(1.0+tanh(x));
  end
```

We use this sigmoid to do a simple graph evaluation. This is done in the code below. We simply evaluate the sigmoid functions at each node using the current synaptic interaction values.

Listing 18.43: The evaluation

```
   function NodeVals = evaluation(Y,W,B,BE)
   %
   % g is the graph
4  % Y is node vector
   % B is the global backward node set information
   % BE is the global backward edge set information
   %
   % get size
9  sizeV = length(Y);
   for i = 1:sizeV
     % get backward node information for neuron i
     BF = B{i};
     % get backward edge information for neuron i
14   BEF = BE{i};
     lenB = length(BF);
     lenBEF = length(BEF);
     sum = 0.0;
     for j = 1:lenBEF
19     link = BEF(j);
       pre = BF(j);
       sum = sum + W(link)*Y(pre);
     end
     NodeVals(i) = sigmoid(sum);
24 end
```

A simple edge update function is then given by in the file below. This uses what is called a Hebbian update to change the scalar weight associated with each edge. If the value of the *post* neuron i is *high* and the value of the edge $E_{j \to i}$ is also high, the value of this edge is increased by a multiplier such as 1.05 or 1.1. The code to implement this is below.

Listing 18.44: The Hebbian code: HebbianUpdateSynapticValue

```
 1 function  Wts  =  HebbianUpdateSynapticValue (Y,W,B,BE)
   %
   % link  is  edge  number
   % pre  is  the  pre  neuron
   % post  is  the  post  neuron
 6 % W is  the  edge  vector
   % Y is  the  new  value  of  edge
   %

   % get  size
11 sizeV  =  length (Y);
   sizeE  =  length (W);

   scale  =  1.1;
   for  i  =  1:sizeV
16   % get  backward  node   information  for  neuron  i
     BF  =  B{i};
     % get  backward  edge  information  for  neuron  i
     BEF  =  BE{i};
     lenB  =  length (BF);
21   lenBEF  =  length (BEF);
     for  j  =  1:lenBEF
        link  =  BEF(j);
        pre  =  BF(j);
        post  =  i;
26      hebb  =  Y(post) *W(link);
        weight  =  W(link);
        if  abs (hebb)  >  .36
           weight  =  scale *W(link);
        end
31      Wts(link)  =  weight;
     end
   end
 end
```

A simple example is shown below for the evaluation step for a small OCOS model. First, we set up a node value vector Y the size of the nodes and a weight value vector W the size of edges of the brain model. We fill the weight vector randomly with numbers from -1 to 1. Then we compute the needed backward set information and use it to perform an evaluation step. Then we do a synaptic weight Hebbian update. At this point, we are just showing the basic ideas of this sort of computation. We still have to restrict certain edges to have only negative values as they correspond to inhibitory synaptic contacts. Also, we have not yet implemented the cable equation update step afforded us with the graph's Laplacian. That is to come. Here is an example computation.

Listing 18.45: An OCOS Hebbian Update

```
   vOCOS = {1,2,3,4,5,6,7,8};
 2 eOCOS = {[1;2],[2;3],[2;4],[2;5],[3;6],[4;7],[5;8],[2;7],[1;7]};
   VOCOS = vertices(vOCOS);
   EOCOS = edges(eOCOS);
   OCOS = graphs(VOCOS,EOCOS);
   KOCOS = incidence(OCOS);
 7 [m,n] = size(KOCOS);
   Y = zeros(1,m);
   W = -1+2*rand(1,n);
   [B,F,BE] = BFsets0(OCOS,KOCOS);
   Y = evaluation(OCOS,Y,W,B,BE);
12 W = HebbianUpdateSynapticValue(Y,W,B,BE);
```

We initialize all the node values to zero; hence, all of the input values to the nodes will be zero as well and so the first evaluation will return a constant 0.5.

Listing 18.46: First evaluation

```
   Y
   Y =
 3    0.50000    0.50000    0.50000    0.50000    0.50000    0.50000
          0.50000    0.50000
   W
   W =
      0.6294    0.8927    -0.8206    0.9094    0.2647    -0.8854
         -0.4430    0.0938    1.0065
```

If we now evaluate again, we will see the change in Y.

Listing 18.47: Second evaluation

```
   Y = evaluation(OCOS,Y,W,B,BE);
   Y
   Y =
 4    0.5000    0.6524    0.7095    0.3056    0.7129    0.5658
          0.5535    0.3910
```

We can also do a simulation loop. Consider this code

Listing 18.48: A simple simulation: DoSim

```
 1 function [Y,W] = DoSim(NumIters,g,Kg,B,BE,Y,W)
   %
   for t=1:NumIters
     Y = evaluation(g,Y,W,B,BE);
     W = HebbianUpdateSynapticValue(Y,W,B,BE);
 6 end
```

Let's evaluate for say 20 steps.

Listing 18.49: A 20 step evaluation

```
[Y,W]  =  DoSim(20,OCOS,KOCOS,B,BE,Y,W);
```

After 20 iterations, we find

Listing 18.50: Second evaluation

```
Y
Y =
      0.5000       0.9792      1.0000      0.1689      1.0000      0.6293
                   0.9888      0.2919
 ₄ W
   W =
      4.2346       6.0059     -0.8206      6.1182      0.2647     -5.9566
                  -0.4430      0.0938      6.7713
```

If the graph was a feed forward architecture, we could also implement a link update strategy using the partial derivatives of an error function with respect to the mutable parameters of the graph architecture. But that is another story!

18.3   Training

We are now ready to attempt to train our networks to match input to output data. We begin with the code to train a feedforward architecture to do that. We have already done this in Chap. 16 for the classical matrix-vector feedforward network architecture using gradient descent. The chain feedforward network architecture is more flexible and we have discussed it carefully in Chap. 17 and derived the new algorithms for evaluation and gradient update. So now we will implement them. First, let's redo the evaluation code so we can handle node functions with gains and offsets. Also, we want to be able to have different node functions for input and output nodes. We do this in the new evaluation code here. We will do it a bit differently here. Now we code the needed derivatives as follows. We need a lot of σ' terms. So, for this type of σ, we note

$$\sigma(y, o, g) = \frac{1}{2}\left((H + L) + (H - L)\tanh\left(\frac{y - o}{g}\right)\right)$$

$$\frac{\partial \sigma}{\partial y} = \frac{H - L}{2g}\left(\mathrm{sech}\left(\frac{y - o}{g}\right)\right)^2$$

and so

$$\frac{\partial \sigma}{\partial o} = -\frac{H-L}{2g}\left(\mathrm{sech}\left(\frac{y-o}{g}\right)\right)^2$$

$$= -\frac{\partial \sigma}{\partial y}$$

and

$$\frac{\partial \sigma}{\partial o} = -\frac{H-L}{2g^2}\left(\mathrm{sech}\left(\frac{y-o}{g}\right)\right)^2$$

$$= -\frac{1}{g}\frac{\partial \sigma}{\partial y}.$$

So in the following code, we will define the $\frac{\partial \sigma}{\partial y}$ functions.

Listing 18.51: Initializing the Node Functions for the Chain FFN

```
    nodefunction = {};
    nodefunctionprime = {};
    %
4   sigma = @(y,offset ,gain) 0.5*(1 + tanh( (y - offset)/gain) );
    sigmainput = @(y,offset ,gain) (y - offset)/gain;
    SL = -1.2;
    SH = 1.2;
    % ouput nodes range from SL to SH
9   sigmaoutput = @(y,offset ,gain) 0.5*( (SH+SL) + (SH-SL)* tanh( (y -
        offset)/gain) );
    %
    nodefunction{1} = sigma;
    nodefunction{2} = sigmainput;
    nodefunction{3} = sigmaoutput;
14  %
    sigmaprime = @(y,offset ,gain) (1/(2*gain))* sech((y - offset)/gain)
        .^2;
    sigmainputprime = @(y,offset ,gain) 1/gain;
    sigmaoutputprime = @(y,offset ,gain) ( (SH-SL)/(2*gain) )* sech((y -
        offset)/gain).^2;
    %
19  nodefunctionprime{1} = sigmaprime;
    nodefunctionprime{2} = sigmainputprime;
    nodefunctionprime{3} = sigmaoutputprime;
```

Note again **nodefunction{1}** is the main node processing function,
nodefunction{2} is the input node function and **nodefunction{3}** is the
input node function. You see this used in the evaluation code as follows

Listing 18.52: Graph evaluation code: evaluation2.m

```
function [InVals,OutVals] = evaluation2(nodefunction,g,I,IN,OUT,W,O,G
   ,B,BE)
%
% nodefunction are the node functions as a cell
% g is the graph
5 % I is the input vector
% IN is the list of input nodes as a cell
% OUT is the list of output nodes as a cell
% Y is node vector
% W is the link vector
10 % O is the offset vector
% G is the gain vector
% B is the global backward node set information
% BE is the global backward edge set information
%
15 % OutVals is the new node output vector
% InVals is the node input vector
%

% get sizes
20 sizeIN = length(IN);
sizeOUT = length(OUT);
Nodes = g.v;
[m,sizeY] = size(Nodes.v);
InVals = zeros(1,sizeY);
25
sigmainput = nodefunction{1};
sigmamain = nodefunction{2};
sigmaoutput = nodefunction{3};
sigma = {};
30 % set all node functions to sigmamain
for i = 1:sizeY
   sigma{i} = sigmamain;
end
% reset input node functions to sigmainput
35 for j = 1:sizeIN
   sigma{IN{j}} = sigmainput;
end
% reset output node functions to sigmaoutput
for j = 1:sizeOUT
40   sigma{OUT{j}} = sigmaoutput;
end
for i = 1:sizeY
   % get backward node  information for neuron i
   BF = B{i};
   BEF = BE{i};
   lenB = length(BF);
   lenBEF = length(BEF);
   sum = 0.0;
50 % add input if this is an input node
   for j = 1:sizeIN
      if i == IN{j}
         sum = sum + I(j);
      end
55   end
   for j = 1:lenBEF
      link = BEF(j);
      pre = BF(j);
      sum = sum + W(link)*Y(pre);
60   end
   InVals(i) = sum;
   Y(i) = sigma{i}(sum,O(i),G(i));
end

65 OutVals = Y;
end
```

We use the code similar to what we did in the matrix ffn work.

Listing 18.53: Testing the new evaluation code

```
   nodefunction = {};
   nodefunctionprime = {};
   sigma = @(y,offset,gain) 0.5*(1 + tanh( (y - offset)/gain) );
 4 nodefunction{1} = sigma;
   sigmainput = @(y,offset,gain) (y - offset)/gain;
   nodefunction{2} = sigmainput;
   SL = -1.2;
   SH = 1.2;
 9 % ouput nodes range from SL to SH
   sigmaoutput = @(y,offset,gain) 0.5*( (SH+SL) + (SH-SL)* tanh( (y -
       offset)/gain) );
   nodefunction{3} = sigmaoutput;

   sigmaprime = @(y,offset,gain) (1/(2*gain))* sech((y - offset)/gain)
       .^2;
14 nodefunctionprime{1} = sigmaprime;
   sigmainputprime = @(y,offset,gain) 1/gain;
   nodefunctionprime{2} = sigmainputprime;
   sigmaoutputprime = @(y,offset,gain) ( (SH-SL)/(2*gain) )* sech((y -
       offset)/gain).^2;
   nodefunctionprime{3} = sigmaoutputprime;
19 %
   % construct a 1-5-1 MFFN as a cffn
   v = [1;2;3;4;5;6;7];
   e = {[1;2],[1;3],[1;4],[1;5],[1;6],[2;7],[3;7],[4;7],[5;7],[6;7]};
   E = edges(e);
24 V = vertices(v);
   G = graphs(V,E);
   KG = incidence(G);
   [NodeSize,EdgeSize] = size(KG);
   Y = zeros(1,NodeSize);
29 OL = -1.4;
   OH = 1.4;
   GL = 0.8;
   GH = 1.2;
   Offset = OL+(OH-OL)*rand(1,NodeSize);
34 Gain = GL+(GH-GL)*rand(1,NodeSize);
   W = -1+2*rand(1,EdgeSize);
   [B,F,BE] = BFsets0(G,KG);
   IN = {1};
   OUT = {7};
39 I = 0.3;
   [y,Y] = evaluation2(nodefunction,G,I,IN,OUT,W,Offset,Gain,B,BE);
   Y =

       0.060865    0.247482    0.223323    0.522128    0.863697    0.962500
          -0.627949
```

Now to do gradient updates, we need to calculate the error of the chain ffn on the
input and training data. This is done as usual with the energy function.

Listing 18.54: The energy calculation: energy.m

```
   function [E,yI,YI,OI,EI] = energy(nodefunction,g,I,IN,OUT,D,Y,W,O,G,B
       ,BE)
 2 %
   % g is the graph
   % Y is node vector
   % W is the link vector
   % O is the offset vector
 7 % G is the gain vector
   % B is the global backward node set information
   % BE is the global backward edge set information
   %
   % I collection of input data
12 % I is a matrix  S rows, Number of Inputs = cols
   % IN input nodes
   % D collection of target data
   % D is a matrix  S rows, Number of outputs = cols
   % OUT output nodes
17 %
   % YI is Y vector for each input
   % OI is Output vector for each input
   % EI is error vector for each input
   %
22 S = length(I);
   sizeY = length(Y);
   sizeD = length(D);
   sizeOUT = length(OUT);
   %
27 YI = zeros(S,sizeY);
   yI = zeros(S,sizeY);
   OI = zeros(S,sizeOUT);
   EI = zeros(S,sizeOUT);

32 sum = 0;
   for i=1:S
     [yI(i,:),YI(i,:)] = evaluation2(nodefunction,g,I(i,:),IN,OUT,W,O,G,
         B,BE);
     error = 0;
     for j = 1:sizeOUT
37     OI(i,j) = YI(i,OUT{j});
       EI(i,j) = YI(i,OUT{j}) - D(i,j);
       error = error + (EI(i,j))^2;
     end
     sum = sum + error;
42 end
   E = .5*sum;
   end
```

We can test our code with the following session.

Listing 18.55: Testing the Energy Code

```
1 X = linspace(0,2*pi,41);
  Input = X';
  U = cos(X);
  Target = U';
  %setup 1-5-1 CFFN as before
6 %setup nodefunction as before
  ...
  [E,yI,YI,OI,EI] = energy(nodefunction,G,Input,IN,OUT,Target,Y,W,
      Offset,Gain,B,BE);
  % see what the energy is
  E =   35.0738
```

Now we are ready to train. However, to do gradient updates, we need forward edge information. So we need to update the **BFsets0** code to return that information. Here is the changed code which is now **BFsets**.

Listing 18.56: Returning forward edge information: Changed BFsets code

```
function [BackGlobal,ForwardGlobal,BackEdgeGlobal,ForwardEdgeGlobal]
    = BFsets(g,Kg)
%
% g is the graph
% Kg is the incidence matrix of the graph
5 % g.e is the edge object of the graph
%
% get edges
Eg = g.e;
% get edge array
10 E = Eg.e;

[Kgrows,Kgcols] = size(Kg);

BackGlobal = {};
15 ForwardGlobal = {};
BackEdgeGlobal = {};
ForwardEdgeGlobal = {};
for i = 1:Kgrows
  BackGlobal{i} = [];
20  ForwardGlobal{i} = [];
  BackEdgeGlobal{i} = [];
  ForwardEdgeGlobal{i} = [];
  for j = 1:Kgcols
    d = E(:,j);
25    u = d(1);
    v = d(2);
    if Kg(i,j) == 1
      ForwardGlobal{i} = [ForwardGlobal{i},v];
      ForwardEdgeGlobal{i} = [ForwardEdgeGlobal{i},j];
30    else if Kg(i,j) == -1
      BackGlobal{i} = [BackGlobal{i},u];
      BackEdgeGlobal{i} = [BackEdgeGlobal{i},j];
    end
  end
35 end
end
```

Now we are ready to look at the gradient update code. We think it is interesting to be able to see how large the gradient norm is and to find at what component this maximum occurs. So we wrote a utility function to do this. Given a vector, it takes the absolute value of all the entries and finds the location of the maximum one. With that, we can return this location in the variable **LocMaxGrad**. Here is the code for the utility.

Listing 18.57: Find the location of the absolute maximum of a vector

```
  function imax = getargmax(V,tol)
  %
  % V is a vector
  %
5 LengthV = length(V);
  % remove minus signs
  W = abs(V);
  M = max(W);
  imax = 1;
10 for i=1:LengthV
     if W(i) > M - tol
        imax = i;
        break;
     end
15 end
```

Now here is the update code. Note we start the ξ loop at the largest node in the **OUT** set as the nodes after that are irrelevant to the gradient calculations.

Listing 18.58: Gradient Update Code

```
  function [WUpdate,OUpdate,GUpdate,NormGrad,LocMaxGrad] =
     GradientUpdate(nodefunction,nodefunctionprime,g,Y,W,O,G,B,BE,F,FE,
     I,IN,D,OUT,r)
  %
  % nodefunction is the cell of nodefunctions
  % nodefunctionprime is the cell of derivatives
5 % g is the graph
  % Y is the node vector
  % W is the link vector
  % O is the offset vector
  % G is the gain vector
10 % B is the backward set information
  % BE is the backward edge information
  % F is the forward set information
  % FE is the global forward edge information
  %
15 % I is the list of input vectors
  % IN is the list of input nodes
  % D is the list of target vectors
  % OUT is the list of output nodes
  %
```

```
20  % r is the amount of gradient update
    %
    % turn gain updates on or off: DOGAIN = 0
    % turns them off
    DOGAIN = 0;
25
    sizeY = length(Y);
    sizeW = length(W);
    sizeI = length(I);
    sizeD = length(D);
30  sizeIN = length(IN);
    sizeOUT = length(OUT);
    xi = zeros(sizeI, sizeY);
    DEDW = zeros(1, sizeW);
    DEDO = zeros(1, sizeY);

    sigmainputprime = nodefunctionprime{1};
    sigmamainprime = nodefunctionprime{2};
    sigmaoutputprime = nodefunctionprime{3};
40  sigmaprime = {};
    % set all node functions to sigmamain
    for i = 1:sizeY
      sigmaprime{i} = sigmamainprime;
    end
45  % reset input node functions to sigmainput
    for j = 1:sizeIN
        sigmaprime{IN{j}} = sigmainputprime;
    end
    % reset output node functions to sigmaoutput
50  for j = 1:sizeOUT
        sigmaprime{OUT{j}} = sigmaoutputprime;
    end

    %find maximum output node
55  OMax = OUT{1};
    for i=2:sizeOUT
        if OUT{i} > OMax
            OMax = OUT{i};
        end
60  end

    for alpha = 1:sizeI
        for i = OMax :-1:1
        % get forward edge information for neuron i
65      FF  = F{i};
        FEF = FE{i};
        lenFF = length(FEF);
        %find out if i is a target node
        for j = 1:sizeOUT
70          if OUT{j} == i
                k = j;
                IsTargetNode = 1;
              end
        end
75      if IsTargetNode == 1
            xi(alpha,i) = (Y(alpha,i) - D(alpha,k));
        else
            adder = 0.0;
```

```
80          link  =  FEF(j);
            % F(j) is a node which i goes forward to
            post  =  FF(j);
            adder  =  adder  +  xi(alpha,post)*W(link)*sigmaprime{post}(yI(
                alpha,post),O(post),G(post));
          end
85          xi(alpha,i)  =  adder*YI(alpha,i);
        end % xi calculation
      end % loop on nodes
    end % loop on data

90  for  i  =  1  :  sizeY
      % FE{i} is the global node information
      % where i the ith weight
      % FE{i} = a set of (pre, post)
      FF  =  F{i};
95    FEF  =  FE{i};
      lenFF  =  length(FF);
      for  j  =  1  :  lenFF
        % get the weight index
        link  =  FEF(j);
100       % W(link) = E_{pre -> post} where
        pre  =  i;
        post  =  FF(j);
        adder  =  0.0;
        for  k  =  1:sizeI
              G(post))*YI(k,pre);
        end
        DEDW(link)  =  adder;
      end
    end
110
    for  i  =  1  :  sizeY
      adder  =  0.0;
      for  k  =  1:sizeI
        adder  =  adder  +  xi(k,i)*sigmaprime{i}(yI(k,i),O(i),G(i));
115   end
      DEDO(i)  =  -  adder;
      adder  =  0.0;
      for  k  =  1:sizeI
        adder  =  adder  +  xi(k,i)*sigmaprime{i}(yI(k,i),O(i),G(i))*(yI(k,i)
          -O(i));
120   end
      DEDG(i)  =  -  adder/G(i);
    end

    % find norm
125 if  DOGAIN  ==  1
      Grad  =  [DEDW DEDO DEDG];
    else
      Grad  =  [DEDW DEDO];
    end
130 NormGrad  =  sqrt(  sum(Grad.*Grad)  );
    LocMaxGrad  =  getargmax(Grad,.001);
    gradtol  =  0.05;

    if  NormGrad  >=  gradtol
135   UnitGradW  =  DEDW/NormGrad;
      UnitGradO  =  DEDO/NormGrad;
      UnitGradG  =  DEDG/NormGrad;
      WUpdate  =  W- UnitGradW  *  r;
      OUpdate  =  O  -  UnitGradO  *  r;
```

```
140    %GUpdate = G - UnitGradG * r;
       GUpdate = G;
     else
       WUpdate = W - DEDW * r;
       OUpdate = O - DEDO * r;
145    %GUpdate = G - DEDG * r;
       GUpdate = G;
     end

     end
```

We can then use this to write a training loop.

Listing 18.59: Chain FFN Training Code

```
1 function [W,G,O,Energy] = chainffntrain(nodefunction,
      nodefunctionprime,g,Input,IN,Target,OUT,Y,W,O,G,B,BE,F,FE,lambda,
      NumIters)
  %
  %   We train a chain ffn
  %
  % Input is input matrix of S rows of input data
6 %        S x number of input nodes
  % target is desired target value matrix of S rows of target data
  %        S x number of output nodes
  %
  % g is the chain ffn graph
11 % G is the gain parameters
  % O is the offset parameters
  % Y is the node output data
  %
  Energy = [];
16 [E,yI,YI,OI,EI] = energy(nodefunction,g,Input,IN,OUT,Target,Y,W,O,G,B
      ,BE);
  Energy = [Energy E];
  for t = 1:NumIters
     [W,O,G] = GradientUpdate(nodefunction,nodefunctionprime,g,Y,W,O,G,B
        ,BE,F,FE,Input,IN,Target,OUT,lambda);
     [E,yI,YI,OI,EI] = energy(nodefunction,g,Input,IN,OUT,Target,Y,W,O,G
        ,B,BE);
21   Energy = [Energy E];
  end
  plot(Energy);
  end
```

Here is a sample session in its entirety.

Listing 18.60: Sample Training Session

```
1 nodefunction = {};
  nodefunctionprime = {};
  sigma = @(y,offset,gain) 0.5*(1 + tanh( (y - offset)/gain) );
  sigmainput = @(y,offset,gain) (y - offset)/gain;

6 SL = -1.2;
  SH = 1.2;
  % ouput nodes range from SL to SH
  sigmaoutput = @(y,offset,gain) 0.5*( (SH+SL) + (SH-SL)* tanh( (y -
     offset)/gain) );
  nodefunction{1} = sigmainput;
```

```
11 nodefunction{2} = sigma;
   nodefunction{3} = sigmaoutput;

   sigmaprime = @(y,offset ,gain) (1/(2*gain))* sech((y - offset)/gain)
       .^2;
   sigmainputprime = @(y,offset ,gain) 1/gain;
16 sigmaoutputprime = @(y,offset ,gain) ( (SH-SL)/(2*gain) )* sech((y -
       offset)/gain).^2;
   nodefunctionprime{1} = sigmainputprime;
   nodefunctionprime{2} = sigmaprime;
   nodefunctionprime{3} = sigmaoutputprime;
   % setup training data
21 X = linspace(0,3,31);
   Input = X';
   U = cos(X);
   Target = U';
   % construct a 1-5-1 MFFN as a graph
26 v = [1;2;3;4;5;6;7];
   e = {[1;2],[1;3],[1;4],[1;5],[1;6],[2;7],[3;7],[4;7],[5;7],[6;7]};
   E = edges(e);
   V = vertices(v);
   G = graphs(V,E);
31 KG = incidence(G);
   [NodeSize ,EdgeSize] = size(KG);
   Y = zeros(1,NodeSize);
   OL = -1.4;
   OH = 1.4;
36 GL = 0.8;
   GH = 1.2;
   Offset = OL+(OH-OL)*rand(1,NodeSize);
   Gain = GL+(GH-GL)*rand(1,NodeSize);
   W = -1+2*rand(1,EdgeSize);
41 [B,F,BE,FE] = BFsets(G,KG);
   IN = {1};
   OUT = {7};
   [E, yI , YI, OI, EI] = energy(nodefunction ,G, Input ,IN ,OUT, Target ,Y,W,
       Offset , Gain ,B,BE);
   E =
46     18.1698
   % Start Training
   lambda = .0005;
   NumIters = 10;
   [W, Gain , Offset , Energy] = chainffntrain(nodefunction , nodefunctionprime
       ,G, Input ,IN , Target ,OUT,Y,W, Offset , Gain ,B,BE,F , FE, lambda , NumIters);
51 [W, Gain , Offset , Energy] = chainffntrain(nodefunction , nodefunctionprime
       ,G, Input ,IN , Target ,OUT,Y,W, Offset , Gain ,B,BE,F , FE,.01 ,200);
   Energy(200)
   ans =
       0.1263
```

This first **chainffntrain** code contains a built in plot of the energy shown in Fig. 18.5.

Fig. 18.5 The 1-5-1 cos energy training results

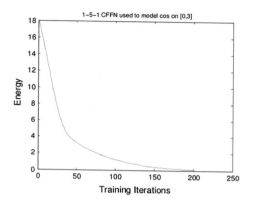

18.4 Polishing the Training Code

Now that we have working code, let's revisit it and see if we can make it a bit more compact. We clearly need some sort of initialization method so we don't have to type so much. Let's handle the initialization of all the node functions in one place. Consider the code below which sets up all the **sigma** functions and then returns them for use. We are experimenting here with allowing every single node function to be unique. We return a **cell** for both **sigma** and **sigmaprime**.

Listing 18.61: Node function Initialization

```
 function [sigma,sigmaprime] = SigmoidInit(NodeSize,IN,OUT)
 %
 % Initialize node transfer functions.
 %
5 nodefunction = {};
 nodefunctionprime = {};
 transferfunction = @(y,offset,gain) 0.5*(1 + tanh( (y - offset)/gain)
    );
 nodefunction{1} = transferfunction;
 transferfunctioninput = @(y,offset,gain) (y - offset)/gain;
10 nodefunction{2} = transferfunctioninput;
 SL = -1.2;
 SH = 1.2;
 % ouput nodes range from SL to SH
 transferfunctionoutput = @(y,offset,gain) 0.5*( (SH+SL) + (SH-SL)*
    tanh( (y - offset)/gain) );
15 nodefunction{3} = transferfunctionoutput;

 transferfunctionprime = @(y,offset,gain) (1/(2*gain))* sech((y -
    offset)/gain).^2;
 nodefunctionprime{1} = transferfunctionprime;
 transferfunctioninputprime = @(y,offset,gain) 1/gain;
```

```
20  nodefunctionprime{2} = transferfunctioninputprime;
    transferfunctionoutputprime = @(y,offset ,gain) ( (SH–SL)/(2*gain) )*
        sech((y - offset)/gain).^2;
    nodefunctionprime{3} = transferfunctionoutputprime;
    %
    sigmamain = nodefunction{1};
25  sigmainput = nodefunction{2};
    sigmaoutput = nodefunction{3};
    sigma = {};

    % get sizes
30  sizeIN = length(IN);
    sizeOUT = length(OUT);

    % set all node functions to sigmamain
    for i = 1:NodeSize
35      sigma{i} = sigmamain;
    end
    % reset input node functions to sigmainput
    for j = 1:sizeIN
        sigma{IN{j}} = sigmainput;
40  end
    % reset output node functions to sigmaoutput
    for j = 1:sizeOUT
        sigma{OUT{j}} = sigmaoutput;
    end
45  %

    sigmamainprime = nodefunctionprime{1};
    sigmainputprime = nodefunctionprime{2};
    sigmaoutputprime = nodefunctionprime{3};
50  sigmaprime = {};
    % set all node functions to sigmamain
    for i = 1:NodeSize
        sigmaprime{i} = sigmamainprime;
    end
55  % reset input node functions to sigmainput
    for j = 1:sizeIN
        sigmaprime{IN{j}} = sigmainputprime;
    end
    % reset output node functions to sigmaoutput
60  for j = 1:sizeOUT
        sigmaprime{OUT{j}} = sigmaoutputprime;
    end

    end
```

We might also want to send in **SL** and **SH** as arguments. This leads to the variant **SigmoidInit2**.

Listing 18.62: Adding upper and lower bounds to the node function initialization

```
    function [sigma,sigmaprime] = SigmoidInit2(NodeSize,SL,SH,IN,OUT)
    %
    % Initialize node transfer functions.
    %
5   ....
    all is the same except SL and SH are arguments so
    no need to initialize them in the code
    ....
    end
```

We then have to alter the code for evaluation, energy calculation, updates and training to reflect the new way we can access the node functions and their derivatives. First, the evaluation. We took out the node function initialization and changed the argument list.

Listing 18.63: The new evaluation function: evaluation3.m

```
 1 function [InVals,OutVals] = evaluation3(sigma,g,I,IN,OUT,Y,W,O,G,B,BE
      )
   %
   % sigma is the nodefunction as a cell
   % g is the graph
   % I is the input vector
 6 % IN is the list of input nodes as a cell
   % OUT is the list of output nodes as a cell
   % Y is node vector
   % W is the link vector
   % O is the offset vector
11 % G is the gain vector
   % B is the global backward node set information
   % BE is the global backward edge set information
   %
   % OutVals is the new node output vector
16 % InVals is the node input vector
   %

   % get sizes
   sizeIN = length(IN);
21 sizeOUT = length(OUT);
   Nodes = g.v;
   [m,sizeY] = size(Nodes.v);
   InVals = zeros(1,sizeY);

26 for i = 1:sizeY
       % get backward node  information for neuron i
       BF = B{i};
       % get backward edge information for neuron i
       BEF = BE{i};
31     lenB = length(BF);
       lenBEF = length(BEF);
       sum = 0.0;
       % add input if this is an input node
       for j = 1:sizeIN
36         if i == IN{j}
               sum = sum + I(j);
           end
       end
       for j = 1:lenBEF
41         link = BEF(j);
           pre = BF(j);
           sum = sum + W(link)*Y(pre);
       end
       InVals(i) = sum;
46     Y(i) = sigma{i}(sum,O(i),G(i));
   end

   OutVals = Y;
   end
```

Then, we use the new evaluation code to calculate the energy.

Listing 18.64: The updated energy code: energy2.m

```
   function  [E,yI,YI,OI,EI]  =  energy2(sigma,g,I,IN,OUT,D,W,O,G,B,BE)
   %
   % g  is  the  graph
   % Y  is  node  vector
 5 % W  is  the  link  vector
   % O  is  the  offset  vector
   % G  is  the  gain  vector
   % B  is  the  global  backward  node  set  information
   % BE  is  the  global  backward  edge  set  information
10 %
   % I  collection  of  input  data
   % I  is  a  matrix  S  rows,  Number  of  Inputs  =  cols
   % IN  input  nodes
   % D  collection  of  target  data
15 % D  is  a  matrix  S  rows,  Number  of  outputs  =  cols
   % OUT  output  nodes
   %
   % YI  is  Y  vector  for  each  input
   % OI  is  Output  vector  for  each  input
20 % EI  is  error  vector  for  each  input
   %
   S  =  length(I);
   Nodes  =  g.v;
   [m,sizeY]  =  size(Nodes);
25 sizeD  =  length(D);
   sizeOUT  =  length(OUT);
   %
   YI  =  zeros(S,sizeY);
   yI  =  zeros(S,sizeY);
30 OI  =  zeros(S,sizeOUT);
   EI  =  zeros(S,sizeOUT);

   sum  =  0;
   for  i=1:S
35    [yI(i,:),YI(i,:)]  =  evaluation3(sigma,g,I(i,:),IN,OUT,W,O,G,B,BE);
      error  =  0;
      for  j  =  1:sizeOUT
         OI(i,j)  =  YI(i,OUT{j});
         EI(i,j)  =  YI(i,OUT{j})  −  D(i,j);
40       error  =  error  +  (EI(i,j))^2;
      end
      sum  =  sum  +  error;
   end
   E  =  .5*sum;
45
   end
```

Next, we change the update code. In addition to removing the node function setup
code, we also streamlined the gradient norm calculation.

Listing 18.65: The new update code: GradientUpdate2.m

```
function [WUpdate,OUpdate,GUpdate,NormGrad,LocMaxGrad] =
    GradientUpdate2(sigma,sigmaprime,g,y,Y,W,O,G,B,BE,F,FE,I,IN,D,OUT,
    r)
%
% g is the graph
% y is the matrix of input vectors for each exemplar
% Y is the matrix of output vectors for each exemplar
% W is the link vector
% O is the offset vector
% G is the gain vector
% FE is the global forward edge set information
% F is the global forward link set
%
% I is the list of input vectors
% IN is the list of input nodes
% D is the list of target vectors
% OUT is the list of output nodes
%
% r is the amount of gradient update
%
% turn gain updates on or off: DOGAIN = 0
% turns them off
DOGAIN = 0;

Nodes = g.v;
[m,sizeY] = size(Nodes);
sizeW = length(W);
S = length(I);
sizeD = length(D);
SN = length(IN);
sizeOUT = length(OUT);
xi = zeros(S,sizeY);
DEDW = zeros(1,sizeW);
DEDO = zeros(1,sizeY);
DEDG = zeros(1,sizeY);

%find maximum output node
OMax = OUT{1};
for i=2:sizeOUT
    if OUT{i} > OMax
        OMax = OUT{i};
    end
end

for alpha = 1:S
    for i = OMax :-1:1
        % get forward edge information for neuron i
        FF = F{i};
        FEF = FE{i};
        lenFF = length(FEF);
        %find out if i is a target node
        for j = 1:sizeOUT
            if OUT{j} == i
                k = j;
                IsTargetNode = 1;
            end
        end
```

```
        if IsTargetNode == 1
          xi(alpha,i) = (Y(alpha,i) - D(alpha,k));
        else
59        adder = 0;
          for j = 1 : lenFF
            link = FEF(j);
            % F(j) is a node which i goes forward to
            post = FF(j);
64          adder = adder + xi(alpha,post)*W(link)*sigmaprime{post}(y(
               alpha,post),O(post),G(post)) ;
          end
          xi(alpha,i) = adder;
        end % test on nodes: output or not
      end % loop on nodes
69 end % loop on data
   %xi

   for alpha = 1:S
     for i = sizeY : -1 : 1
74     % FE{i} is the global node information
       % where i the ith weight
       % FE{i} = a set of (pre, post)
       FF = F{i};
       FEF = FE{i};
79     lenFF = length(FF);
       for j = 1 : lenFF
         % get the weight index
         link = FEF(j);
         % W(link) = E_{pre -> post} where
84       pre = i;
         post = FF(j);
         DEDW(link) = DEDW(link) + xi(alpha,post)*sigmaprime{post}(y(alpha
            ,post),O(post),G(post))*Y(alpha,pre);
       end% forward link loop
       DEDO(i) = DEDO(i) - xi(alpha,i)*sigmaprime{i}(y(alpha,i),O(i),G(i))
          ;
89     %DEDG(i) = DEDG(i) - xi(alpha,i)*sigmaprime{i}(y(alpha,i),O(i),G(i)
          )*(y(alpha,i)-O(i))/G(i);
       end% node loop
     end% exemplar loop

   if DOGAIN == 1
94   Grad = [DEDW DEDO DEDG];
   else
       Grad = [DEDW DEDO];
   end
   NormGrad = sqrt( sum(Grad.*Grad) );
99 %LocMaxGrad = getargmax(Grad,.001);
   LocMaxGrad = 1;
   gradtol = 0.05;
   %
   if NormGrad < gradtol
104    scale = r;
   else
       scale = r/NormGrad;
   end
   %Cscale = scale
109 % do updates
   WUpdate = W- DEDW * scale;
   OUpdate = O - DEDO * scale;
   %GUpdate = G - DEDG * scale;
   GUpdate = G;
114
   end
```

Finally, we need to alter the training loop code also.

Listing 18.66: The altered training loop code: chainffntrain2.m

```
function [W,G,O,Energy,Norm,Loc] = chainffntrain2(sigma,sigmaprime,g,
    Input,IN,Target,OUT,W,O,G,B,BE,F,FE,lambda,NumIters)
%
%   We train a chain ffn
%
5 % Input is input matrix of S rows of input data
%       S x number of input nodes
%   target is desired target value matrix of S rows of target data
%       S x number of output nodes
%
10 % g is the chain ffn graph
%   G is the gain parameters
%   O is the offset parameters
%   Y is the node output data
%
15 Energy = [];
    [E,y,Y,OI,EI] = energy2(sigma,g,Input,IN,OUT,Target,W,O,G,B,BE);
    %E
    Energy = [Energy E];
    for t = 1:NumIters
20     [W,O,G,Norm,Loc] = GradientUpdate2(sigma,sigmaprime,g,y,Y,W,O,G,B,
        BE,F,FE,Input,IN,Target,OUT,lambda);
        [E,y,Y,OI,EI] = energy2(sigma,g,Input,IN,OUT,Target,W,O,G,B,BE);
        Energy = [Energy E];
    end
    plot(Energy);
25 %E
    end
```

Let's look at a sample training session.

Listing 18.67: Setup the training session

```
% setup training data
X = linspace(0,3.14,45);
Input = X';
4 U = cos(X);
Target = U';
% setup 1-5-1 MFFN as a chain FFN
v = [1;2;3;4;5;6;7];
e = {[1;2],[1;3],[1;4],[1;5],[1;6],[2;7],[3;7],[4;7],[5;7],[6;7]};
9 E = edges(e);
V = vertices(v);
G = graphs(V,E);
KG = incidence(G);
[NodeSize,EdgeSize] = size(KG);
14 % initialize tunable parameters
Y = zeros(1,NodeSize);
OL = -1.4;
OH = 1.4;
GL = 0.8;
19 GH = 1.2;
WL = -2.1;
WH = 2.1;
Offset = OL+(OH-OL)*rand(1,NodeSize);
Gain = GL+(GH-GL)*rand(1,NodeSize);
24 W = WL+(WH-WL)*rand(1,EdgeSize);
% get link information
[B,F,BE,FE] = BFsets(G,KG);
% set input and output index sets
IN = {1};
29 OUT = {7};
% initialiaze node functions and derivatives
[sigma,sigmaprime] = SigmoidInit2(NodeSize,-1.2,1.2,IN,OUT);
```

Listing 18.68: Do the training

```
% Get Initial Energy
[E,yI,YI,OI,EI] = energy2(sigma,G,Input,IN,OUT,Target,Y,W,Offset,Gain
   ,B,BE);
E = 18.2366
% Start Training
[W,Gain,Offset,Energy,Norm,Loc] = chainffntrain2(sigma,sigmaprime,G,
   Input,IN,Target,OUT,W,Offset,Gain,B,BE,F,FE,0.0005,100);
E 16.0724
...
many steps
ans = 0.1
```

We can see the results of our approximation in Fig. 18.6 by generating a test file.

Listing 18.69: A Test File

```
X = linspace(0,3.14,101);
Input = X'; U = cos(X);
Target = U'; Z = [];
for i=1:101
   [y,Y] = evaluation3(sigma,G,Input(i),IN,OUT,W,Offset,Gain,B,BE);
   Z = [Z Y(7)];
end
plot(X,U,X,Z);
```

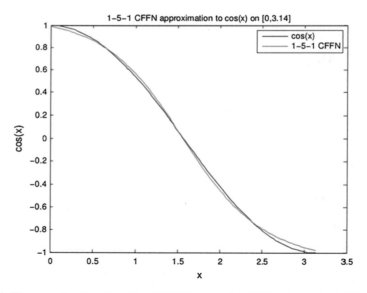

Fig. 18.6 The approximation of cos(t) on [0,3.14] using a 1-5-1 FFN graph: error is 0.1

18.5 Comparing the CFFN and MFFN Code

Let's take a moment to check to see if our two ways of building approximations using network architectures give the same results. To do this, we will setup a particular 1-5-1 network as both a CFFN and a MFFN and calculate the starting energy and one gradient step to see if they match. This is a bit harder than it sounds as the data structures are a bit different. First, here is the setup. This code is careful to set up both versions of the 1-5-1 problem the same. The parameters are chosen randomly as before and we map the MFFN choices to their CFFN counterparts so that both techniques start the same. We use simple node computation initialization functions too. We added one for the MFFN which we show here.

Listing 18.70: Initializing the MFFN node functions

```
   function [nodefunction , nodefunctionprime] = SigmaMFFNInit(SL,SH)
 2 %
   % Initialize  sigmoids  needed  for  MFFN
   %
   nodefunction = {};
   nodefunctionprime = {};
 7 sigma = @(y, offset , gain)  0.5*(1 + tanh( (y - offset )/gain) );
   nodefunction{1} = sigma;
   sigmainput = @(y, offset , gain)  (y - offset )/gain;
   nodefunction{2} = sigmainput;
   % ouput nodes range from SL to SH
12 sigmaoutput = @(y, offset , gain)  0.5*( (SH+SL) + (SH−SL)* tanh( (y −
       offset )/gain) );
   nodefunction{3} = sigmaoutput;

   sigmaprime = @(y, offset , gain)  (1/2)* sech((y − offset )/gain).^2;
   nodefunctionprime{1} = sigmaprime;
17 sigmainputprime = @(y, offset , gain)  1;
   nodefunctionprime{2} = sigmainputprime;
   sigmaoutputprime = @(y, offset , gain)  ((SH−SL)/2)* sech((y − offset )/
       gain).^2;
   nodefunctionprime{3} = sigmaoutputprime;

22 end
```

The function **SigmoidInit2** does a similar job for the CFFN node function initializations. Note the code to set the parameters in the CFFN to match the MFFN ones is a bit cumbersome as it has to been manually since the data structures don't match up nicely.

Listing 18.71: Stetting up the CFFN versus MFFN test

```
     function  testcffneqmffn ()
     %
     X = linspace (0 ,3 ,21) ;
     Input = X' ;
   5 U = sin (X) .^2;
     Target = U' ;
     %setup  the  1-5-1  MFFN
     LayerSizes = [1;5;1] ;
     OL = -1.4;
  10 OH =  1.4;
     GL =  0.8;
     GH =  1.2;
     TL = -2.1;
     TH =  2.1;
  15 [G,O,T] = mffninit (GL,GH,OL,OH,TL,TH, LayerSizes ) ;
     % initialize  the  nodefunctions  for  the  MFFN  which
     % are  off  by  a  gain  factor  from  the  ones  used  in  the  CFFN
     [ nodefunction , nodefunctionprime ] = SigmaMFFNInit ( -0.5 ,1.5 ) ;
     %now  initialize  the  nodefunctions  for  the  cffn
  20 IN={1};
     OUT={7};
     [ sigma , sigmaprime ] = SigmoidInit2 (7 , -0.5 ,1.5 ,IN ,OUT) ;
     %setup  1-5-1  CFFN
     v = [1;2;3;4;5;6;7] ;
  25 %this  sets  up  T^1_1 = E1,  T^1_2 = E2,  T^1_3 = E3
     %                           T^1_4 = E4,  T^1_5 = E5,
     %                           T^2_1 = E6,  T^2_2 = E7,  T^2_3 = E8
     %                           T^2_4 = E9,  T^2_5 = E10
     e = { [1;2] ,[1;3] ,[1;4] ,[1;5] ,[1;6] ,[2;7] ,[3;7] ,[4;7] ,[5;7] ,[6;7] } ;
  30 E = edges (e) ;
     V = vertices (v) ;
     GG = graphs (V,E) ;
     KGG = incidence (GG) ;
     [ NodeSize , EdgeSize ] = size (KGG) ;
  35 % setup  parameters
     GC = zeros (1 ,NodeSize) ;
     OC = zeros (1 ,NodeSize) ;
     WC = zeros (1 ,EdgeSize) ;
     for  i =1:5
  40    WC( i )     = T{1}( i ) ;
        WC( i +5) = T{2}( i ) ;
     end
     GC(1) = G{1}(1) ;
     OC(1) = O{1}(1) ;
  45 for  i = 1:5
        GC( i +1) = G{2}(1 ,i ) ;
        OC( i +1) = O{2}(1 ,i ) ;
     end
     GC(7) = G{3}(1) ;
  50 OC(7) = O{3}(1) ;
     %get  link  information
     [B,F,BE,FE] = BFsets (GG,KGG) ;
     %get  mffn  energy
     [y ,Y,RE,EMFFN] = mffneval2 (Input , Target , nodefunction ,G,O,T, LayerSizes
        ) ;
  55 EMFFN
     %get  cffn  energy
     [ECFFN, yCFFNI , YCFFNI , OCFFNI , ECFFNI ] = energy2 (sigma ,GG, Input , IN ,OUT,
        Target ,WC,OC,GC,B,BE) ;
     ECFFN

  60 end
```

We finish this initialization code by doing the energy evaluations for both versions.
We show this result below. Note they are the same value so it seems the two energy
evaluations return the same number.

Listing 18.72: The initial energy for the CFFN and MFFN code

```
EMFFN =
    2.9173
ECFFN =
    2.9173
```

Next, we test one gradient update. To do this, we created new update and training functions for both the CFFN and the MFFN which simply add diagnostic material. These are the functions **chainffntraindiagnostic, Gradient Updatediagnostic, mffntraindiagnostic** and **mffnupdatediagnostic**. If you look at them, you can see the changes we made are pretty straightforward; we leave that exploration to you! The testing and the results are shown next.

Listing 18.73: One gradient step for both CFFN and MFFN

```
%train mffn 1 step
[G,O,T,EMFFN] = mffntraindiagnostic(Input,Target,nodefunction,
    nodefunctionprime,G,O,T,LayerSizes,y,Y,0.0005,1);
%Train cffn 1 step
[WC,GC,OC,ECFFN,Norm,Loc] = chainffntraindiagnostic(sigma,sigmaprime,
    GG,Input,IN,Target,OUT,WC,OC,GC,B,BE,F,FE,0.0005,1);
GradMFFN =
    -1.3319      0.6797     -0.2617      0.1119     -0.4563      2.7344
     4.2227      0.5844      0.0905      1.5443      0.6065
     2.0345     -1.2520      0.2508     -0.3006      0.6916     -4.8874
NormGradMFFN =
    7.8205
GradCFFN =
    -1.3319      0.6795     -0.2618      0.1118     -0.4566      2.7349
     4.2220      0.5844      0.0905      1.5442      0.6078
     2.0354     -1.2515      0.2509     -0.3005      0.6919     -4.8868
NormGradCFFN =
    7.8202
```

18.6 Handling Feedback

The evaluation algorithm as shown in Table 17.1 must be interpreted carefully in the case of feedback connections. If the backward set $\mathcal{B}(i)$ contains feedback, then how should one do the evaluation? An example will make this clear. Take a standard **Folded Feedback Pathway** cortical circuit model from Raizada and Grossberg (2003) as shown in Fig. 18.7a. This is the usual biological circuit model and the nodes are labeled backwards for our purposes—as we have mentioned before. We relabel them in with the nodes starting at $N1$ in the redone figure on the left. There are 11 nodes here and 11 edges with $E_{11 \to 10}$, $E_{10 \to 9}$ and $E_{10 \to 9}$ being explicit feedback.

Let's assume external input comes into N_8 from the thalamus and into N_{11} from the cortical column above it. The input for the nodal calculation for N_7 in the graph in Fig. 18.7b is then

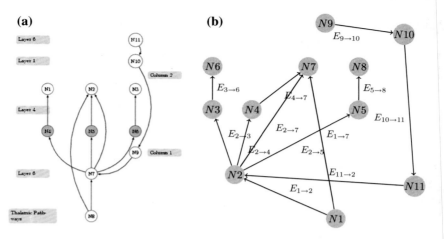

Fig. 18.7 A Folded Feedback Pathway cortical circuit in two forms: the relabeled graph on the right has external Input into Node 1 from the thalamus and into Node 9 from the cortical column above. **a** The Folded Feedback Pathway DG. **b** The Relabeled Folded Feedback Pathway DG

$$y_7^{new} = E_{8\to 7}Y_8^{old} + E_{9\to 7}Y_9^{old}$$

which in terms of a global clock would be implemented as

$$y_7(t+1) = E_{8\to 7}Y_8(t) + E_{9\to 7}Y_9(t)$$

where we can initialize the nodal outputs a variety of ways: for example, by calculating all the outputs without feedback initially. and hence, setting $Y_9(0) = Y_{10}(0) = 0$.

When we relabel this graph as shown in Fig. 18.7b, it is now clear this a feed forward chain of computational nodes and the only feedback is the relabeled edge $E_{11\to 2}$. The external input comes into N_1 from the thalamus and into N_9 from the cortical column above it. The input for the nodal calculation for N_2 is in terms of time ticks on our clock

$$y_2^{new} = E_{1\to 2}Y_1^{old} + E_{11\to 2}Y_{11}^{old}$$

or

$$y_2(t+1) = E_{1\to 2}Y_1(t) + E_{11\to 2}Y_{11}(t)$$

However we choose to implement the graph, there will be feedback terms and so we must interpret the evaluation and update equations appropriately. Then, once a graph structure is chosen, we apply the usual update equations in the form given by Table 18.1a and the Hebbian updates using Table 18.1b. In this version of a Hebbian update algorithm, we differ from the one we discussed earlier in this chapter. In the

Table 18.1 Feedback evaluation and Hebbian update algorithms

(a) Feedback Evaluation	(b) Feedback Hebbian Update
$\text{for}(i = 0;\ i < N;\ i++)\ \{$ $\quad \text{if }(i \in \mathcal{U})$ $\qquad y^i(t+1) = x^i(t) + \sum_{j \in \boldsymbol{\mathcal{B}}(i)}\ \boldsymbol{E}_{j \to i}(t) f_j(t)$ $\quad \text{else}$ $\qquad y^i(t+1) = \sum_{j \in \boldsymbol{\mathcal{B}}(i)}\ \boldsymbol{E}_{j \to i}(t) f_j(t)$ $\qquad f_i(t+1) = \sigma^i(y^i(t+1), p)$ $\}$	$\text{for}(i = 0;\ i < N;\ i++)\ \{$ $\quad \text{for }(j \in \boldsymbol{\mathcal{B}}(i))\ \{$ $\qquad y_p(t) = f_i(t)\,\boldsymbol{E}_{j \to i}(t)$ $\qquad \text{if }(y_p(t) > \epsilon)$ $\qquad\quad \boldsymbol{E}_{j \to i}(t) = \zeta\,\boldsymbol{E}_{j \to i}(t)$ $\qquad \}$ $\}$

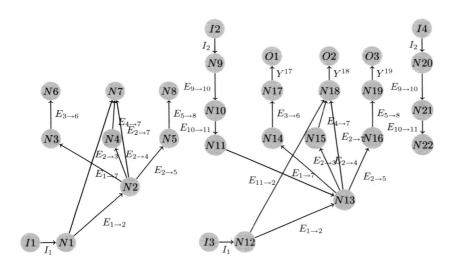

Fig. 18.8 The Folded Feedback Pathway as a lagged architecture

first version, we checked to see if the value of the *post* neuron is *high* and the value of the edge $E_{pre \to post}$ is also high, the value of this edge is increased by a multiplier ξ. To check this, we just looked at the value of $Y^{post} \times E_{pre \to post}$ and if it was high enough we increased the weight $E_{pre \to post}$. Here, the idea is similar but this time we look at the value of $Y^{pre} \times E_{pre \to post}$ and if this values is high enough, we increase the edge. There are other versions too. The idea is to somehow increase the edge weight when the pre signal and the post signal are both high.

Another approach is to handle feedback terms using a **lag** as shown in Fig. 18.8. We make two copies of the graph which are here drawn using nodes N_1 through N_{11} on the left and these correspond to what is happening at time t. On the right is the copy of the graph with the nodes labeled N_{12} to N_{22} which corresponds to time $t + 1$. The feedback connection is then the link between N_{11} and N_{13} with edge weight $E_{11 \to 2}$; hence the feedback has now become a feedforward link in this new configuration. The nodes N_{12} and N_{22} retain the usual edge weights from time t as you can see in the figure. The number of tunable parameters is still the same as in the original

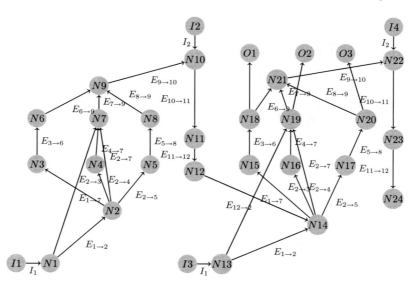

Fig. 18.9 Adding column to column feedback to the original Folded Feedback Pathway as a lagged architecture

network with feedback but using this doubling procedure, we have converted the network into a feedforward one. Once the graph is initialized, we apply evaluation and update algorithms modified a bit to take into account this new doubled structure. Note here there are inputs into Nodes N_1 and N_{12} of value I_1 and inputs into Nodes N_9 and N_{20} of value I_2. The outputs here are N_{17}, N_{18} and N_{19} rather than N_6, N_7 and N_8. For a given lagged problem, we need to decide how to handle this doubling of the inputs and outputs.; we simply show the possibilities here. Now Fig. 18.8 shows a portion of a cortical circuit where the OCOS subcircuit appears disconnected from the FFP portion. Let's make it a bit more realistic by adding a node N_9 in the column above this column circuit which is connected to N_6, N_7 and N_8. The output from N_9 then connects to the FFP originally given by nodes N_9, N_{10} and N_{11}. We will shift the FFP nodes up by one making them N_{10}, N_{11} and N_{12}. We draw this in Fig. 18.9. In this new architecture, the calculations through the copy that corresponds to time t are generating the N_{12} output and the second copy of the network is necessary to take into account the feedback from N_{12} to N_2.

Let's look at some MatLab implementations of these ideas. First, we need to redo how we parse the incidence matrix of a graph so that we find self and non self feedback terms. Consider the new incidence code below. There are now two new possibilities we focus on. For a given node **ii**, we set up the entries in its columns. We loop through the edges using the index **jj**. Recall each edge consists of a **in** and **out** pair. The term **e2(1,jj)** gives the **in** value of the edge. If in value matches the given node **ii**, we check to see if the in value **ii** is less than the out value. If this is true, this means this edge goes out from the node **ii** and so we set the incidence matrix value **IncidMat(ii,jj)** **=** **1**. However, if the reverse inequality holds, we know this means the in value is bigger than or equal to the out value. This means we have feedback, and so we set **IncidMat(ii,jj)** **=** **2**. The tests for incoming edges is similar: we are in the case where the node **ii** is now the out value. if the in value is smaller than **ii**, we set **IncidMat(ii,jj)** **=** **-1** as this is an incoming edge. If the inequality goes the other way, this is an incoming feedback edge.

Listing 18.74: The incidence matrix calculation for a CFFN with feedback

```
   function IncidMat = incidenceFB(G)
   %
   % g is a graph having
   % vertices g.v
 5 % edges g.e
   %
   % Get the edge object from g
   E = G.e;
   % get the edges from the edge object
10 e2 = E.e;
   % get the vertices object from g
   V = G.v;
   % get the vertices from the vertices object
   v2 = V.v;
15 % find out how many vertices and edges there are
   [m, sizeV] = size(v2);
   [m, sizeE] = size(e2);
   IncidMat = zeros(sizeV, sizeE);
   %
20 % setup incidence matrix
   %
   for ii = 1:sizeV
       for jj = 1:sizeE
           if e2(1,jj) == ii
25             if e2(1,jj) < e2(2,jj)
                   IncidMat(ii,jj) = 1;
               else
                   IncidMat(ii,jj) = 2;
               end
30         else if e2(2,jj) == ii
               if e2(1,jj) < e2(2,jj)
                   IncidMat(ii,jj) = -1;
               else
                   IncidMat(ii,jj) = -2;
35             end
           end
       end
   end
40 end
```

Once we have the new incidence matrix, we need to convert this into a dot file for printing. If we have self feedback, the sum of the entries in a given column will be positive and so we set the to and from nodes of that edge to be the same. If the we have a $+1$ and a -1 in the column for a given edge, this corresponds to our usual forward edge. If we have a $+2$ and a -2 in the same column, this is a feedback edge.

Listing 18.75: Converting the incidence matrix for a CFFN with feedback into a dot file

```
    function incToDotFB(inc,width,height,aspect,filename)
    [N,E] = size(inc);
    [fid,msg] = fopen(filename, 'w');
    fprintf(fid, 'digraph G {\n');
5   fprintf(fid, 'size=\"%f,%f\";\n',width,height);
    fprintf(fid, 'rankdir=TB;\n');
    fprintf(fid, 'ratio=%f;\n',aspect);
    for ii = 1:N
        fprintf(fid, '%d [shape=\"rectangle\" fontsize=20 fontname=\"
            Times new Roman\" label=\"%d\"];\n', ii, ii);
10  end

    fprintf(fid,'\n');

    for jj = 1:E
15      a = 0;
        b = 0;
        for ii = 1:N
            switch inc(ii,jj)
                case 1
20                  a = ii;
                case -1
                    b = ii;
                case 2
                    a = ii;
25                  if abs(sum(inc(:,jj))) > 0
                        b = ii;
                    end
                case -2
                    b = ii;
30          end
        end
        if (a ~= 0) && (b ~= 0)
            fprintf(fid, '%d->%d;\n', a, b);
        end
35  end
    fprintf(fid, '\n}');
    fclose(fid);
    end
```

Let's see how this works on a modified OCOS that has some feedback and self feedback.

Listing 18.76: Testing the CFFN feedback incidence matrix

```
    V= [1;2;3;4;5;6;7;8];
2   E = {[1;2],[2;3],[2;4],[2;5],[3;6],[4;7],[5;8],[2;7],[1;7],...
    [8;4],[6;6]};
    v= vertices(V);
    e = edges(E);
    G = graphs(v,e);
7   KG = incidenceFB(G);
    KOCOS =
        1    0    0    0    0    0    0    0    1    0    0
       -1    1    1    1    0    0    0    1    0    0    0
        0   -1    0    0    1    0    0    0    0    0    0
12      0    0   -1    0    0    1    0    0    0   -2    0
        0    0    0   -1    0    0    1    0    0    0    0
        0    0    0    0   -1    0    0    0    0    0    2
        0    0    0    0    0   -1    0   -1   -1    0    0
        0    0    0    0    0    0   -1    0    0    2    0
17  incToDotFB(KG,6,6,1,'OCOSwithFB.dot');
```

Fig. 18.10 A modified
OCOS with self feedback
and feedback

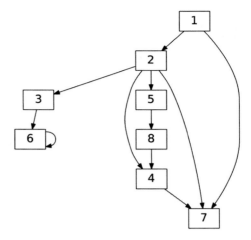

This generates the graphic we see in Fig. 18.10. As an exercise, let's setup the
OCOS/FFP with column to column interaction we have shown as a lagged FFN
in code. We start by constructing the original graph of this circuit with feedback.

Listing 18.77: Constructing the OCOS/FFP with column to column feedback

```
V = [1;2;3;4;5;6;7;8;9;10;11;12];
E =   {[1;2],[2;3],[2;4],[2;5],[3;6],[4;7],[5;8],[2;7],[1;7],...
3           [6;9],[7;9],[8;9],[9;10],[10;11],[11;12],[12;2]};
v = vertices(V);
e = edges(E);
G = graphs(v,e);
KG = incidenceFB(G);
8 incToDotFG(KG,10,10,1,'OCOSFFPandFB.dot');
```

We generate the plot as usual which is shown in Fig. 18.11. Note the incidence
matrix tells us where the feedback edges are: $KG(12, 16) = 2$ and $KG(2, 16) = -2$.
We need to extract the feedback edges and build the new graph which is two copies of
the old one. There is only one feedback edge here, so we find **FBedge(1) = 16**.

Listing 18.78: Finding the feedback edges

```
%get the edges of the graph
eG = G.e;
[rows,cols] = size(KG);
%find the feedback edges
5 FBedge = [];
for i = 1:rows
    for j = 1:cols
        if KG(i,j) == 2
            %edge j is feedback
10          FBedge = [FBedge, j];
        end
    end
endFBedge(1)
    ans =
15      16
```

Fig. 18.11 The plot of the
OCOS/FFP with column
feedback graph

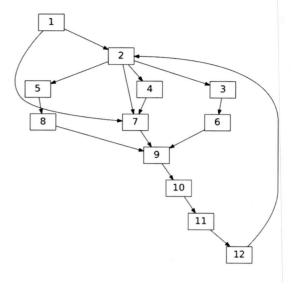

We now know the feedback edge is 16 and so we can remove it. We had to
write remove edge functions to do this. We show them next. There is the function **subtract.m** in the **@edges** directory that removes an edge and then we
use this function in the **@graphs** directory to remove an edge from the graph with
subtractedge function. We show these functions next.

Listing 18.79: The edges object subtract function

```
    function W = subtract(E,p)
    %
    % edge  p  is  removed  from  edge  list
    %
  5 n = length(E.e);
    out = {};
    for i=1:n
      if(i ~=p)
        temp = {[E.e(i).in;E.e(i).out]};
 10     out = [out,temp];
      end
    end
    W = edges(out);
    end
```

Listing 18.80: The graphs object subtract function

```
 1 function W = subtractedge(g,p)
   %
   % remove  an  edge  from  an  existing  graph
   %
   E = subtract(g.e,p);
 6 W = graphs(g.v,E);
   end
```

We remove edge 16 and then copy the resulting graphs. We have not automated this yet and we know the new edge goes from 12 to 14 so we add this new edge to the double copy.

Listing 18.81: Subtract the feedback edge and construct the double copy

```
   G = subtractedge(G,FBedge(1));
   G.e
 3 ans =
            1   2   2   2   3   4   5   2   1   6   7   8    9   10  11
            2   3   4   5   6   7   8   7   7   9   9   9   10   11  12
   W = addgraph(G,G,{[12;14]});
   W.v
 8          1   2   3   4   5   6   7   8   9  10  11  12
           13  14  15  16  17  18  19  20  21  22  23  24
   W.e
            1   2   2   2   3   4   5   2   1   6   7   8   9
            2   3   4   5   6   7   8   7   7   9   9   9  10
13
           10  11  13  14  14  14  15  16  17  14  13
           11  12  14  15  16  17  18  19  20  19  19

           18  19  20  21  22  23  12
18         21  21  21  22  23  24  14
   KW = incidenceFB(W);
   incToDot(KW,10,10,1,'DoubledOCOSFFP.dot');
```

The new dot file generates Fig. 18.12. We see it is fairly easy to construct the doubled graph although it is clear a lot of work is required to automate this process. Also,

Fig. 18.12 The plot of the doubled OCOS/FFP with column feedback turned into a feedforward edge

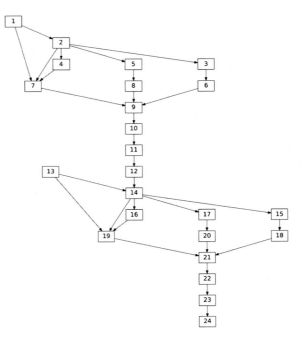

we will apply the standard evaluation, update and training algorithms for the CFFN
to this network and they need to be modified. First, only edges 1–15 and the old
feedback edge which is now edge 31 correspond to tunable weight values. Also the
offsets and gains of only nodes 1 through 12 are modifiable. Since we usually don't
tune the gains, this network has 16 tunable edges and 12 tunable offsets for a total of
28 parameters. We will explore these architectures in the next volume which will be
devoted to building realistic models of cognitive dysfunction. Just for fun, let's list
what we would need to do:

- Suppose the number of original nodes in the graph G is N and the edges for G
  consist of M feedforward edges and Q feedback edges.
- We double G to the new graph W which has $2N$ nodes and $2P$ edges. We then add
  the Q feedback edges as new feedforward edges from the original graph G to its
  copy as you have seen us do in the example.
- We have to make a decision about the input and output nodes. If **IN** and **OUT**
  are those sets, a reasonable thing to do is add the input nodes in the original part of the doubled graph to the copy; i.e. **NewIN = {IN, IN+N}** and
  to drop the output nodes in the original part and use those nodes in the copy;
  i.e. **NewOUT = OUT + N** which adds N to each index in **IN** and **OUT** as
  appropriate.
- The nodes $N + 1$ to $2N$ use the same node function as the original nodes so we
  have to set them up that way.
- The only tunable parameters are the values for the edges 1 to P from the original
  or the values from $P + 1$ to $2P$ in the copy and the new feedforward edges we
  created $2P + 1$ to $2P + Q$ and the offsets and gains for the original nodes 1 to N
  or the offsets and gains in the copy. It makes more sense to do the updates in the
  copy as we know there is a big problem with attenuation in the size of the partial
  derivatives the farther we are removed from the raw target error. Hence, we will
  update the link weights for $P + 1$ to $2P$ and the offsets and gains in the copy as
  well. These updated values will then be used to set the parameter values in the
  original graph part of the lagged graph.
- we adjust the update code so that we only calculate $\frac{\partial E}{\partial E(j)}$ for the appropriate j and
  the $\frac{\partial E}{\partial O(k)}$ and $\frac{\partial E}{\partial G(k)}$ for the right k.
- We can then train as usual.

18.7   Lagged Training

We have added a new function **removeedges** to both **@edges** and **@graphs**.
These are nice functions that make it easier for us to setup the lagged graphs.

Listing 18.82: Removing a list of edges from an edges object

```
function W = removeedges(E, List)
%
% the edges in List is removed from edges object
%
N = length(List);
%sort List
SL = sort(List);
%remove last edge
W = subtract(E, SL(N));
%remove remaining edges
for p = N-1:-1:1
    W = subtract(W, SL(p));
end
end
```

Then we use this function to remove edges from a graphs object.

Listing 18.83: Removing a list of edges from a graph

```
function W = removeedges(g, List)
%
% remove list of edges from an existing graph
%
E = removeedges(g.e, List);
W = graphs(g.v, E);
end
```

Once we have removed edges, we want to find and keep the list of edges removed and the edge information itself. We do this with the functions **ExtractFBEdges** and **ExtractNewFFEdgeInfo**. The function **ExtractFBEdges** stores the link numbers associated with the feedback edges we find in the original graph.

Listing 18.84: Extracting the feedback edge indices from the incidence matrix

```
function FBedge = ExtractFBEdges(KG)
%
% KG is incoming incidence matrix
%
%
[rows, cols] = size(KG);
FBedge = [];
for i = 1:rows
    for j = 1:cols
        if KG(i,j) == 2
            %edge j is feedback
            FBedge = [FBedge, j];
        end
    end
end

end
```

Each of the feedback edges has an integer pair of the form $[a, b]$ where $a > b$ because it is a feedback edge. The function below sets up the new feedforward edges we need in the lagged graph to have the form $[a, b + N]$ where N is the number of nodes in the original graph.

Listing 18.85: Extracting the new feedforward edges constructed from the old feed-back ones

```
function  NewFFedges = ExtractNewFFEdgeInfo(G, List)
%
3 % G  is  original  graph  with  feedback
% List  is  list  of  FB  edges
%
NewFFedges = {};
M = length(List);
8 N = length(G.v);
ge = G.e;
for  i = 1:M
     NewFFedges{i} = [ ge(1,List(i)); ge(2,List(i))+N];
end
13
end
```

We have a new gradient update algorithm **GradientUpdateLag** and a new train-ing algorithm **chainffntrainlag.m**. We are going to implement gradient scal-ing here which takes the form

$$
\nabla^{scaled} E = \begin{bmatrix} \lambda_1 & 0 & 0 & \dots & 0 \\ 0 & \lambda_2 & 0 & \dots & 0 \\ \vdots & \vdots & \ddots & \vdots & \vdots \\ 0 & \dots & 0 & \dots & \lambda_N \end{bmatrix} \nabla E
$$

where we set the multiplier $\lambda_i = 1$ is the absolute value of the ith component of the gradient is too small (say less than 0.01) and otherwise $\lambda_i = 1/\lambda_{max}$ where λ_{max} is the largest absolute value of the components of the gradient. This works reasonably well as long as the gradient has components which are reasonably uniform in scale. If the gradient has only a few components that are big, this will still result in a scaled gradient that only updates a few components. We can make some general comments about the code which are in the source, but here we have taken them out and put them in the main body. First, this is a lagged graph so there are 2 copies of the original graph giving us a lagged graph with nodes 1 to N from the original and nodes $N + 1$ to $2N$ from the copy. The node functions of the first N nodes are repeated for the second set of nodes: so input nodes are in **IN** and **IN+N**. The output nodes are in **OUT** and **OUT+N** and we use those output nodes to set the node functions for the lagged graph. For training purposes, we do use the double input, **IN, IN+N** but the outputs are just the nodes in the copy, **OUT + N**.

If the number of edges in the original graph was M we know $M = P + Q$, where P were feedforward edges and Q were feedback edges which are redone as feedforward edges in the lagged graph. The parameters **w** from $P + 1$ to $2P$, and **w** from $2P + 1$ to $2P + Q$ are updateable as well as the gains and offsets from $N + 1$ to $2N$. So the original code for the gradient updates must be altered to accommodate the fact that half of the parameters of the lagged network are not to be updated. The ξ calculations are the same as we want to include all the contributions from all nodes and links.

Note as usual we start the ξ loops at **OMax**, the largest node in the **OUT** set as the nodes after that do not contribute to the energy. Next, we have the structure of **w** is

$W = $ [link weights for the old graph without feedback,

link weights for the copy of the graph without feedback,

link weights for all feedback edges]

which is the same as

$$W = [W_1, \ldots, W_P, W_1, \ldots, W_P, W_{2P+1}, \ldots, W_{2P+Q}]$$

In this code **NGE** is P and **NFB** is Q. We generate all partials and then set some of them to zero like this. We use **DEDW** to be the vector of partials with respect to link weights, **DEDO** to be partials for the offsets and **DEDG**, the partials for the gains.

$$DEDW = [\text{set to 0 partials for 1 to } NGE,$$

$$\text{keep partials for } NGE + 1 \text{ to } 2* NGE,$$

$$\text{keep partials for } 2* NGE + 1 \text{ to } 2* NGE + NFB]$$

$$DEDO = [\text{set to 0 partials for 1 to } NGN,$$

$$\text{keep partials for } NGN + 1 \text{ to } 2* NGN]$$

$$DEDG = [\text{set to 0 partials for 1 to } NGN,$$

$$\text{keep partials for } NGN + 1 \text{ to } 2* NGN]$$

Of course, it is more complicated if we use gradient scaling, but the idea is essentially the same. The last part of the code implements gradient descent using either the raw gradient information or the scaled gradients.

Listing 18.86: New Lagged Gradient Update Code

```
function [WUpdate,OUpdate,GUpdate,Grad,ScaledGrad,NormGrad] = ...
GradientUpdateLag(sigma,sigmaprime,g,y,Y,W,O,G,B,BE,F,FE,I,IN,D,OUT,r
      ,NGN,NGE,NFB)
%
% g is the lagged graph
% y is the matrix of input vectors for each exemplar
% Y is the matrix of output vectors for each exemplar
% W is the link vector
% O is the offset vector
% G is the gain vector
% FE is the global forward edge set information
% F is the global forward link set
%
% I is the list of input vectors
% IN is the list of input nodes
% D is the list of target vectors
% OUT is the list of output nodes
%
% r is the amount of gradient update
% NGN is the size of the original graph nodes
% NGE is the size of the original graph FF edges
% NFB is the number of feedback edges
%
% turn gain updates on or off: DOGAIN = 0
% turns them off
```

```
25 DOGAIN = 0;

   Nodes = g.v;
   [m, sizeY] = size(Nodes);
   sizeW = length(W);
30 S = length(I);
   sizeD = length(D);
   SN = length(IN);
   sizeOUT = length(OUT);
   xi = zeros(S, sizeY);
35 DEDW = zeros(1, sizeW);
   DEDO = zeros(1, sizeY);
   DEDG = zeros(1, sizeY);

   %find maximum output node
40 OMax = OUT{1};
   for i=2:sizeOUT
       if OUT{i} > OMax
           OMax = OUT{i};
       end
45 end
   %We start the xi loops at OMax, the largest node in the OUT set
   %as the nodes after that do not contribute to the energy.
   for alpha = 1:S
       for i = OMax :-1:1
50         % get forward edge information for neuron i
           FF  = F{i};
           FEF = FE{i};
           lenFF = length(FEF);
           %find out if i is a target node
55         for j = 1:sizeOUT
               if OUT{j} == i
                   k = j;
                   IsTargetNode = 1;
               end
60         end
           if IsTargetNode == 1
               xi(alpha, i) = (Y(alpha, i) - D(alpha, k));
           else
               adder = 0;
65             for j = 1 : lenFF
                   link = FEF(j);
                   % F(j) is a node which i goes forward to
                   post = FF(j);
                   adder = adder + xi(alpha, post)*W(link)*sigmaprime{post}(y(
                       alpha, post), O(post), G(post)) ;
70             end
               xi(alpha, i) = adder;
           end % test on nodes: output or not
       end % loop on nodes
   end % loop on data
75 %xi

   for alpha = 1:S
       for i = OMax : -1 : 1
       % FE{i} is the global node information
80     % where i the ith weight
       % FE{i} = a set of (pre, post)
       FF = F{i};
       FEF = FE{i};
       lenFF = length(FF);
85     for j = 1 : lenFF
           % get the weight index
           link = FEF(j);
           % W(link) = E_{pre -> post} where
           pre = i;
90         post = FF(j);
           DEDW(link) = DEDW(link) + xi(alpha, post)*sigmaprime{post}(y(alpha
               , post), O(post), G(post))*Y(alpha, pre);
           end% forward link loop
```

```
       DEDO( i ) = DEDO( i ) - xi(alpha,i)*sigmaprime{i}(y(alpha,i),O(i),G(i))
         ;
       DEDG( i ) = DEDG( i ) - xi(alpha,i)*sigmaprime{i}(y(alpha,i),O(i),G(i))
         *(y(alpha,i)-O(i))/G(i);
95    end% node loop
     end% exemplar loop

     %we have found all the gradients
     %now set the right pieces to 0
100  DEDW(1,1:NGE) = 0;
     DEDO(1,1:NGN) = 0;
     DEDG(1,1:NGN) = 0;

     %usual gradient
105  if DOGAIN == 1
       Grad = [DEDW DEDO DEDG];
     else
       Grad = [DEDW DEDO];
     end
110  NormGrad = sqrt( sum(Grad.*Grad) );
     %
     %we implement gradient scaling
     Scale_Tol = 0.01;
     WScale = ones(1,length(W));
115  OScale = ones(1,length(O));
     GScale = ones(1,length(G));
     ADEDW = max(abs(DEDW));
     ADEDO = max(abs(DEDO));
     ADEDG = max(abs(DEDG));
120  for i = 1: length(W)
         if abs(DEDW(i)) > Scale_Tol
           WScale(i) = 1/ADEDW;
         end
     end
125  for i = 1: length(O)
         if abs(DEDO(i)) > Scale_Tol
           OScale(i) = 1/ADEDO;
         end
     end
130  for i = 1: length(G)
         if abs(DEDG(i)) > Scale_Tol
           GScale(i) = 1/ADEDG;
         end
     end
135  if DOGAIN == 1
       SW = diag(WScale)*DEDW';
       SO = diag(OScale)*DEDO';
       SG = diag(GScale)*DEDG';
       ScaledGrad = [SW' SO' SG'];
140  else
       SW = diag(WScale)*DEDW';
       SO = diag(OScale)*DEDO';
       ScaledGrad = [SW' SO'];
     end
145  gradtol = 0.05;
     %
     if NormGrad < gradtol
       scale = r;
       %update all the links.
150    WUpdate = W- DEDW * scale;
       %set copies to updates
       WUpdate(1,1:NGE) = WUpdate(1,NGE+1:2*NGE);

       %update all offsets. This does not alter
155    %offsets 1 to NGN
       OUpdate = O - DEDO * scale;
       OUpdate(1,1:NGN) = OUpdate(1,NGN+1:2*NGN);
       if DOGAIN == 1
           %update all gains. This does not alter
```

```
160         %gains  1  to  NGN
            GUpdate = G - DEDG * scale;
            GUpdate(1,1:NGN) = GUpdate(1,NGN+1:2*NGN);
          else
            GUpdate = G;
165       end
        else% normgrad  is  larger  than  0.05
          NormScaledGrad = sqrt( sum(ScaledGrad.*ScaledGrad) );
          %update  all  the  links.
          WUpdate = W- r*DEDW*diag(WScale);
170       %set  copies  to  updates
          WUpdate(1,1:NGE) = WUpdate(1,NGE+1:2*NGE);

          %update  all  offsets.   This  does  not  alter
          %offsets  1  to  NGN
175       OUpdate = O - r*DEDO*diag(OScale);
          OUpdate(1,1:NGN) = OUpdate(1,NGN+1:2*NGN);

          if DOGAIN == 1
            %update  all  gains.   This  does  not  alter
180         %gains  1  to  NGN
            GUpdate = G - r*DEDG*diag(EScale);
            GUpdate(1,1:NGN) = GUpdate(1,NGN+1:2*NGN);
          else
            GUpdate = G;
185     end
      end

      end
```

Finally, we put it all together to write a simple training loop.

Listing 18.87: New lagged training code

```
    function [W,G,O,Energy,Grad,ScaledGrad,Norm] = ...
  2 chainffntrainlag(sigma,sigmaprime,g,Input,IN,Target,...
    OUT,W,O,G,B,BE,F,FE,lambda,NumIters,NGN,NGE,NFB)
    %
    %   We  train  a  chain  ffn
    %
  7 % Input  is  input  matrix  of  S  rows  of  input  data
    %       S  x  number  of  input  nodes
    % target  is  desired  target  value  matrix  of  S  rows  of  target  data
    %       S  x  number  of  output  nodes
    %
 12 % g  is  the  chain  ffn  graph
    % G  is  the  gain  parameters
    % O  is  the  offset  parameters
    % Y  is  the  node  output  data
    %
 17 Energy = [];
    [E,y,Y,OI,EI] = energy2(sigma,g,Input,IN,OUT,Target,W,O,G,B,BE);
    E
    Energy = [Energy E];
    for t = 1:NumIters
 22   [W,O,G,Grad,ScaledGrad,Norm] = GradientUpdateLag(sigma,sigmaprime,g
          ,y,Y,W,O,G,B,BE,F,FE,Input,IN,Target,OUT,lambda,NGN,NGE,NFB);
      [E,y,Y,OI,EI] = energy2(sigma,g,Input,IN,OUT,Target,W,O,G,B,BE);
      Energy = [Energy E];
    end
    Energy(NumIters)
 27 plot(Energy);
    end
```

We need to test this code and see how it does. First, we setup the lagged graph.

Listing 18.88: Example: Setting up the original graph with feedback

```
   V =  [1;2;3;4;5;6;7;8;9;10;11;12];
2  E =  {[1;2],[2;3],[2;4],[2;5],[3;6],[4;7],[5;8],[2;7],[1;7],...
          [6;9],[7;9],[8;9],[9;10],[10;11],[11;12],[12;2],...
          [11,3],[7;7]};
   v =  vertices(V);
   e =  edges(E);
7  G =  graphs(v,e);
   KG =  incidenceFB(G);
   eG = G.e;
   incToDotFB(KG,10,10,1,'PracticeFB.dot');
```

Note, we generate the original dot file so we can see the original graph. Then we get the information we need to build the lagged graph.

Listing 18.89: Example: Setting up the lagged graph

```
   %Find the feedback edges and subtract them
   FBedge = ExtractFBEdges(KG);
   NewFFedges = ExtractNewFFEdgeInfo(G,FBedge);
   % subtract the feedback edges
5  W = removeedges(G,FBedge);
   %double the graph
   LaggedW = addgraph(W,W,NewFFedges);
   KLaggedW = incidence(LaggedW);
   incToDot(KLaggedW,10,10,1,'NewDoubledFFGraph.dot');
```

We then do some further setup. Note we first set **OUT = {6,7,8,9,18, 19,20,21}** as we need to set all the node functions correctly. Later, we will reset to **OUT = {18,19,20,21}**.

Listing 18.90: Example: Further setup details and input and target sets

```
    IN = {1,13};
    OUT = {6,7,8,9,18,19,20,21};
    NGN = length(G.v);
    NGE = length(G.e) - length(FBedge);
5   NFB = length(FBedge);
    NodeSize = length(LaggedW.v);
    SL = -0.5;
    SH = 1.5;
    %setup training
10  X = linspace(0,1,101);
    U = [];
    for i = 1:101
        if X(i) >= 0 && X(i) < 0.25
           U =   [U,[1;0;0;1]];
15      elseif X(i) >= 0.25 && X(i) < 0.50
           U = [U,[0;1;0;0]];
        elseif X(i) >= 0.5 && X(i) < 0.75
           U = [U,[0;0;1;1]];
        else
20         U = [U,[0;0;1;0]];
        end
    end
    %randomly permute the order of the input and target set.
    P = randperm(101);
```

```
25 XX = X(P);
   UU = [];
   for i = 1:101
       j = P(i);
       UU = [UU,U(:,j)];
30 end
   Input = [XX',XX'];
   Target = UU';
   %set all node functions as usual; the copies will
   %be the same as the original
35 [sigma,sigmaprime] = SigmoidInit2(NodeSize,SL,SH,IN,OUT);
```

We designed this input and output data with a more biological situation in mind. The inputs are scalars and so are perhaps a light intensity value that has been obtained from additional layers of neural circuitry. The outputs are taken from the top of the OCOS/FFP stack and so are reminiscent of outputs from the top layer of visual cortex (albeit in a very simplistic way) and the binary patterns of the target—i.e. [1; 0; 0; 1] might represent a coded command to active a certain motor response or a set of chromatophore activations resulting in a specific pattern on the skin of a cephalopod. So this circuit, if we can train it to match the inputs to outputs, is setup to code for four specific motor or chromatophore activations. The feedback edges $11 \rightarrow 3$ and $7 \rightarrow 7$ are not really biological and were added just to help provide a nice test for the code. However, the edge $12 \rightarrow 2$ is part of the usual FFP circuit and so is biologically plausible. The top of the OCOS was allowed to feedforward into a summation node 9 which then fed up to node 10 in the cortical column above the one which contains our OCOS/FFP. That node 10 then provides feedback from the top cortical column into the bottom of our OCOS/FFP stack. So this example is a small step in the direction of modeling something of interest in cognitive modeling!

Listing 18.91: Example: parameters and backward and forward sets

```
   %reset OUT
   OUT = {18,19,20,21};
   %reset node size to that of G
   NodeSize = length(G.v);
 5 OL = -1.4;
   OH = 1.4;
   GL = 0.8;
   GH = 1.2;
   WL = -2.1;
10 WH = 2.1;
   Offset = OL+(OH-OL)*rand(1,NodeSize);
   O = [Offset,Offset];
   Gain = GL+(GH-GL)*rand(1,NodeSize);
   G = [Gain,Gain];
15 Weights1 = WL+(WH-WL)*rand(1,NGE);
   Weights2 = WL+(WH-WL)*rand(1,NFB);
   W = [Weights1,Weights1,Weights2];
   [BG,FG,BEG,FEG] = BFsets(LaggedW,KLaggedW);
```

We can do a simple evaluation.

Listing 18.92: Example: A simple evaluation and energy calculation

```
  [InVals , OutVals] = evaluation3 (sigma , LaggedW , Input (1 ,:) ,IN ,OUT,W,O,G,
      BG,BEG) ;
2 %find error
  [E , yI , YI , OI , EI] = energy2 (sigma , LaggedW , Input , IN , OUT, Target ,W,O,G,BG,
      BEG) ;
  E =
  87.3197
```

Finally, we can do some training.

Listing 18.93: Example: 10 training steps

```
  [W,G,O, Energy , Grad , ScaledGrad , Norm] = ...
    chainffntrainlag (sigma , sigmaprime , LaggedW , Input , IN , Target ,OUT, ...
    W,O,G,BG,BEG,FG,FEG,0.001 ,10 ,NGN,NGE,NFB) ;
```

We list the regular gradient and the scaled gradient. Here there are 15 original feed-forward edges and the first 15 values in Grad are therefore 0. The next 15 correspond to the link weights for the copy and the last 3 are the link weights for the rewritten feedback edges. The values for 28, 29 and 30 are 0 because those edges are after the last output node and so are irrelevant to the energy calculation. Then, there are the 3 partials for the feedback link weights. Finally, there are then 12 offset partials that are zero as they are in the original graph and 9 offsets partials from the copy that are nonzero with the last 3 zero as they are irrelevant. The listing below that is the scaled version of the gradient.

Listing 18.94: Example: 10 training steps gradient information

```
     Grad
2  Grad =
     Columns 1 through 15
          0          0         0          0         0          0
                0          0         0         0          0          0
                0                0         0
     Columns 16 through 30
         -3.2022     0.8050     0.8867     0.4867    -17.4126     2.2200
                -5.1833     2.6164    -0.3360    -0.3775     3.6818     -3.4156
                0                0         0
7  Columns 31 through 45
        4.0001      1.1957     1.6520          0         0          0
        0          0          0         0          0          0          0
                0                0
     Columns 46 through 57
       -103.4473     4.2735    -6.7604   -14.8478   -12.6224     18.4797
                -3.7118     8.8448    -7.7773         0         0          0

12 ScaledGrad =
     Columns 1 through 15
          0          0         0          0         0          0
                0          0         0         0          0          0
                0                0         0
     Columns 16 through 30
        -0.1839      0.0462     0.0509     0.0280    -1.0000     0.1275
                -0.2977     0.1503    -0.0193    -0.0217     0.2114     -0.1962
                0                0         0
```

17	Columns 31 through 45						
	0.2297	0.0687	0.0949	0	0	0	
	0	0	0	0	0	0	0
		0	0				
	Columns 46 through 57						
	−1.0000	0.0413	−0.0654	−0.1435	−0.1220	0.1786	
		−0.0359	0.0855	−0.0752	0	0	0

Finally, let's talk about thresholding and recognition. A given target here has binary values in it and the output from the model is not going to look like that. For example, after some training, we might have a model output of $[0.85; -0.23; 0.37; 0.69]$ with the target supposed to be $[1; , 0; , 0; 1]$. To see if the model has recognized or classified this input, we usually take the model output and apply a threshold function. We set low and high threshold values; say $c_L = 0.4$ and $c_H = 0.6$ and perform the test $V_i = 1$ if $Y_i > c_H$, $V_i = 0$ if $Y_i < c_L$ and otherwise $V_i = 0.5$ where V represents the thresholded output of the model. Then if the error between the thresholded value V and the target is 0, we can say we have classified this input correctly. We do this in code with a pair of functions. First, we write the code to do the thresholding.

Listing 18.95: The thresholding algorithm

```
function R = Recognition (X, LoTol , HiTol )
%
R = zeros ( length (X) ,1) ;
for  i = 1: length (X)
    if  X( i ) > HiTol
        R( i ) = 1;
    elseif  X( i ) < LoTol
        R( i ) = 0;
    else
        R( i ) = 0.5;
    end
end
end
```

Then, we want to run the thresholding algorithm over all our results. This is done in **GetRecognition**.

Listing 18.96: Finding out how many inputs have been classified

```
function [ Error ,sumE,R,Z, Success , Classified ] = GetRecognition (OUT,Y,
    Target , LoTol , HiTol )
%
%
%
[ rows , cols ] = size (Y) ;
R = zeros ( rows , length (OUT) ) ;
Z = zeros ( rows , length (OUT) ) ;
Error = zeros ( rows , length (OUT) ) ;
Success = zeros ( rows ,1) ;
k = 1;
for  j = 1: length (OUT)
    for  i = 1: cols
        if  i == OUT{ j }
            %we have a column that is an output
            Z (: , k) = Y(: , i ) ;
            k = k+1;
            break ;
        end
    end
```

```
20 end
   %Z  now  contains  the  cols  that  match
   % targets
   for  i = 1:rows
        R(i,:)  =  Recognition(Z(i,:),LoTol,HiTol);
25      Error(i,:)  =  (R(i,:)  -  Target(i,:)).^2;
        test  =  sum(Error(i,:));
        if  test == 0
           Success(i)  =  1;
        end
30 end
   sumE  =  0.5*(norm(Error,'fro'))^2;
   Classified  =  sum(Success)/rows*100;
   end
```

For example, after suitable training, we might have results like this

Listing 18.97: Sample classification results

```
   Z(97:101,:)
2
   ans =

        -0.0095       0.1987       0.6900       0.3421
         0.9875      -0.1334      -0.0370       0.9726
7       -0.0095       0.3317       0.6900       0.3205
         1.0231       0.0014      -0.0630       0.9737
         1.0222      -0.0296      -0.0623       0.9746

   Target(97:101,:)
12
   ans =

         0       0       1       1
         1       0       0       1
17       0       0       1       1
         1       0       0       1
         1       0       0       1
```

Recall Z here is the four outputs we are monitoring. We see entry 97 maps to $[0; 0; 1; 0]$ which does not match Target 97 but the output 98 does map to the correct output.

18.8   Better Lagged Training!

It doesn't take long before we all get annoyed with gradient descent. We have played a bit with making it better by adding a type of gradient scaling, but we can do one more trick. Let's look at a technique called **line search**. We have a function of many variables here called E. For convenience, let all the parameters be called the vector p. Any vector D for which the derivative of E in the direction of the vector D, $\nabla E \cdot D < 0$ is a direction in which the value of E goes down is called a descent vector. Our various strategies for picking vectors based on $\nabla E(p^{old})$ all give us descent vectors. Look at a slice through the multidimensional surface of E given by $g(\xi) = E(p^{old} + \xi D)$ where D need not be a unit vector. At a given step in our gradient descent algorithm, we know $g(0) = E(p^{old})$ and if λ is our current choice of step to try, we can calculate $g(\lambda) = E(p^{old} + \lambda D)$. From the chain rule of calculus of more than one variable—ha, you will have to remember what we discussed in Peterson (2015)—we find $g'(\xi) = < \nabla E(p^{old} + \xi D), D >$. Hence, since we know the gradient at the current step, we see $g'(0) = < \nabla E(p^{old}), D >$. We thus have enough information to fit a quadratic approximation of $g(\xi)$ given by $g^{approx}(\xi) = A + B\xi + C\xi^2$. We easily find

$$A = E(p^{old})$$
$$B = - < \nabla E(p^{old}), D >$$
$$C = \frac{E(p^{old} + \lambda D) - E(p^{old}) - B\lambda}{\lambda^2}$$

The new version of the code has been reorganized a bit and is a drop in replacement for the old code. It has a few toggles you can set in the code: if you set **dolinesearch = 0** and leave **dogradscaling = 1**, you get the old version of the gradient update code for the lagged networks.

Listing 18.98: Activating Line Search and Gradient Scaling

```
dogradscaling = 1;
dolinesearch  = 1;
```

The structure of the code has been reorganized as we show in the skeleton here.

Listing 18.99: New GradientUpdateLag skeleton code

```
    if dolinesearch == 1
        % get the first energy value we will need to build the quadratic
3       % approximation to the 1D surface slice using the descent vector
        [EStart,yI,YI,OI,EI] = energy2(sigma,g,I,IN,OUT,D,W,O,G,B,BE);
    end
    % we need the original gradient
    if DOGAIN == 1
8      BaseGrad = [DEDW DEDO DEDG];
    else
       BaseGrad = [DEDW DEDO];
    end
    NormBase = sqrt( sum(BaseGrad.*BaseGrad) );
13
    if dogradscaling == 1
        % implement gradient scaling
        % calculate the gradient vector using gradient
        % scaling and call it ScaledGrad
```

```
18    % set UseGrad
      UseGrad = ScaledGrad;
    else
      % without gradient scaling,
      % we use the original gradient
23    UseGrad = Grad;
    end%dogradscaling loop

    %now get norm of gradient
    gradtol = 0.05;
28  NormGrad = sqrt( sum(UseGrad.*UseGrad) );

    %now do descent step
    if dogradscaling == 1
      % do a descent step using lambda = scale
33    % using either the original descent vector
      % or its normalization
      if NormGrad < gradtol
        %not normalized
        NormDescent = NormGrad;
38    else% normgrad is larger than 0.05
        %normalized
        NormDescent = 1;
      end
    else% not using grad scaling
43    %
      ScaledGrad = Grad;
      if NormGrad < gradtol
        %not normalized
        NormDescent = NormGrad;
48    else%
        % normalized
        NormDescent = 1;
      end
    end
53
    %now we can do a line search
    if dolinesearch == 1
      dolinesearchstep = 1;
      %find the energy for the full step of lambda = scale
58    [EFull,yI,YI,OI,EI] = energy2(sigma,g,I,IN,OUT,D,WUpdate,OUpdate,
        GUpdate,B,BE);
      %set A, BCheck = B and C as discussed in the theory of the
        quadratic approximation
      %of the slice
      A = EStart;
      BCheck = -dot(BaseGrad,UseGrad);
63    C = ( EFull- A - BCheck*scale )/(scale^2);
      %get the optimal step
      lambdastar = -BCheck/(2*C);
      %if C is negative this corresponds to a maximum so reject
      if(C<0 || lambdastar > scale)
68      % we are going to a maximum on the line search; reject
        lambdastar = lambdastar+2*scale;
        %dolinesearchstep = 0;
      end
      %now we have the optimal lambda to use
73    if dolinesearchstep == 1
        %do new step using lamdastar
        scale = lambdastar;
        %do it differently if doing gradient scaling or not
        ...
78      end%dolinsearchstep loop
    end% dolinesearch loop

    %copy updates to the original block of parameters
    WUpdate(1,1:NGE) = WUpdate(1,NGE+1:2*NGE);
83  OUpdate(1,1:NGN) = OUpdate(1,NGN+1:2*NGN);
```

```
    if DOGAIN == 1
       GUpdate(1,1:NGN) = GUpdate(1,NGN+1:2*NGN);
    else
       GUpdate = G;
88 end
```

We can try out our new version of **GradientUpdateLag** which includes line search and see how it does. Note it still takes quite a while, but without line search the iteration count can reach 50,000 or more and you reach about 50 % recognition and stall. The example before used the odd feedbacks of **11 -> 3** and **7 -> 7** which we will now remove for this second example. Also, we remove the output node **10** as it is not very biological. Other than that, the setup is roughly the same, so we only show the construction of the original graph in the results below. We start with a ten step run using gradient scaling and line search. We print the optimal line search step size to show you how it is going. Then we do another and also print the initial recognition results.

Listing 18.100: A Sample Run

```
    V = [1;2;3;4;5;6;7;8;9;10;11;12];
    E = {[1;2],[2;3],[2;4],[2;5],[3;6],[4;7],[5;8],[2;7],[1;7],...
         [6;9],[7;9],[8;9],[9;10],[10;11],[11;12],[12;2]};
    v = vertices(V);
  5 e = edges(E);
    G = graphs(v,e);
    KG = incidenceFB(G);
    %construct lagged graph as before
    %set the IN and OUT so we can initialize the node functions
 10 IN = {1,13};
    OUT = {6,7,8,18,19,20};
    % remove the extra ouput value
    X = linspace(0,1,101);
    U = [];
 15 for i = 1:101
        if X(i) >= 0 && X(i) < 0.25
           U = [U,[1;0;0]];
        elseif X(i) >= 0.25 && X(i) < 0.50
           U = [U,[0;1;0]];
 20     else
           U = [U,[0;0;1]];
        end
    end
    %Initial energy
 25 [E,yI,YI,OI,EI] = ...
    energy2(sigma,LaggedW,Input,IN,OUT,Target,W,O,G,BG,BEG);
    E =
        149.276
    %Train for 10 steps with lambda = .1
 30 [W,G,O,Energy,Grad,ScaledGrad,Norm] = ...
    chainffntrainlag(sigma,sigmaprime,LaggedW,Input,IN,...
    Target,OUT,W,O,G,BG,BEG,FG,FEG,0.001,...
    ,10,NGN,NGE,NFB);
    lambdastar = 0.0572
 35 lambdastar = 0.0585
```

```
      lambdastar  =  0.0600
      lambdastar  =  0.0620
      lambdastar  =  0.0645
      lambdastar  =  0.0678
40    lambdastar  =  0.0724
      lambdastar  =  0.0790
      lambdastar  =  0.0891
      lambdastar  =  0.1057
      Estart  =  149.276 Estop  128.1830
45
      %another 100 steps
      [W,G,O,Energy,Grad,ScaledGrad,Norm] = ...
      chainffntrainlag(sigma,sigmaprime,LaggedW,Input,IN,...
      Target,OUT,W,O,G,BG,BEG,FG,FEG,0.1,...
50    100,NGN,NGE,NFB);
      Estart  =   121.7921 Estop  =  20.4412

      %another 100 steps: eneregy oscillates some
      [W,G,O,Energy,Grad,ScaledGrad,Norm] = ...
55    chainffntrainlag(sigma,sigmaprime,LaggedW,Input,IN,...
      Target,OUT,W,O,G,BG,BEG,FG,FEG,0.1,...
      100,NGN,NGE,NFB);
      Estart  =  20.4315   Estop  =  20.1791

60    %another 100 steps: energy oscillates some
      %cut scale to .01
      [W,G,O,Energy,Grad,ScaledGrad,Norm] = ...
      chainffntrainlag(sigma,sigmaprime,LaggedW,Input,IN,...
      Target,OUT,W,O,G,BG,BEG,FG,FEG,0.01,...
65    100,NGN,NGE,NFB);
      Estart  =  20.1533   Estop  =  20.0652

      %check recognition
      [E,yI,YI,OI,EI] = ...
70    energy2(sigma,LaggedW,Input,IN,OUT,Target,W,O,G,BG,BEG);
      [Error,sumE,R,Z,Success,Classified] = GetRecognition(OUT,YI,Target
        ,0.4,0.6);
      Classified , sumE
      Classified  =  39.6040
      sumE  =  21.4062
```

Now run for an additional 11,000 steps after which we check recognition and we see we are about 50 %.

Listing 18.101: Our example run continued

```
1 Classified , sumE
  Classified =
     49.5050
  sumE =
     18.1875
```

This is clearly not very good. We are using excessive iteration and still not getting good recognition. Let's revisit the gradient update algorithm and line search and see if we can improve it.

18.9   Improved Gradient Descent

Let's tackle the training code by making it more modular. We are going to use the functions **GradientUpdateLagTwo** to handle the gradient stuff and **chainffn trainlagtwo** for the iterative training. We rewrote the update code as seen below; some portions are left out as they are identical. We set up the gradient vectors based on whether we are doing gradient scaling or not with a nice block.

Listing 18.102: Set up gradient vectors

```
    ScaleTol = 0.01;
    if dogradscaling == 1
      [WScale, OScale, GScale, UseGrad] = GetScaledGrad (DOGAIN, ScaleTol ,W,O,
          G,DEDW,DEDO,DEDG);
    else
5     UseGrad = BaseGrad;
    end
    NormGrad = sqrt ( sum ( UseGrad.*UseGrad ) );
```

We use the new function **GetScaledGrad** to find the scaled gradient vector. This function contains all the code we used before, but by pulling it out we make the gradient update code easier to follow and debug.

Listing 18.103: Finding the scaled gradient

```
    function [WScale, OScale, GScale, UseGrad] = GetScaledGrad (DOGAIN,
        ScaleTol ,W,O,G,DEDW,DEDO,DEDG)
    %
    WScale = ones (1, length (W));
    OScale = ones (1, length (O));
5   GScale = ones (1, length (G));
    ADEDW = max ( abs (DEDW));
    ADEDO = max ( abs (DEDO));
    ADEDG = max ( abs (DEDG));
    for i = 1: length (W)
10    if abs (DEDW( i )) > ScaleTol
          WScale ( i ) = 1/ADEDW;
      end
    end
    for i = 1: length (O)
15    if abs (DEDO( i )) > ScaleTol
          OScale ( i ) = 1/ADEDO;
      end
    end
    for i = 1: length (G)
20    if abs (DEDG( i )) > ScaleTol
          GScale ( i ) = 1/ADEDG;
      end
    end
    if DOGAIN == 1
25    SW = diag ( WScale )*DEDW';
      SO = diag ( OScale )*DEDO';
      SG = diag ( GScale )*DEDG';
      ScaledGrad = [SW'  SO' SG'];
    else
```

```
30    SW = diag(WScale)*DEDW';
      SO = diag(OScale)*DEDO';
      ScaledGrad = [SW' SO'];
    end
    UseGrad = ScaledGrad;
35
    end
```

Once we have found the appropriate gradient, we must find the updates. If we don't do a line search, we simply do a usual descent step using our descent vector. We have placed the descent code into two new function **DoDescentStepGradScaling** which handles the scaled gradient case and **DoDescentStepRegular** which uses just a straight gradient without scaling.

Listing 18.104: Doing a descent step

```
    gradtol = 0.05;
    %now do descent step
    if dogradscaling == 1
       [WUpdate, OUpdate, GUpdate] = DoDescentStepGradScaling (DOGAIN,
          NormBase, NormGrad,W,O,G, ...
5      WScale, OScale, GScale,DEDW,DEDO,DEDG, gradtol ,r);
    else
       [WUpdate, OUpdate, GUpdate, NormDescent] = ...
       DoDescentStepRegular (DOGAIN, NormGrad,W,O,G,DEDW,DEDO,DEDG, gradtol ,
          r);
    end
```

It is much easier to see the structure of the code now as all of the complicated computations are placed in their own modules. These modules are lifted straight from the old update code. First, look at the scaled descent.

Listing 18.105: The Scaled Gradient Descent Step

```
1 function [WUpdate, OUpdate, GUpdate, NormDescent] = ...
    DoDescentStepGradScaling (DOGAIN, NormBase, NormGrad,W,O,G, WScale, OScale
       , GScale , ...
    DEDW,DEDO,DEDG, gradtol , scale )
    %
    %
6 if NormBase < gradtol
       WUpdate = W- DEDW * scale ;
       OUpdate = O - DEDO * scale ;
       if DOGAIN == 1
          GUpdate = G - DEDG * scale ;
11     else
          GUpdate = G;
       end
       NormDescent = NormBase;
    else% NormBase is larger than 0.05
```

```
16      WUpdate = W- scale*DEDW*diag(WScale)/NormGrad;
        OUpdate = O - scale*DEDO*diag(OScale);
        if DOGAIN == 1
           GUpdate = G - scale*DEDG*diag(EScale);
        else
21         GUpdate = G;
        end
        NormDescent = NormGrad;
     end

26 end
```

Note the regular descent is much simpler. Before all of this mess was in the main code block which was making the gradient update code hard to read.

Listing 18.106: The Regular Gradient Descent Step

```
   function [WUpdate, OUpdate, GUpdate, NormDescent] = ...
   DoDescentStepRegular (DOGAIN, NormGrad, W, O, G, DEDW, DEDO, DEDG, gradtol ,
       scale )
   %
 4 if NormGrad < gradtol
      WUpdate = W- DEDW * scale;
      OUpdate = O - DEDO * scale;
      if DOGAIN == 1
         GUpdate = G - DEDG * scale;
 9    else
         GUpdate = G;
      end
      NormDescent = NormGrad;
   else% normgrad is larger than 0.05
14    WUpdate = W- scale*DEDW/NormGrad;
      OUpdate = O - scale*DEDO/NormGrad;
      if DOGAIN == 1
         GUpdate = G - scale*DEDG/NormGrad;
      else
19       GUpdate = G;
      end
      NormDescent = 1;
   end

24 end
```

Finally, we want to tackle the line search code. The previous version did not take advantage of what we did at a current line search step and instead kept resetting the initial step size. So for example, we could run 2000 steps and each one would start line search at whatever λ value we used as the start. This meant we did not take advantage of what we were learning. So now we will return two things from our update code when we use line search. First, the current value of **lambdastar** we find and second, a minimum value of the line search value. If we used the current **lambdastar** we find to reset the next iteration, we could progressively shrink our calculated **lambdastar** to zero and effectively stop our progress. So we return a good estimate of how small we want our new start of the line search to be in the variable **lambdastart**. The new line search looks the same in principle. The only difference is we set the value of **lambdastart** at the end using the value of **lambdamin**. This will prevent our line search from using too small of search values.

Listing 18.107: The new update code

```
function [WUpdate,OUpdate,GUpdate,UseGrad,NormBase,NormGrad,
    lambdastar,lambdastart] = ...
GradientUpdateLag(sigma,sigmaprime,g,y,Y,W,O,G,B,BE,F,FE,I,IN,D,OUT,r
    ,NGN,NGE,NFB)
%
% gain update toggle, grad scaling toggle, linesearch toggle
DOGAIN = 0;
dogradscaling = 1;
dolinesearch = 1;

%set up weight and derivative vectors
Nodes = g.v;
[m,sizeY] = size(Nodes);
sizeW = length(W);
S = length(I);
sizeD = length(D);
SN = length(IN);
sizeOUT = length(OUT);
xi = zeros(S,sizeY);
DEDW = zeros(1,sizeW);
DEDO = zeros(1,sizeY);
DEDG = zeros(1,sizeY);

% find the xi's
....code here

%find the partials
.....code here

%we have found all the gradients
%now set the right pieces to 0
DEDW(1,1:NGE) = 0;
DEDO(1,1:NGN) = 0;
DEDG(1,1:NGN) = 0;

%get base gradient
if DOGAIN == 1
    BaseGrad = [DEDW DEDO DEDG];
else
    BaseGrad = [DEDW DEDO];
end
NormBase = sqrt( sum(BaseGrad.*BaseGrad) );
```

```
    ScaleTol = 0.01;
    if dogradscaling == 1
      [WScale, OScale, GScale, UseGrad] = GetScaledGrad(DOGAIN, ScaleTol ,W,O,
        G,DEDW,DEDO,DEDG);
45 else
      UseGrad = BaseGrad;
    end
    NormGrad = sqrt( sum(UseGrad.*UseGrad) );

50 gradtol = 0.05;
    %now do descent step
    if dogradscaling == 1
      [WUpdate, OUpdate, GUpdate] = DoDescentStepGradScaling(DOGAIN,
        NormBase, NormGrad,W,O,G, ...
      WScale, OScale, GScale ,DEDW,DEDO,DEDG, gradtol ,r);
55 else
        [WUpdate, OUpdate, GUpdate, NormDescent] = ...
        DoDescentStepRegular(DOGAIN, NormGrad ,W,O,G,DEDW,DEDO,DEDG, gradtol ,
            r);
    end

60 %linesearch
    lambdamin = 0.5;
    if dolinesearch == 1
      [EStart , yI , YI, OI, EI] = energy2(sigma ,g, I ,IN ,OUT,D,W,O,G,B,BE);
      [EFull , yI , YI, OI, EI]   = energy2(sigma ,g, I ,IN,OUT,D, WUpdate, OUpdate
        , GUpdate ,B,BE);
65    A = EStart ;
      B = -dot(BaseGrad , UseGrad);
      C = ( EFull- A - B*r )/(r^2);
      lambdastar = -B/(2*C);
      if (C<0)
70        % we are going to a maximum on the line search ;
          lambdastar = lambdastar+2*r ;
      end
      %do new optimal step
      if dogradscaling == 1
75        [WUpdate, OUpdate, GUpdate, NormDescent] = ...
          DoDescentStepGradScaling(DOGAIN, NormBase, NormGrad ,W,O,G, WScale ,
            OScale , GScale , ...
          DEDW,DEDO,DEDG, gradtol , lambdastar);
      else% not using grad scaling
          [WUpdate, OUpdate, GUpdate, NormDescent] = ...
80        DoDescentStepRegular(DOGAIN, NormGrad ,W,O,G,DEDW,DEDO,DEDG,
            gradtol ,lambdastar);
      end
      % don't return too small a lambdastar
      if lambdastar < lambdamin
        lambdastart = lambdamin;
85    else
        lambdastart = lambdastar;
      end
    else
        lambdastar = r;
90      lambdastart = lambdastar;
    end

    %copy updates to the original block of parameters
    WUpdate(1,1:NGE) = WUpdate(1,NGE+1:2*NGE);
95 OUpdate(1,1:NGN) = OUpdate(1,NGN+1:2*NGN);
```

```
     if DOGAIN == 1
         GUpdate(1,1:NGN) = GUpdate(1,NGN+1:2*NGN);
     else
         GUpdate = G;
100  end

     end
```

The new training code is very similar although some of the return variables are different. Note we have added code to do a plot of our calculated line search values as well as the energy using the **subplot** command in MatLab.

Listing 18.108: The new training code

```
    function [W,G,O,Energy,Grad,NormGrad,lambda,lambdastart] = ...
    chainffntrainlagtwo(sigma,sigmaprime,g,Input,IN,Target,OUT,W,O,G,B,BE
        ,F,FE,lambda,NumIters,NGN,NGE,NFB)
    %
    Energy = [];
  5 [E,y,Y,OI,EI] = energy2(sigma,g,Input,IN,OUT,Target,W,O,G,B,BE);

    scale = [];
    Energy = [Energy E];
    lambdastart = lambda;
 10 for t = 1:NumIters
        [W,O,G,Grad,NormBase,NormGrad,lambda,lambdastart] =
            GradientUpdateLagTwo(sigma,sigmaprime,...
            g,y,Y,W,O,G,B,BE,F,FE,Input,IN,Target,OUT,lambdastart,NGN,NGE,
                NFB);
        [E,y,Y,OI,EI] = energy2(sigma,g,Input,IN,OUT,Target,W,O,G,B,BE);
        Energy = [Energy E];
 15     scale = [scale lambda];
    end
    Energy(1), Energy(NumIters)
    subplot(2,1,1), plot(Energy);
    subplot(2,1,2), plot(scale);
 20 end
```

Let's try this out on a model. The model is similar to what we have been testing our code on. We changed the training data a bit and altered the node function setup some. We'll let you compare the examples and find the changes.

Listing 18.109: Another example setup

```
    V = [1;2;3;4;5;6;7;8;9;10;11;12];
    E = {[1;2],[2;3],[2;4],[2;5],[3;6],[4;7],[5;8],[2;7],[1;7],...
        [6;9],[7;9],[8;9],[9;10],[10;11],[11;12],[12;2]};
    v = vertices(V);
  5 e = edges(E);
    G = graphs(v,e);
    KG = incidenceFB(G);
    eG = G.e;

 10 %Find the feedback edges and subtract them
    FBedge = ExtractFBEdges(KG);
    NewFFedges = ExtractNewFFEdgeInfo(G,FBedge);

    % subtract the feedback edges
```

```
15 W = removeedges (G, FBedge);

   %double the graph
   LaggedW = addgraph (W,W, NewFFedges);
   KLaggedW = incidence (LaggedW);
20
   IN = {1,13};
   OUT = {6 ,7 ,8 ,18 ,19 ,20};
   NGN = length (G.v);
   NGE = length (G.e) − length (FBedge);
25 NFB = length (FBedge);
   NodeSize = length (LaggedW.v);
   SL = −0.2;
   SH = 1.2;

30 %setup training
   X = linspace (0 ,3 ,101);
   U = [];
   for i = 1:101
       if X(i) >= 0 && X(i) < .33
35         U = [U,[1;0;0]];
       elseif X(i) >= .33 && X(i) < .67
           U = [U,[0;1;0]];
       else
           U = [U,[0;0;1]];
40     end
   end

   %randomly permute the order of the input and target set.
   P = randperm (101);
45 XX = X(P);
   UU = [];
   for i = 1:101
       j = P(i);
       UU = [UU,U(: ,j)];
50 end

   Input = [XX' ,XX'];
   Target = UU';

55 %set all node functions as usual; the copies will
   %be the same as the original
   [sigma , sigmaprime] = SigmoidInit2 (NodeSize ,SL ,SH ,IN ,OUT);

   IN = {1,13};
60 OUT = {18 ,19 ,20};

   %reset node size to that of G
   EdgeSize = length (G.e);
   OL = −1.4;
65 OH = 1.4;
   GL = 0.8;
   GH = 1.2;
   WL = −2.1;
   WH = 2.1;
70 Offset = OL+(OH−OL)*rand (1 ,NGN);
   O = [Offset , Offset];
   Gain = GL+(GH−GL)*rand (1 ,NGN);
   G = [Gain , Gain];
   Weights = WL+(WH−WL)*rand (1 , EdgeSize);
```

```
75 W = [ Weights , Weights ] ;
   [BG,FG,BEG,FEG] = BFsets (LaggedW ,KLaggedW ) ;
   [InVals , OutVals] = evaluation3 (sigma , LaggedW , Input (1 ,:) ,IN ,OUT,W,O,G,
       BG,BEG) ;
   %find error
   [E , yI , OI , EI ] = energy2 (sigma , LaggedW , Input , IN ,OUT, Target ,W,O,G,BG,
       BEG) ;
80 E
```

We start the training and get the following results.

Listing 18.110: Starting the training

```
   %initial energy
   E = 74.4614
   %set starting lambda value
   rstart = 0.5;
 5 %train for 200 steps
   [W,G,O, Energy , Grad , NormGrad , r , rstart ] = ...
   chainffntrainlagtwo (sigma , sigmaprime , LaggedW , Input , IN , ...
   Target ,OUT,W,O,G,BG,BEG, FG,FEG, rstart ,200 ,NGN,NGE,NFB) ;
   Estart = 74.4614 Estop = 5.8951
10
   %significant progess: check recognition
   [E , yI , YI , OI , EI ] = ...
   energy2 (sigma , LaggedW , Input , IN ,OUT, Target ,W,O,G,BG,BEG) ;
   [Error ,sumE,R,Z, Success , Classified ] = ...
15 GetRecognition (OUT, YI , Target ,0.4 ,0.6 ) ;
   Classified , sumE
   Classified = 85.1485
   sumE = 4.0312
```

Only 200 steps and 85 % recognition!! We are doing much better with the new line search code.

Listing 18.111: Some more training

```
   %650 more steps
 2 [E , yI , YI , OI , EI ] = energy2 (sigma , LaggedW , Input , IN ,OUT, Target ,W,O,G,BG,
       BEG) ;
   [Error ,sumE,R,Z, Success , Classified ] = GetRecognition (OUT, YI , Target
       ,0.4 ,0.6 ) ;
   Classified , sumE
   Classified = 91.0891
   sumE = 3.7187
 7
   %200 more steps
   [E , yI , YI , OI , EI ] = energy2 (sigma , LaggedW , Input , IN ,OUT, Target ,W,O,G,BG,
       BEG) ;
   [Error ,sumE,R,Z, Success , Classified ] = GetRecognition (OUT, YI , Target
       ,0.4 ,0.6 ) ;
   Classified , sumE
12 Classified = 91.0891
   sumE = 3.6875
```

We might be able to get to 100 % but this is enough to show you how the training works. Finally, to see how our model works on new data, consider the following test. We set up new input and corresponding target data. These inputs are created using a different **linspace** command so the inputs will not be the same as before. You can see recognition is still fine.

Listing 18.112: Testing the model

```
  %setup testing
  X = linspace(0,3,65);
  U = [];
  for i = 1:65
5     if X(i) >= 0 && X(i) < .33
        U =  [U,[1;0;0]];
      elseif X(i) >= .33 && X(i) < .67
        U = [U,[0;1;0]];
      else
10       U = [U,[0;0;1]];
      end
  end
  %randomly permute the order of the input and target set.
  P = randperm(65);
15 XX = X(P);
  UU = [];
  for i = 1:65
     j = P(i);
     UU = [UU,U(:,j)];
20 end

  Input = [XX',XX'];
  Target = UU';

25 [E,yI,YI,OI,EI] = energy2(sigma,LaggedW,Input,IN,OUT,Target,W,O,G,BG,
     BEG);
  [Error,sumE,R,Z,Success,Classified] = GetRecognition(OUT,YI,Target
     ,0.4,0.6);
  Classified
  Classified = 89.23
```

Note on the training data, we classified 91.0891 % which is 92 out of the original 101 and on the testing, we got 89.23 % which is 58 out of 65 correct. For many purposes, this level of recognition is pretty good.

References

J. Bower, D. Beeman, *The Book of Genesis: Exploring Realistic Neural Models with the GEneral NEural SImulation System*, 2nd edn. (Springer TELOS, New York, 1998)

D. Harland, R. Jackson, Portia perceptions: the umwelt of an sraneophagic jumping spider, in *Complex Worlds from Simpler Nervous Systems*, ed. by F. Prete (MIT Press, Cambridge, 2004), pp. 5–40. A Bradford Book

M. Hines, *Neuron* (2003). http://www.neuron.yale.edu

K. Kral, F. Prete, In the mind of a hunter: the visual world of the Praying Mantis, in *Complex Worlds from Simpler Nervous Systems*, ed. by F. Prete (MIT Press, Cambridge, 2004), pp. 75–116. A Bradford Book

S. North, E. Ganser, *Graphviz: Graph Visualization Software* (AT & T, 2010). http://www.graphviz. org

J. Peterson, *Calculus for Cognitive Scientists: Derivatives, Integration and Modeling*, Springer Series on Cognitive Science and Technology (Springer Science+Business Media Singapore Pte Ltd, Singapore, 2015 In press)

R. Raizada, S. Grossberg, Towards a theory of the laminar architecture of cerebral cortex: computational clues from the visual system. Cereb. Cortex **13**, 100–113 (2003)

S. Zhang, M. Srinivasan, Exploration of cognitive capacity in honeybees: higher functions emerge from small brain, in *Complex Worlds from Simpler Nervous Systems*, ed. by F. Prete (MIT Press, Cambridge, 2004), pp. 41–74. A Bradford Book

Chapter 19
Address Based Graphs

In order to create complicated directed graph structures for our neural modeling, we need a more sophisticated graph object. Create a new folder called **Graphs** and in it create the usual **@graphs**, **@vertices** and **@edges** subfolders. We will then modify the code we had in the previous **@edges** and **@vertices** to reflect our new needs. Let's look at how we might build a cortical column. Each **can** in a cortical column is a six layer structure which consists of the OCOS, FFP and Two/Three building blocks. The OCOS will have addresses [0; 0; 0; 0; 0; 1 − 7] as there are 7 neurons. The FFP is simpler with addresses [0; 0; 0; 0; 0; 1 − 2] as there are just 2 neurons, The six neuron Two/Three circuit will have 6 addresses, [0; 0; 0; 0; 0; 1 − 6]. To distinguish these neurons in different circuits from one another, we add a unique integer in the fifth component. The addresses are now OCOS ([0; 0; 0; 0; 1; 1 − 7]), OCOS ([0; 0; 0; 0; 2; 1 − 2]) and Two/Three ([0; 0; 0; 0; 3; 1 − 6]), So a single **can** would look like what is shown in Eq. 19.1.

$$
\begin{aligned}
&\text{can } L1 \;.\;.\;.\;.\; y\;.\;. \qquad\quad \text{FFP, y's} \qquad \text{Address } [0; 0; 0; 0; 2; 1-2]\\
&\qquad\ L2\ z\;.\;.\;z\;.\;.\;z \ \ \text{Two/Three, z's} \ \ \text{Address } [0; 0; 0; 0; 3; 1-6]\\
&\qquad\ L3\;.\;.\;z\;.\;.\;.\;.\\
&\qquad\ L4\ z\ x\;.\;x\;.\;x\ z \qquad \text{OCOS x's} \qquad \text{Address } [0; 0; 0; 0; 1; 1-7]\quad (19.1)\\
&\qquad\quad\ .\;x\;.\;x\;.\;x\;.\\
&\qquad\ L5\;.\;.\;.\;.\;y\;.\;.\\
&\qquad\ L6\;.\;.\;.\;x\;.\;.\;.
\end{aligned}
$$

We are not showing the edges here, for convenience. We can then assemble **can**s into **column**s by stacking them vertically. This leads to the structure in Eq. 19.2. The

© Springer Science+Business Media Singapore 2016

J.K. Peterson, *BioInformation Processing*, Cognitive Science and Technology,
DOI 10.1007/978-981-287-871-7_19

fourth component of each address now indicates the **can** number: 1, 2 or 3. We also have to add a link between **can**s that provides the vertical processing.

$$
\begin{array}{llllll}
\text{can 3 } L1 & & y & \text{FFP y's} & [0;0;0;3;2;1-2] \\
L2\ z & z & z & \text{Two/Three z's} & [0;0;0;3;3;1-6] \\
L3 & z \\
L4\ z\ x & x & x\ z & \text{OCOS x's} & [0;0;0;3;1;1-7] \\
& x & x & x \\
L5 & & y \\
L6 & x & & & \text{1 interconnection} \\
\text{can 2 } L1 & & y & \text{FFP y's} & [0;0;0;2;2;1-2] \\
L2\ z & z & z & \text{Two/Three z's} & [0;0;0;2;3;1-6] \\
L3 & z \\
L4\ z\ x & x & x\ z & \text{OCOS x's} & [0;0;0;2;1;1-7] \\
& x & x & x \\
L5 & & y \\
L6 & x & & & \text{1 interconnection} \\
\text{can 1 } L1 & & y & \text{FFP y's} & [0;0;0;1;2;1-2]\text{2 edges} \\
L2\ z & z & z & \text{Two/Three z's} & [0;0;0;1;3;1-6]\text{7 edges} \\
L3 & z \\
L4\ z\ x & x & x\ z & \text{OCOS x's} & [0;0;0;1;1;1-7]\text{7 edges} \\
& x & x & x \\
L5 & & y \\
L6 & x \\
Th & t & & \text{Thalamus t's}
\end{array}
\tag{19.2}
$$

We can label the address of this **column** as [0; 0; 1; ·; ·; ·]. We could then use this as a model of cortex or assemble a rectangular **sheet** of **columns**. The resulting model of cortex would then be addresses as [0; 1; ·; ·; ·; ·]. Eventually, we will have at least 7 modules in our small brain model. We will the following address scheme.

Sensory Model One	[0; 1; ·; ·; ·; ·]
Sensory Model Two	[0; 2; ·; ·; ·; ·]
Associative Cortex	[0; 3; ·; ·; ·; ·]
Motor Cortex	[0; 4; ·; ·; ·; ·]
Thalamus	[0; 5; ·; ·; ·; ·]
MidBrain	[0; 6; ·; ·; ·; ·]
Cerebellum	[0; 7; ·; ·; ·; ·]

Then the small brain model itself would have the address **[1; ·; ·; ·; ·; ·]**. To use the brain model, we add an **input** and **output** module and construct a full model as follows:

Input	**[1; ·; ·; ·; ·; ·]**
Brain	**[2; ·; ·; ·; ·; ·]**
Output	**[3; ·; ·; ·; ·; ·]**

We can also construct an asymmetric brain model consisting of a left and right half brain connected by a corpus callosum model. This would be addressed as follows:

Input	**[1; ·; ·; ·; ·; ·]**
Left Brain	**[2; ·; ·; ·; ·; ·]**
Corpus Callosum	**[3; ·; ·; ·; ·; ·]**
Right Brain	**[4; ·; ·; ·; ·; ·]**
Thalamus	**[5; ·; ·; ·; ·; ·]**
MidBrain	**[6; ·; ·; ·; ·; ·]**
Cerebellum	**[7; ·; ·; ·; ·; ·]**
Output	**[8; ·; ·; ·; ·; ·]**

which can be expanded into a more detailed listing easily.

Input		**[1; ·; ·; ·; ·; ·]**
Left Brain		**[2; ·; ·; ·; ·; ·]**
	Sensory Model One	**[2; 1; ·; ·; ·; ·]**
	Sensory Model Two	**[2; 2; ·; ·; ·; ·]**
	Associative Cortex	**[2; 3; ·; ·; ·; ·]**
	Motor Cortex	**[2; 4; ·; ·; ·; ·]**
Corpus Callosum		**[3; ·; ·; ·; ·; ·]**
Right Brain		**[4; ·; ·; ·; ·; ·]**
	Sensory Model One	**[4; 1; ·; ·; ·; ·]**
	Sensory Model Two	**[4; 2; ·; ·; ·; ·]**
	Associative Cortex	**[4; 3; ·; ·; ·; ·]**
	Motor Cortex	**[4; 4; ·; ·; ·; ·]**
Thalamus		**[5; ·; ·; ·; ·; ·]**
MidBrain		**[6; ·; ·; ·; ·; ·]**
Cerebellum		**[7; ·; ·; ·; ·; ·]**
Output		**[8; ·; ·; ·; ·; ·]**

For example, if we used a single **column** with three **cans** for a cortex module, we could add a simple thalamus model using one reversed OCOS circuit as shown in Eq. 19.3. In this picture, we have addressed the thalamus as $[0; 5; 0; 1; 1; 1 - 7]$ but depending on how we set up the full model, this could change.

can 3	$L1$	y	FFP y's 3	$[0; 0; 0; 3; 2; 1 - 2]$
	$L2\ z$ $\quad z \quad z$	Two/Three z's 3	$[0; 0; 0; 3; 3; 1 - 6]$
	$L3 \quad z$		
	$L4\ z \quad x \quad x \quad x\ z$	OCOS x's 3	$[0; 0; 0; 3; 1; 1 - 7]$
	$\quad x \quad x \quad x$		
	$L5 \quad\quad y$		
	$L6 \quad x$		1 interconnection
can 2	$L1 \quad\quad y$	FFP y's 2	$[0; 0; 0; 2; 2; 1 - 2]$
	$L2\ z \quad z \quad z$	Two/Three z's 2	$[0; 0; 0; 2; 3; 1 - 7]$
	$L3 \quad z$		
	$L4\ z \quad x \quad x \quad x\ z$	OCOS x's 2	$[0; 0; 0; 2; 1; 1 - 7]$
	$\quad x \quad x \quad x$		
	$L5 \quad\quad y$		
	$L6 \quad x$		1 interconnection
can 1	$L1 \quad\quad y$	FFP y's 1	$[0; 0; 0; 1; 2; 1 - 2]$ 2 edges
	$L2\ z \quad z \quad z$	Two/Three z's 1	$[0; 0; 0; 1; 3; 1 - 6]$ 7 edges
	$L3 \quad z$		
	$L4\ z \quad x \quad x \quad x\ z$	OCOS x's 1	$[0; 0; 0; 1; 1; 1 - 7]$ 7 edges
	$\quad x \quad x \quad x$		
	$L5 \quad\quad y$		
	$L6 \quad x$		2 interconnections
thalamus	$\quad t$	Thalamus t's	$[0; 5; 0; 1; 1; 1 - 7]$ 6 edges
	$t \quad t \quad t$		
	$t \quad t \quad t$		

$$(19.3)$$

We see **can 1** is one OCOS, one FFP and one Two/Three for 15 neurons and so here, a three **can column** has 45 neurons.

19.1   Graph Class Two

We need to extend our **graph** class implementation so that we can use six dimensional vectors as addresses. We also want to move seamlessly back and forth from the global node numbers needed for the incidence matrix and the Laplacian and the full address information. Let's start with the vertices or nodes of the graph.

19.1.1   Vertices Two

We have a new constructor as vertices are now six dimensional addresses.

Listing 19.1: Address based vertices

```
function V = vertices(a)
%
% create nodes from the structure a
%
% a has the form
% { [n5;n4;n3;n2;n1;n0], [], [], ... , []}
% where
% n0 = local node number          ie OCOS, FFP, 2/3
% n1 = which local module
% n2 = subsubsubmodule number  ie can
% n3 = subsubmodule number     ie column
% n4 = submodule number        ie sensory cortex
% n5 = module number           ie small brain model
%

%
% s
%
s = struct();
t = struct();
n = length(a);
for i=1:n
    b = a{i};
    t.number = i;
    t.address = a{i};
    s.node5 = b(1);
    s.node4 = b(2);
    s.node3 = b(3);
    s.node2 = b(4);
    s.node1 = b(5);
    s.node0 = b(6);
    V.v(i) = s;
    V.n(i) = t;
end
V = class(V,'vertices');
end
```

There are more options in the **subsref.m** code.

Listing 19.2: Address based vertices subsref

```
function out = subsref(V,s)
%
% give access to private elements
% of the vertices object
%
switch s.type
    case "."
        fld = s.subs;
        if(strcmp(fld,"v"))
            n = length(V.v);
            out = [V.v(1).node5;V.v(1).node4;V.v(1).node3;...
                   V.v(1).node2;V.v(1).node1;V.v(1).node0];
            for i=2:n
                out = [out [V.v(i).node5;V.v(i).node4;V.v(i).node3;...
                       V.v(i).node2;V.v(i).node1;V.v(i).node0]];
            end
        elseif(strcmp(fld,"n"))
            n = length(V.n);
            for i=1:n
```

```
20            out(i).number = V.n(i).number;
              out(i).address    = V.n(i).address;
          end
       else
          error("invalid property for vertices object");
25        end
       otherwise
          error("@vertices/subsref: invalid subscript type for vertices")
              ;
      end
  end
```

We can add a node one at a time. Again, this inefficiency arises from the fact that we can't simply glue the incoming data to the end of the existing node list.

Listing 19.3: Add a single node with address based vertices add

```
1 function W = add(V,u)
  %
  % node is added to vertex list
  %
  s = struct();
6 W = V.v;
  n = length(W);
  for i=1:n
     s.node5 = W(i).node5;
     s.node4 = W(i).node4;
11   s.node3 = W(i).node3;
     s.node2 = W(i).node2;
     s.node1 = W(i).node1;
     s.node0 = W(i).node0;
     out{i} = [s.node5;s.node4;s.node3;s.node2;s.node1;s.node0];
16 end
     s.node5 = u(1);
     s.node4 = u(2);
     s.node3 = u(3);
     s.node2 = u(4);
21   s.node1 = u(5);
     s.node0 = u(6);
     out{n+1} = [s.node5;s.node4;s.node3;s.node2;s.node1;s.node0];
     W = vertices(out);
     end
```

Given that we can't concatenate, it would be more efficient to add lists of node simultaneously.

Listing 19.4: Add a node list address based vertices addv

```
  function W = addv(V,nodes)
  %
  % group of nodes are added to node list
  %
5 s = struct();
  W = V.v;
  n = length(W);
  [m,sizenodes] = size(nodes);
  for i=1:n
10   s.node5 = W(i).node5;
     s.node4 = W(i).node4;
     s.node3 = W(i).node3;
     s.node2 = W(i).node2;
     s.node1 = W(i).node1;
```

```
15      s.node0  = W( i ).node0;
        out{i}  =  [s.node5;s.node4;s.node3;s.node2;s.node1;s.node0];
      end
      for  i = 1:sizenodes
        s.node5  =  nodes{i}(1,1);
20      s.node4  =  nodes{i}(2,1);
        s.node3  =  nodes{i}(3,1);
        s.node2  =  nodes{i}(4,1);
        s.node1  =  nodes{i}(5,1);
        s.node0  =  nodes{i}(6,1);
25    out{n+i}  =  [s.node5;s.node4;s.node3;s.node2;s.node1;s.node0];
      end
    W =  vertices(out);
    end
```

19.1.2   Edges Two

We now have edges which are pairs of six dimensional addresses. Here is the constructor.

Listing 19.5: Address based edges

```
    function  E =  edges(a)
2 %
  % create  edges  from  the  vector  of  pairs  a
  % of  the  form  [in, out]
  %
  %
7 % The  edges  come  in  as  a  pair  of  addresses
  % of  the  form
  % {  [n5;n4;n3;n2;n1;n0]_in ,  [n5;n4;n3;n2;n1;n0]_out }
  %
    s =  struct();
12 t =  struct();
   n =  length(a);
   for  i=1:n
     s.in  =  a{i}(:,1);
     s.out =  a{i}(:,2);
17   u =  [s.in,s.out];
     E.e{i}    = u;
     t.number = i;
     t.edge   = u;
     E.n{i}    = t;
22 end
   E =  class(E,'edges');
   end
```

The overloaded **subsref.m** function is similar to what we did before.

Listing 19.6: Address based edges subsref

```
function out = subsref(E,s)
%
% give access to private elements
% of edges
5 %
    switch s.type
      case "."
        fld = s.subs;
        if(strcmp(fld,"e"))
10        n = length(E.e);
          out{1} = [E.e{1}(:,1),E.e{1}(:,2)];
          for i=2:n
            out{i} = [E.e{i}(:,1),E.e{i}(:,2)];
          end
15      elseif(strcmp(fld,"n"))
          n = length(E.n);
          for i=1:n
            out(i).number = i;
            out(i).edge = [E.e{i}(:,1),E.e{i}(:,2)];
20        end
        else
          error("invalid property for edges object");
        end
      otherwise
25      error("@edges/subsref: invalid subscript type for edges");
    end
end
```

We can add a single edge at a time.

Listing 19.7: Add a single edge address based edge add

```
function W = add(E,u,v)
%
3 % edge is added to edge list
%
s = struct();
F = E.e;
n = length(F);
8 for i=1:n
    s.in  = F{i}(:,1);
    s.out = F{i}(:,2);
    out{i} = [s.in,s.out];
  end
13 out{n+1} = [u,v];
  W = edges(out);
end
```

However, it is more efficient to add a list of edges at one time.

Listing 19.8: Add an edge list address based edges addv

```
function W = addv(E,links)
%
% group of edges are added to edge list
%
5 s = struct();
F = E.e;
[m,sizeedges] = size(F);
[m,sizelinks] = size(links);
out = {};
```

```
10 for i=1:sizeedges
      s.in  = F{i}(:,1);
      s.out = F{i}(:,2);
      out{i} = [s.in,s.out];
   end
15 for i = 1:sizelinks
      s.in  = links{i}(:,1);
      s.out = links{i}(:,2);
      out{sizeedges+i} = [s.in,s.out];
   end
20 W = edges(out);
   end
```

19.1.3   Graph Class

The graph class is quite similar to what we had before but we have added some options.

- **g.v** returns the vertices object as before. The difference is that the vertices are now six dimensional addresses.
- **g.e** returns the usual edge object. Here, the difference is that the edge pairs are not pairs of six dimensional addresses.
- **g.n** returns the nodes data structure obtained from the **nodes = V.n** expression. This gives us essentially a structure where **nodes(i)** is a structure whose **number** field is the global node number and whose address field is the node's six dimensional address.
- **g.l** returns the edges data structure obtained from the **links = E.n;** expression. This gives us essentially a structure like the one above: the **number** field is the global node number and the address field is the node's six dimensional address.

For example, if we had the code snippet

Listing 19.9: Setting global locations

```
vOCOS = { [0;0;0;0;0;1],[0;0;0;0;0;2],[0;0;0;0;0;3],...
          [0;0;0;0;0;4],[0;0;0;0;0;5],[0;0;0;0;0;6],...
          [0;0;0;0;0;7]};
eOCOS = {  [[0;0;0;0;0;1],[0;0;0;0;0;2]],...
5          [[0;0;0;0;0;1],[0;0;0;0;0;3]],...
           [[0;0;0;0;0;1],[0;0;0;0;0;4]],...
           [[0;0;0;0;0;2],[0;0;0;0;0;5]],...
           [[0;0;0;0;0;3],[0;0;0;0;0;6]],...
           [[0;0;0;0;0;4],[0;0;0;0;0;7]],...
10         [[0;0;0;0;0;1],[0;0;0;0;0;6]]};
VOCOS = vertices(vOCOS);
EOCOS = edges(eOCOS);
OCOS = graphs(VOCOS,EOCOS);
```

Then, here is what we would find for the various parts of the graph object **OCOS**. The **OCOS.v** data is a matrix of addresses. Each column is a node address. In the code above, we first construct an OCOS graph whose nodes are

Listing 19.10: Initial OCOS node addresses

```
    vOCOS = OCOS.v
    vOCOS =

        0    0    0    0    0    0    0
  5     0    0    0    0    0    0    0
        0    0    0    0    0    0    0
        0    0    0    0    0    0    0
        0    0    0    0    0    0    0
        1    2    3    4    5    6    7
```

Note that our OCOS circuit now does not contain a node 8 as that is a thalamus neuron and we will model it separately later. Then, we use the methods **addlocationtonodes** to relabel the node addresses and and **addlocation toedges** to relabel the edge addresses with the fundamental building block address [0; 0; 0; 0; 1; ·]. This gives

Listing 19.11: OCOS addresses after updated location

```
    locationOCOS = [0;0;0;0;1;0];
    OCOS=addlocationtonodes(OCOS,locationOCOS);
    OCOS=addlocationtoedges(OCOS,locationOCOS);
    vOCOS = OCOS.v
  5 vOCOS =

        0    0    0    0    0    0    0
        0    0    0    0    0    0    0
        0    0    0    0    0    0    0
 10     0    0    0    0    0    0    0
        1    1    1    1    1    1    1
        1    2    3    4    5    6    7
```

For example, the address of the third node is

Listing 19.12: Updated node 3 address

```
    vOCOS(:,3)
    ans =
  3
        0
        0
        0
        0
  8     1
        3
```

which reflects what we said before. The OCOS circuit is a basic building block whose address is [0; 0; 0; 0; 1; ·]. This neuron is the third one in that circuit and so its address is [0; 0; 0; 0; 1; 3]. Since we want to maintain a list of global nodes, the next data structure maintains a table whose first column is the global node number and whose second column is the address of that node.

Listing 19.13: OCOS vertices data

```
1 nOCOS = OCOS.n
  nOCOS =
  {
      1x7  struct  array  containing  the  fields:

6         number
          address
  }
```

Hence, to access the third node, we type

Listing 19.14: Accessing node 3 address

```
  nOCOS(3)
2 ans =
  {
        number  =   3
        address =

7         0
          0
          0
          0
          1
12        3

  }
```

Then we can ask what the global node number is as well as the six dimensional address.

Listing 19.15: Find OCOS global node numbers

```
1 lOCOS = OCOS.l
  lOCOS =
  {
      1x7  struct  array  containing  the  fields:
```

```
 6        number
          edge
     }

    lOCOS(3)
11 ans =
       scalar   structure   containing   the   fields :
          number  =   3
          edge  =
               0     0
16             0     0
               0     0
               0     0
               1     1
               1     4
21 }
```

The **OCOS.1** command returns a data structure whose fields are the global edge number and an edge. Recall an edge is between the **in node** and the **out node**. So, here edge is a pair of six dimensional addresses with the first address begin the the address of the **in node** the second one being the address of the **out node**. For example, the information on the third edge is

Listing 19.16: OCOS global node 3 addresses

```
    lOCOS(3) . number
    ans = 3
```

To extract the edges, we use the **.edge** command.

Listing 19.17: OCOS global node 3 addresses

```
    lOCOS(3) . edge
    ans =
         0     0
         0     0
 5       0     0
         0     0
         1     1
         6     3
```

The **in** and **out** parts of the edge correspond to the first and second column, respectively.

Listing 19.18: Accessing OCOS edge 3 in and out data

```
    lOCOS(3).edge(:,1)
    ans =
        0
        0
5       0
        0
        1
        6
    lOCOS(3).edge(:,2)
10  ans =
        0
        0
        0
        0
15      1
        3
```

With all the above said, we can now look at the constructor and see how we implemented this.

Listing 19.19: Address based graph constructor graph

```
    function g = graphs(V,E)
    %
    % constructor for graph g
4   %
    % V = vertices object
    % E = edges object
    %
    g.v = V;
9   g.e = E;
    nodes = V.n;
    g.n = nodes;
    links = E.n;
    g.l = links;
14  g = class(g,'graphs');
    end
```

We overload **subsref.m** as follows. There are the usual cases.

Listing 19.20: Address based graph constructor graph subsref

```
    function out = subsref(g,s)
    %
    % give access to private elements
    % of the graph
5   %
    switch s.type
       case "."
          fld = s.subs;
          if(strcmp(fld,"v"))
10           Nodes = g.v;
             out = Nodes.v;
          elseif (strcmp(fld,"e"))
             Edges = g.e;
             out = Edges.e;
```

```
15        elseif (strcmp(fld,"n"))
            out = g.n;
          elseif (strcmp(fld,"l"))
            out = g.l;
          else
20          error("invalid property for graph object");
          end
        otherwise
          error("@vertices/subsref: invalid subscript type for graphs");
      end
25 end
```

19.2   Class Methods Two

First, let's look at how we add the *mask* or *global address* to a given graph.

19.2.1   Add Location Methods

We begin with the code to add a *mask* or *location* to each node and vertex of a graph.

Listing 19.21: Address based addlocationtonodes

```
  function g = addlocationtonodes(g,vector)
  %
  %
  %
5 nodes = g.v;
  V = nodes.v;
  [n,nodesize] = size(V);
  for i=1:nodesize
    B{i} = V(:,i)+vector;
10 end
  W = vertices(B);
  g.v = W;
  nodes2 = W.n;
  g.n = nodes2;
15 end
```

We then add locations to edges.

Listing 19.22: Address based addlocationtoedges

```
  function g = addlocationtoedges(g,vector)
  %
  %
  %
5 u = g.e;
  E = u.e;
  [n,edgesize] = size(E);
    for i = 1:edgesize
      B{i} = [E{i}(:,1) + vector ,E{i}(:,2) + vector];
10   end
  F = edges(B);
  g.e = F;
  links = F.n;
  g.l = links;
15 end
```

19.2.2 Add Edge Methods

We can add one edge as follows.

Listing 19.23: Address based addedge

```
  function W = addedge(g,u,v)
  %
  % adjoin an edge to an existing graph
  %
5 E = add(g.e,u,v);
  W = graphs(g.v,E);
  end
```

We can also add a list of edges.

Listing 19.24: Address based addedgev

```
  function W = addedgev(g,links)
  %
3 % adjoin an edge to an existing graph
  %
  E = addv(g.e,links);
  W = graphs(g.v,E);
  end
```

19.2.3 Add Node Methods

We can add a single node.

Listing 19.25: Address based addnode

```
  function W = addnode(g,j)
  %
3 % adjoin a node to an existing graph
  %
  V = add(g.v,j);
  W = graphs(V,g.e);
  end
```

We can also add a list of nodes.

Listing 19.26: Address based addnodev

```
  function W = addnodev(g,nodes)
  %
3 % adjoin nodes to an existing graph
  %
  F = addv(g.v,nodes);
  W = graphs(F,g.e);
  end
```

19.2.4   Finding the Incidence Matrix

This is now harder than you might think as the incidence matrix requires us to use global node and edge numbers and all of our addresses are six dimensional vectors. The basic idea is that once we create our graph with six dimensional addresses, we want to create a data structure B which links the pairs (**global node number**, **node address**) somehow. We will do this by creating a vector B. We come up with a function ϕ which assigns to each vector address a and *unique* integer $\phi(a)$. We then set the entry $B(\phi(a))$ in the matrix to have the value of the global node number that is linked to this address. So for example, say the global node of $a = [0; 2; 3; 0; 1; 2]$ is 8 and $\phi(a) = 46$. Then we set $B(46) = 8$. The other rows of B are given values of 0. So it would be nice if we kept the size of B small and had as few 0's in it as possible. Hence, $B(\phi(a))$ is the global node number corresponding to the address a. A standard way to do this in other programming languages is to create what is called an **associative array** C which would allows us to use the vector address as a *key* whose value is the corresponding node i; hence, we could write $A([0; 2; 3; 0; 1; 2])$ if that address is in the graph and this would return the global node for that address. This is called a reverse lookup on the six dimensional address. Now **MatLab/Octave** does not have *associative containers* so we cannot use the six dimensional address as a *key*. Now once, we have created our function ϕ (called a hashing function), we can calculate the associative array B and write a function to do the reverse lookups. Now in our graph data structure for a graph **g, nodes = g.n** gives us a structure where **nodes(i).number** gives the global node number i and **nodes(i).address** is the corresponding vector address. The function **link2global** will take a six dimensional address and find the global node number. It has the following code.

Listing 19.27: Link2global

```
    function node = link2global(nodes,address)
    %
    % find global node number for a given address
    % in the data structure nodes where
  5 % nodes(i).address(j) is the value of the jth slot of
    % the vector address.
    %
    [sizeaddress,n] = size(address);
    [m,sizeA] = size(nodes);
 10 b = zeros(sizeaddress,1);
    % loop through the data structure nodes.
    % The only way we can match address is if
    % all the entries of b sum to zero.
    for i = 1:sizeA
 15   for j = 1:sizeaddress
        b(j) = (address(j) - nodes(i).address(j))^2;
      end
      if sum(b) == 0
        node = i;
 20     break;
      end
    end

    end
```

The most straightforward way to build the incidence matrix is then to do a loop as follows.

Listing 19.28: Straightforward incidenceOld

```
 1 function K = incidenceOld(g)
   %
   % g.v is the vertices object of the graph
   % g.e is the edge object of the graph
   %
 6 % get node associative array from the vertices
   V = g.v;
   nodes = g.n;

   % get link associative array from the edges
11 E = g.e;
   edges = g.l;

   % get sizes
   [row,sizeV] = size(nodes);
16 [row,sizeE] = size(edges);
   K = zeros(sizeV,sizeE);

   for i = 1:sizeV
     for j = 1:sizeE
21     % associative array from V gives correspondence
       % global node number -> actual hierarchical address
       a = nodes(i).number;
       b = nodes(i).address;
       % get global edge number
26     c = edges(j).number;
       % get address pair for this global edge number
       d = edges(j);
       %
       % this is expensive as in general
31     % it is a linear search
       %
       u = link2global(nodes,d(:,1));
       v = link2global(nodes,d(:,2));
       if u == i
36       K(i,j) = 1;
       else if v == i
         K(i,j) = -1;
       end
     end
41 end

   end
```

Let's count this out. For a typical small brain model of 500 nodes and 3000 edges, a **link2global** call takes about 500 or less iterations. We need two calls for each edge. So we have about 1000 iterations for each edge. But we have 3000 edges, so that puts us up to 3 million iterations to loop through the edges for each node. Since there are 500 nodes, this leaves us with an iteration count of 1.5 billion. **So when we use this method on a small brain model, it takes a long time to return with the incidence matrix**!

A better way to to use the method **Address2Global** which loops through the nodes of the graph and assigns to each node six dimensional address its unique global node number via our choice of hash function ϕ. One way to do this is to take the six dimensional address, convert it to a string and then convert that string to an integer. This creates a list of unique integers. This conversion is what is called a *hashing function*. We then set up a large vector B as follows. We initialize B to all zeroes. If a six dimensional address corresponded to the unique integer N and global node number I, we would set $B(N) = I$. For example, if the address is [0; 2; 3; 2; 1; 2]— this neuron is in sensory module 2, column 3, can 2 and is neuron 2 in the fundamental OCOS building block. This address converts to the string **'023212'** which then converts to some integer; the simplest conversion would be to set the integer to be 23212. However, we can do other conversions too. In this simplest conversion choice, if the global node of this neuron was, say, 67, we would then set $B(23212) = 67$. This uses storage to offset the search cost; in fact, the size of B becomes enormous and we will find out we can actually run out of memory! But this method is straightforward to code and it is instructive to see how we can refine our ideas and make them better with reflection! The code for this hashing function is given below. As mentioned, this is clearly inefficient as there are mostly zeros in this vector. The loop cost here is the size of the node list—for a small brain model about 500.

Listing 19.29: Address2Global

```
   function [Vertex,Address,B] = Address2Global(g)
   %
   % this function assigns to the unique
   % integer link corresponding to a node address
 5 % its global node number.
   % g.v is the vertices object of the graph
   %
   % get node associative array from the vertices
   V = g.v;
10 nodes = g.n;
   [row,sizeV] = size(nodes);
   %
   Vertex = zeros(1,sizeV);
   Address = zeros(1,sizeV);
15 for i=1:sizeV
      p = nodes(i).number;
      Vertex(1,i) = nodes(i).number;
      A = nodes(i).address;
      a = str2num( Address2String(A ));
20    Address(1,i) = a;
   end
   biggest = max(Address);
   B = zeros(1,biggest);
   for i = 1:sizeV
25    B(Address(1,i)) = Vertex(1,i);
   end
   end
```

The code above uses the utility function **Address2String**. This converts a six dimensional address into a string. Now we have to be careful here. Here is an example: suppose we have 3 nodes with addresses [0, 2, 32, 1, 1, 1], [0, 2, 3, 2, 11, 1] and [0, 2, 3, 2, 11, 1]. Then all 3 addresses will give the same integer 0232111 which is a hash collision. Hence, the code in this **Address2String** can generate hash collisions.

Listing 19.30: Address2String

```
function S = Address2String(A)
%
n = length(A);
S = [num2str(A(1))];
for i=2:n
    S = [S,num2str(A(i))];
end

end
```

We can fix this by finding the maximum number of nodes for each level. Our addresses have the form $[n_5, n_4, n_3, n_2, n_1, n_0]$ which in Matlab is encoded as **[n6,n5,n4,n3,n2,n1]**. We can easily find the maximum number of possibilities in each address entry. We use this function.

Listing 19.31: GetAddressMaximums

```
function [M,T] = GetAddressMaximums(G)
%
% G is the graph
%
nG = G.v;
[m,n] = size(nG);
M = zeros(m,1);
for i=1:m
    M(i) = max(nG(i,:));
end
T = zeros(m,1);
T(1) = M(1)+1;
for i=2:m
    T(i) = (M(i)+1)*T(i-1);
end
```

Then, once we have these numbers, we add an appropriate offset to the hash calculations. For example, if the n_0 address had a maximum range of $2-2$, we would add $22 + 1$ to all the entries in the n_1 slot. If the n_1 slot has a maximum range of 9, we would add $(22 + 1)(9 + 1)$ to all the n_1 entries. This will ensure that each address is unique. We then use these altered addresses to compute the hash value. We will still use the original **Address2String** but we will now alter the addresses we send in. However, using the string conversions generates very large addresses. So let's try it again. Instead of using the string conversions, we will use the address maximums in each slot directly with the function **FindGlobalAddress** which calculates the number of unique addresses we need more directly. This code is shown below

Listing 19.32: FindGlobalAddresses

```
function a = FindGlobalAddress(A,gT)
%
n = length(A);
a = A(1);
for i=2:n
    a = a + A(i)*gT(i-1);
end

end
```

The code for the new incidence matrix calculation then has this form.

Listing 19.33: Incidence function form

```
function K = incidence(g)

end
```

Then we get the address maximums for each slot.

Listing 19.34: Get address slot maximums

```
% get maximum values for each address component
[gM,gT] = GetAddressMaximums(g);
```

The first thing we do is get size information and set up a blank incidence matrix.

Listing 19.35: Set up blank incidence matrix

```
%
% g.v is the vertices object of the graph
% g.e is the edge object of the graph
%
% get node associative array from the vertices
V = g.v;
nodes = g.n;

% get link associative array from the edges
E = g.e;
edges = g.l;

% get sizes
[row,sizeV] = size(nodes);
[row,sizeE] = size(edges);
K = zeros(sizeV,sizeE);
```

Then, we loop through the edges and convert all the six dimensional address pairs to integers as we described. For a small brain model, this takes about 3000 iterations. These integer pairs are stored in the structure **out**.

Listing 19.36: Convert address to unique hash values

```
  % setup incidence matrix
  %
  d = edges(1).edge;
  %
5 u = FindGlobalAddress(d(:,1),gT);
  v = FindGlobalAddress(d(:,2),gT);
  out{1} = [u,v];
  for i=2:sizeE
    d = edges(i).edge;
10   u = FindGlobalAddress(d(:,1),gT);
    v = FindGlobalAddress(d(:,2),gT);
    out{i} = [u,v];
  end
```

We then call the **Address2Global** function which will set up a link between the node address and the global node number. This is stored in a matrix B; i.e. $B(A_i) = V_i$ where A_i is the node address and V_i is the global node number. The code also uses the address slot maximums.

Listing 19.37: Address2Global New

```
  function [Vertex,Address,B] = Address2Global(g)
  %
  % this function assigns to the unique
  % integer link corresponding to a node address
5 % its global node number.
  % g.v is the vertices object of the graph
  % get maximum values for each address component
  [gM,gT] = GetAddressMaximums(g);
  %
10 % get node associative array from the vertices
  V = g.v;
  nodes = g.n;
  [row,sizeV] = size(nodes);
  %
15 Vertex = zeros(1,sizeV);
  Address = zeros(1,sizeV);
  for i=1:sizeV
    Vertex(1,i) = nodes(i).number;
    A = nodes(i).address;
20   a = FindGlobalAddress(A,gT);
    Address(1,i) = a;
  end
  biggest = max(Address);
  B = zeros(1,biggest);
25 for i = 1:sizeV
    B(Address(1,i)) = Vertex(1,i);
  end

  end
```

This costs about 500 more iterations for a small brain model. We call this function as follows:

Listing 19.38: Calling Address2Global

```
1 [Vertex, Address, B] = Address2Global(g);
```

Finally, we set up the incidence matrix. This is always a long double loop which cost 500×3000 iterations for a small brain model even without the reverse lookup problem. The vector **B** contains our needed reverse lookup information.

Listing 19.39: Setting up the incidence matrix and closing out

```
    for i = 1:sizeV
      for j = 1:sizeE
        d = out{j};
        u = out{j}(1,1);
5       v = out{j}(1,2);
        if B(1,u) == i
          K(i,j) = 1;
          else if B(1,v) == i
            K(i,j) = -1;
10      end
      end
    end

    end
```

The full code for the **incidence** function is then

Listing 19.40: Address based incidence

```
    function K = incidence(g)
    %
    % g.v is the vertices object of the graph
    % g.e is the edge object of the graph
5   %
    % get maximum values for each address component
    [gM,gT] = GetAddressMaximums(g);
    %
    % get node associative array from the vertices
10  V = g.v;
    nodes = g.n;

    % get link associative array from the edges
    E = g.e;
15  edges = g.l;

    % get sizes
    [row, sizeV] = size(nodes);
    [row, sizeE] = size(edges);
20  K = zeros(sizeV, sizeE);
    %
    % setup incidence matrix
    %
    d = edges(1).edge;
25  u = FindGlobalAddress(d(:,1),gT);
    v = FindGlobalAddress(d(:,2),gT);
    out{1} = [u,v];
    for i=2:sizeE
      d = edges(i).edge;
```

```
30    u = FindGlobalAddress(d(:,1),gT);
      v = FindGlobalAddress(d(:,2),gT);
     out{i} = [u,v];
   end

35 [Vertex,Address,B] = Address2Global(g);

   for i = 1:sizeV
     for j = 1:sizeE
       % associative array from V gives correspondence
40     % global node number -> actual hierarchical address
       d = out{j};
       u = out{j}(1,1);
       v = out{j}(1,2);
       if B(1,u) == i
45       K(i,j) = 1;
       else if B(1,v) == i
         K(i,j) = -1;
       end
     end
   end
50 end

   end
```

For example, here is a simple OCOS incidence calculation. We begin by building the OCOS graph.

Listing 19.41: First, setup the OCOS graphj

```
vOCOS = { [0;0;0;0;0;1] ,[0;0;0;0;0;2] ,[0;0;0;0;0;3] ,...
  [0;0;0;0;0;4] ,[0;0;0;0;0;5] ,[0;0;0;0;0;6] ,...
3 [0;0;0;0;0;7]};
  eOCOS = {  [[0;0;0;0;0;1] ,[0;0;0;0;0;2]] ,...
  [[0;0;0;0;0;1] ,[0;0;0;0;0;3]] ,...
  [[0;0;0;0;0;1] ,[0;0;0;0;0;4]] ,...
  [[0;0;0;0;0;2] ,[0;0;0;0;0;5]] ,...
8 [[0;0;0;0;0;3] ,[0;0;0;0;0;6]] ,...
  [[0;0;0;0;0;4] ,[0;0;0;0;0;7]] ,...
  [[0;0;0;0;0;1] ,[0;0;0;0;0;7]]};
  VOCOS = vertices(vOCOS);
  EOCOS = edges(eOCOS);
13 OCOS = graphs(VOCOS,EOCOS);
  locationOCOS = [0;0;0;0;1;0];
  OCOS=addlocationtonodes(OCOS,locationOCOS);
  OCOS=addlocationtoedges(OCOS,locationOCOS);
```

Next, although it is part of the **incidence** code, we can explicitly find the associative array *B* to see what it looks like.

Listing 19.42: The Address2Global Calculation

```
[Vertex,Address,B] = Address2Global(OCOS);
gM =
   0
   0
5  0
   0
   1
   7
gT =
```

```
10      1
        1
        1
        1
        2
15     16
     row  =   1
     sizeV  =   7
     biggest  =   15
```

We see the associative array *B* has 15 rows (the value of **biggest**). There are only two slots in the address vectors that have entries: slot 5 and 6 with corresponding maxima 1 and 7. The algorithm we have described for assigning a global number to each address then finds the associative array *B*. Note we have

Listing 19.43: The actual B matrix

```
   > Vertex =

        1   2   3   4   5   6   7

 5 > Address
   Address =

        3   5   7   9  11  13  15
      > B
10 B =

        0   0   1   0   2   0   3   0   4   0   5   0   6   0   7
```

This is how it works. if **nodes = OCOS.n**, then **nodes(1).number** is 1, the global node. This is stored as **Vertex(1)**. Then **nodes(1).address** is the nodes address which is $[0; 0; 0; 0; 1; 1]'$ which is converted into an integer via our hashing function and stored as **Address(1)** or 3. Thus $B(3) = 1$ and we can see how we assign these unique integers to the global node numbers. The actual incidence matrix is then calculated to be

Listing 19.44: Sample OCOS incidence calculation

```
   KOCOS = incidence(OCOS)
   KOCOS =
        1   1   1   0   0   0   1
       -1   0   0   1   0   0   0
 5      0  -1   0   0   1   0   0
        0   0  -1   0   0   1   0
        0   0   0  -1   0   0   0
        0   0   0   0  -1   0  -1
        0   0   0   0   0  -1   0
```

19.2.5 Get the Laplacian

Once we know how to find the incidence matrix, we can find the Laplacian of the graph easily.

Listing 19.45: Address based laplacian

```
1 function L = laplacian(g)
  %
  % g is a graph having
  % vertices g.v
  % edges g.e
6 %
  K = incidence(g);
  L = K*K';
  end
```

19.3 Evaluation and Update Strategies in Graphs

At any neuron i in our graphs, the neurons which interact with it via a synaptic contact are in the backward set for neuron i, $\mathcal{B}(i)$. Hence, the input to neuron i is the sum

$$\sum_{j \in \mathcal{B}(i)} E_{j \to i} Y(j)$$

where $E_{j \to i}$ is value we assign the edge between j and i in our graph and $Y(j)$ is the output we assign to neuron j. We also know the neurons that neuron i sends its output signal to are in its forward set $\mathcal{F}(i)$. As usual, both of these sets are readily found by looking at the incidence matrix for the graph. We can do this in the code **BFsets** which is a new graph method placed in the **@graphs** directory. This will be more complicated code than the version we have for global node and link numbers. In it, we find the backward and forward sets for each neuron in both the six dimensional address form and the global node form. In a given row of the incidence matrix, the positive 1's tells us the edges corresponding to the forward links and the negative 1's give us the backward edges. So we look at each row of the incidence matrix and calculate the needed nodes for the backward and forward sets.

Listing 19.46: BFsets

```
function [BackA , BackGlobal , ForwardA , ForwardGlobal , ...
            BackEdgeGlobal , ForwardEdgeGlobal ] = BFsets ( g , Kg )
%
% g is the graph
% Kg is the incidence matrix of the graph
% g.e is the edge object of the graph
% g.l is the edge array of the graph
% BackA is the Backward sets with addresses
% BackGlobal is the Backward sets with global node numbers
% ForwardA is the Forward sets with addresses
% ForwardGlobal is the Forward sets with global node numbers
% BackEdgeGlobal is the Bakward edges with global numbers
% ForwardEdgeGlobal is the Forward edges with global numbers
%
% get maximum values for each address component
[gM, gT] = GetAddressMaximums ( g ) ;

% get link associative array from the edges
E = g.e;
edges = g.l ;

% get sizes
[ row , sizeE ] = size ( edges ) ;
%
% setup out matrix
%
d = edges (1) . edge ;
%u = str2num ( Address2String ( d ( : , 1 ) ) ) ;
%v = str2num ( Address2String ( d ( : , 2 ) ) ) ;
u = FindGlobalAddress ( d ( : , 1 ) , gT ) ;
v = FindGlobalAddress ( d ( : , 2 ) , gT ) ;
out {1} = [ u , v ] ;
for  i =2: sizeE
   d = edges ( i ) . edge ;
   %u = str2num ( Address2String ( d ( : , 1 ) ) ) ;
   %v = str2num ( Address2String ( d ( : , 2 ) ) ) ;
   u = FindGlobalAddress ( d ( : , 1 ) , gT ) ;
   v = FindGlobalAddress ( d ( : , 2 ) , gT ) ;
   out { i } = [ u , v ] ;
end

[ Vertex , Address , B ] = Address2Global ( g ) ;

[ Kgrows , Kgcols ] = size ( Kg ) ;

BackA = { } ;
ForwardA = { } ;
BackGlobal = { } ;
ForwardGlobal = { } ;
BackEdgeGlobal = { } ;
for  i = 1: Kgrows
   BackA { i } = [] ;
   ForwardA { i } = [] ;
   BackGlobal { i } = [] ;
   ForwardGlobal { i } = [] ;
   BackEdgeGlobal { i } = [] ;
   ForwardEdgeGlobal { i } = [] ;
   for  j = 1: Kgcols
      d = edges ( j ) . edge ;
      u = d ( : , 1 ) ;
      v = d ( : , 2 ) ;
      a = out { j } ( 1 , 1 ) ;
      b = out { j } ( 1 , 2 ) ;
      if  Kg ( i , j ) == 1
```

```
65      ForwardA{i} = [ForwardA{i}, v];
        ForwardGlobal{i} = [ForwardGlobal{i}, B(1,b)];
        ForwardEdgeGlobal{i} = [ForwardEdgeGlobal{i},j];
      else if Kg(i,j) == -1
        BackA{i} = [BackA{i},u];
70      BackGlobal{i} = [BackGlobal{i}, B(1,a)];
        BackEdgeGlobal{i} = [BackEdgeGlobal{i},j];
      end
    end
  end
75 end
```

Let's see how this works with a simple example. We again build an OCOS module
with 7 nodes and 7 edges.

Listing 19.47: Build an OCOS Graph

```
  vOCOS = {[0;0;0;0;0;1],[0;0;0;0;0;2],[0;0;0;0;0;3],...
    [0;0;0;0;0;4],[0;0;0;0;0;5],[0;0;0;0;0;6],...
    [0;0;0;0;0;7]};
  eOCOS = { [[0;0;0;0;0;1],[0;0;0;0;0;2]],...
5 [[0;0;0;0;0;1],[0;0;0;0;0;3]],...
  [[0;0;0;0;0;1],[0;0;0;0;0;4]],...
  [[0;0;0;0;0;2],[0;0;0;0;0;5]],...
  [[0;0;0;0;0;3],[0;0;0;0;0;6]],...
  [[0;0;0;0;0;4],[0;0;0;0;0;7]],...
10 [[0;0;0;0;0;1],[0;0;0;0;0;6]]};
  VOCOS = vertices(vOCOS);
  EOCOS = edges(eOCOS);
  OCOS = graphs(VOCOS,EOCOS);
  locationOCOS = [0;0;0;0;1;0];
15 OCOS=addlocationtonodes(OCOS,locationOCOS);
  OCOS=addlocationtoedges(OCOS,locationOCOS);
  KOCOS = incidence(OCOS);
  [BackA, BackGlobal, ForwardA, ForwardGlobal,...
    BackEdgeGlobal, ForwardEdgeGlobal] = BFsets(OCOS,KOCOS);
```

Let's generate the graphical file for the OCOS graph.

Listing 19.48: Build OCOS.dot

```
1 incToDot(KOCOS,8,6,1.0,'ocos.dot');
```

Now create the graphical file which we show in Fig. 19.1.

Listing 19.49: Create AddressOCOS.pdf

```
dot -Tpdf -o AddressOCOS.pdf OCOS.dot
```

Now the backward global information here is $\mathcal{B}\{1\} = \{\}$, $\mathcal{B}\{2\} = \{1\}$, $\mathcal{B}\{3\} = \{1\}$,
$\mathcal{B}\{4\} = \{1\}$, $\mathcal{B}\{5\} = \{2\}$, $\mathcal{B}\{6\} = \{3, 1\}$ and $\mathcal{B}\{7\} = \{4\}$. This information is stored
in **BackGlobal** which looks like

Fig. 19.1 Typical OCOS
graph

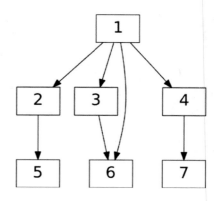

Listing 19.50: BackGlobal

```
BackGlobal
BackGlobal =
{
     [1 ,1]  =  [](0 x0)
5    [1 ,2]  =   1
     [1 ,3]  =   1
     [1 ,4]  =   1
     [1 ,5]  =   2
     [1 ,6]  =
10
          3    1

     [1 ,7]  =   4
}
```

To access $B\{i\}$ in code, we would then use

Listing 19.51: Access BackGlobal Entries

```
1 BackGlobal{2}
  ans =   1
  BackGlobal{6}
  ans =
     3  1
```

and so on. The forward sets are similar and the other data structures give the same
information but list everything using the full vector addresses. Thus, **BackA{2}**
gives the same information as **BackGlobal{2}** but gives the addresses instead of
global node numbers.

Listing 19.52: BackA{6}

```
BackA{2}
ans =
     0     0
     0     0
 5   0     0
     0     0
     1     1
     3     1
```

The simplest evaluation strategy then is to let each neuron have an output value
determined by the usual simple sigmoid function with code in **sigmoid.m**. We use
this sigmoid to do a simple graph evaluation. We then evaluate the sigmoid functions
at each node using the current synaptic interaction values just as we did in the global
graph code.

Listing 19.53: A Simple Evaluation Loop

```
   function [NodeVals] = evaluation(Y,W,B,BE)
   %
   % g is the graph
   % Y is node vector
 5 % B is the global backward node set information
   % BE is the global backward edge set information
   %
   % get size
10 sizeV = length(Y);

   for i = 1:sizeV
     % get backward node information for neuron i
     BF = B{i};
15   % get backward edge information for neuron i
     BEF = BE{i};
     lenB = length(BF);
     lenBEF = length(BEF);
     sum = 0.0;
20   for j = 1:lenBEF
       link = BEF(j);
       pre = BF(j);
       sum = sum + W(link)*Y(pre);
     end
25   NodeVals(i) = sigmoid(sum);
   end
```

A simple edge update function is then given by in the file below. This uses what is
called a Hebbian update to change the scalar weight associated with each edge. If
the value of the *post* neuron i is *high* and the value of the edge $E_{j \to i}$ is also high,
the value of this edge is increased by a multiplier such as 1.05 or 1.1. The code to
implement is the same as the global graph code and is not shown.

A simple example is shown below for the evaluation step for a brain model such
as we will build in Chap. 20. This discussion is a little out of place, but we think it is
worth the risk. The next chapter shows how we build a human brain model consisting
of many modules linked together and in subsequent chapters we also build models
of a cuttlefish and pigeon brain. Hence the function **brain** here is a generic place

keeper for any such brain model we build. Of course, to make this really sink in, you'll have to read the next chapter also and go back and forth a bit to get it straight!

So to set up the evaluation, first, we set up a node value vector Y the size of the nodes and a weight value vector W the size of edges of the brain model. Here we will also use an offset and gain vector. For the evaluation, we still assume the neuron or node value is given by the equation

$$\sum_{j \in B(i)} E_{j \to i} \, Y(j)$$

where we now posit that each edge $E_{j \to i}$ is simply a value called a *weight*, W_{ji} and the neuron evaluation uses a sigmoid calculation. If x is the input into a neuron and if o is the offset of the neuron and g is its gain, we find the neuronal output is given by

$$Y = 0.5 \left(1 + \tanh\left(\frac{x - o}{g} \right) \right).$$

So we will alter the evaluation code a bit to allow us to do this more interesting sigmoid calculation.

Listing 19.54: A Evaluation Loop With Offsets and Gains

```
   function [NodeVals] = evaluation(g,Y,W,O,G,B,BE)
   %
   % g is the graph
   % Y is node vector
 5 % W is edge values
   % O is the offset vector
   % G is the gain vector
   % B is the global backward node set information
   % BE is the global backward edge set information
10 %

   % get size
   sizeV = length(Y);

15 for i = 1:sizeV
      % get backward node information for neuron i
      BF = B{i};
      % get backward edge information for neuron i
      BEF = BE{i};
20    lenB = length(BF);
      lenBEF = length(BEF);
      sum = 0.0;
      for j = 1:lenBEF
         link = BEF(j);
25       pre = BF(j);
         sum = sum + W(link)*Y(pre);
      end
      X = ( sum - O(i) )/ G(i);
      Y(i) = sigmoid(X);
30 end

   NodeVals = Y;

   end
```

We need to set up the offset vector and gain vector to be the size of the graphs vertices and initialize them in some fashion. We fill the weight vector randomly with numbers from -1 to 1. Then we compute the needed backward set information and use it to perform an evaluation step. Then we do a synaptic weight Hebbian update. At this point, we are just showing the basic ideas of this sort of computation. We still have to restrict certain edges to have only negative values as they correspond to inhibitory synaptic contacts. Also, we have not yet implemented the cable equation update step afforded us with the graph's laplacian. That is to come. The function **HebbianUpdateSynapticValue** is the same as the one we used in **GraphsGlobal** because it is written using only global graph information which has been extracted from the address based graph. Here is how we would evaluation the values (i.e. neuron values) at each node in a brain model. We are jumping ahead of course, as we have not yet discussed how to build such a brain model, but this code snippet will give you the idea. We start by building our brain model.

Listing 19.55: Build A Brain Model

```
   [ NodeSizes , EdgeSizes , Brain ]  =  buildbrain (1 ,1 ,1) ;
   Build  SensoryOne
   Neurons  1  to  45  are  Sensory  One
   Links  1  to  61  are  Sensory  One
 5 Build  SensoryTwo
   Neurons  46  to  90  are  Sensory  Two
   Links  62  to  122  are  Sensory  Two
   Build  AssociativeCortex
   Neurons  91  to  270  are  Associative  Cortex
10 Links  123  to  372  are  Associative  Cortex
   Build  MotorCortex
   Neurons  271  to  315  are  Motor  Cortex\n
   Links  373  to  433  are  Associative  Cortex\n
   Build  Thalamus
15 Neurons  316  to  329  are  Thalamus
   Links  434  to  447  are  Thalamus
   Build  MidBrain
   have  dopamine  edges
   add  serotonin  edges
20 add  norepinephrine  edges
   Neurons  330  to  335  are  MidBrain
   Links  448  to  450  are  MidBrain
   Build  Cerebellum
   Neurons  336  to  380  are  Cerebellum
25 Links  451  to  511  are  Cerebellum
```

Then we construct the brain model's incidence matrix.

Listing 19.56: Construct the Brain Incidence Matrix

```
   KBrain  =  incidence ( Brain ) ;
   [m, n ]  =  size ( KBrain )
   m =  380
   n =  707
 5 W =  HebbianUpdateSynapticValue (Y,W, BGlobal , BEGlobal ) ;
```

Next, get the backward and forward information for this graph.

Listing 19.57: Get Forward and Backward Information

```
[BA, BGlobal , FA, FGlobal , BEGlobal , FEGlobal ]  =...
    BFsets ( Brain , KBrain ) ;
```

Now, initialize the output vector **Y**, the offset and gain vectors, **O** and **G** and the edge weight vector **W**. Here we initialize **W** to be randomly chosen in the range $[-1, 1]$, the offset **O** to be randomly in the range $[-1, 1]$ and the Gain **G** to be randomly in the range $[0.45, 0.9]$.

Listing 19.58: Initializing Y, O, G and W

```
Y = zeros (1 ,380);
O = -1+2*rand (1 ,380);
G = 0.5*(0.9+rand (1 ,380));
W = -1+2*rand (1 ,380);
```

We start with

Listing 19.59: First 5 Initial Node Values

```
1 Y(1:5)
  ans =

      0    0    0    0    0
```

Now evaluate the graph for the first time.

Listing 19.60: First Graph Evaluation

```
1 Y = evaluation ( Brain ,Y,W,O,G, BGlobal , BEGlobal ) ;
```

We have done the first evaluation and now the node values have changed.

Listing 19.61: First 5 Node Values

```
  Y(1:5)
  ans =

4     0.466747    0.729127    0.451896    0.178188    0.066558
```

Now do the first Hebbian Update which updates the weight values on the edges. Follow that by another evaluation to see what happened.

Listing 19.62: The First Hebbian Update

```
1 W = HebbianUpdateSynapticValue(Y,W,BGlobal,BEGlobal);
  Y = evaluation(Brain,Y,W,O,G,BGlobal,BEGlobal);
```

We can see how the new weight values have effected the nodal computations!

Listing 19.63: First 5 Nodal Values After One Hebbian Update

```
  Y(1:5)
  ans =
3
      0.446827    0.704026    0.391186    0.433918    0.051564
```

19.4 Adding Inhibition

Now let's consider how we might train a given graph to have certain target node values for specified inputs. We will illustrate this with a very simple OCOS graph and one input at global node 1 of value 0.3. The specified output is in nodes 5, 6 and 7 of value 0, 1 and 0. We will continue to use simple sigmoid node evaluation engines for now. We encode the input and target information as follows. First the inputs:

Listing 19.64: Setting Up the Input Data

```
1 Input = [0.3;0;0;0;0;0;0]
  Input =

      0.00000
      0.00000
6     0.00000
      0.00000
      0.00000
      0.00000
      0.30000
```

Then the targets:

Listing 19.65: Setting Up Training Data

```
  Target = struct();
  DesiredOutput = {};
  Target.in = 5;  Target.value = 0.0;
  DesiredOutput{1} = [Target.in,Target.value];
5 Target.in = 6;  Target.value = 1.0;
  DesiredOutput{2} = [Target.in,Target.value];
  Target.in = 7;  Target.value = 0.0;
  DesiredOutput{3} = [Target.in,Target.value];
```

We usually call this sort of information *Training Data*. We can check how we did with a few simple tests.

Listing 19.66: Testing the Training Data

```
  DesiredOutput{1}
2 ans =

     5    0

  DesiredOutput{1}(1)
7 ans =   5
  DesiredOutput{1}(2)
  ans = 0
  DesiredOutput{2}(1)
  ans =   6
12 DesiredOutput{2}(2)
  ans =   1
  DesiredOutput{3}(1)
  ans =   7
  DesiredOutput{3}(2)
17 ans = 0
```

Next, set up the OCOS graph.

Listing 19.67: Setup OCOS Graph

```
  vOCOS = {[0;0;0;0;0;1],[0;0;0;0;0;2],[0;0;0;0;0;3],...
           [0;0;0;0;0;4],[0;0;0;0;0;5],[0;0;0;0;0;6],...
3          [0;0;0;0;0;7]};
  eOCOS = { [[0;0;0;0;0;1],[0;0;0;0;0;2]],...
  [[0;0;0;0;0;1],[0;0;0;0;0;3]],...
  [[0;0;0;0;0;1],[0;0;0;0;0;4]],...
  [[0;0;0;0;0;2],[0;0;0;0;0;5]],...
8 [[0;0;0;0;0;3],[0;0;0;0;0;6]],...
  [[0;0;0;0;0;4],[0;0;0;0;0;7]],...
  [[0;0;0;0;0;1],[0;0;0;0;0;6]]};
  VOCOS = vertices(vOCOS);
  EOCOS = edges(eOCOS);
13 OCOS = graphs(VOCOS,EOCOS);
  locationOCOS = [0;0;0;0;1;0];
  OCOS=addlocationtonodes(OCOS,locationOCOS);
  OCOS=addlocationtoedges(OCOS,locationOCOS);
```

We then find the incidence matrix, the backward and forward data and initialize all the graph values.

Listing 19.68: Incidence Matrix, Backward and Forward Data and Initialization

```
  KOCOS = incidence(OCOS);
  [BA,BGlobal,FA,FGlobal,BEGlobal,FEGlobal] = BFsets(OCOS,KOCOS);
  Y = zeros(1,7);
4 O = -1+2*rand(1,7);
  G = 0.5*(0.9+rand(1,7));
  W = rand(1,7)
  W =
      0.527088     0.959165     0.048946     0.374755     0.322442     0.698353
           0.184643
9 % so weights are all positive. Now we set the inhibitory edges.
  W(4) = -W(4);
  W(5) = -W(5);
  W(6) = -W(6);
  W =
14     0.45076    0.81485    0.45086    -0.79840    -0.10357    -0.74836
           0.59966
```

We are setting edges 4, 5 and 6 to be inhibitory as we know this is what the neuro-biology tells us. We do this by manually changes these edge weights to be negative once they are randomly set. We set W to be values randomly chosen between 0 and 1 and then set the inhibitory values. Now we have to do a preliminary evaluation. This uses a new function **evalwithinput**.

Listing 19.69: Preliminary Evaluation With Input

```
Y = evalwithinput (OCOS,Y,W,O,G,Input ,BGlobal ,BEGlobal) ;
Y =
   0.85963     0.67221     0.46610     0.19435     0.63086     0.94294
      0.92217
```

The new function is like the old **evaluation** except it use input data to seed each nodal calculation. Here is the updated function: the only thing that changed is that each sum is initialized to an input value rather than zero.

Listing 19.70: Evaluation With Input Data

```
   function [NodeVals] = evalwithinput (g,Y,W,O,G,Input ,B,BE)
   %
   % g is the graph
   % Y is node vector
5  % W is edge values
   % O is the offset vector
   % G is the gain vector
   % Input is the input vector
   % B is the global backward node set information
10 % BE is the global backward edge set information
   %
   % get size
   sizeV = length (Y) ;
15
   for i = 1: sizeV
      % get backward node information for neuron i
      BF = B{ i };
      % get backward edge information for neuron i
20    BEF = BE{ i };
      lenB = length (BF) ;
      lenBEF = length (BEF) ;
      sum = Input ( i ) ;
      for j = 1: lenBEF
25       link = BEF( j ) ;
         pre = BF( j ) ;
         sum = sum + W( link )*Y( pre ) ;
      end
      X = ( sum - O( i ) )/ G( i ) ;
30    Y( i ) = sigmoid (X) ;
   end

   NodeVals = Y;

35 end
```

Now we need to alter the edge values of the graph so that we can map our input data to our desired target data. We do this with the modified Hebbian training algorithm shown in **HebbianUpdateErrorSignal**. This function performs a single update step. but we still have to write the code!

Listing 19.71: Hebbian Training With an Error Signal

```
Inhibitory = [2;3;4];
scale = 1.1;
Tol = .36;
TargTol = .25;
W = HebbianUpdateErrorSignal(Y,O,G,W,BGlobal,BEGlobal,Inhibitory,
    DesiredOutput,scale,Tol,TargTol)
```

We need to set some tolerances before we do the Hebbian updates: **scale** is the usual multiplier we see in a Hebbian update, **Tol** is the tolerance that determines if an edge weight should be updated because of a correlation with the post nodal value and **TarTol** is the fraction of the error update we use. The new function is listed below. This function is a bit complicated as we are now handling updates differently depending on whether the neuron is inhibitory or excitatory and whether the neuron is an input or a target. We start by taking the given input and target data and use it to set up a data structure which hold this information. The **ValueAndType** structure holds three things: the neuron value **value**; whether it is inhibitory or not **inhibitory = 1** or **inhibitory = 0**; and whether it is a target or not, **target = 1** or **target = 0**. We loop through the neurons and initialize the **ValueAndType** data structure and then store it in a cell array **Neuron** for access later.

Listing 19.72: Initialized the Neuron Data

```
sizeV = length(Y);
sizeE = length(W);
sizeT = length(DesiredOutput);
sizeI = length(Inhibitory);
ValueAndType = struct();
Neuron = {};
for i=1:sizeV
    ValueAndType.value = Y(i);
    ValueAndType.inhibitory = 0;
    ValueAndType.target = 0;
    for k=1:sizeI
        if i == Inhibitory(k);
            ValueAndType.inhibitory = 1;
        end
    end
    for k=1:sizeT
        if i == DesiredOutput{k}(1);
            ValueAndType.target = 1;
            ValueAndType.value = DesiredOutput{k}(2);
        end
    end
    v = ValueAndType.value;
    inh = ValueAndype.inhibitory;
    tar = ValueAndType.target;
    Neuron{i} = [v,inh,tar];
end
```

The update loop is therefore more complicated. In skeleton form, we have

Listing 19.73: Update loop in skeleton form

```
    for i = 1:sizeV
      % get backward node  information for neuron i
      BF = B{i};
4     % get backward edge information for neuron i
      BEF = BE{i};
      lenB = length(BF);
      lenBEF = length(BEF);
      value = Neuron{i}(1);
9     inhibitory = Neuron{i}(2);
      target = Neuron{i}(3);
      % see if we have a target neuron
      TargetMatch = 0;
      IsTarget = 0;
14    if i ==target
         IsTarget = 1;
      end
      %not a target neuron
      if IsTarget == 0
19       for j = 1:lenBEF
           % do usual backward edge evaluation loop
           % if Y post is inhibitory do one thing
           if inhibitory == 1
              ...
24         end
           % if Y post is excitatory do another thing
           if inhibitory == 0
              ...
           end
29         Wts(link) = weight;
         end% backward links loop for non targets
      end
      % a target neuron
      if IsTarget ==1
34       % do the usual backward edge evaluation loop
           % if Y post is inhibitory do one thing
           if inhibitory == 1
              ...
         end
39         % Y post is excitatory do another thing
           if inhibitory == 0
              ...
         end
         Wts(link) = weight;
44       end% backwards links loop for targets
      end % is is a target
    end% neuron loop
```

If we have a target node i, we have the usual evaluation $x_i = \sum_{j \in \mathcal{B}(i)} E_{j \to i} Y(j)$ and the target value is given by $Y(i) = 0.5 \left(1 + \tanh\left(\frac{x(i) - o(i)}{g(i)} \right) \right)$ for the neuron's offset $o(i)$ and gain $g(i)$. The neuron node calculation function is thus given by

$$\sigma(x, o, g) = \left(1 + \tanh\left(\frac{x - o}{g} \right) \right).$$

which has inverse

$$\sigma^{-1}(y, o, g) = \left(o + \frac{g}{2} \ln\left(\frac{y}{1 - y} \right) \right).$$

where $y = \sigma(x, o, g)$. Now if the target value is 1 or 0, this will be unachievable as the inverse will return ∞ and $-\infty$, respectively. So in this case, we replace the inputs by a suitable $1 - \epsilon$ and ϵ for ease of computation. For a given target, we have $\sigma^{-1}(Y(i)) = \frac{x(i) - o(i)}{g(i)}$. This leads to $x(i) = o(i) + g(i)\sigma^{-1}(Y(i))$ and so we have

$$\sum_{j \in B(i)} E_{j \to i} Y(j) = o(i) + g(i)\sigma^{-1}(Y(i)).$$

We then apply Hebbian ideas and choose which of the summands in this sum to update. For example, if the $E_{j \to i} Y(i)$ is sufficiently large, we note

$$E_{j \to i} Y(j) \approx o(i) + g(i)\sigma^{-1}(Y(i))$$

and so we could update using

$$E_{j \to i} \approx \frac{o(i) + g(i)\sigma^{-1}(Y(i))}{Y(j)}.$$

But $Y(j)$ could be zero and we probably don't want to use the full *update*. So we use

$$E_{j \to i} \approx \epsilon_1 \frac{o(i) + g(i)\sigma^{-1}(Y(i))}{Y(j) + \epsilon_2}.$$

The equation above shows how we handle the target updates with slight adjustments for the inhibitory case.

Listing 19.74: Target Updates

```
% set sigmoid inverse function
SigInv = @(y,o,g) ( o + (g/2)*log( (y)./(1-y) ) );
% a target neuron
if IsTarget ==1
    targetinverse = O(i) + G(i)*SigInv(value ,O(i),G(i))
    for j = 1:lenBEF
        link = BEF(j);
        pre = BF(j);
        post = i;
        hebb = Y(post)*W(link);
        weight = W(link);
        inhibitory = Neuron{post}(2);
        % Y post is inhibitory
```

```
14          if inhibitory == 1
            if hebb < -Tol
                weight = TargTol*targetinverse/(Y(pre)+.1);
            end
         end
19       % Y post is excitatory
         if inhibitory == 0
            if hebb > Tol
                weight = TargTol*targetinverse/(Y(pre)+.1);
            end
24       end
         Wts(link) = weight;
      end% backwards links loop for targets
    end % is is a target
```

The full code is below.

Listing 19.75: HebbianUpdateErrorSignal

```
 function Wts = HebbianUpdateErrorSignal(Y,O,G,W,B,BE,Inhibitory ,...
 DesiredOutput ,scale ,Tol ,TargTol)
3 %
 % B is the Backward Global node information
 % BE is the Backward Global Edge information
 % W is the edge vector
 % Y is the new value of edge
8 % Inhibitory is a vector of the nodes that are inhibitory
 % DesiredOutput is a cell of structs
 % each struct is of type Target.in = neuron index
 % and Target.value = neuron value)
 % DesiredOutput{2}(1) gives the index for the second cell entry
13 % DesiredOutput{2}(2) gives the value for the second cell entry
 % scale is the weight rescaling factor
 % Tol is the desired Hebbian tolerance
 % TargTol is the desired Hebbian tolerance for target weight updates
 %
18 % setup sigmoid inverse.
 SigInv = @(y,o,g) ( o + (g/2)*log( (y)./(1-y) ) );

 % get size
 sizeV = length(Y);
23 sizeE = length(W);
 sizeT = length(DesiredOutput);
 sizeI = length(Inhibitory);
 ValueAndType = struct();
 Neuron = {};
28 for i=1:sizeV
     ValueAndType.value = Y(i);
     ValueAndype.inhibitory = 0;
     ValueAndType.target = 0;
     for k=1:sizeT
33      if i == Inhibitory(k);
            ValueAndype.inhibitory = 1;
         end
      end
      for k=1:sizeI
38      if i == DesiredOutput{k}(1);
            ValueAndype.target = 1;
         end
      end
      v = ValueAndType.value;
```

```
43    inh = ValueAndype.inhibitory;
      tar = ValueAndType.target;
      Neuron{i} = [v,inh,tar];
   end

48 for i = 1:sizeV
      % get backward node  information for neuron i
      BF = B{i};
      % get backward edge information for neuron i
      BEF = BE{i};
53    lenB = length(BF);
      lenBEF = length(BEF);
      value = Neuron{i}(1);
      inhibitory = Neuron{i}(2);
      target  = Neuron{i}(3);
58    % see if we have a target neuron
      TargetMatch = 0;
      IsTarget = 0;
      if i ==target
         IsTarget = 1;
63    end
      %not a target neuron
      if IsTarget == 0
         for j = 1:lenBEF
            link = BF(j);
68          pre = BF(j);
            post = i;
            hebb = Y(post)*W(link);
            weight = W(link);
            inhibitory = Neuron{post}(2);
73          % Y post is inhibitory
            if inhibitory == 1
               if hebb < -Tol
                  weight = -scale*abs(W(link))
               end
78          end
            % Y post is excitatory
            if inhibitory == 0
               if hebb > Tol
                  weight = scale*abs(W(link));
83          end
            end
            Wts(link) = weight;
         end% backward links loop for non targets
      end
88    % a target neuron
      if IsTarget ==1
         targetinverse = O(i) + G(i)*SigInv(value,O(i),G(i))
         for j = 1:lenBEF
            link = BEF(j);
93          pre = BF(j);
            post = i;
            hebb = Y(post)*W(link);
            weight = W(link);
            inhibitory = Neuron{post}(2);
98          % Y post is inhibitory
            if inhibitory == 1
               if hebb < -Tol
                  weight = TargTol*targetinverse/(Y(pre)+.1);
               end
```

```
103                end
                   % Y post is excitatory
                   if inhibitory == 0
                     if hebb > Tol
                        weight = TargTol*targetinverse/(Y(pre)+.1);
108                  end
                   end
                   Wts(link) = weight;
              end% backwards links loop for targets
            end % is is a target
113 end% neuron loop

    end
```

Listing 19.76: A Sample One Step Training Session

```
    %
    % Inputs
    %
    Input = [0.3;0;0;0;0;0;0];
 5  %
    % Targets
    %
    Target = struct();
    DesiredOutput = {};
10  Target.in = 5; Target.value = 0.0;
    DesiredOutput{1} = [Target.in,Target.value];
    Target.in = 6; Target.value = 1.0;
    DesiredOutput{2} = [Target.in,Target.value];
    Target.in = 7; Target.value = 0.0;
15  DesiredOutput{3} = [Target.in,Target.value];
    %
    Inhibitory = [2;3;4];
    %
    % Build OCOS
20  %
    vOCOS = {[0;0;0;0;0;1],[0;0;0;0;0;2],[0;0;0;0;0;3],...
             [0;0;0;0;0;4],[0;0;0;0;0;5],[0;0;0;0;0;6],...
             [0;0;0;0;0;7]};
    eOCOS = { [[0;0;0;0;0;1],[0;0;0;0;0;2]],...
25  [[0;0;0;0;0;1],[0;0;0;0;0;3]],...
    [[0;0;0;0;0;1],[0;0;0;0;0;4]],...
    [[0;0;0;0;0;2],[0;0;0;0;0;5]],...
    [[0;0;0;0;0;3],[0;0;0;0;0;6]],...
    [[0;0;0;0;0;4],[0;0;0;0;0;7]],...
30  [[0;0;0;0;0;1],[0;0;0;0;0;6]]};
    VOCOS = vertices(vOCOS);
    EOCOS = edges(eOCOS);
    OCOS = graphs(VOCOS,EOCOS);
    locationOCOS = [0;0;0;0;1;0];
```

```
35 OCOS=addlocationtonodes (OCOS, locationOCOS);
   OCOS=addlocationtoedges (OCOS, locationOCOS);
   %
   % find incidence matrix
   %
40 KOCOS = incidence(OCOS);
   %
   % find backward and forward data
   %
   [BA,BGlobal,FA,FGlobal,BEGlobal,FEGlobal] = BFsets(OCOS,KOCOS);
45 %
   % Initiliazation
   %
   Y = zeros(1,7);
   O = -1+2*rand(1,7);
50 G = 0.5*(0.9+rand(1,7));
   W = rand(1,7);
   % so weights are all positive. Now we set the inhibitory edges.
   W(4) = -W(4);
   W(5) = -W(5);
55 W(6) = -W(6);
   W =
       0.92507    0.18987    0.68864   -0.58308   -0.16078   -0.23091
             0.69714
   %
   % Do first evaluation using inputs
60 %
   Y = evalwithinput (OCOS,Y,W,O,G,Input ,BGlobal,BEGlobal);
   Y =
       0.3740340    0.9484705    0.1426123    0.6130007    0.0063026
             0.2021721    0.1505239
   scale = 1.1;
65 Tol = .05;
   TargTol = .20;
   W = HebbianUpdateErrorSignal (Y,O,G,W,BGlobal,BEGlobal,Inhibitory ,
       DesiredOutput ,scale ,Tol ,TargTol);
   W =
       0.92507    0.18987    0.68864   -0.58308   -0.16078   -0.23091
             0.76685
70 Y = evalwithinput (OCOS,Y,W,O,G,Input ,BGlobal,BEGlobal);
   Y =
       0.3740340    0.9484705    0.1426123    0.6130007    0.0063026
             0.2126961    0.1505239
```

Now if we wish to do these updates for multiple steps, we need a training function. A simple one is shown in **HebbianTrainingWithError** whose code is given below.

Listing 19.77: Multistep Training

```
function [Y,YSeries ,W, WSeries] = HebbianTrainingWithError(Graph,
    KGraph,T,Y,W,O,G,Input ,Inhibitory ,DesiredOutput , scale ,Tol ,TargTol)
% Graph = incoming graph
% T is number of iterations
% Y is the node values
5 % W is the edge weights
% O is the node offsets
% G is the node gains
% Inhibitory is the inhibitory neurons
% Input is the input vector
10 % DesiredOutput is a cell of structs
% each struct is of type Target.in = neuron index
% and Target.value = neuron value)
% DesiredOutput{2}(1) gives the index for the second cell entry
% DesiredOutput{2}(2) gives the value for the second cell entry
15 % scale is the weight rescaling factor
% Tol is the desired Hebbian tolerance
% TargTol is the desired Hebbian tolerance for target weight updates
% scale is the Hebbian update multiplier
% Tol is the Hebbian tolerance
20 % TargTol is the target multiplier
%
[BA,BGlobal ,FA,FGlobal ,BEGlobal ,FEGlobal] = BFsets (Graph ,KGraph ) ;
%
%Y = evalwithinput (Graph ,Y,W,O,G, Input , BGlobal , BEGlobal ) ;
25 %
WSeries = W';
YSeries = Y';
for t = 1:T-1
    % use Hebbian update to reset link weights
30  Wnew = HebbianUpdateErrorSignal (Y,O,G,W,BGlobal ,BEGlobal ,Inhibitory
        , DesiredOutput , scale ,Tol ,TargTol ) ;
    % reevaluate the graph outputs
    Ynew = evalwithinput (Graph ,Y,Wnew,O,G, Input , BGlobal , BEGlobal ) ;
    W = Wnew;
    Y = Ynew;
35  YSeries = [YSeries Y'];
    WSeries = [WSeries W'];
end
```

Here is how it works in a practice session. We will assume all the initialization from the previous sample session and just show the training. Here we use 151 steps–setting the iteration count to $T = 152$ actually does 151 steps.

Listing 19.78: Training for 151 Steps

```
scale = 1.1;
Tol = .05;
3 TargTol = .01;
[Y,YS,W,WS] = HebbianTrainingWithError (OCOS,KOCOS,152 ,Y,W,O,G,...
Input , Inhibitory , DesiredOutput , scale ,Tol ,TargTol ) ;
```

After training, we find

Listing 19.79: Y after 151 steps

```
W =
    9.2507e-01     1.8987e-01     6.8864e-01    -5.8308e-01    -1.6078e-01
   -2.3091e-01     1.3646e+06
Y =
    0.3740340     0.9484705     0.1426123     0.6130007     0.0063026
    1.0000000     0.1505239
```

and we are close to achieving our targets: remember we do not use the target values
of 0 and 1. However, we are making no effort at controlling the sizes of the edge
weights as you can see from the value of $W(7)$!

Chapter 20
Building Brain Models

We are interested in modeling the brains and other neural systems of a variety of creatures. This text has given us the tools to begin to do this and in this chapter, we will show how to build a human brain model using address based graphs in a modular fashion. Some of our modules will have a lot of structure and some will be just sketches. Our intent is to furnish you with a guide to how you could use these ideas to build a model of any neural system of interest. The general process is to look at the literature of the animal in question and try to find the neural circuit diagrams from which a graph model can be built. This is not an easy task and with the help of students, we have built preliminary models of honeybee, spider, pigeon and squid/cuttlefish brains. It is a very interesting journey and in the next volume we will explore it much further.

Now to build a small brain from component modules, we need code to build the neural modules and link them together. We do this in several steps. The first one is to assemble a cortex module. We have already discussed this quite a bit. The plan is to build a cortical can out of OCOS, FFP and Two/Three circuits, assemble three or more cans into a cortical column and then assemble multiple columns into a cortical sheet. The code to do this is very modular.

20.1 Build A Cortex Module

A cortex module is built from cans assembled into columns which are then organized into one or two dimensional sheets. So the process of building a cortex module is pretty intense! However, we can use the resulting module as an isocortex building block as we know all cortex prior to environmental imprinting is the same. Hence, we can build a cortex module for visual cortex, auditory cortex etc and simply label their respective global vector addresses appropriately.

© Springer Science+Business Media Singapore 2016

J.K. Peterson, *BioInformation Processing*, Cognitive Science and Technology,

DOI 10.1007/978-981-287-871-7_20

Fig. 20.1 Two components
of a can circuit. **a** The OCOS
circuit. **b** The FFP circuit

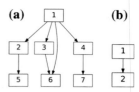

20.1.1   Build A Cortical Can

We will build a cortical can using the OCOS, FFP and Two/Three building blocks.
After we build the OCOS, FFP and Two/Three blocks we set their location addresses
as we discussed. OCOS is [0; 0; 0; 0; 1; ·], FFP is [0; 0; 0; 0; 2; ·] and Two/Three is
[0; 0; 0; 0; 3; ·]. We use the function **buildcan**.

Listing 20.1: Structure of buildcan

```
function  [OCOS,FFP,TwoThree,CanOne]  =  buildcan()

end
```

First,we build the OCOS, FFP and Two/Three circuit blocks, seen in Figs. 20.1a,b
and 20.2a, and set their location masks as discussed above. We have simplified the
FFP so that it consists of only 2 nodes: we just specify there is node 1 in the can
above which connects to node 1 of the OCOS module. Again, note we are not using
the thalamus node in the OCOS circuit.

Listing 20.2: Build the OCOS, FFP and Two/Three circuit blocks

```
   vOCOS = {[0;0;0;0;0;1],[0;0;0;0;0;2],[0;0;0;0;0;3],...
 2         [0;0;0;0;0;4],[0;0;0;0;0;5],[0;0;0;0;0;6],...
           [0;0;0;0;0;7]};
   eOCOS = { [[0;0;0;0;0;1],[0;0;0;0;0;2]],...
             [[0;0;0;0;0;1],[0;0;0;0;0;3]],...
             [[0;0;0;0;0;1],[0;0;0;0;0;4]],...
 7           [[0;0;0;0;0;2],[0;0;0;0;0;5]],...
             [[0;0;0;0;0;3],[0;0;0;0;0;6]],...
             [[0;0;0;0;0;4],[0;0;0;0;0;7]],...
             [[0;0;0;0;0;1],[0;0;0;0;0;6]]};
   VOCOS = vertices(vOCOS);
12 EOCOS = edges(eOCOS);
   OCOS = graphs(VOCOS,EOCOS);
   locationOCOS = [0;0;0;0;1;0];
   OCOS=addlocationtonodes(OCOS,locationOCOS);
   OCOS=addlocationtoedges(OCOS,locationOCOS);
17
   vFFP = {[0;0;0;0;0;1],[0;0;0;0;0;2]};
   eFFP = {[[0;0;0;0;0;1],[0;0;0;0;0;2]]};
   VFFP = vertices(vFFP);
   EFFP = edges(eFFP);
```

```
22 FFP = graphs (VFFP, EFFP) ;
   locationFFP = [0;0;0;0;2;0];
   FFP=addlocationtonodes (FFP, locationFFP ) ;
   FFP=addlocationtoedges (FFP, locationFFP ) ;

27 vTwoThree = {
   [0;0;0;0;0;1] ,[0;0;0;0;0;2] ,[0;0;0;0;0;3] ,...
   [0;0;0;0;0;4] ,[0;0;0;0;0;5] ,[0;0;0;0;0;6]};
   eTwoThree = {
   [[0;0;0;0;0;5] ,[0;0;0;0;0;1]] ,[[0;0;0;0;0;6] ,[0;0;0;0;0;3]] ,...
32 [[0;0;0;0;0;1] ,[0;0;0;0;0;2]] ,[[0;0;0;0;0;1] ,[0;0;0;0;0;4]] ,...
   [[0;0;0;0;0;3] ,[0;0;0;0;0;2]] ,[[0;0;0;0;0;3] ,[0;0;0;0;0;4]] ,...
   [[0;0;0;0;0;4] ,[0;0;0;0;0;2]]};
   VTwoThree = vertices (vTwoThree ) ;
   ETwoThree = edges (eTwoThree ) ;
37 TwoThree = graphs (VTwoThree, ETwoThree ) ;
   locationTwoThree = [0;0;0;0;3;0];
   TwoThree=addlocationtonodes (TwoThree, locationTwoThree ) ;
   TwoThree=addlocationtoedges (TwoThree, locationTwoThree ) ;
```

Then, we add all the individual nodes together into the graph **CanOne**.

Listing 20.3: Construct CanOne

```
%
% CanOne Step 1
% add the nodes of FFP to OCOS to make CanOne Step 1
%
5 nodes = FFP.v;
  [n, nodesize ] = size (nodes ) ;
  CanOne = addnode (OCOS, nodes (: ,1) ) ;
  for i = 2:nodesize
    CanOne = addnode (CanOne, nodes (: , i )) ;
10 end
  %
  % CanOne Step 2
  % add the nodes of TwoThree to CanOne Step 1
  %       to make CanOne Step 2
15 %
  nodes = TwoThree.v;
  [n, nodesize ] = size (nodes ) ;
  for i = 1:nodesize
    CanOne = addnode (CanOne, nodes (: , i )) ;
20 end
```

Now that **CanOne** has all the nodes, we add the edges. **CanOne** already has the OCOS edges, so we just need to add the FFP and Two/Three edges.

Listing 20.4: Add the FFP and Two/Three edges

```
% Now add all the edges from the components to CanOne
% add edges from FFP
%
  links = FFP.e;
5 CanOne = addedgev (CanOne, links ) ;
  % now add edges from TwoThree
  links = TwoThree.e;
  CanOne = addedgev (CanOne, links ) ;
```

Next, we connect the components.

Fig. 20.2 The last
component of the can circuit
and the can circuit itself. **a**
Two-three circuit. **b** A can
circuit

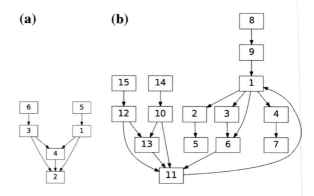

Listing 20.5: Connect the components

```
  links={};
2 links{1} = [[0;0;0;0;2;2],[0;0;0;0;1;1]];
  links{2} = [[0;0;0;0;3;2],[0;0;0;0;1;1]];
  links{3} = [[0;0;0;0;1;6],[0;0;0;0;3;2]];
  CanOne = addedgev(CanOne,links);
```

We can visualize the can easily now. We build the can, its incidence matrix and then
generate the needed dot file. We then can see the assembled components in a can in
Fig. 20.2b.

20.1.2 Build A Cortical Column

Once we can build a **can**, we can glue **cans** together to build **columns**. We will
default to a three **can** size for a column. The function is

Listing 20.6: Structure of buildcolumn

```
  function [CanOne,CanTwo,CanThree,Column] = buildcolumn()

  end
```

First, we build three **cans** and set there location masks: **CanOne** is [0; 0; 0; 1; ·; ·],
CanTwo is [0; 0; 0; 2; ·; ·] and **CanThree** is [0; 0; 0; 3; ·; ·].

Listing 20.7: Building three cans

```
   [OCOS,FFP,TwoThree,CanOne]  =  buildcan();
 2 [OCOS,FFP,TwoThree,CanTwo]  =  buildcan();
   [OCOS,FFP,TwoThree,CanThree] = buildcan();
   %
   locationCanOne    = [0;0;0;1;0;0];
   locationCanTwo    = [0;0;0;2;0;0];
 7 locationCanThree = [0;0;0;3;0;0];
   %
   CanOne=addlocationtonodes(CanOne,locationCanOne);
   CanOne=addlocationtoedges(CanOne,locationCanOne);
   %
12 CanTwo=addlocationtonodes(CanTwo,locationCanTwo);
   CanTwo=addlocationtoedges(CanTwo,locationCanTwo);
   %
   CanThree=addlocationtonodes(CanThree,locationCanThree);
   CanThree=addlocationtoedges(CanThree,locationCanThree);
```

Next, we add the cans to make a column. This uses the add a node at a time approach which is inefficient, but makes it clearer what we are doing.

Listing 20.8: Adding the cans to make a column

```
   %
   % add CanOne, CanTwo and CanThree nodes
   % to make a Column
 4 %
   % Column Step 1
   % add the nodes of CanTwo to CanOne to make Column Step 1
   %
   nodes = CanTwo.v;
 9 [n,nodesize] = size(nodes);
   Column = addnode(CanOne,nodes(:,1));
   for i = 2:nodesize
     Column = addnode(Column,nodes(:,i));
   end
14 %
   nodes = CanThree.v;
   [n,nodesize] = size(nodes);
   Column = addnode(Column,nodes(:,1));
   for i = 2:nodesize
19   Column = addnode(Column,nodes(:,i));
   end
```

Now that **Column** has all the nodes of the three **can**s, we add in all the **can** edges. Since **Column** contains all of **CanOne**, we only have to add the edges of **CanTwo** and **CanThree**. This also uses the inefficient add an edge at a time approach.

Listing 20.9: Add the other can edges

```
%
% Now  add  all  the  edges  from  the  components  to  Column
% edges  from  CanOne  are  already  there
%
5 % now  add  edges  from  CanTwo
%
links = CanTwo.e;
[n,edgesize] = size(links);
for i = 1:edgesize
10    Column = addedge(Column,links{i}(:,1),links{i}(:,2));
end
%
% now  add  edges  from  CanThree
%
15 links = CanThree.e;
[n,edgesize] = size(links);
for i = 1:edgesize
    Column = addedge(Column,links{i}(:,1),links{i}(:,2));
end
```

We then add the connections between the **can**s.

Listing 20.10: Add connections between cans

```
1 %
% add  connecting  edges  between  components
% neuron  1  in  can  2  OCOS  to  neuron  1  in  Can  1  FFP
%        [0;0;0;2;1;1]  ->  [0;0;0;1;2;1]
% neuron  1  in  can  3  OCOS  to  neuron  1  in  Can  2  FFP
6 %        [0;0;0;3;1;1]  ->  [0;0;0;2;2;1]
%
% neuron  2  in  Can  1  TwoThree  to  neuron  1  in  can  2  OCOS
%        [0;0;0;1;3;2]  ->  [0;0;0;2;1;1]
% neuron  2  in  Can  2  TwoThree  to  neuron  1  in  can  3  OCOS
11 %        [0;0;0;2;3;2]  ->  [0;0;0;3;1;1]
%
Column = addedge(Column,[0;0;0;2;1;1],[0;0;0;1;2;1]);
Column = addedge(Column,[0;0;0;3;1;1],[0;0;0;2;2;1]);
Column = addedge(Column,[0;0;0;1;3;2],[0;0;0;2;1;1]);
16 Column = addedge(Column,[0;0;0;2;3;2],[0;0;0;3;1;1]);
```

We can see a typical column in Fig. 20.3. We first generate the dot file and then the associated graphic as follows:

Listing 20.11: Generate the column dot file and graphic

```
[CanOne,CanTwo,CanThree,Column] = buildcolumn();
KColumn = incidence(Column);
incToDot(KColumn,6,6,1.0,'Column.dot');
```

And generate the graphic in another window with

Fig. 20.3   A column circuit

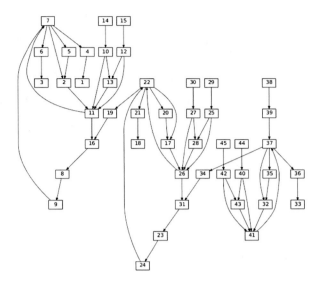

Listing 20.12: Generating the graphic with dot

```
dot −Tpdf −oBrainColumn.pdf  Column.dot
```

20.1.3   Build A Cortex Sheet

To build the model of cortex, we then glue together as many cortex columns as we want. We use two choices: a type of **'single'** which means just one column is used and a type of **'sheet'** which means we want a cortex model that is a rectangular sheet of *r* rows and *c* columns. We do this with a *switch* statement. The function is then

Listing 20.13: The structure of the buildcortex function

```
   function  Cortex = buildcortex(r,c,type)
   %
   switch type
     case ("single")
5      % here there is only one column
       %
       % so only one column name is needed
       % so just call it Cortex
       %
10     % build column = Cortex
       ...
     case ("sheet")
       % here the columns are in a non degenerate sheet;
       % ie at least two columns.
```

```
15    %   names  start  at  the  bottom  row:  for  convenience
      %   say  r = 2  and  c = 3
      %
      %   Column11 ,... , Column13
      %   Column21 ,... , Column23
20    %
         ...
      end
   end
```

The **'single'** case is easy. We build a single column and set its location to [0; 0; 1; ·; ·; ·].

Listing 20.14: The single case

```
   case  (" single ")
     %  here  there  is  only  one  columns
     %
     %  so  only  one  column  name  is  needed
5    %  so  just  call  it  Cortex
     %
     %  build  column = Cortex
     [ c1 , c2 , c2 , Cortex ] = buildcolumn () ;
     %  set  location
10   location   = [0;0;1;0;0;0];
     %  add  location  to  nodes  of  Cortex
     Cortex=addlocationtonodes ( Cortex , location ) ;
     %  add  location  to  edges  of  Cortex
     Cortex=addlocationtoedges ( Cortex , location ) ;
15   %
```

In the case of a **'sheet'**, we simply build as many columns as the architecture requires. We build names for the columns in the sheet as we go so we can refer to them.

Listing 20.15: Constructing the column case

```
     %   here  the  columns  are  in  a  non  degenerate  sheet;
     %   ie  at  least  two  columns .
     %   names  start  at  the  bottom  row:  for  convenience
     %   say  r = 2  and  c = 3
5    %
     %   Column11 ,... , Column13
     %   Column21 ,... , Column23
     %
     base = 'Column';
10   %  loop  through  sheet
     for  a = 1:r
       for  b = 1:c
         u = (a−1)∗c+b;
         %  create  name  string
15       name{u} = [ base , num2str ( a ) , num2str ( b ) ];
         %  build  column
         [ c1 , c2 , c2 , name{u}] = buildcolumn () ;
         %  set  location
         location   = [0;0;u;0;0;0];
20       %  add  nodes  to  column
         name{u}=addlocationtonodes ( name{u} , location ) ;
         %  add  edges  to  column
         name{u}=addlocationtoedges ( name{u} , location ) ;
       end
25   end
```

We then build the cortex model by gluing the columns together. We start by adding nodes.

Listing 20.16: Adding nodes to the cortex module

```
     %
     % We have now constructed an array of r x c
     % columns.  Now we assemble into a cortical module.
     %
5    % add Column nodes in name{u} to make the Cortex module
     %
     % Cortex Step 1
     % add the nodes of name{2} to name{1} to make Column Step 1
     %
10   nodes = name{2}.v;
     [n,nodesize] = size(nodes);
     Cortex = addnode(name{1},nodes(:,1));
     for i = 2:nodesize
        Cortex = addnode(Cortex,nodes(:,i));
15   end
     %
     % now add the other columns
     %
     for u = 3:r*c
20      nodes = name{u}.v;
        [n,nodesize] = size(nodes);
        for i = 1:nodesize
           Cortex = addnode(Cortex,nodes(:,i));
        end
25   end
```

Now add the edges. The cortex module already has the edges of the first column, so we only add the edges from column two on. We use the more efficient add a list of edges here.

Listing 20.17: Adding edges to the cortex module

```
     %
     % Now add all the edges from the components to Cortex
     % edges from name{1} are already there
     %
5    % now add edges from the other columns
     %
     for u = 2:r*c
        links = name{u}.e;
        Cortex = addedgev(Cortex,links);
10   end
     %
```

Then, we add the inter column connections. We do this by building a list of edges to add called **links{ }**. The code shows the old single edge command commented out above the addition to the list of edges.

Listing 20.18: Add the inter column connections

```
      % now add intercolumn connections
      %
      links = {};
      i = 0;
5     for a = 1:r
        for b = 1:c−1
          u = (a−1)*c+b;
          i = i+1;
          %Cortex = addedge(Cortex,[0;0;u;1;3;6],[0;0;u+1;1;3;5]);
10        links{i} = [[0;0;u;1;3;6],[0;0;u+1;1;3;5]];
        end
      end
      for a = 1:r
        for b = 1:c−1
15        u = (a−1)*c+b;
          i = i+1;
          %Cortex = addedge(Cortex,[0;0;u;2;3;6],[0;0;u+1;2;3;5]);
          links{i} = [[0;0;u;2;3;6],[0;0;u+1;2;3;5]];
        end
20    end
      for a = 1:r
        for b = 1:c−1
          u = (a−1)*c+b;
          i = i+1;
25        %Cortex = addedge(Cortex,[0;0;u;3;3;6],[0;0;u+1;3;3;5]);
          links{i} = [[0;0;u;3;3;6],[0;0;u+1;3;3;5]];
        end
      end
      Cortex = addedgev(Cortex,links);
```

We can graph a typical 2×2 cortex by generating its dot file. We can see this in Fig. 20.4. We build the cortex, its incidence matrix and generate the dot file with these lines:

Listing 20.19: The cortex dot file and the graphical image

```
1 Cortex = buildcortex(2,2,'sheet');
  KCortex = incidence(Cortex);
  incToDot(KCortex,6,6,1,'Cortex.dot');
```

And, as usual, generate the graphic in another window with

Listing 20.20: Generating the Cortex figure with dot

```
  dot −Tpdf −oBrainCortex.pdf Cortex.dot
```

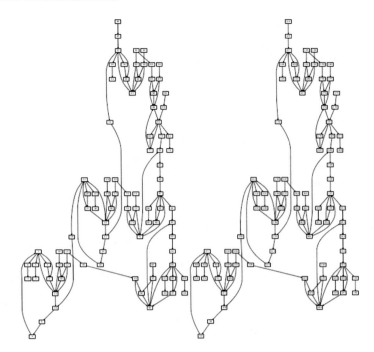

Fig. 20.4   A cortex circuit

20.2   Build A Thalamus Module

In this sample code, we build a thalamus module by gluing together reversed OCOS blocks. Ours will have just two, but the pattern is easily repeated. We can build much better models of thalamus by following the ideas in (Sherman and Guillery 2006) and their modification in (Sherman and Guillery 2013). But that will be for another time. Right now we are building a simple placeholder for the thalamus functions in the full brain model. Since these networks are reversed OCOS's (ROCOS's), they form a new fundamental block. The function is as follows:

Listing 20.21: Structure of the buildthalamus function

```
function  Thalamus  =  buildthalamus ( thalamusSize )

end
```

First, we build the reverse OCOS node and edge objects.

Listing 20.22: Build the reverse OCOS nodes and edges

```
       %
     2 % Use  size  Reverse  OCOS  modules
       %
       vROCOS  =  {[0;0;0;0;0;1],[0;0;0;0;0;2],[0;0;0;0;0;3],...
                  [0;0;0;0;0;4],[0;0;0;0;0;5],[0;0;0;0;0;6],...
                  [0;0;0;0;0;7]};
     7 eROCOS  =  {
           [[0;0;0;0;0;4],[0;0;0;0;0;1]],[[0;0;0;0;0;5],[0;0;0;0;0;2]],...
                  [[0;0;0;0;0;6],[0;0;0;0;0;3]],...
                  [[0;0;0;0;0;7],[0;0;0;0;0;4]],...
                  [[0;0;0;0;0;7],[0;0;0;0;0;5]],...
                  [[0;0;0;0;0;7],[0;0;0;0;0;6]],...
    12            [[0;0;0;0;0;7],[0;0;0;0;0;2]]};
       VROCOS  =  vertices(vROCOS);
       EROCOS  =  edges(eROCOS);
```

We then build the needed reverse OCOS blocks and assemble into one module. We
construct individual names for the reverse OCOS graph objects and set their location
masks to be [0; 0; 0; 0; 1, ·] to [0; 0; 0; 0; 7, ·]. The case of **thalamusSize == 1**
is easy. But if there are more than one, we have to assemble more carefully.

Listing 20.23: Building with one reversed OCOS block

```
       if  thalamusSize  ==  1
          Thalamus  =  graphs(VROCOS,EROCOS);
          location  =  [0;0;0;0;1;0];
          Thalamus  =  addlocationtonodes(Thalamus,location);
     5    Thalamus  =  addlocationtoedges(Thalamus,location);
       else
          ....
       end
```

In the more than one case, we do this. We create names for each thalamus module.
Then we create thalamus objects and set their addresses.

Listing 20.24: Building with more than one reversed OCOS blocks

```
       base  =  'ROCOS';
     2 for  i=1:thalamusSize
          % create  name  string
          name{i}  =  [base,num2str(i)];
          name{i}  =  graphs(VROCOS,EROCOS);
          location  =  [0;0;0;0;i;0];
     7    name{i}  =  addlocationtonodes(name{i},location);
          name{i}  =  addlocationtoedges(name{i},location);
       end
```

Next glue all the nodes together.

Listing 20.25: Glue the thalamus nodes together

```
 1   %
     % Now add all the nodes from the components to Thalamus
     % nodes from name{1} are already there
     %
     nodes = name{2}.v;
 6   [n,nodesize] = size(nodes);
     Thalamus = addnode(name{1},nodes(:,1));
     for i = 2:nodesize
       Thalamus = addnode(Thalamus,nodes(:,i));
     end
11   %
     % now add the other ROCOS's
     %
     for i = 3:thalamusSize
       nodes = name{i}.v;
16     [n,nodesize] = size(nodes);
       for i = 1:nodesize
         Thalamus = addnode(Thalamus,nodes(:,i));
       end
     end
```

Then add the edges. Note our simple thalamus model that consists of multiple reverse OCOS

Listing 20.26: Add the thalamus edges

```
     %
     % Now add all the edges from the components to Thalamus
     % edges from name{1} are already there
     %
 5   % now add edges from the other columns
     %
     for i = 2:thalamusSize
       links = name{i}.e;
       Thalamus = addedgev(Thalamus,links);
10   end
```

At this point, we don't have intermodule connections between the different ROCOS modules so it truly is a simple model! We then build a thalamus graph as usual and construct its dot file to visualize it.

Listing 20.27: Build a Thalamus Module with Two Pieces

```
Thalamus = buildthalamus(2);
KThalamus = incidence(Thalamus);
incToDot(KThalamus,6,6,1,'Thalamus.dot');
```

We see this in Fig. 20.5.

Fig. 20.5 A two copy
ROCOS Thalamus circuit

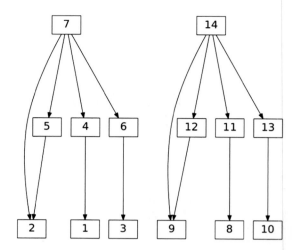

20.3   Build A MidBrain Module

The midbrain module controls how neurotransmitters are used in our brain model.
Again, this model will be quite simple just to illustrate the points. You can easily
imagine many ways to extend and that will be necessary to build useful models
for various purposes. We build a simple midbrain model by assembling **NumDop**
dopamine, **NumSer** serotonin and **NumNor** norepinephrine neurons into two neuron
sets. The function template is

Listing 20.28: Structure of the buildmidbrain function

```
function  MidBrain  =  buildmidbrain (NumDop, NumSer , NumNor)

end
```

First, we build the neurotransmitter node and edge lists. Each neurotransmitter is
modeled as two nodes with a simple connection between them. First, we build nodes
and edges for each neurotransmitter.

Listing 20.29: Build nodes and edges for each neurotransmitter

```
  %
2 %  Neurotransmitter  Modules
  %
  %  Build  Dopamine  neurons
  N = NumDop;
  vDopNT = {};
```

```
 7 for i = 1:N
     vDopNT{i} = [0;0;0;0;0;i];
   end
   for i = 1:N
     vDopNT{N+i} = [0;0;0;0;0;N+i];
12 end
   eDopNT = {};
   for i = 1:N
     eDopNT{i} = [[0;0;0;0;0;i],[0;0;0;0;0;N+i]];
   end
17
   % Build Serotonin neurons
   N = NumSer;
   vSerNT = {};
   for i = 1:N
22   vSerNT{i} = [0;0;0;0;0;i];
   end
   for i = 1:N
     vSerNT{N+i} = [0;0;0;0;0;N+i];
   end
27 eSerNT = {};
   for i = 1:N
     eSerNT{i} = [[0;0;0;0;0;i],[0;0;0;0;0;N+i]];
   end
32 % Build Norepinephrine neurons
   N = NumNor;
   vNorNT = {};
   for i = 1:N
     vNorNT{i} = [0;0;0;0;0;i];
37 end
   for i = 1:N
     vNorNT{N+i} = [0;0;0;0;0;N+i];
   end
   eNorNT = {};
42 for i = 1:N
     eNorNT{i} = [[0;0;0;0;0;i],[0;0;0;0;0;N+i]];
   end
```

Then, we build vertices and edges objects and then their corresponding neurotransmitter objects.

Listing 20.30: Build neurotransmitter objects

```
 1 VDopNT = vertices(vDopNT);
   EDopNT = edges(eDopNT);
   Dopamine = graphs(VDopNT,EDopNT);
   location = [0;0;0;0;1;0];
   Dopamine = addlocationtonodes(Dopamine,location);
 6 Dopamine = addlocationtoedges(Dopamine,location);

   VSerNT = vertices(vSerNT);
   ESerNT = edges(eSerNT);
   Serotonin = graphs(VSerNT,ESerNT);
11 location = [0;0;0;0;2;0];
   Serotonin = addlocationtonodes(Serotonin,location);
   Serotonin = addlocationtoedges(Serotonin,location);
```

```
   VNorNT = vertices(vNorNT);
16 ENorNT = edges(eNorNT);
   Norepinephrine  = graphs(VNorNT,ENorNT);
   location = [0;0;0;0;3;0];
   Norepinephrine = addlocationtonodes(Norepinephrine,location);
   Norepinephrine = addlocationtoedges(Norepinephrine,location);
```

Then, we finish by gluing the nodes together into a MidBrain object and add the edges.

Listing 20.31: Glue neurotransmitter modules into the midbrain

```
   nodes = Serotonin.v;
   [n,nodesize] = size(nodes);
   MidBrain = addnode(Dopamine,nodes(:,1));
   for i = 2:nodesize
5      MidBrain = addnode(MidBrain,nodes(:,i));
   end

   nodes = Norepinephrine.v;
   [n,nodesize] = size(nodes);
10 for i = 1:nodesize
      MidBrain = addnode(MidBrain,nodes(:,i));
   end

   links = Serotonin.e;
15 disp('add serotonin edges');
   MidBrain = addedgev(MidBrain,links);

   links = Norepinephrine.e;
   disp('add norepinephrine edges');
20 MidBrain = addedgev(MidBrain,links);
```

We build the MidBrain module and its dot file as usual.

Listing 20.32: Generate the midbrain and its the dot file

```
   MidBrain = buildmidbrain(2,2,2);
   KMidBrain = incidence(MidBrain);
   incToDot(KMidBrain,6,6,1.0,'MidBrain.dot');
```

We then plot the image and we can see the midbrain object in Fig. 20.6.

Fig. 20.6 A MidBrain circuit

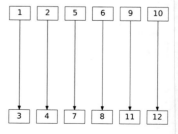

20.4   Building the Brain Model

Our simple brain model will consist of two sensory modules, an associative cortex, a motor cortex, a thalamus model, a midbrain model and a cerebellum model. We are not modeling memory or any motor output functions. We can add these later but this will be a nice simple model with a fair bit of structure. Note the model is full of feedback and input and output nodes are scattered all through the model. So all of our hard work at trying to understand how to build input to output maps using derivative based tools and Hebbian techniques as well as the conversion of a feedback graph into a lagged feedforward are very relevant. Later, we will briefly introduce another training technique that we think will be useful in efficiently building models of cognitive function and dysfunction. The generic model then looks like what we show in Fig. 20.7. We are not showing input/output models here.

The template we use to build the brain model is given below. It will return how many nodes and edges are in the model as well as the graph that represents our model of this brain.

Listing 20.33: Structure of the buildbrain function

```
function [NodeSizes, EdgeSizes, Brain] = buildbrain(NumDop, NumSer,
    NumNor)

end
```

where in the call to **buildbrain** we can choose how many of each neurotransmitter we want to use and we return a vector of size for neurons and edges so that we can build a better visualization of this graph. There are seven modules here and each has its own number of neurons and edges. So the global neuron and edge numbering scheme

Fig. 20.7 A simple brain model not showing input and output connections. The edge connections are labeled $E_{\alpha\beta}$ where α and β are the abbreviations for the various modules

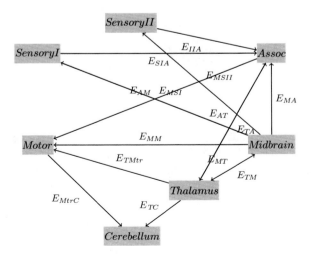

can be broken up into pieces relevant to each module. The vectors **NodeSizes** and **EdgeSizes** contain this information. In the original **incToDot** code, we just drew a graph simply using the global node and edge numbers. We will now automate the construction of the dot file differently. We will call each module its own cluster in the digraph and allow for us to set individual colors for each module and both internal links and intermodule links. Now let's discuss how we build our brain model. This will still not have an input or output section. First, we initialize the size counters.

Listing 20.34: Initialize counters

```
NodeSizes = [];
EdgeSizes = [];
```

We then build the **SensoryOne** cortical module and set its mask to [0; 1; ·; ·; ·; ·]. This allows us to set the node size and edge of the first module in the counters.

Listing 20.35: Build sensory cortex module one

```
    disp('Build SensoryOne');
    SensoryOne = buildcortex(1,1,'single');
    location = [0;1;0;0;0;0];
    SensoryOne=addlocationtonodes(SensoryOne,location);
  5 SensoryOne=addlocationtoedges(SensoryOne,location);

    N1 = length(SensoryOne.v);
    E1 = length(SensoryOne.e);
    NodeSizes(1) = N1;
 10 EdgeSizes(1) = E1;
    disp(['Neurons 1 to ',num2str(N1),' are Sensory One']);
    disp(['Links 1 to ',num2str(E1),' are Sensory One']);
```

We then build the **SensoryTwo** cortical module, set its mask to [0; 2; ·; ·; ·; ·] and set counters.

Listing 20.36: Build sensory cortex module two

```
    disp('Build SensoryTwo');
    SensoryTwo = buildcortex(1,1,'single');
    location = [0;2;0;0;0;0];
    SensoryTwo=addlocationtonodes(SensoryTwo,location);
  5 SensoryTwo=addlocationtoedges(SensoryTwo,location);

    N2 = length(SensoryTwo.v);
    E2 = length(SensoryTwo.e);
    NodeSizes(2) = N2;
 10 EdgeSizes(2) = E2;
    I = N1+1;
    disp(['Neurons ',num2str(I),' to ',num2str(N1+N2),' are Sensory Two'
        ]);
    disp(['Links ',num2str(E1+1),' to ',num2str(E1+E2),' are Sensory Two'
        ]);
```

We will model the associative cortex as a 2×2 sheet and build the object **AssociativeCortex** with mask address [0; 3; ·; ·; ·; ·].

Listing 20.37: Build the associative cortex module

```
    disp('Build AssociativeCortex');
    AssociativeCortex = buildcortex(2,2,'sheet');
    location = [0;3;0;0;0;0];
    AssociativeCortex=addlocationtonodes(AssociativeCortex,location);
5   AssociativeCortex=addlocationtoedges(AssociativeCortex,location);

    N3 = length(AssociativeCortex.v);
    E3 = length(AssociativeCortex.e);
    NodeSizes(3) = N3;
10  EdgeSizes(3) = E3;
    I = N1+N2+1;
    disp(['Neurons ',num2str(I),' to ',num2str(N1+N2+N3),'
      are Associative Cortex']);
    disp(['Links ',num2str(E1+E2+1),' to ',num2str(E1+E2+E3),'
15    are Associative Cortex']);
```

We choose to model the motor cortex as a single cortical column and set its mask to [0; 4; ·; ·; ·; ·].

Listing 20.38: Build the motor cortex module

```
    disp('Build MotorCortex');
    MotorCortex = buildcortex(1,1,'single');
    location = [0;4;0;0;0;0];
    MotorCortex=addlocationtonodes(MotorCortex,location);
5   MotorCortex=addlocationtoedges(MotorCortex,location);

    N4 = length(MotorCortex.v);
    E4 = length(MotorCortex.e);
    NodeSizes(4) = N4;
10  EdgeSizes(4) = E4;
    I = N1+N2+N3+1;
    disp(['Neurons ',num2str(I),' to ',num2str(N1+N2+N3+N4),'
      are Motor Cortex']);
    disp(['Links ',num2str(E1+E2+E3+1),' to ',num2str(E1+E2+E3+E4),'
15    are Associative Cortex']);
```

We then build the thalamus model, **Thalamus** with mask [0; 5; ·; ·; ·; ·].

Listing 20.39: Build the thalamus module

```
    disp('Build Thalamus');
    Thalamus = buildthalamus(2);
    location = [0;5;0;0;0;0];
    Thalamus=addlocationtonodes(Thalamus,location);
5   Thalamus=addlocationtoedges(Thalamus,location);

    N5 = length(Thalamus.v);
    E5 = length(Thalamus.e);
    NodeSizes(5) = N5;
10  EdgeSizes(5) = E5;
    I = N1+N2+N3+N4+1;
    disp(['Neurons ',num2str(I),' to ',num2str(N1+N2+N3+N4+N5),'
      are Thalamus']);
    disp(['Links ',num2str(E1+E2+E3+E4+1),' to ',num2str(E1+E2+E3+E4+E5),
      '
15    are Thalamus']);
```

Next, we model the midbrain as we discussed with 25 each of three different neurotransmitters. We set the mask now to [0; 6; ·; ·; ·; ·].

Listing 20.40: Build the midbrain module

```
disp('Build MidBrain');
MidBrain = buildmidbrain(NumDop,NumSer,NumNor);
location = [0;6;0;0;0;0];
MidBrain=addlocationtonodes(MidBrain,location);
MidBrain=addlocationtoedges(MidBrain,location);

N6 = length(MidBrain.v);
E6 = length(MidBrain.e);
NodeSizes(6) = N6;
EdgeSizes(6) = E6;
I = N1+N2+N3+N4+N5+1;
disp(['Neurons ',num2str(I),' to ',num2str(N1+N2+N3+N4+N5+N6),'
    are MidBrain']);
disp(['Links ',num2str(E1+E2+E3+E4+E5+1),' to ',num2str(E1+E2+E3+E4+
    E5+E6),'
    are MidBrain']);
```

Finally, we model the cerebellum as a single cortical column for now and set its mask to $[0; 7; \cdot; \cdot; \cdot; \cdot]$.

Listing 20.41: Build the cerebellum module

```
disp('Build Cerebellum');
Cerebellum = buildcortex(1,1,'single');
location = [0;7;0;0;0;0];
Cerebellum=addlocationtonodes(Cerebellum,location);
Cerebellum=addlocationtoedges(Cerebellum,location);

N7 = length(Cerebellum.v);
E7 = length(Cerebellum.e);
NodeSizes(7) = N7;
EdgeSizes(7) = E7;
I = N1+N2+N3+N4+N5+N6+1;
disp(['Neurons ',num2str(I),' to ',num2str(N1+N2+N3+N4+N5+N6+N7),'
    are Cerebellum']);
disp(['Links ',num2str(E1+E2+E3+E4+E5+E6+1),' to ',num2str(E1+E2+E3+
    E4+E5+E6+E7),'
    are Cerebellum']);
```

We can then build a simple brain model object, **Brain**. We begin by gluing together all the module nodes.

Listing 20.42: Glue brain modules together

```
% Brain Step 1
% add the nodes of SensoryTwo to SensoryOne make Brain Step 1
%
nodes = SensoryTwo.v;
[n,nodesize] = size(nodes);
Brain = addnode(SensoryOne,nodes(:,1));
for i = 2:nodesize
    Brain = addnode(Brain,nodes(:,i));
end
```

```
10 %
   % Brain Step 2
   % add the nodes of Associative to Brain
   %
   nodes = AssociativeCortex.v;
15 [n,nodesize] = size(nodes);
   for i = 1:nodesize
      Brain = addnode(Brain,nodes(:,i));
   end
   %
20 % Brain Step 3
   % add the nodes of MotorCortex to Brain
   %
   nodes = MotorCortex.v;
   [n,nodesize] = size(nodes);
25 for i = 1:nodesize
      Brain = addnode(Brain,nodes(:,i));
   end
   % Brain Step 4
   % add the nodes of Thalamus to Brain
30 %
   nodes = Thalamus.v;
   [n,nodesize] = size(nodes);
   for i = 1:nodesize
      Brain = addnode(Brain,nodes(:,i));
35 end
   % Brain Step 5
   % add the nodes of MidBrain to Brain
   %
   nodes = MidBrain.v;
40 [n,nodesize] = size(nodes);
   for i = 1:nodesize
      Brain = addnode(Brain,nodes(:,i));
   end
   % Brain Step 6
45 % add the nodes of Cerebellum to Brain
   %
   nodes = Cerebellum.v;
   [n,nodesize] = size(nodes);
   for i = 1:nodesize
50    Brain = addnode(Brain,nodes(:,i));
   end
```

Then we add in the edges. We already have the edges from Sensory Cortex One.

Listing 20.43: Add the brain module edges

```
   %
   %
   % now add edges from components to Brain
 4 %
   % Now add all the edges from SensoryTwo
   %
   links = SensoryTwo.e;
   [n,edgesize] = size(links);
 9 for i = 1:edgesize
      Brain = addedge(Brain,links{i}(:,1),links{i}(:,2));
   end
   %
   % Now add all the edges from AssociativeCortex
14 %
   links = AssociativeCortex.e;
   [n,edgesize] = size(links);
   for i = 1:edgesize
      Brain = addedge(Brain,links{i}(:,1),links{i}(:,2));
```

```
19 end
   %
   % Now add all the edges from MotorCortex
   %
   links = MotorCortex.e;
24 [n,edgesize] = size(links);
   for i = 1:edgesize
     Brain = addedge(Brain,links{i}(:,1),links{i}(:,2));
   end
   %
29 % Now add all the edges from Thalamus
   %
   links = Thalamus.e;
   [n,edgesize] = size(links);
   for i = 1:edgesize
34   Brain = addedge(Brain,links{i}(:,1),links{i}(:,2));
   end
   %
   % Now add all the edges from Midbrain
   %
39 links = MidBrain.e;
   [n,edgesize] = size(links);
   for i = 1:edgesize
     Brain = addedge(Brain,links{i}(:,1),links{i}(:,2));
   end
44 %
   % Now add all the edges from Cerebellum
   %
   links = Cerebellum.e;
   [n,edgesize] = size(links);
49 for i = 1:edgesize
     Brain = addedge(Brain,links{i}(:,1),links{i}(:,2));
   end
```

The next part is harder to write down as we have to carefully add all the edges between neurons in various components.

We are going to setup the connections now for all the modules as shown in Fig. 20.7. There are lot of them and in general, this is pretty time consuming and intellectually demanding to setup. In the text below, we go through all the steps. We can show you an intermediate step in Fig. 20.8.

This figure only shows the E_{SIA} and E_{SIIA} connections which we setup below. We are connecting the FFP circuits in all three cans of each sensory cortex to the Two/Three circuits in the associative cortex model. Our associative cortex consists of four columns each having three cans. We connect the sensory cortex to neuron 5 of the Two/Three circuit in each column and each can in the associative cortex. So for example, we connect [0; 1; 1; 3; 2; 1] to [0; 3; i; j; 3; 5] for i running from 1 to 4 (these are the associative cortex column indices) and for j running from 1 to 3 (these are the can indices). This gives 4×3 such connections. Now the address [0; 1; 1; 3; 2; 1] is for sensory cortex one, column one (there is only one column in each sensory cortex) and node 1 in the FFP circuit (that is the number 2 in the address's fifth column and the number 1 in the sixth column). We have to do that for the other two column in each sensory cortex. So we have a total of 3×12 such

Fig. 20.8 The two sensory
cortex modules and the
associative cortex module
with intermodule links

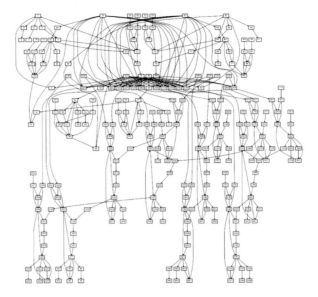

connections for each sensory cortex and two sensory cortexes giving a total of 72
connections to code. It is hard to say how to do this. We have experimented with
graphic user interfaces which allow us to connect nodes by clicking on the pre node
and the post node to establish the link, but if you look at Fig. 20.8, you can see the
complexity of the drawing starts making this very hard to do. We still find it easier
to code this sort of stuff, but then our minds might be different from yours! So feel
free to experiment yourself and find a convenient way to set these links up you can
live with!

Now, let's get started with all the setup. We start with the simplest interconnections.
First, we setup 9 links from sensory cortex one to associative cortex.

Listing 20.44: Connections sensory cortex one and associative cortex

```
  %
  % add connections from sensory cortex one to associative cortex
  % Sensory Cortex      address [0;1;0;-;-;-]
4 % Sensory Cortex One has 1 Column so 3 cans
  %          Column     address [0;1;1;-;-;-]
  %          Can        address [0;1;1;1;-;-]
  %          Can        address [0;0;1;2;-;-]
  %          Can        address [0;0;1;3;-;-]
9 % neuron 1 of ffp can 3 in sensory one to right neurons
  % of each 2/3 circuit in associative cortex
  % can 1, can 2, can 3
  links{1} = [[0;1;1;3;2;1],[0;3;1;1;3;5]];
  links{2} = [[0;1;1;3;2;1],[0;3;1;2;3;5]];
14 links{3} = [[0;1;1;3;2;1],[0;3;1;3;3;5]];
  % neuron 1 of ffp can 2 in sensory cortex one to right neurons
  % of each 2/3 circuit in associative cortex
  % can 1, can 2, can 3
  links 4  = [[0;1;1;2;2;1],[0;3;1;1;3;5]];
```

```
19  links{5} = [[0;1;1;2;2;1],[0;3;1;2;3;5]];
    links{6} = [[0;1;1;2;2;1],[0;3;1;3;3;5]];
    % neuron 1 of ffp can 1 in sensory one to right neurons
    % of each 2/3 circuit in associative cortex
    % can 1, can 2, can 3
24  links{7} = [[0;1;1;1;2;1],[0;3;1;1;3;5]];
    links{8} = [[0;1;1;1;2;1],[0;3;1;2;3;5]];
    links{9} = [[0;1;1;1;2;1],[0;3;1;3;3;5]];
```

Second, we encode in the 9 links from sensory cortex two to associative cortex.

Listing 20.45: Connections sensory cortex two and associative cortex

```
    %
    % neuron 1 of ffp can 3 in sensory two to right neurons
    % of each 2/3 circuit in associative cortex
4  % can 1, can 2, can 3
    links{10} = [[0;2;1;3;2;1],[0;3;1;1;3;5]];
    links{11} = [[0;2;1;3;2;1],[0;3;1;2;3;5]];
    links{12} = [[0;2;1;3;2;1],[0;3;1;3;3;5]];
    % neuron 1 of ffp can 2 in sensory two to right neurons
9  % of each 2/3 circuit in associative cortex
    % can 1, can 2, can 3
    links{13} = [[0;2;1;2;2;1],[0;3;1;1;3;5]];
    links{14} = [[0;2;1;2;2;1],[0;3;1;2;3;5]];
    links{15} = [[0;2;1;2;2;1],[0;3;1;3;3;5]];
14 % neuron 1 of ffp can 1 in sensory two to right neurons
    % of each 2/3 circuit in associative cortex
    % can 1, can 2, can 3
    links{16} = [[0;2;1;1;2;1],[0;3;1;1;3;5]];
    links{17} = [[0;2;1;1;2;1],[0;3;1;2;3;5]];
19  links{18} = [[0;2;1;1;2;1],[0;3;1;3;3;5]];
```

We then assign 9 links from associative cortex to motor cortex.

Listing 20.46: Connections associative cortex and motor cortex

```
1  %
    % neuron 1 of ffp can 3 in associative cortex to right neurons
    % of each 2/3 circuit in motor cortex
    % can 1, can 2, can 3
    links{19} = [[0;3;1;3;2;1],[0;4;1;1;3;5]];
6  links{20} = [[0;3;1;3;2;1],[0;4;1;2;3;5]];
    links{21} = [[0;3;1;3;2;1],[0;4;1;3;3;5]];
    % neuron 1 of ffp can 2 in associative cortex to right neurons
    % of each 2/3 circuit in motor cortex
    % can 1, can 2, can 3
11 links{22} = [[0;3;1;2;2;1],[0;4;1;1;3;5]];
    links{23} = [[0;3;1;2;2;1],[0;4;1;2;3;5]];
    links{24} = [[0;3;1;2;2;1],[0;4;1;3;3;5]];
    % neuron 1 of ffp can 1 in associative cortex to right neurons
    % of each 2/3 circuit in motor cortex
16 % can 1, can 2, can 3
    links{25} = [[0;3;1;1;2;1],[0;4;1;1;3;5]];
    links{26} = [[0;3;1;1;2;1],[0;4;1;2;3;5]];
    links{27} = [[0;3;1;1;2;1],[0;4;1;3;3;5]];
```

We then add 10 links from thalamus to sensory cortex one and two.

Listing 20.47: Connections thalamus and sensory cortex one

```
1 %
  % add thalamus to sensory cortex one can 1 OCOS
  links{28} = [[0;5;0;0;1;4],[0;1;1;1;1;7]];
  links{29} = [[0;5;0;0;1;5],[0;1;1;1;1;7]];
  links{30} = [[0;5;0;0;1;6],[0;1;1;1;1;7]];
6 links{31} = [[0;5;0;0;1;2],[0;1;1;1;1;7]];
  links{32} = [[0;5;0;0;1;2],[0;1;1;1;1;2]];
  % add thalamus to sensory cortex two can 1 OCOS
  links{33} = [[0;5;0;0;1;4],[0;2;1;1;1;7]];
  links{34} = [[0;5;0;0;1;5],[0;2;1;1;1;7]];
11 links{35} = [[0;5;0;0;1;6],[0;2;1;1;1;7]];
  links{36} = [[0;5;0;0;1;2],[0;2;1;1;1;7]];
  links{37} = [[0;5;0;0;1;2],[0;2;1;1;1;2]];
```

Next are 10 more links from thalamus to associative cortex and motor cortex.

Listing 20.48: Connections thalamus to associative and motor cortex

```
  % add thalamus to associative cortex can 1 OCOS
2 links{38} = [[0;5;0;0;1;4],[0;3;1;1;1;7]];
  links{39} = [[0;5;0;0;1;5],[0;3;1;1;1;7]];
  links{40} = [[0;5;0;0;1;6],[0;3;1;1;1;7]];
  links{41} = [[0;5;0;0;1;2],[0;3;1;1;1;7]];
  links{42} = [[0;5;0;0;1;2],[0;3;1;1;1;2]];
7 % add thalamus to motor cortex can 1 OCOS
  links{43} = [[0;5;0;0;1;4],[0;4;1;1;1;7]];
  links{44} = [[0;5;0;0;1;5],[0;4;1;1;1;7]];
  links{45} = [[0;5;0;0;1;6],[0;4;1;1;1;7]];
  links{46} = [[0;5;0;0;1;2],[0;4;1;1;1;7]];
12 links{47} = [[0;5;0;0;1;2],[0;4;1;1;1;2]];
```

Finally, we add connections from thalamus to the cerebellum.

Listing 20.49: Connections Thalamus to cerebellum

```
  %
  % now add connections from thalamus to cerebellum
3 %
  links{48} = [[0;5;0;0;1;4],[0;7;1;3;2;1]];
  links{49} = [[0;5;0;0;1;4],[0;7;1;2;2;1]];
  links{50} = [[0;5;0;0;1;4],[0;7;1;1;2;1]];
```

Then add connections from cerebellum to motor cortex.

Listing 20.50: Connections cerebellum to motor cortex

```
  %
  % now add connections from cerebellum to motor cortex
4 %
  % add cerebellum to motor cortex two/three can 3
  links{51} = [[0;7;1;3;2;1],[0;4;1;3;3;5]];
  links{52} = [[0;7;1;2;2;1],[0;4;1;3;3;5]];
  links{53} = [[0;7;1;1;2;1],[0;4;1;3;3;5]];
```

```
 9 % add cerebellum to motor cortex two/three can 2
   links{54} = [[0;7;1;3;2;1],[0;4;1;2;3;5]];
   links{55} = [[0;7;1;2;2;1],[0;4;1;2;3;5]];
   links{56} = [[0;7;1;1;2;1],[0;4;1;2;3;5]];
   % add cerebellum to motor cortex two/three can 1
14 links{57} = [[0;7;1;3;2;1],[0;4;1;1;3;5]];
   links{58} = [[0;7;1;2;2;1],[0;4;1;1;3;5]];
   links{59} = [[0;7;1;1;2;1],[0;4;1;1;3;5]];
```

This gives us a link list of 59 elements we add to the **Brain** object.

Listing 20.51: Add intermodule connections to brain

```
Brain = addedgev(Brain,links);
```

Next, we add in the dopamine links. This is done is a loop and for ease of understanding, we set up all the dopamine links in the link list **doplinks**. First, we set
neurotransmitter sizes.

Listing 20.52: Set neurotransmitter sizes

```
sizeDopamine = NumDop;
sizeSerotonin = NumSer;
sizeNorepinephrine = NumNor;
```

Then, we set the dopamine links.

Listing 20.53: Set the dopamine connections

```
   %
 2 % add dopamine connections
   %
   sizeDopamine = NumDop;
   sizeSerotonin = NumSer;
   sizeNorepinephrine = NumNor;
 7 for i=1:sizeDopamine
       %N = 51+(i-1)*45;
       N = (i-1)*45;
       In = [0;6;0;0;1;i];
       % sensory cortex one
12     doplinks{N+1} = [In,[0;1;1;1;3;5]];
       doplinks{N+2} = [In,[0;1;1;1;3;6]];
       doplinks{N+3} = [In,[0;1;1;2;3;5]];
       doplinks{N+4} = [In,[0;1;1;2;3;6]];
       doplinks{N+5} = [In,[0;1;1;3;3;5]];
17     doplinks{N+6} = [In,[0;1;1;3;3;6]];
       % sensory cortex two
       doplinks{N+7}  = [In,[0;2;1;1;3;5]];
       doplinks{N+8}  = [In,[0;2;1;1;3;6]];
       doplinks{N+9}  = [In,[0;2;1;2;3;5]];
22     doplinks{N+10} = [In,[0;2;1;2;3;6]];
       doplinks{N+11} = [In,[0;2;1;3;3;5]];
       doplinks{N+12} = [In,[0;2;1;3;3;6]];
       % associative cortex
       doplinks{N+13} = [In,[0;3;1;1;3;5]];
```

```
27    doplinks{N+14} = [In ,[0;3;1;1;3;6]];
      doplinks{N+15} = [In ,[0;3;1;2;3;5]];
      doplinks{N+16} = [In ,[0;3;1;2;3;6]];
      doplinks{N+17} = [In ,[0;3;1;3;3;5]];
      doplinks{N+18} = [In ,[0;3;1;3;3;6]];
32    % motor cortex
      doplinks{N+19} =    [In ,[0;4;1;1;3;5]];
      doplinks{N+20} =    [In ,[0;4;1;1;3;6]];
      doplinks{N+21} =    [In ,[0;4;1;2;3;5]];
      doplinks{N+22} =    [In ,[0;4;1;2;3;6]];
37    doplinks{N+23} =    [In ,[0;4;1;3;3;5]];
      doplinks{N+24} =    [In ,[0;4;1;3;3;6]];
      % thalamus
      for j = 1:7
        u = (j−1)*3;
42      Out1 = [0;5;0;0;j;1];
        Out2 = [0;5;0;0;j;2];
        Out3 = [0;5;0;0;j;3];
        doplinks{N+24+u+1} = [In ,Out1];
        doplinks{N+24+u+2} = [In ,Out2];
47      doplinks{N+24+u+3} = [In ,Out3];
      end
    end
    Brain = addedgev(Brain ,doplinks);
    [a,b] = size(doplinks);
```

We then handle serotonin links with the list **serlinks**.

Listing 20.54: Set the serotonin connections

```
    for i=1:sizeSerotonin
      N = (i−1)*45;
      In = [0;6;0;0;2;i];
      % sensory cortex one
5     serlinks{N+1} = [In ,[0;1;1;1;3;5]];
      serlinks{N+2} = [In ,[0;1;1;1;3;6]];
      serlinks{N+3} = [In ,[0;1;1;2;3;5]];
      serlinks{N+4} = [In ,[0;1;1;2;3;6]];
      serlinks{N+5} = [In ,[0;1;1;3;3;5]];
10    serlinks{N+6} = [In ,[0;1;1;3;3;6]];
      % sensory cortex two
      serlinks{N+7}  = [In ,[0;2;1;1;3;5]];
      serlinks{N+8}  = [In ,[0;2;1;1;3;6]];
      serlinks{N+9}  = [In ,[0;2;1;2;3;5]];
15    serlinks{N+10} = [In ,[0;2;1;2;3;6]];
      serlinks{N+11} = [In ,[0;2;1;3;3;5]];
      serlinks{N+12} = [In ,[0;2;1;3;3;6]];
      % associative cortex
      serlinks{N+13} = [In ,[0;3;1;1;3;5]];
20    serlinks{N+14} = [In ,[0;3;1;1;3;6]];
      serlinks{N+15} = [In ,[0;3;1;2;3;5]];
      serlinks{N+16} = [In ,[0;3;1;2;3;6]];
      serlinks{N+17} = [In ,[0;3;1;3;3;5]];
      serlinks{N+18} = [In ,[0;3;1;3;3;6]];
25    % motor cortex
      serlinks{N+19} =    [In ,[0;4;1;1;3;5]];
      serlinks{N+20} =    [In ,[0;4;1;1;3;6]];
      serlinks{N+21} =    [In ,[0;4;1;2;3;5]];
      serlinks{N+22} =    [In ,[0;4;1;2;3;6]];
```

```
30    serlinks{N+23} =   [In,[0;4;1;3;3;5]];
      serlinks{N+24} =   [In,[0;4;1;3;3;6]];
      % thalamus
      for j = 1:7
        u = (j-1)*3;
35      Out1 = [0;5;0;0;j;1];
        Out2 = [0;5;0;0;j;2];
        Out3 = [0;5;0;0;j;3];
        serlinks{N+24+u+1} = [In,Out1];
        serlinks{N+24+u+2} = [In,Out2];
40      serlinks{N+24+u+3} = [In,Out3];
      end
    end
    Brain = addedgev(Brain,serlinks);
```

Finally, we handle norepinephrine links with the list **norlinks**.

Listing 20.55: Set the norepinephrine connections

```
    %
    % add norepinephrine connections
    %
    for i=1:sizeNorepinephrine
5     N = (i-1)*45;
      In = [0;6;0;0;3;i];
      % sensory cortex one
      norlinks{N+1} = [In,[0;1;1;1;3;5]];
      norlinks{N+2} = [In,[0;1;1;1;3;6]];
10    norlinks{N+3} = [In,[0;1;1;2;3;5]];
      norlinks{N+4} = [In,[0;1;1;2;3;6]];
      norlinks{N+5} = [In,[0;1;1;3;3;5]];
      norlinks{N+6} = [In,[0;1;1;3;3;6]];
      % sensory cortex two
15    norlinks{N+7}  = [In,[0;2;1;1;3;5]];
      norlinks{N+8}  = [In,[0;2;1;1;3;6]];
      norlinks{N+9}  = [In,[0;2;1;2;3;5]];
      norlinks{N+10} = [In,[0;2;1;2;3;6]];
      norlinks{N+11} = [In,[0;2;1;3;3;5]];
20    norlinks{N+12} = [In,[0;2;1;3;3;6]];
      % associative cortex
      norlinks{N+13} = [In,[0;3;1;1;3;5]];
      norlinks{N+14} = [In,[0;3;1;1;3;6]];
      norlinks{N+15} = [In,[0;3;1;2;3;5]];
25    norlinks{N+16} = [In,[0;3;1;2;3;6]];
      norlinks{N+17} = [In,[0;3;1;3;3;5]];
      norlinks{N+18} = [In,[0;3;1;3;3;6]];
      % motor cortex
      norlinks{N+19} =   [In,[0;4;1;1;3;5]];
30    norlinks{N+20} =   [In,[0;4;1;1;3;6]];
      norlinks{N+21} =   [In,[0;4;1;2;3;5]];
      norlinks{N+22} =   [In,[0;4;1;2;3;6]];
      norlinks{N+23} =   [In,[0;4;1;3;3;5]];
      norlinks{N+24} =   [In,[0;4;1;3;3;6]];
35    % thalamus
      for j = 1:7
        u = (j-1)*3;
        Out1 = [0;5;0;0;j;1];
        Out2 = [0;5;0;0;j;2];
```

```
         norlinks{N+24+u+1}  =  [In,Out1];
         norlinks{N+24+u+2}  =  [In,Out2];
         norlinks{N+24+u+3}  =  [In,Out3];
      end
45 end
   Brain = addedgev(Brain,norlinks);
```

We can now build a small brain model.

Listing 20.56: Build a simple brain model

```
   [NodeSizes,EdgeSizes,Brain] = buildbrain(2,2,2);
   Build  SensoryOne
   Neurons 1 to 45 are Sensory One
 4 Links 1 to 58 are Sensory One
   Build  SensoryTwo
   Neurons 46 to 90 are Sensory Two
   Links 59 to 116 are Sensory Two
   Build  AssociativeCortex
 9 Neurons 91 to 270 are Associative Cortex
   Links 117 to 354 are Associative Cortex
   Build  MotorCortex
   Neurons 271 to 315 are Motor Cortex
   Links 355 to 412 are Motor Cortex
14 Build  Thalamus
   Neurons 316 to 329 are Thalamus
   Links 413 to 426 are Thalamus
   Build  MidBrain
   have dopamine edges
19 add  serotonin  edges
   add  norepinephrine  edges
   Neurons 330 to 341 are MidBrain
   Links 427 to 432 are MidBrain
   Build  Cerebellum
24 Neurons 342 to 386 are Cerebellum
   Links 433 to 490 are Cerebellum
   Edges 491 to 873 are Intermodule  connections
```

Then get the incidence matrix and the graph's dot file.

Listing 20.57: Generate brain incidence matrix and dot file

```
   KBrain = incidence(Brain);
   incToDot(KBrain,6,6,1.0,'Brain.dot');
```

We next generate the graphic.

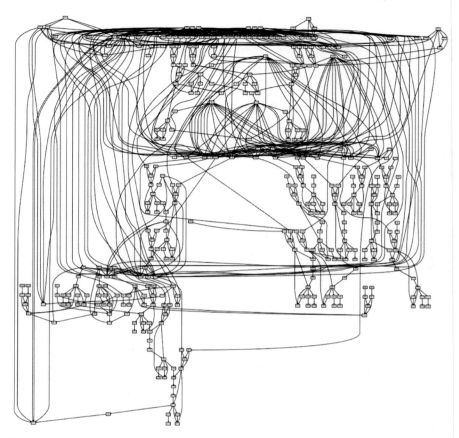

Fig. 20.9 A Brain model

Listing 20.58: Generate the brain graphic

```
dot -Tpdf -oBrainBrain.pdf Brain.dot
```

This generates the graph shown in Fig. 20.9. However, this visualization is less than ideal. It would be better to be able to color portions of the graph differently as well as organize modules into graph sub clusters. This can be done, although it is tedious. We have written such code but how we do it depends a great deal on what type of brain model we are building so it is very hard to automate it. This kind of code is available on request; we will be handle to share our pain with you! Also, there are standard ways to generate the many files you can use to make a simple movie from your simulation. The dot code is setup to allow color changes based on link weight intensities and so forth. We have also not included that code here. Again, it is hard to make it generic; we tend to write such code as needed for a project at hand. Still, contact us if you want to talk about it!

References

S.M. Sherman, R. Guillery, *Exploring the Thalamus and Its Role in Cortical Function* (The MIT Press, Cambridge, 2006)
S.M. Sherman, R. Guillery, *Functional Connections of Cortical Areas: A New View of the Thalamus* (The MIT Press, Cambridge, 2013)

Part VII
Models of Cognition Dysfunction

Chapter 21
Models of Cognitive Dysfunction

To finish this text, let's outline a blueprint for the creation of a model of cognitive function and/or dysfunction which is *relatively* small computationally.

21.1 Cognitive Modeling

We are interested in doing this for several important reasons. First, in order to make judicious statements about both cognitive dysfunction and policies that ameliorate the problem, we require that there is a functioning model of **normal brain function**. This is very hard to do, yet if we attempt it and develop a reasonable approximation, we can then use the normal brain model to help us understand what happens if neural modules and their interconnections change due to disease and trauma. Also, the computational capability of stand alone robotic platforms is always less than desired, so a model which can perhaps add higher level functions such as emotions to decision making algorithms would be quite useful. Hence, we will deliberately search out appropriate approximations to neural computation and deploy them in a robust, yet small scale scalable architecture in the quest for our models. However, the real reason is that in order to make progress on many different fronts in cognitive research, we need an overarching model of neural computation and it engenders high level things like associative learning and cognition. In our minds, this quest is quite similar to understanding how to answer difficult questions such as *"Which pathway to cancer is dominant in a colon cancer model?"* or *"How do we understand the spread of altruism throughout a population?"* and so on. In our earlier sections, we went through the arguments that can be brought to bear to bring insight into these important questions and our intent is similar here. How can we phrase and answer questions of this sort:

- *"How is associative learning done?"* Answering this sheds light on fundamental problems in **deep learning** and how to build better algorithms to accomplish it. It also is at the heart of understanding **shape**, **handwriting** and many other things.

© Springer Science+Business Media Singapore 2016
J.K. Peterson, *BioInformation Processing*, Cognitive Science and Technology,
DOI 10.1007/978-981-287-871-7_21

- *"What is a good painting?"* How do we decide a painting is good in some sense as a composition?
- *"How do we decide a musical composition is good?"* There are many ways notes can be strung together. We interpret some combinations as more pleasing than others. Why?
- *"What are emotions?"* Are emotions by products of sufficiently complicated neural wiring? Are they pragmatic ways to reduce search spaces in neural optimization algorithms?
- *"What is a normal brain?"* This is similar to asking how we know in a simulation if a simulated host dies—a question we asked in the West Nile Virus model of infection. We want to explain survival curve data but explaining our simulation results requires we understand how to flag a host's history as indicative of death or life at the end of the simulation. If we could design a model of a normal brain, it would open the door to lesion studies and much more.
- *"How do we generate good music or good art?"* This is part of associative learning really. If we could encode a bunch of examples of good music or art into a brain architecture, how could this be used to take start sequences of notes to generate new compositions that are interesting and labeled as good? Note this requires the model can supply emotional labellings!
- *"How do we understand the need for brain asymmetry?"* If an organism has known asymmetry in two halves of their brain, the question is why? What is the advantage of that?

There are many more questions we are interested in. Note building simulations at various levels of detail for how a neuron generates its axonal pulse, how second messenger triggers influence that output and so forth do not answer these types of questions. We will always make model error anyway so we think we should focus on the bare bones of the functional brain model. First, all brain models connect computational nodes (typically neurons) to other computational nodes using edges between the nodes. This architecture of nodes plus edges is really quite fixed. The details of the nodal and edge processing can change a lot, but this connectivity architecture remains the same. This means each brain model has an associated graph of nodes with edges. The edges have a direction, so the graph must be a directed graph. At the minimum, we need to generate directed graph models of brain function with at least cortical, thalamus and midbrain modules having norepinephrine, serotonin and dopamine neurotransmitter modulation. We have already discussed the full details of how to generate neutral and emotionally labeled music and painting data in this text in Chaps. 11 and 12. So let's use it to discuss how we might build a model of normal brain function. We won't actually build this yet as that is a very detailed process and it will be left for the next volume. However, in this text all the tools have been developed and we are ready to begin the building of such models both for human targets and a variety of small brained animals.

But let's get back to modeling human cognitive dysfunction using the music and painting data sets. Recall from Lang et al. (1998), it was shown that people respond to emotionally tagged or affective images in a semi-quantitative manner.

Human volunteers were shown various images and their physiological responses were recorded in two ways. One was a skin galvanic response and the other a fMRI parameter. The data shows that the null responses are associated with images that have no emotional tag. Further, the images cleanly map to distinct 2D locations in skin response and fMRI space, the emotional grid, when the emotional contents of the images differ. Hence, we will assume that if a database of music and paintings were separated into states of anger, sadness, happiness and neutrality, we would see a similar separation of response. We have designed emotionally labeled data using a grammatical approach to composition using an approach to assembling data known as a Würfelspiel matrix which consists of P rows and three columns. In the first column are placed the **nouns**; in the third column, are placed the **objects**; and in the second column, are placed the **verbs**. Each sentence fragment $S_{ijk} =$ **Noun**$_i$ + **Verb**$_j$ + **Object**$_k$ constitutes a composed phrase and there are P^3 possible combinations.

$$A = \begin{bmatrix} \textbf{Noun}_0 & \textbf{Verb}_0 & \textbf{Object}_0 \\ \textbf{Noun}_1 & \textbf{Verb}_1 & \textbf{Object}_1 \\ \vdots & \vdots & \vdots \\ \textbf{Noun}_{P-1} & \textbf{Verb}_{P-1} & \textbf{Object}_{P-1} \end{bmatrix} \qquad (21.1)$$

We have used this technique in our development of emotionally tagged music and painting data in Chaps. 11 and 12. For auditory data, we use musical fragments where the nouns become opening phrases, the verbs transitions and the objects, the closing. For the visual data, we use painting compositions made from foreground (the noun), midground (the verb) and background (the object). Hence, for neutral data plus three emotionally labeled sets of grammars, we have $4P^3$ input sequences for our model of visual and auditory cortex. We assume each musical and painting data presentation corresponds to a two dimensional emotional grid location which we assume has nice separation properties. We thus have a set of data which can be used to imprint a brain model with *correct* or *normal* emotional responses.

Each of our directed graphs also has node and edge functions associated with it and these functions are time dependent as what they do depends on first and second messenger triggers, the hardware structure of the output neuron on so forth. We therefore model the neural circuitry of a brain using a directed graph architecture consisting of computational nodes N and edge functions E which mediate the transfer of information between two nodes. Hence, if N_i and N_j are two computational nodes, then $E_{i \to j}$ would be the corresponding edge function that handles information transfer from node N_i and node N_j. For our purposes, we will assume here that the neural circuitry architecture we describe is fixed, although dynamic architectures can be handled as sequence of directed graphs. We organize the directed graph using interactions between neural modules (visual cortex, thalamus etc.) which are themselves subgraphs of the entire circuit. Once we have chosen a direct graph to represent our neural circuitry, note the addition of a new neural module is easily handled by adding it and its connections to other modules as a subgraph addition.

Hence, at a given level of complexity, if we have the graph $\mathcal{G}(N, E)$ that encodes the connectivity we wish to model, then the addition of a new module or modules simply generates a new graph $\mathcal{G}'(N', E')$ for which there are straightforward equations for explaining how G' relates to G which are easy to implement. The update equations for a given node then are given as an input/output pair. For the node N_i, let y_i and Y_i denote the input and output from the node, respectively. Then we have

$$y_i(t+1) = I_i + \sum_{j \in \mathcal{B}(i)} E_{j \to i}(t)\, Y_j(t)$$
$$Y_i(t+1) = \sigma_i(t)(y_i(t)).$$

where I_i is a possible external input, $\mathcal{B}(i)$ is the list of nodes which connect to the input side of node N_i and $\sigma_i(t)$ is the function which processes the inputs to the node into outputs. This processing function is mutable over time t because second messenger systems are altering how information is processing each time tick. Hence, our model consists of a graph $\mathcal{G}$ which captures the connectivity or topology of the brain model on top of which is laid the instructions for information processing via the time dependent node and edge processing functions. A simple look at edge processing shows the nodal output which is perhaps an action potential which is transferred without change to a synaptic connection where it initiates a spike in Ca^{++} ions which results in neurotransmitter release. The efficacy of this release depends on many things, but we can focus on four: $r_u(i, j)$, the rate of reuptake of neurotransmitter in the connection between node N_i and node N_j; the neurotransmitter is destroyed via an appropriate oxidase at the rate $r_d(i, j)$; the rate of neurotransmitter release, $r_r(i, j)$ and the density of the neurotransmitter receptor, $n_d(i, j)$. The triple $(r_u(i, j), r_d(i, j), r_r(i, j)) \equiv T(i, j)$ determines a net increase or decrease of neurotransmitter concentration between the two nodes: $r_r(i, j) - r_u(i, j) - r_d(i, j) \equiv r_{net}(i, j)$. The efficacy of a connection between nodes is then proportional to the product $r_{net}(i, j) \times n_d(i, j)$. Hence, each triple is a determining signature for a given neurotransmitter and the effectiveness of the neurotransmitter is proportional to the new neurotransmitter flow times the available receptor density. A very simple version of this is to simply assign the value of the edge processing function $E_{i \to j}$ to be the weight $W_{i,j}$ as is standard in a simple connectionist architecture. Of course, it is more complicated as our graphs allow feedback easily by simply defining the appropriate edge connections. All of these parameters are also time dependent, so we can add a (t) to all of the above to indicate that, but we have not done so as we do not want too much clutter. We have worked out how to use approximations to nodal computation in Chaps. 6 and 9 and more details of how these approximations can be used in asynchronous computation environments are given in Peterson (2015).

Table 21.1 Evaluation and Hebbian update algorithms

(a) Evaluation	(b) Hebbian Update
$\text{for}(i = 0; \, i < N; \, i++) \, \{$	$\text{for}(i = 0; \, i < N; \, i++) \, \{$
$\quad \text{if } (i \in \mathcal{U})$	$\quad \text{for } (j \in \mathcal{B}(i)) \, \{$
$\qquad y^i = x^i + \sum_{j \in \mathcal{B}(i)} \mathbf{E}_{j \to i} f_j$	$\qquad y_p = f_i \, \mathbf{E}_{j \to i}$
$\quad \text{else}$	$\qquad \text{if } (y_p > \epsilon)$
$\qquad y^i = \sum_{j \in \mathcal{B}(i)} \mathbf{E}_{j \to i} f_j$	$\qquad\quad \mathbf{E}_{j \to i} = \zeta \, \mathbf{E}_{j \to i}$
$\quad f_i = \sigma^i(y^i, p)$	$\quad \}$
$\}$	$\}$

21.1.1 Training Algorithms

Recall, we assume there is a vector function f which assigns an output to each node. In the simplest case, this vector function takes the form of $f(i) = \sigma_i$ and, of course, these computations can be time dependent. Given an arbitrary input vector $\vec{x}$, the DG computes node outputs via an iterative process. We let $\mathcal{U}$ denote the nodes which receive external inputs and $\mathcal{B}(i)$, the backward set of node N_i. The edge from node N_i to node N_j is assigned an edge function $\mathbf{E}_{i \to j}$ which could be implemented as assigning a scalar value to this edge. Then, the evaluation is shown in Table 21.1a, where N is the number of nodes in the DG and the ith node function is given by the function σ^i which processes the current node input y^i.

We can also adjust parameters in the DG using the classic idea of a Hebbian update which would be implemented as shown in Table 21.1b: choose a tolerance ϵ and at each node in the backward set of the post neuron i, update the edge value by the factor ζ if the product of the post neuron node value f_i and the edge value $\mathbf{E}_{j \to i}$ exceeds ϵ. The update algorithm then consists of the paired operations: sweep through the DG to do an evaluation and then use both Hebbian updates and the graph flow equations to adjust the edge values.

21.2 Information

21.2.1 Laplacian Updates

For a given directed graph, the *gradient* of f is defined by $\nabla f = K^T f$ and the Laplacian, by $\nabla^2 f$. We posit the flow of information through $\mathcal{G}$ is given by the graph based partial differential equation $\nabla^2 f - \alpha \frac{\partial f}{\partial t} - \beta f = -\mathcal{I}$, where $\mathcal{I}$ is the external input. This is similar to a standard cable equation. We interpret the $\frac{\partial f}{\partial t}$ using a standard forward difference $\mathbf{\Delta} f$ which is defined at each iteration time t by $\mathbf{\Delta} f(0) = 0$ and otherwise $\mathbf{\Delta} f(t) = f(t) - f(t - 1)$. This gives, using a finite difference for the $\frac{\partial f}{\partial t}$ term, Eq. 21.2, where we define the finite difference $\mathbf{\Delta} f_n(t)$ as $f_n(t + 1) - f_n(t)$.

$$\nabla^2 f - \alpha \, \mathbf{\Delta} f - \beta f = -\mathbf{\mathcal{I}}. \tag{21.2}$$

The update equation is then

$$\mathbf{KK}^T f - \alpha \, \mathbf{\Delta} f - \beta f = -\mathbf{\mathcal{I}}.$$

For iteration t, we then have the update equation

$$\mathbf{KK}^T f(t + 1) - \alpha \, \mathbf{\Delta} f(t) - \beta f(t + 1) = -\mathbf{\mathcal{I}}(t + 1).$$

which gives

$$\left(\mathbf{KK}^T - (\alpha + \beta)\mathbf{Id}\right) f(t + 1) + \beta f(t) = -\mathbf{\mathcal{I}}(t + 1).$$

where $\mathbf{Id}$ is the appropriate identity matrix. Let $\mathcal{H}_{\alpha,\beta}$ denote the operator $\mathbf{KK}^T - (\alpha + \beta)\mathbf{Id}$. We thus have the iterative equation

$$\mathcal{H}_{\alpha,\beta} f(t + 1) = -\beta f(t) - \mathbf{\mathcal{I}}(t + 1).$$

or since $\mathcal{H}_{\alpha,\beta}$ is invertible,

$$f(t + 1) = -\mathcal{H}_{\alpha,\beta}^{-1} \beta f(t) - \mathcal{H}_{\alpha,\beta}^{-1} \mathbf{\mathcal{I}}(t + 1).$$

For convenience, let $\mathbf{\Lambda}(t + 1) = -\mathcal{H}_{\alpha,\beta}^{-1} \mathbf{\mathcal{I}}(t + 1)$. Then, we have

$$f(t + 1) = \mathbf{\Lambda}(t + 1).$$

Now let's switch to a more typical nodal processing formulation. Recall, each node N_i has an input $y_i(t)$ which is processed by the node using the function σ_i as $Y_i(t) = \sigma_i(y_i(t))$. The node processing could also depend on parameters, but here we will only assume an offset and a gain is used in each computation. Thus, the node values f are equivalent to the node outputs Y. This allows us to write the update as

$$Y_i(t + 1) = \sigma_i(y_i(t)) = \mathbf{\Lambda}_i(t + 1).$$

or

$$y_i(t) = \sigma_i^{-1} \left(\mathbf{\Lambda}_i(t + 1)\right).$$

For simplicity, let's assume a sigmoid nodal computation for each node:

$$\sigma_i(x) = 0.5\left(1 + \tanh\left(\frac{x-o}{g}\right)\right).$$

where o and g are traditional offset and gain parameters for the computation. It is straightforward to find

$$\sigma_i^{-1}(\Lambda_i(t+1)) = O_i + \frac{g_i}{2}\ln\left(\frac{1+\Lambda_i(t+1)}{1-\Lambda_i(t+1)}\right)$$

where O_i is the offset to node i and g_i is the gain. We also know the graph evaluation equations for y_i then give

$$y_i(t) = \sum_{j\in B(i)} E_{j\to i}(t)\, Y_j(t) = O_i + \frac{g_i}{2}\ln\left(\frac{1+\Lambda_i(t+1)}{1-\Lambda_i(t+1)}\right)$$

where we don't need to add a nodal input as that is taken care of in the external input I. Pick the maximum $E_{j\to i}(t)\, Y_j(t)$ component and label its index j^M. Then set

$$E_{j^M\to i}(t)\, Y_j^M(t) = O_i + \frac{g_i}{2}\ln\left(\frac{1+\Lambda_i(t+1)}{1-\Lambda_i(t+1)}\right)$$

and solve for $E_{j^M\to i}(t)$ to find

$$E_{j^M\to i}(t) = \frac{1}{Y_j^M(t)}\left(O_i + \frac{g_i}{2}\ln\left(\frac{1+\Lambda_i(t+1)}{1-\Lambda_i(t+1)}\right)\right)$$

We could also decide to do this update for a larger subset of the indices contributing to the nodal output value in which case several edge weights could be updated for each iteration. For example, this update could be undertaken for all synaptic edges that meet a standard Hebbian update tolerance criterion. Implementing this update model into Matlab is quite similar to what we have already shown in the Hebbian case.

21.2.2 Module Updates

Given the update strategies, above it is worth mentioning that once we have a given graph structure for our model, we will probably want to update it by adding module interconnections, additional modules and so forth. We find this is our normal pathway to building a better model as our understanding of the neural computation we need to approximate changes as we study and we then need to add additional complexity to our model. However, just to illustrate what can happen, let's just look at what happens when we add a new subgraph to an existing graph. We want to see what happens in the update strategies. Hence, we assume we have a model $\mathcal{G}_\infty(N_1, E_1)$ with incidence matrix K_1. We then add to that model the subgraph $\mathcal{G}_\in(N_2, E_2)$ with incidence matrix K_2. For convenience of exposition, let's assume the first graphs incidence matrix is 9×12 and the second graphs incidence matrix is 20×30. When we combine these neural modules, we must decide on how to connect the neurons of $\mathcal{G}_1$ to the neurons of $\mathcal{G}_2$. The combined graph has $N_1 + N_2$ nodes and $E_1 + E_2$ edges plus the additional edges from the connections between the modules. The combined graph has an incidence matrix K and the connections between $\mathcal{G}_1$ to $\mathcal{G}_2$ will give rise sub matrices in K of the form

$$
K =
\begin{bmatrix}
K_1 & O_1 & C \\
(9 \times 12) & (9 \times 30) & (9 \times N) \\
\hline
O_2 & K_2 & D \\
(20 \times 12) & (20 \times 30) & (20 \times N)
\end{bmatrix}
$$

where N is the number of intermodule edges we have added. The new incidence matrix is thus $29 \times (42 + N)$. The matrix K^T is then

$$
K^T =
\begin{bmatrix}
K_1^T & O_2^T \\
(12 \times 9) & (12 \times 20) \\
\hline
O_1^T & K_2 \\
(30 \times 9) & (30 \times 20) \\
\hline
C^T & D^T \\
(N \times 9) & (N \times 20)
\end{bmatrix}
$$

and the new Laplacian is

$$
KK^T =
\begin{bmatrix}
K_1 & O_1 & C \\
(9 \times 12) & (9 \times 30) & (9 \times N) \\
\hline
O_2 & K_2 & D \\
(20 \times 12) & (20 \times 30) & (20 \times N)
\end{bmatrix}
\begin{bmatrix}
K_1^T & O_2^T \\
(12 \times 9) & (12 \times 20) \\
\hline
O_1^T & K_2 \\
(30 \times 9) & (30 \times 20) \\
\hline
C^T & D^T \\
(N \times 9) & (N \times 20)
\end{bmatrix}
$$

After multiplying, we have (multiplies involving our various zero matrices O_1 and O_2 vanish)

$$KK^T = \left[\begin{array}{c|c} \dfrac{K_1K_1^T + CC^T}{(9 \times 9)} & \dfrac{CD^T}{(9 \times 20)} \\ \hline \dfrac{DC^T}{(20 \times 9)} & \dfrac{K_1K_1^T + CC^T}{(20 \times 20)} \end{array}\right]$$

This can be rewritten as

$$KK^T = \begin{bmatrix} K_1K_1^T & O \\ O & K_2K_2^T \end{bmatrix} + \begin{bmatrix} CC^T & O \\ O & DD^T \end{bmatrix} + \begin{bmatrix} O & CD^T \\ DC^T & O \end{bmatrix}$$

for zero sub matrices of appropriate sizes in each matrix all labeled O. Recall the Laplacian updating algorithm is given by

$$KK^T f - \alpha \frac{\partial f}{\partial t} - \beta f = -\mathcal{I}.$$

which can be written in finite difference form using $f(t)$ for the value of the node values at time t as

$$KK^T f(t) - \alpha \, \Delta f(t) - \beta f(t) = -\mathcal{I}(t).$$

where $I(t)$ is the external input at time t and $\Delta f(0) = 0$ and otherwise $\Delta f(t) = f(t) - f(t-1)$. Now divide the node vectors f and $\mathcal{I}$ into the components f_I, f_{II}, $\mathcal{I}_I$ and $\mathcal{I}_{II}$ to denote the nodes for subgraph $\mathcal{G}_1$ and $\mathcal{G}_2$, respectively. Then, we see we can write the Laplacian for the full graph in terms of the components used to assemble the graph from its modules and module interconnections.

$$KK^T \begin{bmatrix} f_I \\ f_{II} \end{bmatrix} = \begin{bmatrix} K_1K_1^T & O \\ O & K_2K_2^T \end{bmatrix} \begin{bmatrix} f_I \\ f_{II} \end{bmatrix} + \begin{bmatrix} CC^T & O \\ O & DD^T \end{bmatrix} \begin{bmatrix} f_I \\ f_{II} \end{bmatrix}$$
$$+ \begin{bmatrix} O & CD^T \\ DC^T & O \end{bmatrix} \begin{bmatrix} f_I \\ f_{II} \end{bmatrix}$$

This can be rewritten as

$$KK^T \begin{bmatrix} f_I \\ f_{II} \end{bmatrix} = \begin{bmatrix} K_1K_1^T + CC^T \\ K_2K_2^T + DD^T \end{bmatrix} \begin{bmatrix} f_I \\ f_{II} \end{bmatrix} + \begin{bmatrix} CD^T \\ DC^T \end{bmatrix} \begin{bmatrix} f_{II} \\ f_I \end{bmatrix}$$

Now substitute this into the update equation to find

$$\begin{bmatrix} K_1K_1^T + CC^T \\ K_2K_2^T + DD^T \end{bmatrix} \begin{bmatrix} f_I \\ f_{II} \end{bmatrix} + \begin{bmatrix} CD^T \\ DC^T \end{bmatrix} \begin{bmatrix} f_{II} \\ f_I \end{bmatrix} - \alpha \begin{bmatrix} \Delta f_I \\ \Delta f_{II} \end{bmatrix} - \beta - \alpha \begin{bmatrix} f_I \\ f_{II} \end{bmatrix} = - \begin{bmatrix} \mathcal{I}_I \\ \mathcal{I}_{II} \end{bmatrix}$$

This can be then further rewritten as follows:

$$K_1 K_1^T f_I - \alpha \, \Delta f_I - \beta f_I + CC^T f_I + CD^T f_{II} = -\mathcal{I}_I$$
$$K_2 K_2^T f_{II} - \alpha \, \Delta f_{II} - \beta f_{II} + DD^T f_{II} + DC^T f_I = -\mathcal{I}_{II}$$

Hence, the terms $\left(CC^T + CD^T\right) f_I$, $\left(DD^T + DC^T\right) f_{II}$, $CD^T f_{II}$ and $DC^T f_I$ represent the mixing of signals between modules. We see if we have an existing model we can use these update equations to add to an existing graph a new graph. Thus, we can take a trained brain module and add a new cortical sub module and so forth and to some extent retain the training effort we have already undertaken. Now add time indices and expand the difference terms to get

$$K_1 K_1^T f_I^t - \alpha \, (f_I^t - f_I^{t-1}) - \beta f_I^t + CC^T f_I^t + CD^T f_{II}^t = -\mathcal{I}_I^t$$
$$K_2 K_2^T f_{II}^t - \alpha \, (f_{II}^t - f_{II}^{t-1}) - \beta f_{II}^t + DD^T f_{II}^t + DC^T f_I^t = -\mathcal{I}_{II}^t$$

If we let $\mathcal{L}f = KK^T f - \alpha f - \beta f$, for the graph with incidence matrix K, we can rewrite the update as

$$\mathcal{L}_I f_I^t + CC^T f_I^t + CD^T f_{II}^t = \alpha f_I^{t-1} - \mathcal{I}_I^t$$
$$\mathcal{L}_{II} f_{II}^t + DD^T f_{II}^t + DC^T f_I^t = \alpha f_{II}^{t-1} - \mathcal{I}_{II}^t$$

21.2.3 Submodule Two Training

If we wanted to train a portion of the cortex to match sensory data, we can organize our brain graph model as two modules. The first graph is all of the model except the portion of the cortex we want to train. Hence, the two module update equations derived in the previous section are applicable. Let's assume the cortical submodule we are interested in is the second graph and let E_{II} be the orthonormal basis consisting of the eigenvectors for the matrix $\mathcal{L}_{II}$ with λ_{II} the corresponding vector of eigenvalues. Then using the representation of vectors with respect to the basis E_{II}, we have $f_{II}^t = \sum_j f_{II}^{tj} E_{II}^j$ and $\mathcal{I}_{II}^t = \sum_j \mathcal{I}_{II}^{tj} E_{II}^j$ and so

$$\sum_j f_{II}^{tj} \mathcal{L}_{II} E_{II}^j + \sum_j f_{II}^{tj} DD^T E_{II}^j + DC^T f_I^t = \sum_j f_{II}^{t-1,j} \alpha E_{II}^j - \sum_j \mathcal{I}_{II}^{tj} E_{II}^j.$$

However, we also know $\mathcal{L}_{II} E_{II}^j = \lambda_{II}^j E_{II}^j$. Thus, we obtain

$$\sum_j f_{II}^{tj} \lambda_{II}^j E_{II}^j + \sum_j f_{II}^{tj} DD^T E_{II}^j + DC^T f_I^t = \sum_j f_{II}^{t-1,j} \alpha E_{II}^j - \sum_j \mathcal{I}_{II}^{tj} E_{II}^j.$$

Define the new vectors $F^j_{II} = DD^T E^j_{II}$ and $g^t_{II} = DC^T f^t_I$. Substituting these into our expression, we find

$$\sum_j \left(f^{tj}_{II} \chi^j_{II} - \alpha f^{t-1,j}_{II} + \mathcal{I}^{tj}_{II} \right) E^j_{II} + \sum_j f^{tj}_{II} F^j_{II} + g^t_{II} = 0.$$

We can expand F^j_{II} and g^t_{II} also in the basis E_{II} giving $F^j_{II} = \sum_k F^{jk}_{II} E^j_{II}$ and $g^t_{II} = \sum_j g^{tj}_{II} E^j_{II}$. From these expansions, we find

$$\sum_j \left(f^{tj}_{II} \chi^j_{II} - \alpha f^{t-1,j}_{II} + \mathcal{I}^{tj}_{II} + g^{tj}_{II} \right) E^j_{II} + \sum_j \sum_k f^{tj}_{II} F^{jk}_{II} E^k_{II} = 0.$$

Now reorganize the double sum to write

$$\sum_j \left(f^{tj}_{II} \chi^j_{II} - \alpha f^{t-1,j}_{II} + \mathcal{I}^{tj}_{II} + g^{tj}_{II} + \sum_k f^{tk}_{II} F^{kj}_{II} \right) E^j_{II} = 0.$$

This immediately implies that for all indices j, we must have

$$f^{tj}_{II} \chi^j_{II} - \alpha f^{t-1,j}_{II} + \mathcal{I}^{tj}_{II} + g^{tj}_{II} + \sum_k f^{tk}_{II} F^{kj}_{II} = 0.$$

This can be rewritten as

$$f^{tj}_{II} \chi^j_{II} - \alpha f^{t-1,j}_{II} + \mathcal{I}^{tj}_{II} + g^{tj}_{II} + < f^t_{II} F^k_{II} > = 0.$$

We now have a set of equations for the unknowns $f^{t,j}_{II}$. We can rewrite this as a matrix equation. Let Λ_{II} denote the diagonal matrix

$$\Lambda_{II} = \begin{bmatrix} \lambda^1_{II} & 0 & 0 & \cdots & 0 \\ 0 & \lambda^1_{II} & 0 & \cdots & 0 \\ \vdots & \vdots & \vdots & \vdots & \vdots \\ 0 & \cdots & 0 & 0 & \lambda^P_{II} \end{bmatrix}$$

where P is the number of nodes in the second module. Further, let $\mathcal{F}_{II}$ be the matrix

$$\mathcal{F}^T_{II} = \begin{bmatrix} F^{11}_{II} & F^{21}_{II} & \cdots & F^{P1}_{II} \\ F^{12}_{II} & F^{22}_{II} & \cdots & F^{P2}_{II} \\ \vdots & \vdots & \vdots & \\ F^{1P}_{II} & F^{2P}_{II} & \cdots & F^{PP}_{II} \end{bmatrix}$$

Then, we can rewrite the update equation more succinctly as

$$(\mathbf{\Lambda}_{II} + \mathcal{F}_{II}^T)f_{II}^t + (\mathcal{I}_{II}^t + g_{II}^t - \alpha f_{II}^{t-1}) = 0.$$

Finally, let $\mathcal{D}_{II}^{t,t-1} = -(\mathcal{I}_{II}^t + g_{II}^t - \alpha f_{II}^{t-1})$. Then we have $(\mathbf{\Lambda}_{II} + \mathcal{F}^T)f_{II}^t = \mathcal{D}_{II}^{t,t-1}$. It then follows that we can update via

$$f_{II}^t = (\mathbf{\Lambda}_{II} + \mathcal{F}^T)^{-1}\mathcal{D}_{II}^{t,t-1} = \mathcal{H}_{II}^t$$

where we know the matrix inverse exists because CC^T is invertible. This is the same form as the calculations we did in Sect. 21.2.1. Let's assume we have a sigmoidal nodal computation $\sigma_i(x) = 0.5\left(1 + \tanh\left(\frac{x-o}{g}\right)\right)$ and so $f_{II}^{ti} = \sigma_i(y_{II}^{ti}) = \xi_{II}^{ti}$ or $y_{II}^{ti} = \sigma_i^{-1}(\xi_{II}^{ti})$ where o and g are traditional offset and gain parameters for the computation. We know the graph evaluation equations for y_{II}^{ti} then give

$$y_{II}^{ti} = \sum_{j \in B(i)} E_{j \to i}(t)f_{II}^{tj} = O_i + \frac{g_i}{2}\ln\left(\frac{1 + \xi_{II}^{ti}}{1 - \xi_{II}^{ti}}\right)$$

Pick the maximum $E_{j \to i}(t)f_{II}^{tj}$ component and label its index j^M. Then set

$$E_{j^M \to i}(t)f_{II}^{tM} = O_i + \frac{g_i}{2}\ln\left(\frac{1 + \xi_{II}^{ti}}{1 - \xi_{II}^{ti}}\right)$$

and solve for $E_{j^M \to i}(t)$ to find

$$E_{j^M \to i}(t) = \frac{1}{f_{II}^{tM}}\left(O_i + \frac{g_i}{2}\ln\left(\frac{1 + \xi_{II}^{ti}}{1 - \xi_{II}^{ti}}\right)\right)$$

Thus, we have a method for updating a portion of the parameters that influence each nodal output. We combine this with the standard Hebbian updates with inhibition to complete the training algorithm.

We can also turn this into recursive equation by replacing critical components at time t with time $t-1$ to give

$$f_{II}^{tj}\chi_{II}^j = \alpha f_{II}^{t-1,j} - \mathcal{I}_{II}^{tj} - g_{II}^{t-1,j} + <f_{II}^{t-1}F_{II}^k> .$$

This gives the update

$$f_{II}^{tj} = \frac{1}{\chi_{II}^j}\left(\alpha f_{II}^{t-1,j} - \mathcal{I}_{II}^{tj} - g_{II}^{t-1,j} + <f_{II}^{t-1}F_{II}^k>\right).$$

which allows us to alter the parameters of the jth nodal function using its inverse $(f_{II}^j)^{-1}$ as we did earlier.

21.2.4 Submodule One Training

To update the first graph in the pair, we use the update equation

$$\mathcal{L}_I f_I^t + CC^T f_I^t + CD^T f_{II}^t = \alpha f_I^{t-1} - \mathcal{I}_I^t$$

The arguments we present next are quite similar to the ones presented for submodule two training; hence, we will be briefer in our exposition. The first graph is all of the model except the portion of the cortex we want to train. Let E_I be the orthonormal basis consisting of the eigenvectors for the matrix $\mathcal{L}_I$ with λ_I the corresponding vector of eigenvalues. Then using the representation of vectors with respect to the basis E_I, we have $f_I^t = \sum_j f_I^{tj} E_I^j$ and $\mathcal{I}_I^t = \sum_j \mathcal{I}_I^{tj} E_I^j$ and so $\sum_j f_I^{tj} \mathcal{L}_I E_I^j + \sum_j f_I^{tj} CC^T E_I^j + CD^T f_{II}^t = \sum_j f_I^{t-1,j} \alpha E_I^j - \sum_j \mathcal{I}_I^{tj} E_I^j$. However, we also know $\mathcal{L}_I E_I^j = \lambda_I^j E_I^j$. Thus, we obtain

$$\sum_j f_I^{tj} \lambda_I^j E_I^j + \sum_j f_I^{tj} CC^T E_I^j + CD^T f_{II}^t = \sum_j f_I^{t-1,j} \alpha E_I^j - \sum_j \mathcal{I}_I^{tj} E_I^j.$$

Define the new vectors $F_I^j = CC^T E_I^j$ and $g_I^t = CD^T f_{II}^t$. Substituting these into our expression, we find

$$\sum_j \left(f_I^{tj} \lambda_I^j - \alpha f_I^{t-1,j} + \mathcal{I}_I^{tj} \right) E_I^j + \sum_j f_I^{tj} F_I^j + g_I^t = 0.$$

We can expand F_I^j and g_I^t also in the basis E_I giving $F_I^j = \sum_k F_I^{jk} E_I^j$ and $g_I^t = \sum_j g_I^{tj} E_I^j$. From these expansions, we find $\sum_j \left(f_I^{tj} \lambda_I^j - \alpha f_I^{t-1,j} + \mathcal{I}_I^{tj} + g_I^{tj} \right) E_I^j + \sum_j \sum_k f_I^{tj} F_I^{jk} E_I^k = 0$. Now reorganize the double sum to write $\sum_j \left(f_I^{tj} \lambda_I^j - \alpha f_I^{t-1,j} + \mathcal{I}_I^{tj} + g_I^{tj} + \sum_k f_I^{tk} F_I^{kj} \right) E_I^j = 0$. This immediately implies that for all indices j, we must have $f_I^{tj} \lambda_I^j - \alpha f_I^{t-1,j} + \mathcal{I}_I^{tj} + g_I^{tj} + \sum_k f_I^{tk} F_I^{kj} = 0$. This can be rewritten as $f_I^{tj} \lambda_I^j - \alpha f_I^{t-1,j} + \mathcal{I}_I^{tj} + g_I^{tj} + <f_I^t F_I^k> = 0$. We now have a set of equations for the unknowns f_I^{tj}. We can rewrite this as a matrix equation. Let Λ_I denote the diagonal matrix $diag(\Lambda_I)$ similar to what we did before. Further, let $\mathcal{F}_I$ be the matrix $(\mathcal{F}_I^{ij})$. Then, we can rewrite the update equation more succinctly as

$$(\Lambda_I + \mathcal{F}_I^T) f_I^t + (\mathcal{I}_I^t + g_I^t - \alpha f_I^{t-1}) = 0.$$

Finally, let $\mathcal{D}_I^{t,t-1} = -\left(\mathcal{I}_I^t + g_I^t - \alpha f_I^{t-1}\right)$. Then we have $\left(\Lambda_I + \mathcal{F}_I^T\right) f_I^t = \mathcal{D}_I^{t,t-1}$. It then follows that we can update via

$$f_I^t = \left(\Lambda_I + \mathcal{F}_I^T\right)^{-1} \mathcal{D}_I^{t,t-1} = \mathcal{H}_I^t$$

where we know the matrix inverse exists because CC^T is invertible. Again, assuming we have a sigmoidal nodal computation, we find $f_I^{ti} = \sigma_i(y_I^{ti}) = \xi_I^{ti}$ or $y_I^{ti} = \sigma_i^{-1}\left(\xi_I^{ti}\right)$ where o and g are traditional offset and gain parameters for the computation. We know the graph evaluation equations for y_i then give

$$y_I^{ti} = \sum_{j \in B(i)} E_{j \to i}(t) f_I^{tj} = O_i + \frac{g_i}{2} \ln\left(\frac{1 + \xi_I^{ti}}{1 - \xi_I^{ti}}\right)$$

Pick the maximum $E_{j \to i}(t) f_I^{tj}$ component and label its index j^M. Then set

$$E_{j^M \to i}(t) f_I^{tM} = O_i + \frac{g_i}{2} \ln\left(\frac{1 + \xi_I^{ti}}{1 - \xi_I^{ti}}\right)$$

and solve for $E_{j^M \to i}(t)$ to find

$$E_{j^M \to i}(t) = \frac{1}{f_I^{tM}}\left(O_i + \frac{g_i}{2} \ln\left(\frac{1 + \xi_I^{ti}}{1 - \xi_I^{ti}}\right)\right)$$

Thus, we have a method for updating a portion of the parameters that influence each nodal output. We combine this with the standard Hebbian updates with inhibition to complete the training algorithm.

We can also turn this into recursive equation by replacing critical components at time t with time $t - 1$ to give $f_I^{tj} \chi_I^j = \alpha f_I^{t-1,j} - \mathcal{I}_I^{tj} - g_I^{t-1,j} + < f_I^{t-1} F_I^k >$. This gives the update $f_I^{tj} = \frac{1}{\chi_I^j}\left(\alpha f_I^{t-1,j} - \mathcal{I}_I^{tj} - g_I^{t-1,j} + < f_I^{t-1} F_I^k >\right)$ which allows us to alter the parameters of the jth nodal function using its inverse $(f_I^j)^{-1}$ as we did earlier. Thus we have a mechanism to update the submodule of the rest of the neural model we are interested in using the orthonormal basis its subgraph provides.

21.3 A Normal Brain Model

Let's address the issue of developing a *normal* brain model for lesion studies. Indeed, we can also look at the difference between a *normal* and a *dysfunctional* brain model as a way to determine a palliative strategy that might restore normalcy. Let's focus on three neurotransmitters here: dopamine, serotonin and norepinephrine. It is well know that disturbances in neurotransmitter signalling and processing are related to mood disorders. A good review of this is given in Russo and Nestler (2013). In fact, reward—seeking behavior circuitry is probably common across many animals as described in Barron et al. (2010). Our model here will be a first attempt at quantifying these interactions; simplified but with some explanatory insight. Each has an associated triple $(r_u, r_d, r_r) \equiv T$ which we label with a superscript for each neurotransmitter. Hence, T^D, T^S and T^N are the respective triples for our three neurotransmitters dopamine, serotonin and norepinephrine, Consider a simple graph model $\mathcal{G}(N, E)$ as shown in Fig. 21.1.

The auditory and visual training data are given in four emotional labellings; **neutral, sad, happy** and **angry**. We assume that each emotional state corresponds to neurotransmitter triples (T_N^D, T_N^S, T_N^N) (**neutral**), (T_S^D, T_S^S, T_S^N) (**sad**), (T_H^D, T_H^S, T_H^N) (**happy**) and (T_A^D, T_A^S, T_A^N) (**angry**), In addition, we assume that these triple states can be different in the visual, auditory and associative cortex and hence, we add an additional subscript label to denote that. We let $(T_{\alpha,\beta}^D, T_{\alpha,\beta}^S, T_{\alpha,\beta}^N)$ be the triple for cortex β ($\beta = 0$ is visual cortex, $\beta = 1$ is auditory cortex and $\beta = 2$ is associative cortex) and emotional state α ($\alpha = 0$ is neutral, $\alpha = 1$ is sad, $\alpha = 2$ is happy and $\alpha = 3$ is angry). Hence, we are using our emotionally labeled data as a way of constructing a simplistic normal brain model. Of course, it is much more complicated than this. Emotion in general is something that is hard to quantify and even somewhat difficult to find good measures for what state a person is in. A good review of that problem is in Mauss and Robinson (2009). Cortical activity that is asymmetric probably plays a role here, see Harmon-Jones et al. (2010), and what is nice about our modeling choices is that we have the tools to model such asymmetry using the graph models. However, our first model is much simpler.

The graph encodes the information about emotional states in both the nodal processing and the edge processing functions. Let's start with auditory **neutral** data and consider the graph shown in Fig. 21.2.

The neurotransmitter triples can exist in four states: **neutral, sad, happy** and **angry** which are indicated by the toggle α in the triples $T_{\alpha,\beta}^\gamma$ where γ denotes the neurotransmitter choice ($\gamma = 0$ is dopamine, $\gamma = 1$ is serotonin and $\gamma = 2$ is norepinephrine). For auditory neutral data, there is a choice of triple $T_{0,1}^\gamma$ and $T_{0,2}^\gamma$ for each neurotransmitter which corresponds to how the brain labels this data as emotionally neutral. Hence, we are identifying a collection of six triples or eighteen numbers with neutral auditory data. This is a vector in $\Re^{18}$ which we will call $V_{0,1,2}$. We need to do this for the other emotional states which gives us three additional vectors $V_{1,1,2}$, $V_{2,1,2}$ and $V_{3,1,2}$. Thus, emotional processing for emotionally labeled auditory data requires us to set four vectors in $\Re^{18}$. We need to do the same thing for the processing of emotionally labeled visual data which will give us the four vectors $V_{0,0,2}$, $V_{1,0,2}$, $V_{2,0,2}$ and $V_{3,0,2}$. To process the data for both types of sensory cortex thus requires eight vectors in $\Re^{18}$. Choose 8 orthonormal vectors in $\Re^{18}$ to correspond to these states. To train the full graph $\mathcal{G}(N, E)$ to understand neutral auditory data, it is a question of what internal outputs of the graph should be fixed or clamped and which should be allowed to alter. For neutral auditory data, we clamp the neurotransmitter triples in the auditory cortex and associative cortex using the vector $V_{0,1,2}$ and we force the output of the associative cortex to be the same as the incoming auditory

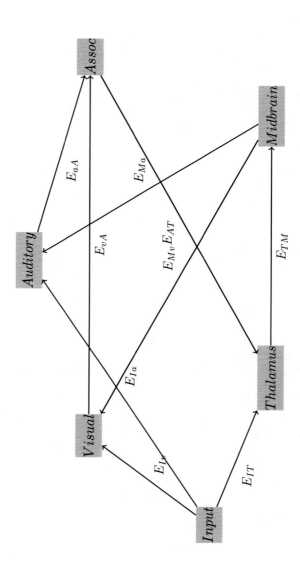

Fig. 21.1 A simple cognitive processing map focusing on a few salient modules. The edge connections are of the form $E_{\alpha,\beta}$ where α and β can be a selection from v (Visual), a (Auditory), A (Associative), T (Thalamus), M (Midbrain) and I (Input)

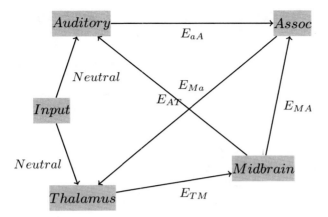

Fig. 21.2   Neural processing for neutral auditory data

Fig. 21.3   Neural processing
for sad auditory data

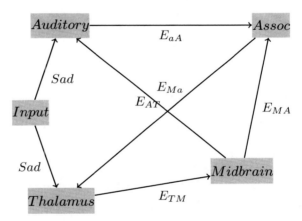

data. The midbrain module makes the neurotransmitter connections to each cortex module, but how these connections are used is determined by the triples and the usual nodal processing. In a sense, we are identifying the neutral auditory emotional state with the vector $V_{0,1,2}$. We assume the Thalamus module can shape processing in the midbrain. Hence, we will let $V_{0,1,2}$ be the desired output for the Thalamus module for neutral auditory data. The Midbrain module accepts the input $V_{0,1,2}$ and uses it to set the triple states in auditory and associative cortex. In effect, this is an abstraction of second messenger systems that affect the neural processing in these cortical modules. We can do this for each emotional state. For example, Fig. 21.3 shows the requisite processing for the cortical submodules.

Processing the sad auditory data clamps the triples in auditory and associative cortex using $V_{1,1,2}$ and clamps the associative cortex output to the incoming sad auditory data. We then train the graph to have a Thalamus output of $V_{1,1,2}$. We do

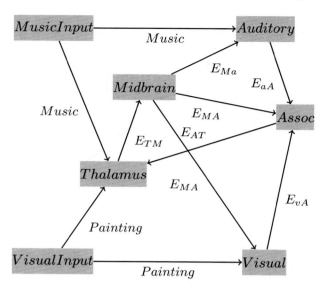

Fig. 21.4 Neural processing for new data

the same thing for the other emotional states for auditory and for all the emotional states for the visual cortex data.

Now consider the graph model accepting new music and painting data. We see the requisite graph in Fig. 21.4.

The music data and painting data will generate an output vector W from the Thalamus module in addition to generating associative cortex outputs corresponding to a music and painting fragment on the basis of the encoded edge and nodal processing that has been set by the training. Hence, we can classify the new combined auditory and visual data as corresponding to an emotional labeling of the form shown in Eq. 21.3.

$$W = \sum_{i=1}^{4} < W, V_{i,1,2} > V_{i,1,2} + \sum_{i=1}^{4} < W, V_{i,0,2} > V_{i,0,2} \qquad (21.3)$$

This is the projection of W to the subspace of $\Re^{18}$ spanned by our neurotransmitter triples. We can then perform standard cluster analysis to determine which emotionally labeling is closest. The training process we have gone through generates what we will call the **normal** brain. This is, of course, an abstraction of many things, but we think it has enough value that it can give us insight into cognitive dysfunction.

21.3.1   The Cognitive Dysfunction Model

Let's summarize what we have discussed. For a given brain model $\mathcal{G}(V, E)$ which consists of auditory sensory cortex, **SCI**, visual cortex, **SCII**, Thalamus, **Th**, Associative Cortex, **AC** and Midbrain, **MB**, we can now build its normal emotional state. We assign neurotransmitter triples $T_{\alpha,\beta}^{\gamma}$ for the four emotional states of neutral, sad, happy and angry for each neurotransmitter. These eighteen values for eight possible states are chosen as orthonormal vectors in $\Re^{18}$ but our choice of values for these triples can also be guided by the biopsychological literature but we will not discuss that here. The auditory and visual input data can be mapped into exclusive regions of skin galvanic response and a fMRI parameter so part of our modeling is to create a suitable abstraction of this mapping from the inputs into the two dimensional skin galvanic response and fMRI space which is $\Re^2$. There are papers that discuss how to model the fMRI response we can use for this; see Friston (2005) and Stephan et al. (2004).

To construct the normal model, we do the following:

- For neutral music data, we train the graph to imprint the emotional states as follows: for neutral music inputs to **SCI**, the clamped **AC** output is the same neutral music input and the clamped **Th** to **MB** output are the neutral triples $T_{0,1}^{\gamma}$ and $T_{0,2}^{\gamma}$ encoded as $V_{0,1,2}$. When training is finished, the neutral music samples are recognized as neutral music with the neutral neurotransmitter triples engaged.
- For neutral painting data, use the combined evaluation/Hebbian update loop to imprint the emotional states by clamping **AC** output to the neutral paint input to **SCII** and clamping the **Th** to **MB** output to the neutral triples $T_{0,1}^{\gamma}$ and $T_{0,2}^{\gamma}$ encoded as $V_{0,0,2}$. When training is finished, the neutral painting samples are recognized as neutral paintings with the neutral neurotransmitter triples engaged.
- Do the same training loop for the sad, happy and angry music and painting data samples using the sad, happy and angry neurotransmitter triples. The **AC** output is clamped to the appropriate data input values and the **Th** to **MB** output is the required neurotransmitter triples.

At this point, the model $\mathcal{G}(V, E)$ assigns known emotionally labeled data in two sensory modalities to their correct outputs in associative cortex. A new music and painting input would then be processed by the model and the **Th** to **MB** connections would assign the inputs to a set of neurotransmitter triple states W and generate a set of **AC** outputs for both the music and painting inputs. These outputs would be interpreted as a new musical sequence and a new painting. We can think of this trained model $\mathcal{G}(V, E)$ as defined the *normal* state. Also, as shown in Eq. 21.3, we can assign an emotional labeling to the new data. We can model dysfunction at this point by allowing changes in midbrain processing.

- Alter T^D, T^S and T^N in the auditory, visual and associative cortex to create a new vector W and therefore a new model $\mathcal{G}^{new}(N, E)$. The nodes and edges do not change but the way computations are processed does change. The change in model can be labeled $\delta \mathcal{G}$. Each triple change $(\delta T^D, \delta T^S, \delta T^N)$ gives a potential mapping to cognitive dysfunction.

- Using the normal model $\mathcal{G}(N, E)$, map all neutral music and painting data to sad data. Hence, we clamp **AC** outputs to sad states but do not clamp the **Th** to **MB** mapping. This will generate new neurotransmitter triple states in **MB** which we can label W^{sad}. The exhibited changes $(\delta T^D, \delta T^S, \delta T^N)$ give us a quantitative model of a cognitive dysfunction having some of the characteristics of depression. Note all 18 parameters from all 3 neurotransmitters are potentially involved. Note also standard drugs such as *Abilify* only make adjustments to 2 or 3 of these parameters which suggest our current pharmacological treatments are too narrow in scope. If we know the normal state, this model would suggest to restore normalcy, we apply drugs which alter neurotransmitter activity by $(-\delta T^D, -\delta T^S, -\delta T^N)$.

- Lesions in a brain can be simulated by the removal of nodes and edges from the graph model $\mathcal{G}(N, E)$. This allows us to study when the normal emotional responses alter and lead to dysfunction.

- If we construct a model $\mathcal{G}_{\mathcal{L}}(N_L, E_L)$ for the left half of the brain, a model $\mathcal{G}_{\mathcal{R}}(N_R, E_R)$ and a model of the corpus callosum, $\mathcal{G}_{\mathcal{C}}(N_C, E_C)$, we can combine these modules using the ideas from Sect. 21.2.2 to create a brain model $\mathcal{G}(N, E)$, which can be used to model problems with right and left half miscommunication problems such as schizophrenia which could be very useful. The interconnections between the right and left half of the brain through the corpus callosum modules and neurotransmitter connections can be altered and the responses of the altered model to a normal brain model can be compared for possible insight.

We think the most important problem in understanding cognition and the working of an full brain model is the issue of how features are extracted from environmental input. There are various brain structures that do this and it appears that cortical structures evolved to handle this task quite some time ago. In Wang et al. (2010), careful studies of the chicken brain have found a region in the telencephalon that is similar to the mammalian auditory cortex and it is also comprised of laminated layers of cells linked by narrow radial columns. Evidence is therefore mounting that the complex circuitry needed to parse environmental input and do feature extraction and associative learning probably evolved in an ancestor common to both mammals and birds at least 300 million years ago. Further, in Sanes and Zipursky (2010), there is a fascinating discussion about the similarities and differences between the mammalian visual cortex and the fly visual cortex. For example, both visual systems use a small number of neuron types divided into many subtypes for specialized computation. They use multiple synaptic contacts to a single presynaptic terminal which connects to multiple postsynaptic sites. Both use multiple cellular layers with regular arrangements of neurons in the layers with an ordered way of mapping computation between layers. Both use lateral interactions through parallel relays to heavily process the raw environmental input received by the photoreceptors in the eyes. They use repeated local modules to handle the global environmental input field and different visual functions are helped by pathways that act in parallel. The main structure for the parallel processing is the organization of the many synaptic connections into parallel laminar regions. The results of this processing is then passed to other portions of the cortex where further feature extraction and associative learning is completed. It may be that the organization of the fly visual system and the mammalian visual system do not share a common evolutionary origin though in contrast to what is now suspected about the common origins of these systems in mammals and birds. Instead, this may appears to be an example of convergent evolution at work: the problems of handing the environmental input shaped the evolution of this type of circuitry. However, we also know that the vertebrate transcription factor Pax6 and its fly orthologue *eyeless* is crucial for eye development so it is still possible that there is a shared evolutionary origin.

There is much work to be done to develop our cognitive dysfunction model and this will be main topic in the next volume. We will use these tools to build both models of human cognitive dysfunction and also models of small brained animals.

References

A. Barron, E. Søvik, J. Cornish, The roles of dopamine and related compounds in reward-seeking behavior across animal phyla. Front. Behav. Neurosci. **4**(163), 1–9 (2010) (Article 163)

K. Friston, Models of brain function in neuroimaging. Annu. Rev. Psychol. **56**, 57–87 (2005) (Annual Reviews)

E. Harmon-Jones, P. Gable, C. Peterson, The role of asymmetric frontal cortical activity in emotion-related phenomena: a review and update. Biol. Psychol. **84**, 451–462 (2010)

P. Lang, M. Bradley, J. Fitzimmons, B. Cuthbert, J. Scott, B. Moulder, V. Nangia, Emotional arousal and activation of the visual cortex: an fMRI analysis. Psychophysiology **35**, 199–210 (1998)

I. Mauss, M. Robinson, Measures of emotion: a review. Cognit. Emot. **23**, 209–237 (2009)

J. Peterson, Nodal computation approximations in asynchronous cognitive models. Comput. Cognit. Sci. (2015) (In Press)

S. Russo, E. Nestler, The brain reward circuitry in mood disorders. Nat. Rev.: Neurosci. **14**, 609–625 (2013)

J. Sanes, S. Zipursky, Design principles of insect and vertebrate visual systems. Neuron **66**(1), 15–36 (2010)

K. Stephan, L. Harrison, W. Penny, K. Friston, Biophysical models of fMRI responses. Curr. Opin. Neurobiol. **14**, 629–635 (2004)

Y. Wang, A. Brzozowsha-Prechtl, H. Karten, Laminary and columnar auditory cortex in avian brain. Proc. Natl. Acad. Sci. **107**(28), 12676–12681 (2010)

Part VIII
Conclusions

Chapter 22
Conclusions

This text has introduced the salient issues that arise in cognitive modeling. We have used relatively simple models of neural computation to construct graph models of brain circuitry at essentially any level of detail we wish. In Chap. 21, we have outlined a simple model of a normal brain and shown how, in principle, we could develop a model of cognitive dysfunction from that. To finish out this book, let's look at some really hard problems that we have been building various tools to address. Let's consider a model of schizophrenia. Using the notations of Chap. 21, a first look could be this. If we were to model schizophrenia with a left and right brain and a connecting corpus callosum, the two modules would be the left and right brain subgraph and the corpus callosum subgraph. Let's look at this two graph system where the first graph contains the left and the right brain and the second graph contains the corpus callosum. Assume we have trained on T samples using the module training equations

$$\left(\Lambda_I + \mathcal{F}_I^T\right) f_I^t + \left(\mathcal{I}_I^t + g_I^t - \alpha \, f_I^{t-1}\right) = 0$$
$$\left(\Lambda_{II} + \mathcal{F}_{II}^T\right) f_{II}^t + \left(\mathcal{I}_{II}^t + g_{II}^t - \alpha \, f_{II}^{t-1}\right) = 0$$

At convergence, we obtain stable values for f_I^t and f_{II}^t we will label as f_I^∞ and f_{II}^∞. Then we must have

$$\left(\Lambda_I + \mathcal{F}_I^T\right) f_I^\infty + \left(\mathcal{I}_I^t + g_I^\infty - \alpha \, f_I^\infty\right) = 0$$
$$\left(\Lambda_{II} + \mathcal{F}_{II}^T\right) f_{II}^\infty + \left(\mathcal{I}_{II}^t + g_{II}^\infty - \alpha \, f_{II}^\infty\right) = 0$$

A change in the connectivity from the Corpus Callosum to the left and right brains and vice-versa implies a change $C \to C'$ and $D \to D'$. This changes terms which are built from the matrices C and D, such as $\mathcal{F}_I^T = DD^T E_{II}$ which changes to $D'(D')^T E_{II}$. After the connectivity changes are made, the graph can evaluate all T inputs to generate a collection of new outputs $S_I' = ((f_I^1)', \ldots, f_I^T)')$ and $S_{II}' = ((f_{II}^1)', \ldots, (f_{II}^T)')$. The original outputs due to the trained graph are S_I and S_{II}. Hence, any measure of the entries of these matrices gives us a way to compare the performance of the graph model of the brain with a *correct* set of corpus callosum

© Springer Science+Business Media Singapore 2016
J.K. Peterson, *BioInformation Processing*, Cognitive Science and Technology,
DOI 10.1007/978-981-287-871-7_22

connectivities to the graph model of a brain with dysfunctional connectivity between the corpus callosum and the two halves of the brain. A simple ratio gives us a useful measure (here, the norm symbols indicate any choice of matrix measurement such as a simple Frobenius norm). Choosing a set of tolerances ϵ_I and ϵ_{II}, we obtain a tool for deciding if there is dysfunction: for example, $\frac{\|S_I'\|}{\|S_I\|} > \epsilon_I$ and $\frac{\|S_{II}'\|}{\|S_{II}\|} > \epsilon_{II}$. In general, if we let Φ denote our tool for measuring the brain model's effectiveness as a function of the measured outputs for a sample set, we would check to see if $\frac{\Phi(S_I')}{\Phi(S_I)} > \epsilon_I$ and $\frac{\Phi(S_{II}')}{\Phi(S_{II})} > \epsilon_{II}$. The choice of Φ is, of course, critical and it is not easy to determine a good choice. Another issue is that we need a way to determine what the *normal* output of a brain model should be so we can make a reasonable comparison. However, despite these obstacles, we see this procedure of submodule training allows us to identify theoretical consequences of changes in corpus callosum communication pathways (i.e. the cross connections to the left and right brain) and perhaps give some illumination to this difficult problem. The above model is a very high level idea for schizophrenia in line with the general comment that some feel schizophrenia is a price we pay for our brains becoming complicated enough to develop language as is discussed in Crow (2000), but even if that is true, it would be much more interesting to develop a dynamic model which responds as a schizophrenic would do. There are many good reasons to build a model first of which is it could give insight. More importantly, most of our data for what is happening in schizophrenics comes from very small data sets and hence a theoretical model of schizophrenia is needed to help shape experiments and give better understanding as to what parts of our interacting neural modules are malfunctioning. A neural simulation could potentially help us with that.

There has been a lot of work which correlates various structural and connectionist problems in the neural circuitry of an individual with schizophrenia when compared to the *normal* population. Most of these studies though are data poor and so we must be careful in drawing conclusions. Synaptic plasticity issues are discussed in Goto et al. (2010) and various neurotransmitter problems appear to be linked to schizophrenia also as you can see the papers Hashimoto (2009) and Domino et al. (2004). The work in Domino et al. (2004) focuses on the N-methyl-D-aspartate (NMDA) receptor. Physical brain changes—again from a small data set—are outlined in (Andreasen et al. 2011) and some of the neuroimaging results about chronic schizophrenia are discussed in Wood et al. (2011). There is developmental hypothesis for schizophrenia outlined in Piper et al. (2012) which is a good read as it helps us with some of the larger issues surrounding our modeling choices. Finally, the review by Fakhoury (2015) is particularly useful in building modeling insight for similar issues that arise in major depressive disorders which often occur with schizophrenia too. In Chap. 7, we present a careful model of free calcium and bound calcium in a cell and show how we can think about a sudden influx of calcium current arising from a second messenger trigger. These ideas are helpful in understanding schizophrenia as there is evidence that abnormalities in these processes are linked to schizophrenia as seen in Eyles et al. (2002). Since we are easily able to model different topological organizations for our brain models, the results in Zhang et al. (2012) help us understand how to set up the graph for a schizophrenia model.

There have been recent and exciting results in treating depressive states, suicidal impulses that are often associated with schizophrenia which are based on using an N-methyl-D-aspartate (NMDA) receptor antagonist using the drug ketamine. A complete review of the use of ketamine for major depressive problems and bipolar depression is given in Lee et al. (2015). How it is used in chronic stress, attention disorders and depressive states is discussed in Li et al. 2011, KNott et al. (2011) and Murrough et al. (2013). Hence, the tools and background given to you in this text will help you on the next journey to develop interesting neural models of cognitive function and dysfunction such as we see in depression and schizophrenia.

Finally, since we know hormones play a major role in brain plasticity as evidenced by Garcia-Segura (2009), we know that our graph models can also be tuned or altered by external hormone based second messenger triggers. We know that in parasitic wasps, the wasp injects a large neurotoxin cocktail into a cockroach's brain which then initiates large scale changes in behavior. In Chap. 9, we discuss toxins that can alter the action potential of a neuron in various ways. The cocktail the wasp injects is very large scale and not a small scale input to the neural model of the cockroach at all. A general outline of parasitic manipulation of a host is given in Weinersmith and Faulkes (2014) and more specific information about how the wasp does this with the cockroach is discussed in Libersat and Gal (2014). Note this cocktail effectively reorganizes the interacting modules of the cockroach brain to perform altered behavior. Thus, there is proof of concept for such a large reorganization of the neural modules of any brain via an appropriate input package. With the graph modeling tools we have developed and suitable approximations, we can hope to build a model of this process also which in the past has been very hard to both conceptualize and to build.

In the next volume, all of these ideas will be used to build real models.

References

N. Andreasen, P. Nopoulos, V. Magnotta, R. Pierson, S. Ziebell, B. Ho, Progressive brain change in schizophrenia: a prospective longitudinal study of first-episode schizophrenia. Biol. Psychiatr. **70**, 672–679 (2011)

T. Crow, Schizophrenia as the price Homo sapiens pays for language: a resolution of the central paradox in the origin of the species. Brain Res. Rev. **31**, 118–129 (2000)

E. Domino, D. Mirzoyan, H. Tsukada, N-methyl-D-aspartate antagonists as drug models of schizophrenia: a surprise link to tobacco smoking. Prog. Neuro-Psychopharmacol. Biol. Psychiatr. **28**, 801–811 (2004)

D. Eyles, J. McGrath, G. Reynolds, Neuronal calcium-binding proteins and schizophrenia. Schizophr. Res. **57**, 27–34 (2002)

M. Fakhoury, New insights into the neurobiological mechanisms of major depressive disorders. Gen. Hosp. Psychiatr. **37**, 172–177 (2015)

L. Garcia-Segura, *Hormones and Brain Plasticity* (Oxford University Press, Oxford, 2009)

Y. Goto, C. Yang, S. Otani, Functional and dysfunctional synaptic plasticity in prefrontal cortex: roles in psychiatric disorders. Biol. Psychiatr. **67**, 199–207 (2010)

K. Hashimoto, Emerging role of glutamate in the pathophysiology of major depressive disorder. Brain Res. Rev. **61**, 105–123 (2009)

V. KNott, A. Millar, J. McIntosh, D. Shah, D. Fisher, C. Blais, V. Ilivitsky, E. Horn, Separate and combined effects of low dose ketamine and nicotine on behavioral and neural correlates of sustained attention. Biol. Psychol. **88**, 83–93 (2011)

E. Lee, M. Della-Selva, A. Liu, S. Himelhoch, Ketamine as a novel treatment for major depressive disorder and bipolar depression: a systematic review and quantitative meta-analysis. Gener. Hosp. Psychiatr. **37**, 178–184 (2015)

N. Li, R. Liu, J. Dwyer, M. Banasr, B. Lee, H. Son, X. Li, G. Aghajanian, R. Duman, Glutamate N-methyl-D-aspartate receptor antagonists rapidly reverse behavioral and synaptic deficits Caused by Chronic stress exposure. Biol. Psychiatr. **69**, 754–761 (2011)

F. Libersat, R. Gal, Wasp voodoo rituals, venom—cocktails, and the zombification of cockroach hosts. Integr. Comp. Biol. **54**, 129–142 (2014)

J. Murrough, D. Iosifescu, L. Chang, R. Al Jurdi, C. Green, A. Perez, S. Iqbal, S. Pillemer, A. Foulkes, A. Shah, D. Charney, S. Mathew, Antidepressant efficacy of ketamine in treatment-resistant major depression: a two-site randomized controlled trial. Am. J. Psychiatr. **170**, 1134–1142 (2013)

M. Piper, M. Beneyto, T. Burne, D. Eyles, D. Lewis, J. McGrath, The neurodevelopmental hypothesis of schizophrenia: convergent clues from epidemiology and neuropathology. Psychiatr. Clin. N. Am. **35**, 571–584 (2012)

K. Weinersmith, Z. Faulkes, Parasitic manipulation of host's phenotype, or how to make a zombie—an introduction the symposium. Integr. Comp. Biol. **54**, 93–100 (2014)

S. Wood, A. Yung, P. McGorry, C. Pantelis, Neuroimaging and treatment evidence for clinical staging in psychotic disorders: from the at-risk mental state to chronic schizophrenia. Biol. Psychiatr. **70**, 619–625 (2011)

Y. Zhang, L. Lin, C. Lin, Y. Zhou, K. Chou, C. Lo, T. Su, T. Jiang, Abnormal topological organization of structural brain networks in schizophrenia. Schizophr. Res. **141**, 109–118 (2012)

Part IX
Background Reading

Chapter 23
Background Reading

We have been inspired by many attempts by people in disparate fields to find meaning and order in the vast compilations of knowledge that they must assimilate. Like them, we have done a fair bit of reading and study to prepare ourselves for necessary abstractions we need to make in our journey. We have learned a lot from various studies of theoretical biology and computation and so forth. You will need to make this journey too, so to help you, here are some specific comments about the sources we have used to learn from. In addition to the books we mentioned in Peterson (2015a, b, c), we have some further recommendations.

23.1 The Central Nervous System

- The functional anatomy of the human brain in Nolte (2002).
- The central nervous system from Brodal (1992).
- Two very useful books which has many diagrams of brain circuitry which you can use to draw your own sketches and color in yourself are Diamond et al. (1985) and Pinel and Edwards (2008). Thinking about these circuits at the same time you draw and color them is a great way to begin to *understand* them. We have used these two texts so much over the last ten years that they are dog-eared and well-thumbed. These are not a stand-alone texts as they have wonderful drawings with extremely terse textual descriptions. But they are nevertheless gold mines of information especially if your way of understanding stuff is visual.
- An assessment of the tools used to understand the function of the brain (fMRI and so on) is presented in Donaldson (2004). We must always remember that evidence for *anything* is always obtained indirectly. There are usually many technical processing steps performed on the raw data collected with many concomitant assumptions before one sees information that is useful. This chain of evidentiary processing must never be forgotten when we build our models.

© Springer Science+Business Media Singapore 2016   525
J.K. Peterson, *BioInformation Processing*, Cognitive Science and Technology,
DOI 10.1007/978-981-287-871-7_23

23.2 Information Theory, Biological Complexity and Neural Circuits

- Biochemical Messengers in Hardie (1991). This is the first book that we began to read about information processing in the nervous system on a technical level. It has useful insights. All in all, we use this one as a technical reference to sometimes dip into. This book explains a lot of the technical detail behind the messaging systems we encounter in our studies.
- Information in the brain in Black (1991). While dated now, this has been an important source as it tries hard to develop a unified theory of information processing. It is an absolutely stunning attempt to understand the *how* and *why* of information processing in the large scale nervous system. As you will see in the body of this report, we have quoted from this work extensively. There is so much interesting speculation in here. It is also extremely technical and unapologetically so!! We used Black's ideas extensively in our previous attempt at developing a good abstraction for biological detail into software design. This is laid out in Peterson (2001).
- The Evolution of Vertebrate Design in Radinsky (1987). This is a wonderful book that attempts to look at the evolution of the vertebrate body plan in very abstract terms.
- Large scale computational theories of the brain in Koch and Davis (1994). This is another useful collection of attempts to model biological complexity and see the *big* picture.
- Neural organization in Arbib et al. (1998). Here Arbib and his coauthors spend a lot of time discussing their approach to seeing the hidden structure in brain circuitry. There are a lot of good ideas in here.
- The paper (Baslow 2011) discusses the modular organization of the vertebrate brain. Pay attention to how these ideas can influence how we build our models using the graph tools in MatLab.
- Two layman's books on how our notions of morality may have evolved and a general introduction to ideas in evolutionary psychology are Buller (2006) and Hauser (2006). These are less technical but provide needed overview.

23.3 Nervous System Evolution and Cognition

We believe there is a lot to learn from studying how organisms have evolved nervous systems. The primitive systems contain a minimal set of structures to allow control, movement and environmental response. We can see in the evolutionary record how as additional structure is added, we not only see a gain in capability but also more constraints on the organism in terms of energy requirements and so forth. Some of the vary useful papers in this area are listed below:

- Catania (2000) analyzes the organization of cortex in the insects. The connection between the development of sensory modules to understand the environment and the fusion required to build higher level precepts for enhanced survival can be studied in many animals. However, there are advantages to studying these things in animals with smaller cognitive systems.
- In Deacon (1990), there is a very interesting review of the state of the art in thinking about brain evolution as of 1990.
- In Miklos (1993), there is even more of an attempt to use the ideas from the evolution of nervous systems to help guide the development of artificial nervous systems for use in agents, artificial life and so forth. Miklos has strong opinions and there is much to think about in this paper.
- In Allman (1999), we can read the most up to date discussion of brain and nervous system evolution. We have been inspired by this book and there are good pointers to additional literature in its reference list.
- Potts (2004) discusses how paleoenvironments may have shaped the evolution of cognition in the great apes while the great ape cognitive systems themselves are outlined in Russon (2004). Further articles can be found in the collection (Russon and Begun 2004).
- Johnson-Frey (2004) studies the neural circuitry that may subserve complex tool use in humans.
- The possible connection between *modularity* of function and the rise of cognition is explored in Coltheart (1999) and in evolutionary terms in Redies and Puelles (2001).
- The appearance of artistic expression in the cave paintings of France ca 25,000 BC (and other clues) has suggested to some that neural reorganization occurred around that time. Other scientists are not so sure. Nevertheless, it is a clue to the relationship between brain structure and higher level processing. Brain endocasts give us clues about the location of sulci on the surface of the brain which allows us to compare the sulci of fossils to those of primates and modern man and look for differences. An interesting article about paleolithic art and its connection to modern humans is presented in Conkey (1999). Holloway (1999) analyzes such sulci location differences in his review of human brain evolution.

23.4 Comparative Cognition

Trying to learn about how cognition might have arisen is made even more difficult when the nervous system studied is as complex as the human brain. There is growing evidence that simpler nervous systems appear to be capable of tasks that would easily be considered as cognitive if seen in a higher vertebrate or primate. These simpler systems occur in several insects and spiders. There appears to be some evidence that parsing and fusing disparate sensory signals into higher level precepts begins to build a cognitive system that at some point *crosses a threshold* to become *aware*. This advanced behavior appears to develop primarily in predators that must assess contradictory and ambiguous information in order to plan an attack. Once these animals include enough varied prey for programmed attacks to be of limited utility, they appear to undergo neural system evolution that allows pre planning of routes that go out of sight and an almost infinite variety of attack strategies. Since these animals have very small nervous systems—say 400,000 neurons—this is forcing many scientists to reconsider almost everything previously said about how cognition evolves. It also means that the software modeling of cognition processes is not as far fetched as once believed as these nervous systems are within our computational grasp.

These general ideas are discussed in the survey (Greenspan and van Swinderen 2004). An introduction to the study of these *simpler nervous systems* begins with Prete (2004). This book has chapters devoted to several interesting small animals. The spider *portia Fimbriata* is studied in Harland and Jackson (2004). Portia is quite amazing as she appears to be self-aware and reacts to its own image in a mirror. The motion perception of amphibians is studied in Ewert (2004). The Praying Mantis also possess impressive cognitive capabilities as detailed in Kral and Prete (2004). Also, the emergence of cognitive function within the small brain of the honeybee is explored in Zhang and Srinivasan (2004). General principles of the evolution of animal communication are studied in Hauser (1996). The evolution of cortical diversity during development in Kornack (2000) and Krubitzer and Huffman (2000). The neural language for words is explored in Pulvermüller (1999).

In addition to these, there are texts that deal explicitly with to problem of comparing the cognitive capacities and potentials of other animals.

- In Butler and Hodos (2005), there is an excellent treatment of how to do these sorts of comparisons. The texts by Wasserman and Zentall (2006) and Shettleworth (2010) are complimentary and are very helpful as if we are interested in understanding how small a neural model can be and still exhibit cognitive behavior, there are clues as to how this can be implemented in the study of alternative neural strategies for solving the problems of integrating environmental signals into useful choices for survival. Also, the papers Edelman et al. (2005) and seth (2005) try to identify how to find evidence of consciousness in non-mammals which again helps widen our perceptions. As all of us who have had pets know, subjective experience in not limited to us as humans; an attempt to find evidence for that in neurobiology is found in Baars (2005). This is another example of high level ideas which are quite difficult to quantify into a model. There is also the question of whether non-humans organize information like we do; i.e. do they have cognitive maps? The paper Bennet (1996), while somewhat old, addresses this question nicely. The general problem of trying to understand another beings neural circuitry is the focus of Saxe et al. (2004).
- There are also sources of information for specific cognitive systems in various animals which really help us see cognition as a general solution to the common problems all animals face. Specifically, Reznikova (2007) is a complete discussion of animal intelligence in general and Panksepp (2005) covers core emotional

feelings in animals. Also, the whole idea of learning in animals is laid out in Zentall et al. (2008). For specific animals, we can turn to the following references:

- A complete treatment of arthropod brains is given in Strausfeld (2012). Navigational issues are explored in Webb et al. (2004).
- the octopus brain is analyzed in Young (1971) which is out of print but a wonderful reference from which we can pull out neural circuitry for cephalopod models. The squid brain's development is outlined in Shigeno et al. (2001) and following this developmental chain helps us contrast it with other neural development pathways.
- Herrold et al. (2011) examines the nidopallium caudolaterale in birds which might be an analogue of a type of cortex in humans. Hence, associations built from environmental signals can be done in multiple ways which gives insight into general cognitive architectures. A review of tool use in birds given in Lefebvre et al. (2002) is also very helpful. In Petkov and Jarvis (2012), a very helpful comparison on bird and primate language origins is presented which is also a way to study how different systems are used to solve similar problems. This paper should be read in conjunction with another survey on human language evolution given in Berwick et al. (2012). Finally, the evolution of avian intelligence is studied in Emery (2006).
- Honeybee neural architectures are studied in Menzel (2012) as a model of cognition. The information processing the honeybee mushroom body does is probed in the papers (Haehnel and Menzel 2010b), (Szyszka et al. 2008) and (Sjöholm et al. 2005). By studying these papers, we can see different ways to implement cortical structure. A more general treatment of insect mushroom bodies is given in Farris and Sinakevitch (2003). Honeybee memory formation is discussed in the two papers (Hadar and Menzel 2010a) and (Roussel et al. 2010).
- The generation of behavior in a tadpole is worked out in detail in Roberts et al. (2010) which helps us understand how low level neuronal information can give rise to high level behavior.
- The fruit fly is capable of very complex cognitive functions which we are just beginning to understand better. An overview of what the fruit fly is capable of is found in Greenspan and van Swinderen (2004) and it makes for good reading. Again, think about generalizations! A nice survey of social learning in insects in general and what that might mean about their neural organizations is given in Giurfa (2012) which is also very interesting.
- The ctenophore Mnemiopsis leidyi has recently had its genome sequenced and we now know its neural architecture was developed in a different ways from other animals. Hence, it also gives us new design principles to build functional small brain architectures. Read Ryan et al. (2013) and Moroz et al. (2014) carefully and take notes.
- Commonalities in the development of the nerve chord in crustaceans is discussed in Harzsch (2003). Comparisons to what happens in arthropods are made which allows generalization.

Some general ideas about brain evolution can also be illuminated by looking at other animals. We note that these issues are explained carefully in Williams and Holland (1998), Marakami et al. (2005) and Sprecher and Reichert (2003).

23.5   Neural Signaling

- A basic treatment of calcium signaling in the cell from Berridge (1997).
- A theoretical discussion of signaling complexes in Bray (1998) which influenced much of the work in Chap. 6.
- Incoming environmental signals must be processed and transduced into other forms. A good treatment of the many transduction cascades employed by a biological system to make sense of raw data is presented in Gomperts et al. (2002a). Of particular interest to the work presented in Chap. 6 is the chapter on *GTP binding proteins* and their role in second messenger triggers (Gomperts et al. 2002b, Chap. 4).

23.6   Gene Regulatory Circuits

A clear theoretical approach to gene regulatory circuits is very helpful in understanding how the complications of cell signaling both evolved and functions. In particular, *gene doubling* is a powerful control metaphor for neural system development. The primary sources we have used in this area are given below. The study of the cis-regulatory module as a multi input/output control device which simultaneously alters structure (hardwire) and signaling (software) helps in the design of useful asynchronous software architectures for cognitive modeling.

- The development and regulation of gene regulatory systems in Davidson (2001a). Of particular interest is the chapter on the *cis-regulatory module* given in Davidson (2001b), Chap. 2.
- The evolution of developmental pathways is discussed in clear terms in Wilkins (2002a). Particularly interesting is the treatment of *genetic pathways and networks in developments* (Wilkins 2002b, Chap. 4).
- The path from embryonic modules to function in the adult nervous system is explored in Redies and Puelles (2004).
- A collection of articles on the theme of modularity in development is presented in Schlosser and Wagner (2004).

23.7 Software

Out of the many books that are available for self-study in all of the areas above, some have proved to be invaluable, while others have been much less helpful. The following annotated list consists of the real gems. To learn to program effectively in an object oriented way in Python, it is helpful to know how to program in a procedural language such as C. Then, learning how to program objects within the constraints of the class syntax of C++ is very useful. This is the route we took in learning how to program in an object oriented way. The final step is to learn how to use a scripting glue language such as Python to build application software. Finally, don't be put off by the publication date of these resources! Many resources are timeless.

C++: The following books need to be on your shelf. Lippman will get you started, but you'll also need Deitel and Deitel and Olshevsky and Ponomarev for nuance.

1. C++ Primer, Lippman (1991). This book is the most basic resource for this area. While very complete, it has shortcomings; for example, it's discussion of call by reference is very unclear and its treatment of dynamic binding in its chapters on OOD is also murky. Nevertheless, it is a good basic introduction. It's biggest problem for us is that all of its examples are so simple (yes, even the *zoo* class is just *too* simple to give us much insight).
2. C++: How to Program, Deitel and Deitel (1994). We have found this book to be of great value. It intermingles excellent C++ coverage with ongoing object oriented design (OOD)material. It is full of practical advice on *software engineering* aspects of OOD design.
3. The Revolutionary Guide to OOP Using C++, Olshevsky and Ponomarev (1994). This book has a wonderful discussion of call by reference and equally good material on dynamic binding.
4. Compiler Design, Wilhem and Maurer (1995). This book has already been mentioned in the text as the source of technical information on how an object-oriented compiler is built. This is an essential resource.

Python: There are two wonderful resources for using Python in computation which are Langtangen (2010, 2012). You can start learning about this now as the transition from using MatLab to using Python is fairly easy. The comments below on object oriented programming are also relevant here, so even though many of the object oriented texts are focusing on C++, don't let that put you off.

Erlang: Erlang is a great choice for our eventual neural simulations as it will allow us to do lesion studies. Of course, nodal computation in Erlang requires we find good approximations to realistic neural computation, but that can be done as we have discussed in this book. Two good references to this language are Armstrong (2013) and Logan et al. (2011). However, the best way to get started is to read (Hébert 2013). We encourage you to start thinking about this way of coding as well as we will start using it in later volumes.

Haskell: Haskell is another language like Erlang which has great potential for writing lesion simulations. You should look at Lipovača (2011) to get started here.

Object Oriented Programming and Design

1. Object-Oriented Analysis and Design with Applications (Booch 1994).
   This is a classic reference to one method of handling large scale OOD. As the
   number of objects in your design grows there is a combinatorial explosion in
   the number of interaction pathways. The Booch method gives a popular software
   engineering tool. This is best to read on a surface level, for impressions and ideas.
2. Designing Object-Oriented C++ Applications Using the Booch Method
   (Martin 1995).
   If you decide to use the Booch method, this book is full of practical advice. It has
   many code examples, but it has a very heavy reliance on *templates*. This is a C++
   language feature we have been avoiding because it complicates the architectural
   details. Hence, the translation of Martin's code fragments into useful insight is
   sometimes difficult, but nonetheless, there is much meat here.
3. Design Patterns for Object-Oriented Software Development (Pree 1995).
   The design of classes and objects is very much an art form. To some extent, like
   all crafts, you learn by doing. As you get more skilled, you realize how little of the
   real knowledge of how to write good classes is written down! This book is full of
   hard won real-world wisdom that comes out of actually being in the programming
   trenches. It is best to surface read and sample.
4. Taming C++: Pattern Classes and Persistence for Large Projects (Soukup 1994).
   We have similar comments for this book. Since our proposed *neural objects* OOD
   project will be a rather massive undertaking, useful insight into the large scale
   OOD is most welcome!
5. Design Patterns: Elements of Reusable Object-Oriented Software (Gamma et al.
   1995).
   As you program, you realize that many classes are essential building blocks of
   many disparate applications. This wonderful book brings together a large number of already worked out OOD solutions to common problems. It is extremely
   important to look at this book carefully. All of the different classes are presented
   in code sketches (not easy to follow, but well worth the effort!).

Neural Simulation Software: To model these things, we can use

1. the **Genesis** modeling language as discussed in **The Book of Genesis: Exploring
   Realistic Neural Models with the GEneral NEural SImulation System**, by
   Bower and Beeman (Bower and Beeman 1998).
2. home grown code written in C or C++.
3. home grown code written in MatLab.
4. home grown code written in Python.
5. home grown code written in Erlang.

23.8 Theoretical Robotics

- In the Ph.D. thesis of Vogt (2000), we see studies and experiments on how to get collections of robots to come up with a language all on their own and to then to communicate using this language. This is extremely fascinating and the mechanisms that Vogt and his colleagues use to implement both software and hardware so as to gain *behavioral* plasticity have great promise.
- The view of situated cognition in Clancey (1997). In this book, Clancey discusses at great length what he considers to be appropriate ways to develop robots that will have the plasticity we need to perform interesting things. There is a lot of food for thought here.

References

J. Allman, *Evolving Brains* (Scientific American Library, New York, 1999)

M. Arbib, P. Erdi, P. Szentagothal, *Neural Organization: Structure, Function and Dynamics* (A Bradford Book, MIT Press, Cambridge, 1998)

J. Armstrong, *Programming Erlang Second Edition: Software for a Concurrent World* (The Pragmatic Bookshelf, Dallas, 2013)

B. Baars, Subjective experience is probably not limited to humans: the evidence from neurobiology and behavior. Conscious. Cogn. **14**, 7–21 (2005)

M. Baslow, The vertebrate brain, evidence of its modular organization and operating system: insights into the brain's basic units of structure, function, and operation and how they influence neuronal signaling and behavior. Front. Behav. Neurosci. **5**, 1–7 (2011). (Article 5)

A. Bennet, Do animals have cognitive maps? J. Exp. Biol. **199**, 219–224 (1996)

M. Berridge, Elementary and global aspects of calcium signalling. J. Physiol. **499**, 291–306 (1997)

R. Berwick, G. Beckers, K. Okanoya, J. Bolhuis, A bird's eye view of human language evolution. Front. Evol. Neurosci. **4**, 1–25 (2012). (Article 5)

I. Black, *Information in the Brain: A Molecular Perspective* (A Bradford Book, MIT Press, Cambridge, 1991)

G. Booch, *Object-Oriented Analysis and Design with Applications*, 2nd edn. (Benjamin/Cummings Publishing Company, Redwood City, 1994)

J. Bower, D. Beeman, *The Book of Genesis: Exploring Realistic Neural Models with the GEneral NEural SImulation System*, 2nd edn. (Springer TELOS, New York, 1998)

D. Bray, Signalling complexes: biophysical constraints on intracellular communication. Annu. Rev. Biophys. Biomol. Struct. **27**, 59–75 (1998)

P. Brodal, *The Central Nervous System: Structure and Function* (Oxford University Press, New York, 1992)

D. Buller, *Adapting Minds: Evolutionary Psychology and the Persistent Quest for Human Nature* (A Bradford Book, The MIT Press, Cambridge, 2006)

A. Butler, W. Hodos, *Comparative Vertebrate Neuroanatomy: Evolution and Adaptation*, 2nd edn. (Wiley-Interscience, New York, 2005)

K. Catania, Cortical organization in insectovoria: the parallel evolution of the sensory periphery and the brain. Brain, Behav. Evol. **55**(6), 311–321 (2000)

W. Clancey, *Situated Cognition: On Human Knowledge and Computer Representations* (Cambridge University Press, Cambridge, 1997)

M. Coltheart, Modularity and cognition. Trends Cogn. Sci. **3**(3), 115–120 (1999)

M. Conkey, A history of the interpretation of European 'paleolithic art': magic, mythogram, and metaphors for modernity, in *Handbook of Human Symbolic Evolution*, ed. by A. Lock, C. Peters (Blackwell Publishers, Massachusetts, 1999)

E. Davidson, *Genomic Regulatory Systems: Development and Evolution* (Academic, San Diego, 2001a)

E. Davidson, Inside the cis-regulatory module: control logic, and how regulatory environment is transduced into spatial patterns of gene expression, *Genomic Regulatory Systems: Development and Evolution* (Academic, San Diego, 2001b), pp. 26–63

T. Deacon, Rethinking mammalian brain evolution. Am. Zool. **30**, 629–705 (1990)

H. Deitel, P. Deitel, C^{++}*: How to Program* (Prentice Hall, Upper Saddle River, 1994)

M. Diamond, A. Scheibel, L. Elson, *The Human Brain Coloring Book* (Barnes and Noble Books, New York, 1985)

D. Donaldson, Parsing brain activity with fMRI and mixed designs: what kind of a state is neuroimaging in? Trends Neurosci. **27**, 442–444 (2004)

D. Edelman, B. Baars, A. Seth, Identifying hallmarks of consciousness in non-mammalian species. Conscious. Cogn. **14**, 169–187 (2005)

N. Emery, Cognitive ornithology: the evolution of avian intelligence. Philos. Trans. R. Soc. B **361**, 23–43 (2006)

J. Ewert, Motion perception shapes the visual world of the amphibians, in *Complex Worlds from Simpler Nervous Systems*, ed. by F. Prete (A Bradford Book, MIT Press, Cambridge, 2004), pp. 117–160

S. Farris, I. Sinakevitch, Development and evolution of the insect mushroom bodies: towards the understanding of conserved developmental mechanisms in a higher brain center. Arthropod Struct. Dev. **32**, 79–101 (2003)

E. Gamma, R. Helm, R. Johnson, J. Vlissides, *Design Patterns: Elements of Reusable Object-Oriented Software* (Addison-Wesley, Reading, 1995)

M. Giurfa, Social learning in insects: a higher-order capacity? Front. Behav. Neurosci. **6**, 1–3 (2012). (Article 57)

B. Gomperts, P. Tatham, I. Kramer, *Signal Transduction* (Academic, San Diego, 2002a)

B. Gomperts, P. Tatham, I. Kramer, GTP-binding proteins and signal transduction, *Signal Transduction* (Academic, San Diego, 2002b), pp. 71–106

R. Greenspan, B. van Swinderen, Cognitive consonance: complex brain functions in the fruit fly and its relatives. Trends Neurosci. **27**(12), 707–711 (2004)

R. Hadar, R. Menzel, Memory formation in reversal learning of the honeybee. Front. Behav. Neurosci. **4**, 1–11 (2010a). (Article 4)

M. Haehnel, R. Menzel, Sensory representation and learning-related plasticity in mushroom body extrinsic feedback neurons of the protocerebral tract. Front. Syst. Neurosci. **4**, 1–13 (2010b). (Article 161)

D. Hardie, *Biochemical Messengers: Hormones, Neurotransmitters and Growth Factors* (Chapman & Hall, London, 1991)

D. Harland, R. Jackson, Portia perceptions: the umwelt of an araneophagic jumping spider, in *Complex Worlds from Simpler Nervous Systems*, ed. by F. Prete (A Bradford Book, MIT Press, Cambridge, 2004), pp. 5–40

S. Harzsch, Ontogeny of the ventral nerve chord in malacostracan crustaceans: a common plan for neuronal development in Crustacea, Hexapoda and other Arthropoda? Arthropod Struct. Dev. **32**, 17–37 (2003)

M. Hauser, *The Evolution of Communication* (A Bradford Book, MIT Press, Cambridge, 1996)

M. Hauser, *Moral Minds: How Nature Designed Our Universal Sense of Right and Wrong* (Harper-Collins, New York, 2006)

F. Hébert, *Learn You Some Erlang for Great Good* (No Starch Press, San Francisco, 2013)

C. Herrold, N. Palomero-Gallagher, B. Hellman, S. Kröner, C. Theiss, O. Güntürkün, K. Zilles, The receptor architecture of the pigeons' nidopallium caudolaterale: an avian analogue to the mammalian prefrontal cortex. Brain Struct. Funct. **216**, 239–254 (2011)

R. Holloway, Evolution of the human brain, in *Handbook of Human Symbolic Evolution*, ed. by A. Lock, C. Peters (Blackwell Publishers, Massachusetts, 1999), pp. 74–116

S. Johnson-Frey, The neural basis of complex tool use in humans. Trends Cogn. Sci. **8**(2), 71–78 (2004)

K. Kral, F. Prete, In the mind of a hunter: the visual world of the praying mantis, in *Complex Worlds from Simpler Nervous Systems*, ed. by F. Prete (A Bradford Book, MIT Press, Cambridge, 2004), pp. 75–116

K. Koch, J. Davis (eds.), *Large Scale Neuronal Theories of the Brain* (A Bradford Book, MIT Press, Cambridge, 1994)

D. Kornack, Neurogenesis and the evolution of cortical diversity: mode, tempo, and partitioning during development and persistence in adulthood. Brain, Behav. Evol. **55**(6), 336–344 (2000)

L. Krubitzer, K. Huffman, A realization of the neocortex in mammals: genetic and epigenetic contributions to the phenotype. Brain, Behav. Evol. **55**(6), 322–335 (2000)

H. Langtangen, *Python Scripting for Computational Science* (Springer, New York, 2010)

H. Langtangen, *A Primer of Scientific Programming with Python* (Springer, New York, 2012)

L. Lefebvre, N. Nicolakakis, D. Boire, Tools and brains in birds. Behaviour **139**, 939–973 (2002)

M. Lipovača, *Learn You a Haskell for Great Good* (No Starch Press, San Francisco, 2011)

S. Lippman, *C++ Primer*, 2nd edn. (Addison-Wesley, Reading, 1991)

M. Logan, E. Merritt, R. Carlsson, *Erlang and OTP in Actions* (Manning, Stamford, 2011)

Y. Marakami, K. Uchida, F. Rijli, S. Kuratani, Evolution of the brain developmental plan: insights from agnathans. Dev. Biol. **280**, 249–259 (2005)

R. Martin, *Designing Object-Oriented C++ Applications Using the Booch Method* (Prentice Hall, Englewood Cliffs, 1995)

R. Menzel, The honeybee as a model for understanding the basis of cognition. Nat. Rev.: Neurosci. **13**, 758–768 (2012)

G. Miklos, Molecules and cognition: the latterday lessons of levels, language and lac: evolutionary overview of brain structure and function in some vertebrates and invertebrates. J. Neurobiol. **24**(6), 842–890 (1993)

L. Moroz, K. Kocot, M. Citrarella, S. Dosung, T. Norekian, I. Povolotskaya, A. Grigorenko, C. Dailey, E. Berezikov, K. Buckely, A. Ptitsyn, D. Reshetov, K. Mukherjee, T. Moroz, Y. Bobkova, F. Yu, V. Kapitonov, J. Jurka, Y. Bobkov, J. Swore, D. Girado, A. Fodor, F. Gusev, R. Sanford, R. Bruders, E. Kittler, C. Mills, J. Rast, R. Derelle, V. Solovyev, F. Kondrashov, B. Swalla, J. Sweedler, E. Rogaev, K. Halancych, A. Kohn, The ctenophore genome and the evolutionary origins of neural systems. Nature **510**, 109–120 (2014)

J. Nolte, *The Human Brain: An Introduction to Its Functional Anatomy* (Mosby, A Division of Elsevier Science, St. Louis, 2002)

V. Olshevsky, A. Ponomarev, *The Revolutionary Guide to OOP Using C++* (WROX Publishers, Birmingham, 1994)

J. Panksepp, Affective consciousness: core emotional feelings in animals and humans. Conscious. Cogn. **14**, 30–80 (2005)

J. Peterson, *A White Paper on Neural Object Design: Preliminaries.* Department of Mathematical Sciences (1995), Revised 1998, Revised 1999, Revised 2001. http://www.ces.clemson.edu/~petersj/NeuralCodes/NeuralObjects3

J. Peterson, *Calculus for Cognitive Scientists: Derivatives, Integration and Modeling*, Springer Series on Cognitive Science and Technology (Springer Science+Business Media Singapore Pte Ltd., Singapore, 2015a in press)

J. Peterson, *Calculus for Cognitive Scientists: Higher Order Models and Their Analysis*, Springer Series on Cognitive Science and Technology (Springer Science+Business Media Singapore Pte Ltd., Singapore, 2015b in press)

J. Peterson, *Calculus for Cognitive Scientists: Partial Differential Equation Models*, Springer Series on Cognitive Science and Technology (Springer Science+Business Media Singapore Pte Ltd., Singapore, 2015c in press)

C. Petkov, E. Jarvis, Birds, primates, and spoken language origins: behavioral phenotypes and neurobiological substrates. Front. Behav. Neurosci. **4**, 1–24 (2012). (Article 12)

J. Pinel, M. Edwards, *A Colorful Introduction to the Anatomy of the Human Brain: A Brain and Psychology Coloring Book* (Pearson, New York, 2008)

R. Potts, Paleoenvironments and the evolution of adaptability in great apes, in *The Evolution of Thought: Evolutionary Origins of Great Ape Intelligence*, ed. by A. Russon, D. Begun (Cambridge University Press, Cambridge, 2004), pp. 237–259

W. Pree, *Design Patterns for Object-Oriented Software Development* (ACM Press Books, Addison-Wesley, New York, 1995)

F. Prete (ed.), *Complex Worlds from Simpler Nervous Systems* (A Bradford Book, MIT Press, Cambridge, 2004)

F. Pulvermüller, Words in the brain's language. Behav. Brain Sci. **22**, 253–336 (1999)

L. Radinsky, *The Evolution of Vertebrate Design* (The University of Chicago Press, Chicago, 1987)

C. Redies, L. Puelles, Modularity in vertebrate brain development and evolution. Bioessays **23**, 1100–1111 (2001)

C. Redies, L. Puelles, Central nervous system development: from embryonic modules to functional modules, in *Modularity in Development and Evolution*, ed. by G. Schlosser, G. Wagner (University of Chicago Press, Chicago, 2004), pp. 154–186

Z. Reznikova, *Animal Intelligence: From Individual to Social Cognition* (Cambridge University Press, Cambridge, 2007)

A. Roberts, W. Li, S. Soffe, How neurons generate behavior in a hatchling amphibian tadpole: an outline. Front. Behav. Neurosci. **4**, 1–11 (2010). Article 16

E. Roussel, J. Sandoz, M. Giurfa, Searching for learning-dependent changes in the antennal lobe: simultaneous recording of neural activity and aversive olfactory learning in honeybees. Front. Behav. Neurosci. **4**, 1–11 (2010). (Article 155)

A. Russon, Great ape cognitive systems, in *The Evolution of Thought: Evolutionary Origins of Great Ape Intelligence*, ed. by A. Russon, D. Begun (Cambridge University Press, Cambridge, 2004), pp. 76–100

A. Russon, D. Begun (eds.), *The Evolution of Thought: Evolutionary Origins of Great Ape Intelligence* (Cambridge University Press, Cambridge, 2004)

J. Ryan, K. Pang, C. Schnitzler, A. Nguyen, R. Moreland, D. Simmons, B. Koch, W. Francis, P. Havlak, S. Smith, N. Putnam, S. Haddock, C. Dunn, T. Wolfsberg, J. Mullikin, M. Martindale, A. Baxevanis, The genome of the ctenophore Mnemiopsis leidyi and its implications for cell type evolution. Science **342**(6164), 1242592-1–12425920-8 (2013)

R. Saxe, S. Carey, N. Kanwisher, Understanding other minds: linking developmental psychology and functional neuroimaging. Annu. Rev. Psychol. **55**, 87–124 (2004). (Annual Reviews)

G. Schlosser, G. Wagner (eds.), *Modularity in Development and Evolution* (University of Chicago Press, Chicago, 2004)

A.K. Seth, Criteria for consciousness in humans and other mammals. Conscious. Cogn. *14*, 119–139 (2005)

S. Shettleworth, *Cognition, Evolution, and Behavior* (Oxford University Press, Oxford, 2010)

S. Shigeno, H. Kidokoro, K. Tsuchiya, S. Segawa, Development of the brain in the oegopsid squid, Todarodes pacificus: an atlas from hatchling to juvenile. Zool. Sci. **18**, 1081–1096 (2001)

M. Sjöholm, I. Sinakevitch, R. Ignell, N. Strausfeld, B. Hansson, Organization of Kenyon cells in subdivisions of the mushroom bodies of a lepidopteran insect. J. Comp. Neurol. **491**, 290–304 (2005)

J. Soukup, *Taming C++: Pattern Classes and Persistence for Large Projects* (Addison-Wesley, Reading, 1994)

S. Sprecher, H. Reichert, The urbilaterian brain: developmental insights into the evolutionary origin of the brain in insects and vertebrates. Arthropod Struct. Dev. **32**, 141–156 (2003)

N. Strausfeld, *Arthropod Brains: Evolution, Functional Elegans, and Historical Significance* (The Belknap Press of Harvard University Press, Massachusetts, 2012)

P. Szyszka, A. Galkin, R. Menzel, Associative and non-associative plasticity in Kenyon cells of the honeybee mushroom body. Front. Syst. Neurosci. **2**, 1–10 (2008). (Article 3)

P. Vogt, Lexicon Grounding on Mobile Robots. Ph.D. thesis, Vrije Universiteit, Brussel, Laboratorium voor Artificiele Intelligie (2000)

E. Wasserman, T. Zentall (eds.), *Comparative Cognition: Experimental Explorations of Animal Intelligence* (Oxford University Press, Oxford, 2006)

B. Webb, R. Harrison, M. Willis, Sensorimotor control of navigation in arthropod and artificial systems. Arthropod Struct. Dev. **33**, 301–329 (2004)

R. Wilhem, D. Maurer, *Compiler Design* (Addison-Wesley, Reading, 1995)

A. Wilkins, *The Evolution of Developmental Pathways* (Sinauer Associates, Sunderland, 2002a)

A. Wilkins, Genetic pathways and networks in development, *The Evolution of Developmental Pathways* (Sinauer Associates, Sunderland, 2002b), pp. 99–126

N. Williams, P. Holland, Molecular evolution of the brain of chordates. Brain, Behav. Evol. **52**, 177–185 (1998)

J. Young, *The Anatomy of the Nervous System of Octopus Vulgaris* (Oxford at the Clarendon Press, Oxford, 1971)

T. Zentall, E. Wasserman, O. Lazareva, R. Thompson, M. Rattermann, Concept learning in animals. Comp. Cogn. Behav. Rev. **3**, 13–45 (2008)

S. Zhang, M. Srinivasan, Exploration of cognitive capacity in honeybees: higher functions emerge from small brain, in *Complex Worlds from Simpler Nervous Systems*, ed. by F. Prete (A Bradford Book, MIT Press, Cambridge, 2004), pp. 41–74

Glossary

A

Address based graph models   The basic idea here is to take the usual graphs, edges and vertices classes and add additional information to them. Each node now has a global node number and a six dimensional address. The usual methods for manipulating these classes are then available. Using these tools, it is easier to build complicated graphs one subgraph at a time. For example, it is straightforward to build cortical cans out of OCOS, FFP and Two/Three subcircuits. The OCOS will have addresses $[0; 0; 0; 0; 0; 1 - 7]$ as there are 7 neurons. The FFP is simpler with addresses $[0; 0; 0; 0; 0; 1 - 2]$ as there are just 2 neurons, The six neuron Two/Three circuit will have 6 addresses, $[0; 0; 0; 0; 0; 1 - 6]$. To distinguish these neurons in different circuits from one another, we add a unique integer in the fifth component. The addresses are now OCOS ($[0; 0; 0; 0; 1; 1 - 7]$), OCOS ($[0; 0; 0; 0; 2; 1-2]$) and Two/Three ($[0; 0; 0; 0; 3; 1-6]$), So a single **can** would look like what is shown below

```
can   L1  .   .   .   .  y  .   .      FFP, y's        Address  [0; 0; 0; 0; 2; 1 − 2]
      L2  z   .   .   z  .   .  z      Two/Three, z's   Address  [0; 0; 0; 0; 3; 1 − 6]
      L3  .   .  z   .   .   .   .
      L4  z  x   .  x   .  x  z      OCOS x's          Address  [0; 0; 0; 0; 1; 1 − 7]
          .  x   .  x   .  x  .
      L5  .   .   .   .  y  .   .
      L6  .   .   .  x   .   .   .
```

We are not showing the edges here, for convenience. We can then assemble **cans** into **columns** by stacking them vertically. There are many details and you should read Chap. 19 carefully. A typical small brain model could then consist of modules with the following addresses, p. 417:

© Springer Science+Business Media Singapore 2016
J.K. Peterson, *BioInformation Processing*, Cognitive Science and Technology,
DOI 10.1007/978-981-287-871-7

Input	$[1; \cdot; \cdot; \cdot; \cdot; \cdot; \cdot]$
Left Brain	$[2; \cdot; \cdot; \cdot; \cdot; \cdot; \cdot]$
Corpus Callosum	$[3; \cdot; \cdot; \cdot; \cdot; \cdot; \cdot]$
Right Brain	$[4; \cdot; \cdot; \cdot; \cdot; \cdot; \cdot]$
Thalamus	$[5; \cdot; \cdot; \cdot; \cdot; \cdot; \cdot]$
MidBrain	$[6; \cdot; \cdot; \cdot; \cdot; \cdot; \cdot]$
Cerebellum	$[7; \cdot; \cdot; \cdot; \cdot; \cdot; \cdot]$
Output	$[8; \cdot; \cdot; \cdot; \cdot; \cdot; \cdot]$

B

Biological Feature Vector of BFV   This is a finite dimensional approximation ζ to an action potential given by

$$\zeta = \begin{cases} (t_0, V_0) & \text{start point} \\ (t_1, V_1) & \text{maximum point} \\ (t_2, V_2) & \text{return to reference voltage} \\ (t_3, V_3) & \text{minimum point} \\ (g, t_4, V_4) & \text{sigmoid model of tail} \\ & V_3 + (V_4 - V_3)\,\tanh(g(t - t_3)) \end{cases}$$

where the model of the tail of the action potential is of the form $V_m(t) = V_3 + (V_4 - V_3)\,\tanh(g(t - t_3))$. Note that $V'_m(t_3) = (V_4 - V_3)\,g$ and so if we were using real voltage data, we would approximate $V'_m(t_3)$ by a standard finite difference. In Sect. 9.3, we derive equations telling us how the various attributes of an action potential are altered when there are changes in these 11 parameters. This information can then be used to link second messenger effects directly to BFV changes, p. 141.

Biological Information Processing   Biological systems process information that uses a computational node paradigm. Inputs are collected in a network of incoming edges called the dendritic arbor, combined in nonlinear ways in the computational unit called the node and then an output is generated. This processing can be modeled in many ways and in the text, we focus on some of them such as matrix feed forward networks and graphs of nodes which we sometimes call chained networks, p. 4.

Brain model   For our purposes, we will consider a brain model to consist of the cerebrum, the cerebellum and the brain stem. Finer subdivisions are then shown in the table below where some structures are labeled with a corresponding number for later reference in figures shown in Chap. 5. The numbering scheme shown below helps us locate brain structures deep in the brain that can only be seen by taking slices. These numbers thus correspond to the brain structures shown in Fig. 5.4 (modules that can be seen on the surface) and the brain slices of Figs. 5.5a, b, and 5.6a, b. A useful model of the processing necessary to combine disparate sensory information into higher level concepts is clearly built on models of cortical processing.

Brain	→	Cerebrum	Cerebellum **1**	Brain Stem
Cerebrum	→	Cerebral Hemisphere	Diencephalon	
Brain Stem	→	Medulla **4**	Pons **3**	Midbrain **2**
Cerebral Hemisphere	→	Amygdala **6**	Hippocampus **5**	
		Cerebral Cortex	Basal Ganglia	
Diencephalon	→	Hypothalamus **8**	Thalamus **7**	
Cerebral Cortex	→	Limbic **13**	Temporal **12**	Occipital **11**
		Parietal **10**	Frontal **9**	
Basal Ganglia	→	Lenticular Nucleus **15**	Caudate Nucleus **14**	
Lenticular Nucleus	→	Global Pallidus **16**	Putamen **15**	

There are many more details you should look over in Chap. 5 including more information about cortical models, p. 67.

C

Calcium trigger event   Some triggers initiate a spike in calcium current which enters the cell and alters the equilibrium between free calcium in the cell and calcium stored in a variety of buffering complexes. This provides useful ways of regulating many biological processes. A second messenger system often involves Ca^{++} ion movement in and out of the cell. The amount of free Ca^{++} ion in the cell is controlled by complicated mechanisms, but some is stored in buffer complexes. The release of calcium ion from these buffers plays a big role in cellular regulatory processes and which protein $P(T_1)$ is actually created from a trigger T_0. Calcium is bound to many proteins and other molecules in the cytosol. This binding is fast compared to calcium release and diffusion. Let's assume there are M different calcium binding sites; in effect, there are different *calcium species*. We label these sites with the index j. Each has a binding rate constant k_j^+ and a disassociation rate constant k_j^-. We assume each of these binding sites is homogeneously distributed throughout the cytosol with concentration B_j. We let the concentration of free binding sites for species j be denoted by $b_j(t, x)$ and the concentration of occupied binding sites is $c_j(t, x)$; where t is our time variable and x is the spatial variable. We let $u(t, x)$ be the concentration of free calcium ion in the cytosol at (t, x). The amount of release/uptake could be a nonlinear function of the concentration of calcium ion. Hence, we model this effect with the nonlinear mapping $f_0(u)$. The diffusion dynamics are then

$$\frac{\partial u}{\partial t} = f_0(u) + \sum_{j=1}^{M} \left(k_j^- c_j - k_j^+ (B_j - c_j)u \right) + D_0 \frac{\partial^2 u}{\partial x^2}$$

The dynamics for c_j are given by

$$\frac{\partial c_j}{\partial t} = -k_j^- c_j + k_j^+ (B_j - c_j)u + D_j \frac{\partial^2 c_j}{\partial x^2}$$

where D_0 is diffusion coefficient for free calcium ion. There are also boundary conditions which we will let you read about in the discussions in Chap. 7. The concentration of total calcium is the sum of the free and the bound. We denote this by $w(t, x)$ and we can show

$$\frac{\partial w}{\partial t} = f_0 \left(w - \sum_{j=1}^{M} c_j \right) + D_0 \frac{\partial^2 w}{\partial x^2} + \sum_{j=1}^{M} (D_j - D_0) \frac{\partial^2 c_j}{\partial x^2}$$

Thus, the total calcium equation is

$$\frac{\partial w}{\partial t} = f_0 \left(w - \sum_{j=1}^{M} c_j \right) + D_0 \frac{\partial^2 w}{\partial x^2} + \sum_{j=1}^{M} (D_j - D_0) \frac{\partial^2 c_j}{\partial x^2}$$

Using several assumptions about how different rates of calcium binding compare we can then reduce this to a new dynamic model for u given by

$$\frac{\partial u}{\partial t} = \frac{f_0(u)}{\Lambda} + \left[\frac{\Lambda D_0 + \sum_{j=1}^{M} D_j \gamma_j}{\Lambda} \right] \frac{\partial^2 u}{\partial x^2}$$

where Λ is a new constant that comes out of the approximations we make. The details are in Chap. 7. Then, by redefining the diffusion constant by $\hat{\mathscr{D}} = \frac{D_0 + \sum_{j=1}^{M} D_j \gamma_j}{\Lambda}$, we arrive at a good approximation for u's dynamics

$$\frac{\partial u}{\partial t} = \frac{f_0(u)}{\Lambda} + \hat{\mathscr{D}} \frac{\partial^2 u}{\partial x^2}$$

This is the model that allows us to estimate the effect of a calcium current spike initiated by a second messenger event which is important in understanding how to approximate nodal computations in simulations, p. 107.

Cellular trigger　　The full event chain for a cellular trigger (you will have to look up the meaning of these terms in Chap. 6) is

T_0	$\rightarrow$ Cell Surface Receptor	$\rightarrow PK/\mathscr{U}$
$PK/\mathscr{U}$	$+ \ T_1/T_1^{\sim}$	$\rightarrow T_1/T_1^{\sim} P$
$T_1/T_1^{\sim} P$	$+ \ $ tagging system	$\rightarrow T_1/T_1^{\sim} P \mathscr{V}_n$
$T_1/T_1^{\sim} P \mathscr{V}_n$	$+ \ fSQ \mathscr{V}_n$	$\rightarrow T_1$
T_1	$\rightarrow$ nucleus	$\rightarrow$ tagged protein transcription $P(T_1)$

where $P(T_1)$ indicates the protein whose construction is initiated by the trigger T_0. Without the trigger, we see there are a variety of ways transcription can be stopped:

- T_1 does not exist in a free state; instead, it is always bound into the complex $T_1/T_1^{\sim}$ and hence can't be activated until the $T_1^{\sim}$ is removed.
- Any of the steps required to remove $T_1^{\sim}$ can be blocked effectively killing transcription:
  - phosphorylation of $T_1^{\sim}$ into $T_1^{\sim}P$ is needed so that tagging can occur. So anything that blocks the phosphorylation step will also block transcription.
  - Anything that blocks the tagging of the phosphorylated $T_1^{\sim}P$ will thus block transcription.
  - Anything that stops the removal mechanism $fSQ\mathcal{V}_n$ will also block transcription.

The steps above can be used therefore to further regulate the transcription of T_1 into the protein $P(T_1)$. Let T_0', T_0'' and T_0''' be inhibitors of the steps above. These inhibitory proteins can themselves be regulated via triggers through mechanisms just like the ones we are discussing. In fact, $P(T_1)$ could itself serve as an inhibitory trigger—i.e. as any one of the inhibitors T_0', T_0'' and T_0'''. Our theoretical pathway is then

$$
\begin{array}{llll}
T_0 & \to \text{Cell Surface Receptor} \to & & PK/\mathcal{U} \\
PK/\mathcal{U} & + \; T_1/T_1^{\sim} & \to \; \text{step i} & T_1/T_1^{\sim}P \\
T_1/T_1^{\sim}P & + \; \text{tagging system} & \to \; \text{step ii} & T_1/T_1^{\sim}P\mathcal{V}_n \\
T_1/T_1^{\sim}P\mathcal{V}_n & + \; fSQ\mathcal{V}_n & \to \; \text{step iii} & T_1 \\
T_1 & \to \text{nucleus} & \to & \text{tagged protein transcription} \\
& & & P(T_1)
\end{array}
$$

where the **step i**, **step ii** and **step iii** can be inhibited as shown below:

$$
\begin{array}{llll}
T_0 & \to \text{Cell Surface Receptor} \to & & PK/\mathcal{U} \\
PK/\mathcal{U} & + \; T_1/T_1^{\sim} & \to \; \text{step i} & T_1/T_1^{\sim}P \\
& & & \uparrow T_0'\text{kill} \\
T_1/T_1^{\sim}P & + \; \text{tagging system} & \to \; \text{step ii} & T_1/T_1^{\sim}P\mathcal{V}_n \\
& & & \uparrow T_0''\text{kill} \\
T_1/T_1^{\sim}P\mathcal{V}_n & + \; fSQ\mathcal{V}_n & \to \; \text{step iii} & T_1 \\
& & & \uparrow T_0'''\text{kill} \\
T_1 & \to \text{nucleus} & \to & \text{tagged protein transcription} \\
& & & P(T_1)
\end{array}
$$

These event chains can then be analyzed dynamically and using equilibrium analysis, approximations to the effects of a trigger can be derived. The full details of that discussion are complex and are given carefully in Chap. 6. The protein $P(T_1)$ can be an important one for the generation of action potentials such a sodium gates. Hence, this is a model of second messenger activity that can alter the hardware of the cell, p. 85.

CFFN back propagation   The CFFN energy function is minimized using gradient descent just as we do in the MFFN case, but the derivation is different as the structure of the problem is in a graph form rather than matrices and vectors. The discussion is technical and we refer you to Sect. 17.2 for the details of how to find the required partial derivatives recursively, p. 319.

CFFN training problem   The output of the CFFN is therefore a vector in $\Re^{no}$ defined by

$$H(x) = \left\{ Y^i \mid i \in \mathcal{V} \right\}$$

and we see that $H : \Re^{n_I} \to \Re^{no}$ is a highly nonlinear function that is built out of chains of feedforward nonlinearities. The parameters that control the value of $H(x)$ are the forward links, the offsets and the gains for each neuron. The cardinality of these parameter sets are given by

$$n_s = \sum_{i=0}^{N-1} | \mathcal{F}(i) |$$
$$n_o = N$$
$$n_g = N$$

where $|\mathcal{F}(i)|$ denotes the size of the forward link set for the ith neuron; n_s denotes the number of synaptic links; n_o, the number of offsets; and n_g, the number of gains. Now let $\mathcal{I} = \{x_\alpha \in R^{n_I} : 0 \le \alpha \le S - 1\}$ and $\mathcal{D} = \{D_\alpha \in R^{no} : 0 \le \alpha \le S - 1\}$ be two given sets of data of size $S > 0$. The set $\mathcal{I}$ is the set of *input exemplars* and the set $\mathcal{D}$ is the set of outputs that are associated with exemplars. Together, the sets $\mathcal{I}$ and $\mathcal{D}$ comprise what is known as the *training* set. Also, from now on, as before in the MFFN case, the subscript notation α indicates the dependence of various variables on the αth exemplar in the sets $\mathcal{I}$ and $\mathcal{D}$. The **training problem** is then to choose the CFFN parameters to minimize an energy function, E, given by

$$E = 0.5 \sum_{\alpha=0}^{S-1} \sum_{i=0}^{n_O-1} f(Y_\alpha^{v_i} - D_\alpha^i),$$
$$= 0.5 \sum_{\alpha=0}^{S-1} \sum_{i \in \mathcal{V}} f(Y_\alpha^i - D_\alpha^{d_i})$$

where f is a nonnegative function of the error term $Y_\alpha^i - D_\alpha^{d_i}$; e.g. using the function $f(x) = x^2$ gives the standard L^2 or least squares energy function, p. 318.

Chained feedforward network or CFFN   We consider a function $H : \mathfrak{R}^{n_I} \rightarrow \mathfrak{R}^{n_O}$ that has a very special nonlinear structure. This structure consists of a string or chain of computational elements, generally referred to as neurons in deference to a somewhat tenuous link to a lumped sum model of post-synaptic potential. Each neuron processes a summed collection of weighted inputs via a saturating transfer function with bounded output range. The neurons whose outputs connect to a given target or **postsynaptic** neuron are called **presynaptic** neurons. Each presynaptic neuron has an output Y which is modified by the synaptic weight $T_{pre,post}$ connecting the presynaptic neuron to the postsynaptic neuron. This gives a contribution $T_{pre,post} Y$ to the input of the postsynaptic neuron. The postsynaptic neuron then processes the inputs using some sort of nodal function which is often a sigmoid as discussed in the MFFN case. The CFFN model consists of a string of N neurons, labeled from 0 to $N - 1$. Some of these neurons can accept external input and some have their outputs compared to external targets. We let

$$\mathcal{U} = \{i \in \{0, \ldots, N - 1\} \mid \text{neuron i is an input neuron}\}$$
$$= \{u_0, \ldots, u_{n_I - 1}\}$$
$$\mathcal{V} = \{i \in \{0, \ldots, N - 1\} \mid \text{neuron i is an output neuron}\}$$
$$= \{v_0, \ldots, v_{n_O - 1}\}$$

We will let n_I and n_O denote the cardinality of $\mathcal{U}$ and $\mathcal{V}$ respectively. The remaining neurons in the chain which have no external role will be called hidden neurons with dimension n_H. Note that $n_H + |\mathcal{U}| = N$. Note that it is possible for an input neuron to be an output neuron; hence $\mathcal{U}$ and $\mathcal{V}$ need not be disjoint sets. The chain is thus divided by function into three possibly overlapping types of processing elements: n_I input neurons, n_O output neurons and n_H internal or hidden neurons. We will let the set of postsynaptic neurons for neuron i be denoted by $\mathcal{F}(i)$, the set of **forward links** for neuron i. Note also that each neuron can be viewed as a postsynaptic neruon with a set of presynaptic neurons feeding into it: thus, each neuron i has associated with it a set of backward links which will be denoted by $\mathcal{B}(i)$. The input of a typical postsynaptic neuron therefore requires summing over the backward link set of the postsynaptic neuron in the following way:

$$y^{post} = x + \sum_{pre \in \mathcal{B}(post)} T_{pre \rightarrow post} Y^{pre}$$

where the term x is the external input term which is only used if the post neuron is an input neuron. Note the CFFN notation is much cleaner than the MFFN and is more general as we can have links in the CFFN model we just cannot handle in the MFFN, p. 315.

E

Evolvable Hardware   Certain types of hardware can be changed due to environmental input. The best example is that of a FPGA—the Field Programmable Gate Array. This consists of a large number of cells which can be thought of a fixed hardware primitives whose interconnections can be shuffled according to real time demand to assemble into large scale hardware modules, p. 4.

F

Fourier Transform   Given a function g defined on the y axis, we define the Fourier Transform of g to be

$$\mathscr{F}(g) = \frac{1}{\sqrt{2\pi}} \int_{-\infty}^{\infty} g(y) \, e^{-j\xi y} \, dy$$

where j denotes the square root of minus 1 and the exponential term is as usual

$$e^{-j\xi y} = \cos(\xi y) - j\sin(\xi y)$$

This integral is well defined if g is what is called square integrable which roughly means we can get a finite value for the integral of g^2 over the y axis, p. 41.

G

Graph model   A model consisting of nodes called vertices and edges. In MatLab, there are three classes we define: the vertices class, the edges class and the graphs class using the particular syntax MatLab requires. This is fairly technical, so you need to look carefully at the details in Sect. 18.1. Roughly speaking, we first create edges and vertices objects and use them to build the graphs objects. To build an OCOS graph, the commands
```
V = [1;2;3;4;5;6;7;8]; v = vertices(V);
```
define the OCOS nodes; the edges are defined by
```
E = [1;2],[2;3],[2;4],[2;5],[3;6],[4;7],[5;8],[2;7],[1;7];
```
and
```
e = edges(E); and OCOS=graphs(v,e);
```
define the OCOS graph. There are many other methods associated with these classes such as adding vertices and edges, adding two graphs together and so forth. The details are in Sect. 18.1. The big thing here is that when two graphs are merged, their individual node and edge numbers are redone and all information about where they originally came from

is lost. Retaining that sort of information requires we add address information, p. 340.

H

Hebbian update   We can also adjust parameters in a graph model using the classic idea of a Hebbian update which would be implemented as shown below: choose a tolerance ϵ and at each node in the backward set of the post neuron i, update the edge value by the factor ζ if the product of the post neuron node value f_i and the edge value $E_{j \to i}$ exceeds ϵ. The update algorithm then consists of the paired operations: sweep through the DG to do an evaluation and then use both Hebbian updates and the graph flow equations to adjust the edge values, p. 499.

$$
\begin{aligned}
&\text{for}(i = 0;\ i < N;\ i++)\ \{ \\
&\quad \text{for } (j \in \mathcal{B}(i))\ \{ \\
&\qquad y_p = f_i\, E_{j \to i} \\
&\qquad \text{if } (y_p > \epsilon) \\
&\qquad\quad E_{j \to i} = \zeta\, E_{j \to i} \\
&\quad \} \\
&\}
\end{aligned}
$$

L

Laplace Transform   Given the function x defined on $s \geq 0$, we define the Laplace Transform of x to be

$$
\mathcal{L}(x) = \int_0^\infty x(s)\, e^{-\beta s}\, ds
$$

The new function $\mathcal{L}(x)$ is defined for some domain of the new variable β. The variable β's domain is called the *frequency* domain and in general, the values of β where the transform is defined depend on the function x we are transforming. Also, in order for the Laplace transform of the function x to work, x must not grow to fast—roughly, x must decay like an exponential function with a negative coefficient, p. 39.

Laplacian graph based information flow   The standard cable equation for the voltage v across a membrane is

$$
\lambda^2\, \nabla^2\, v - \tau\, \frac{\partial v}{\partial t} - v = -r\, \lambda^2\, k
$$

where the constant λ is the space constant, τ is the time constant and r is a geometry independent constant. The variable k is an input source. We assume the flow of information through a graph model G is given by the graph based partial differential equation

$$\nabla_G^2 f - \frac{\tau_G}{\lambda_G^2} \frac{\partial f}{\partial t} - \frac{1}{\lambda_G^2} f = -rk.$$

where f is the vector of node values for the graph and $\nabla_G^2 f = KK' f$ where K is the incident matrix for the graph. Relabel the fraction τ_G / λ_G^2 by μ_1 and the constant $1/\lambda_G^2$ by μ_2 where we drop the G label as it is understood from context. Each computational graph model will have constants μ_1 and μ_2 associated with it and if it is important to distinguish them, we can always add the labellings at that time. This equation gives us the Laplacian graph based information flow model, p. 279.

M

Matrix Feedforward Networks or MFFNs   MFFNs are a specialized network architecture which links computational nodes into a layered structure connected by links or edges. The MFFN is then a function

$$F : R^{n_0} \rightarrow R^{n_{M+1}}$$

that has a special nonlinear structure. The nonlinearities in the FFN are contained in the neurons which are typically modeled by the sigmoid functions,

$$\sigma(y) = 0.5\left(1 + \tanh(y)\right),$$

typically evaluated at

$$y = \frac{x - o}{g},$$

where x is the input, o is the offset and g is the gain, respectively, of each neuron. We can also write the transfer functions in a more general fashion as follows:

$$\sigma(x, o, g) = 0.5\left(1.0 + \tanh\left(\frac{x - o}{g}\right)\right)$$

The feed forward network consists of $M + 1$ layers of neurons connected together with connection coefficients. For each i, $0 \leq i \leq M + 1 - 1$, the ith layer of n_i

neurons is *connected* to the $i + 1$st layer of n_{i+1} neurons by a *connection* matrix, T^i. The feed forward network processes an arbitrary $I \in R^{n_0}$ in the following way. First, for $0 \leq \ell \leq M + 1$ and $1 \leq i \leq n_\ell$ and layer ℓ, to help us organize the notation, we let

σ_i^ℓ denote the *transfer* function, $\qquad$ Y_i^ℓ denote the output,

y_i^ℓ denote the input, $\qquad\qquad\qquad$ O_i^ℓ denote the offset,

g_i^ℓ denote the gain for the ith neuron $\quad$ and X_i^ℓ denote $\left(\frac{y_i^\ell - O_i^\ell}{g_i^\ell} \right)$.

The inputs are processed by the zeroth layer as follows.

$$y_i^0 = x_i$$
$$X_i^0 = (y_i^0 - O_i^0)/g_i^0$$
$$Y_i^0 = \sigma_i^0(X_i^0)$$

At the layer $\ell + 1$, we have

$$y_i^{\ell+1} = \sum_{j=1}^{n_\ell - 1} T_{ji}^\ell Y_j^\ell$$
$$X_i^{\ell+1} = (y_i^{\ell+1} - O_i^{\ell+1})/g_i^{\ell+1}$$
$$Y_i^{\ell+1} = \sigma_i^{\ell+1}(X_i^{\ell+1})$$

So the output from the MFFN at the output layer $M + 1$ is given by p. 287.

$$y_i^{M+1} = \sum_{j=1}^{n_{M+1} - 1} T_{ji}^M Y_j^M$$
$$X_i^{M+1} = (y_i^{M+1} - O_i^{M+1})/g_i^{M+1}$$
$$Y_i^{M+1} = \sigma_i^{M+1}(X_i^{M+1})$$

MFFN back propagation The backpropagation equations for a MFFN are simply a recursive algorithm for computing the partial derivatives of the MFFN error or energy function with respect to each of the tunable parameters. The indicial details for a given training set are fairly intense, so we just refer you to the lengthy and detailed discussion in Sect. 16.3, p. 290.

MFFN training problem With the usual MFFN notation, for any $I \in R^{n_0}$, we know how to calculate the output from the MFFN. It is clearly a process that flows forward from the input layer and it is very nonlinear in general even though it is built out of relatively simple layers of nonlinearities. The parameters that control the value of $F(I)$ are the $n_i n_{i+1}$ coefficients of T^i, $0 \leq i \leq M$; the offsets

O^i, $0 \le i \le M + 1$; and the gains g^i, $0 \le i \le M + 1$. This gives the number of parameters,

$$N = \sum_{i=0}^{M} n_i n_{i+1} + 2 \sum_{i=0}^{M+1} n_i.$$

Now let

$$\mathcal{I} = \left\{ I_\alpha \in R^{n_0} : 0 \le \alpha \le S - 1 \right\}$$

and

$$\mathcal{D} = \left\{ D_\alpha \in R^{n_{M+1}} : 0 \le \alpha \le S - 1 \right\}$$

be two given sets of data of size $S > 0$. The set $\mathcal{I}$ is referred to as the set of *exemplars* and the set $\mathcal{D}$ is the set of outputs that are associated with exemplars. Together, the sets $\mathcal{I}$ and $\mathcal{D}$ comprise what is known as the *training* set. The **training problem** is to choose the N network parameters, $T^0, \ldots, T^M, O^0, \ldots, O^{M+1}$, $g^0, \ldots, g^{M+1}$, such that we minimize,

$$E = 0.5 \sum_{\alpha=0}^{S-1} \| F(I_\alpha) - D_\alpha \|_2^2,$$

$$= 0.5 \sum_{\alpha=0}^{S-1} \sum_{i=0}^{n_{M+1}-1} \left(Y_{\alpha i}^{M+1} - D_{\alpha i} \right)^2$$

where the subscript notation α indicates that the terms correspond to the αth exemplar in the sets $\mathcal{I}$ and $\mathcal{D}$, p. 290.

S

Sigma Three Transition Second messenger events have consequences that can be written in terms of concatenated sigmoid transitions. In many cases of interest, there are three such transformations and this is then called a σ_3 transition. We use the notation

$\sigma_3([T_0], [T_0]_b, g_p;$ inner most sigmoid

$[T_1]_n;$ scale again by $[T_1]_n$

$\frac{r[T_1]_n}{2}, g_e;$ offset, gain of next sigmoid

$[T_1]_n;$ scale again by $[T_1]_n$

     this is input into last sigmoid

$0, g_{Na};$ offset, gain of last sigmoid)

$r;$ scale innermost calculation by r

     this is input to next sigmoid

$s;$ scale results by s

     this is $[P(T_1)]$

where the meanings of the various terms, if not self explanatory, can be found in Chap. 8. Thus, the typical sodium maximum conductance g_{Na} computation can be written as p. 121.

$$g_{Na}(t, V) = g_{Na}^{max} \left(1 + e\delta_{Na}\, h_3([T_0], [T_0]_b, g_p; r; [T_1]_n; \frac{r[T_1]_n}{2}, g_e; s; [T_1]_n; 0, g_{Na})\right)$$
$$\mathcal{M}_{Na}^p(t, V)\mathcal{H}_{Na}^q(t, V)$$

T

Time dependent cable solution   The full cable equation is

$$\lambda_c^2 \frac{\partial^2 v_m}{\partial z^2} = v_m + \tau_m \frac{\partial v_m}{\partial t} - r_o \lambda_c^2 k_e$$

Recall that k_e is current per unit length. Next convert this equation into a diffusion model by making the change of variables $y = \frac{z}{\lambda_c}$ and $s = \frac{t}{\tau_m}$. With these changes, space will be measured in units of space constants and time in units of time constants. We then define the a new *voltage* variable w by

$$w(s, y) = v_m(\tau_m t, \lambda_c z)$$

This gives the scaled cable equation

$$\frac{\partial^2 w}{\partial y^2} = w + \frac{\partial w}{\partial s} - r_o \lambda_c^2 k_e(\tau_m s, \lambda_c y)$$

Make the additional change of variables

$$\Phi(s, y) = w(s, y)\, e^s$$

and we find

$$\frac{\partial^2 \Phi}{\partial y^2} = \frac{\partial \Phi}{\partial s} - r_o \lambda_c^2 \tau_m k_e(\tau_m s, \lambda_c y)\, e^s$$

Then apply to Laplace Transform to both sides to obtain

$$\mathscr{L}\left(\frac{\partial^2 \Phi}{\partial y^2}\right) = \mathscr{L}\left(\frac{\partial \Phi}{\partial s}\right) - r_0 \lambda_c^2 \, \mathscr{L}(P_{nm}(\tau_m s, \lambda_c y) \, e^s)$$

$$\frac{\partial^2 \mathscr{L}(\Phi)}{\partial y^2} = \beta \, \mathscr{L}(\Phi) - \Phi(0, y) - r_0 \lambda_c^2 \, \tau_m \, \mathscr{L}(k_e(\tau_m s, \lambda_c y) \, e^s)$$

Now further assume

$$\Phi(0, y) = 0, \ y \neq 0$$

which is the same as assuming

$$v_m(0, z) = 0, \ z \neq 0;$$

a reasonable physical initial condition. This gives us

$$\frac{\partial^2 \mathscr{L}(\Phi)}{\partial y^2} = \beta \, \mathscr{L}(\Phi) - r_0 \lambda_c^2 \, \tau_m \, \mathscr{L}(k_e(\tau_m s, \lambda_c y) \, e^s)$$

Now apply the Fourier Transform in space to get

$$\mathscr{F}\left(\frac{\partial^2 \mathscr{L}(\Phi)}{\partial y^2}\right) = \beta \, \mathscr{F}(\mathscr{L}(\Phi)) - r_0 \lambda_c^2 \, \tau_m \, \mathscr{F}(\mathscr{L}(k_e(\tau_m s, \lambda_c y) \, e^s))$$

or

$$-\xi^2 \mathscr{F}(\mathscr{L}(\Phi)) = \beta \, \mathscr{F}(\mathscr{L}(\Phi)) - r_0 \lambda_c^2 \, \tau_m \, \mathscr{F}(\mathscr{L}(k_e(\tau_m s, \lambda_c y) \, e^s))$$

For convenience, let

$$\mathscr{T}(\Phi) = \mathscr{F}(\mathscr{L}(\Phi))$$

we see we have

$$-\xi^2 \mathscr{T}(\Phi) = \beta \, \mathscr{T}(\Phi) - r_0 \lambda_c^2 \, \tau_m \, \mathscr{T}(k_e(\tau_m s, \lambda_c y) \, e^s)$$

The rest of the argument is messy but straightforward. The current k_e is replaced by a sequence of current pulses P_{nm} and limit arguments are used to find the value of $\mathscr{T}(k_e(\tau_m s, \lambda_c y) \, e^s)$. The limiting solution for a impulse current I_0 then satisfies

$$-\xi^2 \mathscr{T}(\Phi) = \beta \, \mathscr{T}(\Phi) - r_0 \lambda_c I_0 \implies (\xi^2 + \beta) \mathscr{T}(\Phi) = \frac{r_0 \lambda_c I_0}{\sqrt{2\pi}}$$

We then apply the Laplace and Fourier Transform inverses in the right order to find the solution

$$\Phi(s, y) = (r_0\lambda_c)I_0\left(\lambda_c P_0(x, t)\right)$$

We can then find the full solution w since

$$w(s, y) = \Phi(s, y)\, e^{-s}$$
$$= r_0\lambda_c I_0 \frac{1}{\sqrt{4\pi s}} e^{-\frac{y^2}{4s}} e^{-s}$$

We can write this in the unscaled form at pulse center (t_0, z_0) as

$$v_m(t, z) = r_0\lambda_c I_0 \frac{1}{\sqrt{4\pi\left((t - t_0)/\tau_m\right)}} e^{-\frac{((z-z_0)/\lambda_c)^2}{4((t-t_0)/\tau_m)}} e^{-(t-t_0)/\tau_m}$$

Note this solution does not have time and space separated!, p. 45.

W

Würfelspiel data matrices   There is an 18th century historical idea called The Musicalisches Würfelspiel. In the 1700's, fragments of music could be rapidly prototyped by using a matrix $\mathcal{A}$ of possibilities. It consists of P rows and three columns. In the first column are placed the opening phrases or nouns; in the third column, are placed the closing phrases or objects; and in the second column, are placed the transitional phrases or verbs. Each phrase consisted of L notes and the composer's duty was to make sure that any opening, transitional and closing (or noun, verb and object) was both viable and pleasing for the musical style that the composer was attempting to achieve. The Würfelspiel matrix used in musical composition has the general form below.

$$\mathcal{A} = \begin{bmatrix} \text{Opening 0} & \text{Transition 0} & \text{Closing 0} \\ \text{Opening 1} & \text{Transition 1} & \text{Closing 1} \\ \vdots & \vdots & \vdots \\ \text{Opening P-1} & \text{Transition P-1} & \text{Closing P-1} \end{bmatrix}$$

and the Würfelspiel matrix used in painting composition has the form

$$\mathcal{A} = \begin{bmatrix} \text{Background 0} & \text{Midground 0} & \text{Foreground 0} \\ \text{Background 1} & \text{Midground 1} & \text{Foreground 1} \\ \vdots & \vdots & \vdots \\ \text{Background P-1} & \text{Midground P-1} & \text{Foreground P-1} \end{bmatrix}$$

Thus, a musical or painting composition could be formed by concatenating these fragments together: picking the ith Opening, the jth Transition and the kth Closing phrases would form a musical sentence. Since we would get a different musical sentence for each choice of the indices i, j and k (where each index can take on the values 0 to $P - 1$), we can label the sentences that are constructed by using the subscript i, j, k as follows:

$$S_{i,j,k} = \text{Opening i} + \text{Transition j} + \text{Closing k}$$

Note that there are P^3 possible musical sentences that can be formed in this manner. If each opening, transition and closing fragment is four beats long, we can build P^3 different twelve beat sentences. In a similar way, we can assemble P^3 painting compositions, p. 209.

Index

Printed in the United States
By Bookmasters